ICT Tools and Applications for Accessible Tourism

Celeste Eusébio
University of Aveiro, Portugal

Leonor Teixeira
University of Aveiro, Portugal

Maria João Carneiro
University of Aveiro, Portugal

IGI Global
PUBLISHER of TIMELY KNOWLEDGE

A volume in the Advances in Hospitality, Tourism, and the Services Industry (AHTSI) Book Series

Published in the United States of America by
IGI Global
Business Science Reference (an imprint of IGI Global)
701 E. Chocolate Avenue
Hershey PA, USA 17033
Tel: 717-533-8845
Fax: 717-533-8661
E-mail: cust@igi-global.com
Web site: http://www.igi-global.com

Library of Congress Cataloging-in-Publication Data

Names: Eusebio, Celeste, 1971- editor. | Teixeira, Leonor, 1971- editor. | Carneiro, Maria, 1970- editor.
Title: ICT tools and applications for accessible tourism / Celeste Eusebio, Leonor Teixeira, and Maria Carneiro, editors.
Description: Hershey, PA : Business Science Reference, [2021] | Includes bibliographical references and index. | Summary: "This book provides theoretical and practical contributions pertaining to the field of accessible tourism, which is a growing tourism segment of the market, as social responsibility and inclusivity issues are not only expected but in many cases mandated"-- Provided by publisher.
Identifiers: LCCN 2020020907 (print) | LCCN 2020020908 (ebook) | ISBN 9781799864288 (hardcover) | ISBN 9781799864295 (paperback) | ISBN 9781799864301 (ebook)
Subjects: LCSH: Tourism--Information technology. | Tourism--Social aspects. | Tourism--Management.
Classification: LCC G155.A1 I182 2021 (print) | LCC G155.A1 (ebook) | DDC 338.4791--dc23
LC record available at https://lccn.loc.gov/2020020907
LC ebook record available at https://lccn.loc.gov/2020020908

This book is published in the IGI Global book series Advances in Hospitality, Tourism, and the Services Industry (AHTSI) (ISSN: 2475-6547; eISSN: 2475-6555)

British Cataloguing in Publication Data
A Cataloguing in Publication record for this book is available from the British Library.

For electronic access to this publication, please contact: eresources@igi-global.com.

Advances in Hospitality, Tourism, and the Services Industry (AHTSI) Book Series

Maximiliano Korstanje
University of Palermo, Argentina

ISSN:2475-6547
EISSN:2475-6555

Mission

Globally, the hospitality, travel, tourism, and services industries generate a significant percentage of revenue and represent a large portion of the business world. Even in tough economic times, these industries thrive as individuals continue to spend on leisure and recreation activities as well as services.

The Advances in Hospitality, Tourism, and the Services Industry (AHTSI) book series offers diverse publications relating to the management, promotion, and profitability of the leisure, recreation, and services industries. Highlighting current research pertaining to various topics within the realm of hospitality, travel, tourism, and services management, the titles found within the AHTSI book series are pertinent to the research and professional needs of managers, business practitioners, researchers, and upper-level students studying in the field.

Coverage

- Service Management
- Food and Beverage Management
- Customer Service Issues
- Travel Agency Management
- Service Training
- Tourism and the Environment
- Destination Marketing and Management
- Leisure & Business Travel
- Cruise Marketing and Sales
- Service Design

IGI Global is currently accepting manuscripts for publication within this series. To submit a proposal for a volume in this series, please contact our Acquisition Editors at Acquisitions@igi-global.com or visit: http://www.igi-global.com/publish/.

The Advances in Hospitality, Tourism, and the Services Industry (AHTSI) Book Series (ISSN 2475-6547) is published by IGI Global, 701 E. Chocolate Avenue, Hershey, PA 17033-1240, USA, www.igi-global.com. This series is composed of titles available for purchase individually; each title is edited to be contextually exclusive from any other title within the series. For pricing and ordering information please visit http://www.igi-global.com/book-series/advances-hospitality-tourism-services-industry/121014. Postmaster: Send all address changes to above address.

Titles in this Series

For a list of additional titles in this series, please visit: www.igi-global.com/book-series

Growth of the Medical Tourism Industry and Its Impact on Society Emerging Research and Oportunities
Manjeet Singh (Central University of Jammu, India) and Subbaraman Kumaran (Alliance University, Bangalore, India)
Business Science Reference • ©2021 • 189pp • H/C (ISBN: 9781799834274) • US $195.00

Global Development of Religious Tourism
Emilia Alaverdov (Georgian Technical University, Georgia) and Muhammad Waseem Bari (Government College University, Faisalabad, Pakistan)
Business Science Reference • ©2021 • 366pp • H/C (ISBN: 9781799857921) • US $195.00

Innovation and Entrepreneurial Opportunities in Community Tourism
Jakson Renner Rodrigues Soares (Universidade da Coruña, Spain & Universidade Estadual do Ceará, Brazil)
Business Science Reference • ©2021 • 358pp • H/C (ISBN: 9781799848554) • US $195.00

Handbook of Research on Human Capital and People Management in the Tourism Industry
Vânia Gonçalves Costa (School of Hospitality and Tourism, Polytechnic Institute of Cávado and Ave, Portugal & CiTUR, Portugal & GOVCOPP, University of Aveiro, Portugal) Andreia Antunes Moura (Polytechnic Institute of Coimbra, Portugal & CiTUR, Portugal & GOVCOPP, University of Aveiro, Portugal) and Maria do Rosário Mira (Polytechnic Institute of Coimbra, Portugal & CiTUR, Portugal & GOVCOPP, University of Aveiro, Portugal)
Business Science Reference • ©2021 • 472pp • H/C (ISBN: 9781799843184) • US $285.00

Handbook of Research on the Role of Tourism in Achieving Sustainable Development Goals
Filipa Brandão (GOVCOPP, University of Aveiro, Portugal) Zélia Breda (GOVCOPP, University of Aveiro, Portugal) Rui Costa (GOVCOPP, University of Aveiro, Portugal) and Carlos Costa (GOVCOPP, University of Aveiro, Portugal)
Business Science Reference • ©2021 • 385pp • H/C (ISBN: 9781799856917) • US $285.00

Role and Impact of Tourism in Peacebuilding and Conflict Transformation
Jorge Tavares da Silva (GOVCOPP, University of Aveiro, Portugal) Zélia Breda (GOVCOPP, University of Aveiro, Portugal) and Fabio Carbone (Centre for Trust, Peace, and Social Relations, Coventry University, UK & GOVCOPP, University of Aveiro, Portugal)
Business Science Reference • ©2021 • 402pp • H/C (ISBN: 9781799850533) • US $195.00

IGI Global
PUBLISHER of TIMELY KNOWLEDGE

701 East Chocolate Avenue, Hershey, PA 17033, USA
Tel: 717-533-8845 x100 • Fax: 717-533-8661
E-Mail: cust@igi-global.com • www.igi-global.com

Table of Contents

Detailed Table of Contents

of experience, and weaknesses such as high cost, lack of security. The high cost of both technologies restricts accessibility and marketing. Also, VR and AR have benefits as reducing strategy costs, attracting tourists with gamification strategies, ease of brand promotion, building brand loyalty, collecting data, and customized product development in tourism marketing.

Preface

The book *ICT Tools and Applications for Accessible Tourism* provides theoretical and practical contributions on the role of information and communication technologies (ICT) in the development of accessible tourism. This is an area of utmost relevance in increasing the participation of all people in tourism activities, independent of their characteristics. However, although tourism is one of the most important economic activities worldwide, accounting for 10.3% of the gross domestic product (GDP) and 10% of the total employment generated (WTTC, 2020), there are still several groups in our society that face many constraints to travelling and participating in tourism activities, particularly people with disabilities (PwD), the elderly and others with functional limitations. Nowadays, about 15% of the world's population has some disability (WHO, 2011). This proportion rises to around 40% of the world population when including other categories of people who may have a temporary need for accessible services. In European countries, about 138.6 million people have some kind of specific need for mobility, and by 2020, as a consequence of population ageing, this number is expected to be in the order of 154.6 million (EC, 2014). Accessible tourism enables people with functional and other kind of special needs to have access to tourism activities with equity and dignity (Darcy & Dickson, 2009).

The contribution of tourism to creating an inclusive society requires the adoption of new approaches and strategies that promote the accessibility of tourism destinations, allowing all people, regardless of their health condition, to enjoy tourism experiences. To accomplish this objective, it is of utmost importance to promote the active involvement of all stakeholders in the tourism system (demand, supply, government entities and educational institutions) in the creation of accessible and adapted tourism products. However, the scarce literature in this area suggests that the people working in the tourism industry are not usually aware of the various needs and travel constraints of PwD and that the information delivered by traditional information sources to this market is frequently inadequate, inaccurate or incomplete. Therefore, information and communications technologies (ICTs) may have a crucial role in overcoming the various travel constraints that these people face when planning and carrying out a tourism trip as well as enabling supply agents to develop accessible tourism products. Despite this, and although research regarding accessible tourism has increased in recent years, the number of studies on the contribution of ICTs to the development of accessible research is scarce. Therefore, this book intends to encourage academic discussion and knowledge-sharing regarding this under-researched area in order to: (i) think critically about the challenges of accessible tourism worldwide; (ii) increase knowledge on needs and travel constraints of PwD, the elderly and other people with functional limitations; (iii) promote knowledge of ICTs that can be useful to overcome travel constraints as well as to design accessible tourism products; (iv) promote research that connects theory and practice, enabling the identification of emerging technologies in this field; and (v) encourage future research in a field that has been highly neglected in the literature.

To achieve the above-mentioned objectives, this book incorporates theoretical chapters as well as case studies from different geographical regions, of different groups of people with special needs (PwSN) and of different ICTs. Some examples of the topics discussed in the 13 chapters that are part of this book are the travel constraints of the accessible tourism market, the accessibility of tourism supply agents, the accessibility of tourism websites, assistive technologies to facilitate the participation of PwSN in recreation and tourism activities and to improve their tourism experiences, smart destinations and accessible tourism, tourism information sources used by PwSN and emerging technologies in accessible tourism. A brief description of each chapter is presented below.

A systematic literature review on ICTs and accessible tourism is presented in Chapter 1, "The Impact of ICTs on Accessible Tourism: Evidence Based on a Systematic Literature Review". Based on a systematic literature review, this chapter provides important insights into the technologies that have been used to develop accessible tourism, the topics that have been analysed in the literature, the methodologies used and the areas that have been neglected.

People with visual disabilities are studied in Chapter 2, "The Challenges of Accessible Tourism Information Systems for Tourists With Vision Impairment: Sensory Communications Beyond the Screen". In this chapter, the authors outline accessible information needs of tourists with visual disabilities, drawing upon a small-scale project of nine telephone interviews conducted by the Wales Council of the Blind. It considers user experiences in the context of ICTs to help tourists with visual disabilities navigate information systems such as travel apps, social media and websites, assessing how these technologies meet user needs. Authors also highlight the relevance of designers and tourism providers as facilitators of accessible tourism. Moreover, the authors also remark on the relevance of collaboration across the tourism ecosystem, from digital developers and marketers, alongside disabled people as active stakeholders.

Travel constraints and negotiation strategies used by people with physical disabilities are studied in Chapter 3, "Travel by People With Physical Disabilities: Constraints and Influences in the Decision-Making Process", through a qualitative methodology. This chapter is based on the assumption that access to tourism by PwD continues to be shaped by a variety of constraints that are not being clearly or consistently addressed by those in charge of tourism planning and management. It focuses on the identification of structural constraints and negotiation strategies that influence the decisions and participation of people with physical disabilities in tourism activities. The results presented and discussed in this chapter contribute to extending knowledge about the factors that influence travel choices of PwD, allowing specific recommendations to be identified for improving the tourist supply chain and consequently the overall tourist experience quality.

The use of the Internet by people with special needs (PwSN) (people with disabilities and people with other specific needs) to plan their tourism trips is analysed through a quantitative approach in Chapter 4, "Use of the Internet to Plan Tourism Trips by People With Special Needs". In this chapter, differences in the use of the Internet according to the age and education level are also examined. Based on the results of this study, the authors show that the Internet is the most important information source used by PwSN in planning tourism trips, that PwSN are not very satisfied with the online sources used and that the sources used to plan tourism activities vary according to PwSNs' age and education level.

To increase knowledge on tools and strategies that can be used to develop inclusive cities, a systematic literature review on smart cities and accessible tourism is presented in Chapter 5, "Smart Cities and Accessible Tourism: A Systematic Review". The literature search and the analysis of the articles retrieved is performed according to the Preferred Reporting Items for Systematic Reviews and Meta-Analyses (PRISMA) guidelines. This chapter provides important insights to identify the most significant achievements related to accessible tourism in the context of smart cities, including the types of smart services being developed and their level of maturity.

Two specific case studies concerning the relationship between smart cities and accessible cities are presented in Chapter 6, "Smart and Accessible Cities? Not Always – Accessible Tourism Initiatives in the Autonomous Cities of Buenos Aires and Sydney". In this chapter, the authors discuss disability, ageing and accessibility, and their relationship with technology trends, considering a world full of different kinds of constraints, taking Buenos Aires (Argentina) and Sydney (Australia) as case studies. The authors also examine how a hypothetically smart and accessible city could set the basis to develop tourism for all. In particular, this chapter focuses on how cities that are being designed and planned under the smart city/destination model do not always consider people with disabilities and seniors in their models.

The accessibility of spas is analysed in Chapter 7, "Accessibility to Spa Experiences". More specifically, in this chapter the authors raise awareness of environmental, information, intrinsic and social barriers currently faced by wheelchair-user consumers of spa products and services and the implications these barriers have for consumers and industry alike. The chapter highlights the potential for health and greater mental and social well-being the spa industry has for wheelchair users and how they could mutually benefit each other. It further promotes the case for barrier-free wellness tourism and access to leisure opportunities.

The accessibility of official European tourism websites is examined in Chapter 8, "An Empirical Investigation of the Accessibility of Official European Tourism Websites". In the first part of this chapter, a theoretical background of web accessibility is discussed, including the definition of accessibility, accessibility laws, accessibility recommendations, accessibility automatic tools, and related research. In the second part, the authors present the results of an empirical research, using automatic tools and the Web Content Accessibility Guidelines 2.1 (WCAG 2.1). The most frequent errors are identified, and the authors present relevant recommendations for improving web accessibility.

E-travel agencies and the development of accessible tourism products is analysed in Chapter 9, "Competitiveness Factors of Accessible Tourism E-Travel Agencies". This study seeks to detect the differentiating factors of accessible e-travel agencies that enable them to be competitive in the market and, therefore, to promote the development of accessible products in the tourism sector. To achieve this objective, the authors examine fourteen accessible travel agencies' websites based on a theoretical framework of competitiveness and accessible tourism.

Chapter 10 also presents a study on web accessibility, specifically in the case of museums, "Museum Website Accessibility Examined Using Automatic Diagnostic Tools: A Study in Portugal". To identify improvements that should be introduced in the websites of museums to make information available to all people, the authors, based on the Web Content Accessibility Guidelines (WCAG 2.0), and on the three types of conformance levels (A, AA and AAA), use two web diagnostic tools – AccessMonitor and TAW (Test de Accessibilidad Web) – to examine the website accessibility of museums located in Portugal (N = 575).

The accessibility of museum websites is also examined in Chapter 11, "The Accessibility of Museum Websites: The case of Barcelona". In this chapter, the museums located in Barcelona are analysed in terms of accessibility and accessible information using a holistic approach. The authors use 45 museums in the city of Barcelona as a case study. The accessibility of these websites is assessed through content analysis, taking as reference 4 categories and 36 items that have emerged from a previous literature review.

The focus of Chapter 12, "Assistive Technologies in Museums for People With Visual Impairments", is also museums. However, in this chapter the authors discuss the role of technology, and specifically assistive technology, and identify guidelines to implement this kind of technology to improve the experience of people with visual disabilities during their visits to museums. In this chapter, it is also clear that people with visual disabilities represent a large group of the population but still experience various constraints during their museum visits. Therefore, guidelines are provided to promote more accessible and inclusive museums.

The last chapter of this book, Chapter 13, "Virtual Reality (VR) and Augmented Reality (AR) Technologies for Accessibility and Marketing in the Tourism Industry", presents two important technologies that are increasingly being used in the tourism industry. Specifically, the author investigates the roles of virtual reality (VR) and augmented reality (AR) for accessibility and marketing in tourism. In this chapter, a literature review and examples of VR and AR for accessibility and marketing in tourism industry are presented. The review was conducted with two perspectives: the roles of accessibility and marketing of VR and AR technologies in the tourism industry. Moreover, in this chapter the author also discusses the strengths and weaknesses of these two technologies and their usefulness in promoting accessible tourism.

The editors believe that this book provides theoretical and practical contributions in an area that has been neglected. Accessible tourism is a very important growth market in tourism and should not be ignored by the tourism industry, not only for social responsibility issues but also due to being an excellent business opportunity. Therefore, this book will be useful for all stakeholders of the tourism industry (researchers, students, professionals, public organizations, and international and national organizations that give support to accessible tourism market). Given that this book supplies knowledge to increase the participation of disadvantaged groups in tourism activities by promoting more inclusive societies, it is of benefit to all of society.

Celeste Eusébio
University of Aveiro, Portugal

Leonor Teixeira
University of Aveiro, Portugal

Maria João Carneiro
University of Aveiro, Portugal

REFERENCES

Darcy, S., & Dickson, T. J. (2009). A whole-of-life approach to tourism: The case for accessible tourism experiences. *Journal of Hospitality and Tourism Management*, *16*(1), 32–44. doi:10.1375/jhtm.16.1.32

European Commission. (2014). *Economic Impact and Travel Patterns of Accessible Tourism in Europe – Final Report*. Retrieved from: http://ec.europa.eu/DocsRoom/documents/7221/attachments/1/translations/en/renditions/native

WHO. (2011). *Relatório Mundial sobre a Deficiência* [World Disability Report]. Retrieved from: https://apps.who.int/iris/bitstream/handle/10665/70670/WHO_NMH_VIP_11.01_por.pdf;jsessionid=5BBD98488DC0CF3AD13E7C780D460A2F?sequence=9

World Travel & Tourism Council (WTTC). (2020). *Global Economic Impact & Trends 2020*. Retrieved from: https://wttc.org/Research/Economic-Impact

Acknowledgment

The editors would like to thank all people and organizations involved directly or indirectly in this project. First, our deepest thanks to all people with special needs who participated in the surveys carried out in some chapters of this book. Without their participation the content of this book would be much impoverished. Second, our thanks to all the authors for their excellent contributions to increasing knowledge in an area that is still little explored in the literature. We are also grateful to all reviewers for their valuable suggestions. The editors are also thankful for the support of the research project ACTION - POCI-01-0145-FEDER-030376, funded by FEDER, through COMPETE2020 - Programa Operacional Competitividade e Internacionalização (POCI), and by national funds (OE), through FCT/MCTES. Lastly, we are also grateful to IGI Global Publications for the opportunity to publish this book.

Celeste Eusébio
University of Aveiro, Portugal

Leonor Teixeira
University of Aveiro, Portugal

Maria João Carneiro
University of Aveiro, Portugal

Acknowledgment

Chapter 1
The Impact of ICTs on Accessible Tourism:
Evidence Based on a Systematic Literature Review

Pedro Teixeira
https://orcid.org/0000-0003-4395-4296
DEGEIT, University of Aveiro, Portugal

Leonor Teixeira
https://orcid.org/0000-0002-7791-1932
IEETA, DEGEIT, University of Aveiro, Portugal

Celeste Eusébio
https://orcid.org/0000-0002-2220-5483
GOVCOPP, DEGEIT, University of Aveiro, Portugal

Samuel Silva
IEETA, DETI, University of Aveiro, Portugal

António Teixeira
IEETA, DETI, University of Aveiro, Portugal

ABSTRACT

There is a clear need to address the technological effects of providing conditions for accessibility in tourism. Accessible tourism is becoming very important, and technology may be the tool which will surpass barriers to accessibility in tourism. New information technologies focus not just on the divulgation of information, but also on making sure everything is connected, which in the case of accessible tourism has significant value. Therefore, this chapter aims to share the results of a systematic literature review conducted on the topic of the impact of ICTs on accessible tourism. The research will be done via analysis of a range of bibliometric indicators present in the main academic and research databases. This approach will allow the characterization of the main technologies in the service of accessible tourism and an understanding of how they can enhance the quality of tourism experiences for people with disabilities.

DOI: 10.4018/978-1-7998-6428-8.ch001

INTRODUCTION

Information and communications technologies (ICT) have been responsible for technological growth in many sectors (Sedlar et al., 2018). The *Cambridge* dictionary defines ICT as: "the use of computers and other electronic equipment and systems to collect, store, use, and send data electronically" ("ICT", 2019). Within this perspective, ICT can be described as a combination of "devices, networking components, applications and systems (technologies) that allows people and organizations to interact in the digital world" (Rouse, 2014: 1). ICTs, in their different forms, are responsible for assuring that information and knowledge are spread, playing a central position in modern economies now and in the upcoming years (Byrne & Corrado, 2017). In fact, technological innovations are creating waves of change across many service providers and business strategies, with particular relevance in the tourism sector and in particularly in accessible tourism.

Accessible tourism can be defined as: "a form of tourism that involves collaborative processes between stakeholders that enables people with access requirements, including mobility, vision, hearing and cognitive dimensions of access, to function independently and with equity and dignity through the delivery of universally designed tourism products, services, and environments" (Buhalis & Darcy, 2011: 10-11). Often, tourism products on offer have some problems with accessibility (Avis, Card, & Cole, 2005; Buhalis & Darcy, 2011: 50; Figueiredo et al., 2012; Popiel 2016), implying many barriers for people with disabilities (PwD). **Lack of information** (Waschke, 2004), about the accessibility level of tourism products is a particularly big obstacle because it hinders answers to basic questions related to tourism activity, like how to travel and what to explore. Information is undoubtedly the foundation for accessible tourism, which makes ICTs crucial tools in this area. Accessible tourism is becoming very relevant for the social and economic sector, and ICTs can represent important tools to surpass accessible barriers in this area. The role of ICTs is not just to focus on divulgation of information, but also making sure it is accessible to PwD, since by using ICTs it is possible to connect everything in real time.

With the evolution found across ICTs, a new reality for tourism is starting to appear. The entire process of creating, managing and promoting tourism products is changing radically. Increasingly, the impacts of ICTs are becoming more evident, as networking, dynamic interfaces with consumers and partners and the ability to re-develop tourism products proactively and reactively are critical for the competitiveness of tourism organizations (Buhalis & Law, 2008). ICTs are essential for tourism, as they contribute to improving the sector and make the connection between tourism and technological innovation possible. Therefore, this study aims to share the results of a systematic literature review, conducted on the topic related to the impact of ICTs in accessible tourism. The research will be done by analysing a range of bibliometric indicators, present in one of the main academic and research database – Scopus. This approach will allow the characterization of the main technologies at the service of accessible tourism and understand how they can enhance the quality of tourism experiences for PwD.

The studies selected are related to the use of technologies in tourism. This review analysed a total of 65 documents. The main objective was to identify the main technologies and understand their contribution to the development of accessible tourism. To accomplish these objectives, this chapter is organized into four sections. After this introduction, in section 2, the methodology that was used to select and analyse the papers is explained. The main findings are reported in section 3. Attention was given to the distribution of papers over time, among journals, subject areas, authorship, the geographical contexts, the studied segment of the accessible market analysed, the ICT studied and the research method, as well as the key results found, regarding the impact of ICT in the development of accessible tourism. Finally, in section 4, the main conclusions are summarized, and the main research gaps in the current literature are identified.

Methodology

A systematic literature review was used to conduct this study with the aim of understanding the impact of ICT in the development of accessible tourism, using methods and explanations to identify, select and critically assess relevant research and analyse data from different studies (Siddaway, 2014; Petticrew & Roberts, 2008). The research methodology is divided into two phases. First, the research design was defined to select documents for analysis. Second, a content analysis of the documents selected was performed to meet the objectives of this research.

Document Selection

To select the documents (e.g. journals, proceedings, book chapters) to be analysed in this systematic literature review, the procedure illustrated in Figure 1 was used. The studies were identified with a research in the Scopus database in March 2020, using the following search code:

```
("accessible tourism" OR "inclusive tourism" OR "tourism for all" OR "easy access tourism" OR "barrier free tourism" OR "universal tourism" OR "disabled tourism" OR "disability tourism" OR disabilit* OR "people with disabilit*")
AND (tourism OR visitor* OR tourist* OR "tourism organization*" OR travel*)
AND (ICT OR "information system*" OR "information and communication system*"
OR "information and communication technology" OR "communication technolog*" OR
web*)
```

This code was applied to the title, abstract and keywords, with no restriction of time and subject. The Scopus database was selected due to the fact that it is one of the largest databases of peer-reviewed literature (scientific journals, books and conference proceedings).

Based on this search, a total of 186 records were obtained. Due to language and respective translation constraints, only papers in English, Portuguese and Spanish were included. Based on this first screening, two papers were excluded. Next, studies without author identification were taken out due to scientific coherence. Based on this second screening, 12 records were excluded.

In order to exclude documents that are outside the scope of this research, a manual review of the abstracts and titles of the remaining 172 studies was performed. However, when the abstract and title led to doubts about inclusion, the full document was analysed. Regarding the inclusion criteria, these were determined to only include documents in which the study was based on ICT contributing to the development of accessible conditions in diverse tourism areas. In that sense, medicine and education-based studies, which for example analysed the impact of technology in improving regular life-conditions of PwD in areas other than tourism, were excluded. A large number of documents were focused on eliminating the social exclusion of PwD from society. However, many contexts were not directly related to tourism and explored more social aspects such as education and active participation in daily life. With this third screening, 102 documents were excluded.

Thus, only 70 papers met the objectives set for this study and were therefore selected. Nevertheless, not every paper was available in Scopus. With only 38 documents available in Scopus, further research to obtain the full papers was essential. Hence, 25 studies were obtained by analysing other research sources. However, seven studies were impossible to obtain. In these seven cases, the main author was contacted, which resulted in obtaining two of the missing papers. Therefore, a total of 65 papers were selected for content analysis. The relatively low number of existing papers demonstrates the need to study the potential technological impact in accessible tourism, as it is an extremely important sector for the inclusion of PwD.

Figure 1. PRISMA flow diagram of the selection process

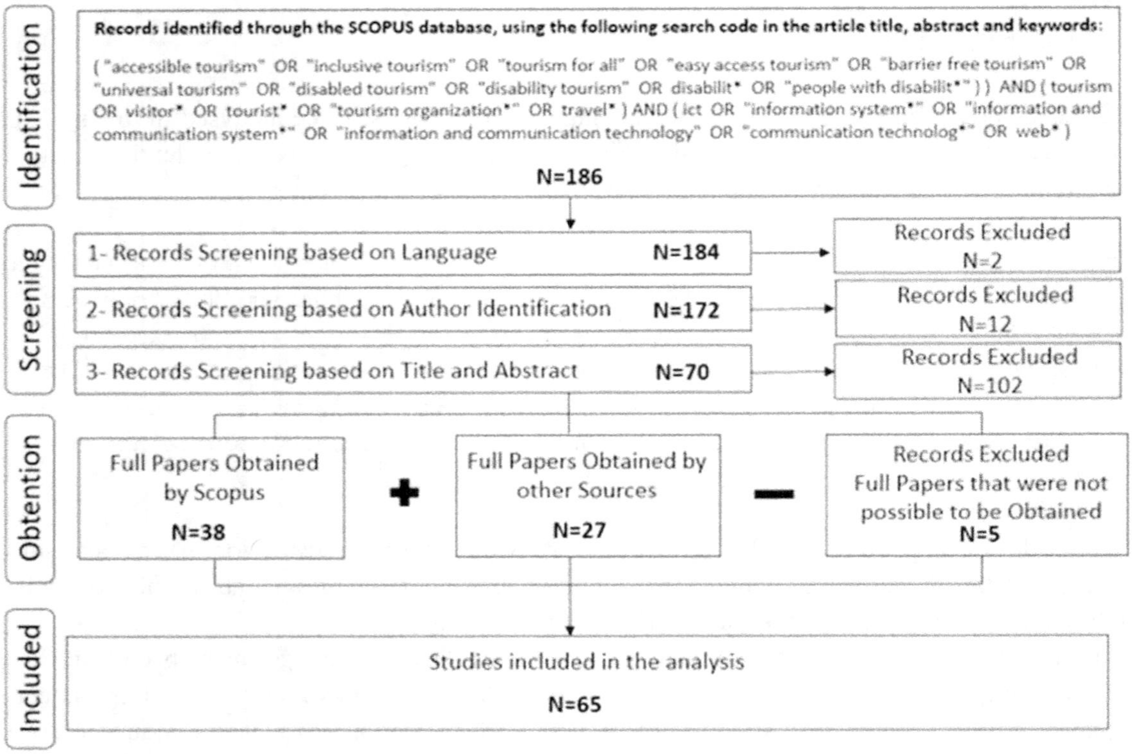

Document Analysis

A total of 65 documents obtained in the previous stage were analysed, using two different approaches. First, based on a bibliometric analysis, several aspects were examined, such as (i) the distribution of scientific documents over time; (ii) distribution of scientific documents by journal and by the authorship; and (iii) distribution of scientific documents by geographical context. Second, considering a qualitative analysis of the papers, some other aspects were examined, such as, (iv) segments of accessible tourism that have been given more attention in the literature concerning ICTs; and (v) the maturity level of the ICT mentioned in these studies. The different types of studies, combined with several types of methodological approaches, allowed the impact of ICTs to be estimated in the development of accessible tourism conditions.

FINDINGS

Distribution of the Papers Over Time

A total of 65 papers were analysed. Since the development of ICTs in tourism has grown over time, the existent literature seems to be following this trend, with most papers being published in the last decade. This trend also illustrates that there has been an increasing idea of helping PwD through the use of technology. Technology development is not stale, and more and more attention has been given to the accessible market. These factors can justify the increase in studies published in recent years. The first publication dates back to 1993 and the second ten years later. 15.4% of all the documents analysed were published in 2019, and the year 2020 is starting to follow this trend (Figure 2). This could mean that, recently, increasing importance has been given to research into both accessible tourism and technological impacts in this particular sector.

Figure 2. Number of papers published by year of publication

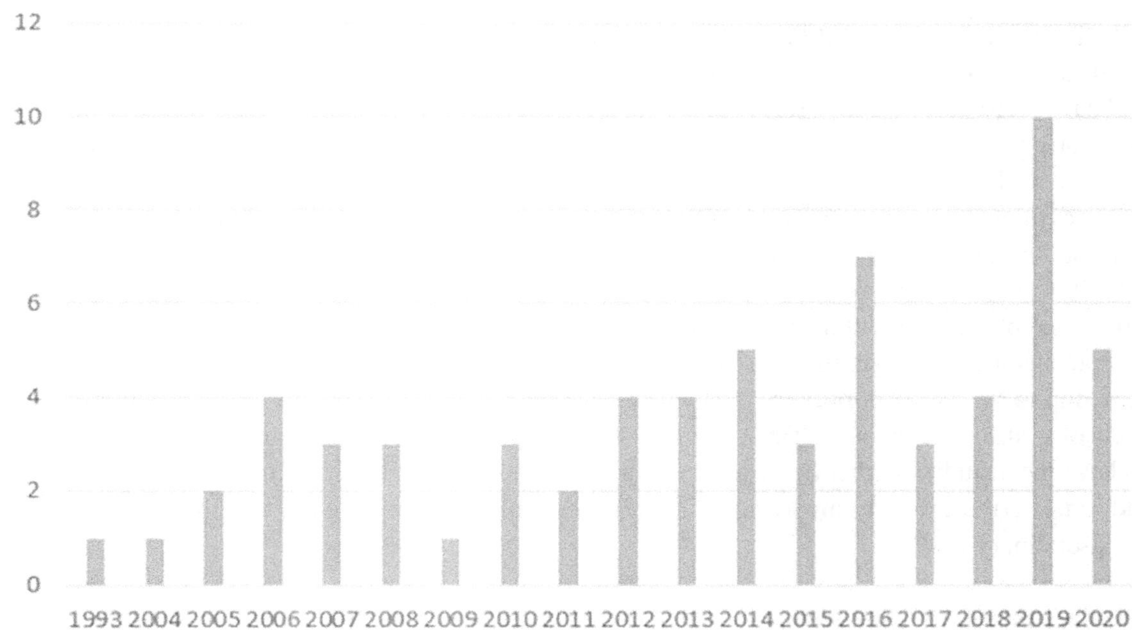

Journals and Authorship

By looking at the analysed sample, it can be seen that the topic is published in a variety of sources and involves different authors. The sources (journals and proceedings) with two or more studies are represented in Table 1, and the authors with two or more published works are shown in Table 1.

Table 1. Sources of the published studies

Journal (J) or Proceedings (P)	Type of Publication	Area of Study	Number of Studies	Percentage
Lecture Notes in Computer Science	P	CS	6	9.23
ACM International Conference Proceedings Series	P	CS	4	6.15
Current Issues in Tourism	J	HL	3	4.62
Advances in Intelligent Systems and Computing	P	IS	2	3.08
Disability and Rehabilitation	J	HD	2	3.08
Iberian Conf. on Information Systems and Technologies	P	IS	2	3.08
Smart Innovation, Systems and Technologies	P	CS	2	3.08
Tourism Management	J	HL	2	3.08

The papers were published in two sources: journals (J) and conference proceedings (P). Based on these results, it can be observed that conferences present a greater number of publications, which may reveal a recent research trend in this area, and that research has not yet reached the journals. The sources with most published studies are proceedings in Lecture Notes in Computer Science, with 10% of all the studies analysed. Different areas are present: i) hospitality and leisure (HL), ii) information systems (IS), iii) computer science (CS); iv) health and disabilities (HD); v) transportation (T); geography (G); mechanics (M); economics (E) and education (EDU). The areas with most published studies are HL (15%), IS (15%) and CS (18%). Despite computer science being the area with most studies published, the presence of the different areas may be an indicator of the variation within the subject, in the areas of both technology and tourism. Most sources and most authors published only one paper, which could correspond to the different backgrounds which each author represents, considering different technical aspects of writing and theories. Table 2 demonstrates that Buhalis, D. and Darcy, S. are the two authors who have most studies in this field, with a total of four each. The variety found in both Tables 1 and 2 could indicate the diversity involving ICTs and accessible tourism, proving that this topic has a high multidisciplinary level.

Table 2. Number of published studies by author

Authors	Number of Studies
Buhalis, D.	5
Darcy, S.	5
Domínguez Vila, T.	4
Wöß, W.	4
Alén González, E.	3
Metrôlho, J.	3
Michopoulou, E.	3
Ribeiro, F.	3
Abreu, A.	2
Bekiaris, E.	2
Daems, J.	2
Dekelver, J.	2
Eusébio, C.	2
Luccio, F.L.	2
Pühretmair, F.	2
Rumetshofer, H.	2
Silveiro, A.	2
Teixeira, L.	2
Zajadacz, A.	2

Geographical Context

Figure 3 illustrates the geographical distribution where the analysed studies took place. There are, however, studies that include more than one country, and 21 do not specify the country where the research took place. A total of 56 studies were performed across countries all around the world. According to the literature review carried out in this chapter, the impact of ICTs on accessible tourism has mainly been studied in Europe (73%) and North America. The USA is the country, where most studies were performed (seven studies). They are followed closely by several European countries like the United Kingdom, Portugal, Spain and Italy. There are studies in every geographic European area (southern, northern and central). On the other hand, there are relatively few studies in South America (4%), Oceania (4%) and Asia (5%). This may imply that developed countries have carried out more research in the fields of ICTs and accessible tourism.

Figure 3. Geographical distribution of studies around the world

Segments of the Accessible Market Analyzed

The accessible market is not homogenous (Buhalis & Michopoulou, 2011). Different types of disabilities have different demands and requirements (Figueiredo et al., 2012). Therefore, it is critical to divide the market into segments with similar needs and wants in order to create customized offerings and suitable products (Buhalis & Michopoulou, 2011). In accordance with the type of disability, the market can be divided into various segments, such as: i) mobility; ii) blind or vision impaired; iii) deaf or hearing impaired; iv) speech; v) cognitive (mental health/intellectual/learning) and vi) hidden (Buhalis & Darcy, 2011: 38). However, the accessible market also includes people with temporary disabilities such as people with crutches during a temporary period, people carrying luggage, small children or people who are big or small in size, people with allergies and also the elderly/seniors/boomers (Blocksidge, 2003; UNWTO, 2017; DEO, 2005; Disabled People's Association, 2015; Buhalis & Darcy, 2011:38; Michopoulou & Buhalis, 2013). Therefore, in order to identify the segments of accessible tourism that have been given more attention in the literature concerning ICTs, seven segments were considered.

Most documents under analysis in the literature review chapter aimed their attention at more than one segment (Figure 4). Indeed many (approximately 43%) approach all segments of the accessible market, meaning that in most cases the same ICT could be used for aiming at all segments. This is mainly explained by the fact that, when developing an accessible ICT, different needs should be taken into consideration. However, it could also be understood that, regarding individual accessible needs, mobility issues are the most explored. There were, however, few studies that only considered speech disabilities, hidden impairments or the elderly population.

Figure 4. Number of studies addressing each accessible market segment

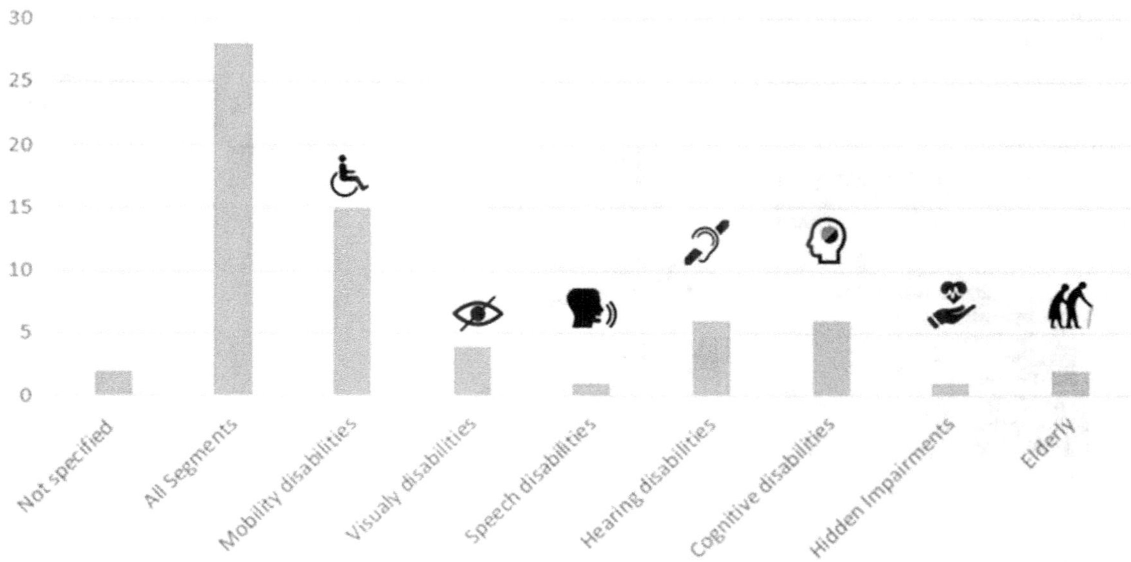

ICT Studied (Technology) and Current Life Cycle

Technologies have a life cycle, which can be described by the S-curve, developed by Khalil (2000). The S-curve classifies technology in three different stages. The first stage is the embryonic state, where improvements on the technology are slow, as it is still in the process of being understood. After that, in the growth period, improvements become faster and faster. When the technology becomes mature, development becomes slower, and a physical limit is reached. Based on the methodologies of the studies analysed, it was possible to understand the rate at which technology is adopted in the tourism marketplace. The S-curve was used to evaluate and classify the current technological innovation state of all ICTs displayed in the studies reviewed (Figure 5).

Figure 5 illustrates the various technologies identified, with the number of studies and the maturity state, according to the S-curve. Most of the technologies identified are still in an embryonic state, which can justify the low number of studies. As expected, the technologies present in the embryonic state are mostly still at the stage of conceptualization and development. In the growth state, there are three ICTs (geographic information systems, tourism information systems and mobile apps), that are clearly in an improvement phase, justifying current investments as regards the tourism sector, as there are cases of technology conceptualization, development and evaluation, especially for mobile apps. At a mature level, there are also three ICTs (global positioning systems, websites and low-technology solutions). These present an important step for current technological analysis, and since scientific knowledge is already present, this is a crucial factor in understanding possible applications, meaning that ICTs in the maturity stage have the most dedicated studies (52%). On an interesting note, most of the studies regarding technologies in a mature state focused on technology analysis, which is particularly true in the case of websites, the ICTs with most studies (38%), with most cases performing an accessibility analysis of the content, especially an evaluation of web accessibility performance.

Figure 5. ICTs identified in the studies, and current position on the technology life cycle spectrum

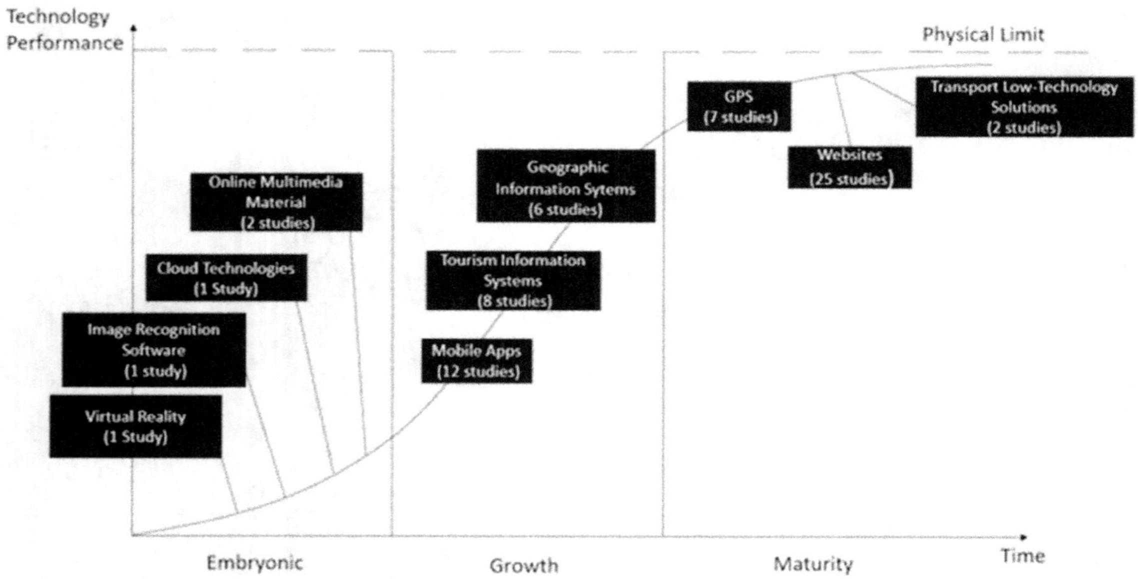

The Impact of ICTs on the Development of Accessible Tourism Conditions

Based on the analysis carried out, the main question of this study can be answered. The main objective of this literature review is to understand the impact of ICTs on the development of accessible tourism conditions for PwD. After careful analysis of the main segments of the accessible market and technological studies in the area of accessible information, essential technological advantages for disabled tourists can now be pointed out. Once again, the interdisciplinary aspect is visible across the review, as some authors analysed more than one technology. A synopsis of the impact of technologies in the promotion of accessible tourism was made (Tables 3, 4 and 5). The identified technologies, which drive, and impact accessible tourism were described, and the main advantages for accessible tourism are pointed out. After studying their fundamental ideas and discoveries, the impact of each technology in accessible tourism was discussed. Due to the high technological content of some papers, the discussion is rather complex. However, since the objective is to study the impact of ICTs on accessible tourism, some trends in each technology can be displayed. Therefore, the following discussion will put a spotlight on each identified ICT, understanding each author's study and its respective impacts on the creation of accessibility features for the tourism sector.

Technologies in an embryonic phase

Table 3 summarizes the impacts of technologies, that are still in an embryonic stage. Image recognition software, virtual reality, and cloud technologies are ICTs which are still in an early stage of technological innovation. All three studies keep the focus on a technological development level. Hamann et al., (2016) developed a smart doorbell that can facilitate the opening of doors, especially for PwD, improving inclusion and security, and which can be a way to unlock the full potential of other ICTs that can use face recognition software to develop accessibility features.

Regarding virtual reality, Thangaraj and Gomathi, (2019) designed a multi-disciplinary ontology engine to help the elderly experience tourism activities. This brings enormous challenges like technology usage, but also a high number of advantages that can be applied to other segments of the accessible market and which can be especially important in planning processes or giving the ability to experience a tourism activity. For cloud computing, a different approach was followed by Chalkia et al. (2015), in which an infrastructure was created that will be able to support the usage of ICTs. This project can be of great value, since it aims to automatically change other ICTs based on the user's needs and preferences, making cloud technology vital for the development of accessible tourism, due to the heterogeneity of the market.

The progress of multimedia material follows a different rationale. Zajadacz and Szmal (2017) developed two proposals for automatic translation of texts into Polish Sign Language, which can greatly

Table 3. Technological impact of ICTs in an embryonic stage

ICT	Description	Author(s)	Impact on Accessible Tourism
Image Recognition Software	Software able to identify and detect objects or people in a digital feature	Hamann et al. (2016)	• Performs the tasks of a doorbell, • Recognizes the tourist, automatically opening doors (for example works as a room key) • Face recognition can be the trigger for various multimedia exhibitions • Promotes social inclusion and security • Facilitates access to potential tourism technological platforms
Virtual Reality	A computer-based simulation of environments or objects, that can be interacted with by a person, providing a realistic experience.	Thangaraj and Gomathi (2019)	• Helps disabled people to perform virtual tours (inaccessible places) • Gives a preview of a tourism offer (helps in making travel decisions). • Combines different ontologies such as text, image, video and simulation • Allows more immersive tourism experiences • Creates virtual realities of several points of interest (e.g. monuments) • Simulates virtual tour maps
Cloud Technologies	Data centres where information can be stored and be made available to many users.	Chalkia et al. (2015)	• Helps to create frameworks for the development of accessible tourism technological tools • Provides communication between diverse ICTs • Shared information and knowledge • Provides help to disabled tourists while using technological devices
Online Multimedia Material	Image, audio and video content available online to different types of users	Zajadacz and Szmal (2017); Lauría (2016)	• Allows multisensory and bilingual guidebooks, available to every tourist • Provides a unique information tool for disabled and non-disabled people • Results in cooperation of different fields of knowledge and skills • Enables education and training of workers in accessible tourism • Enables design and creation of accessible tourism services and products • Strategic formulation can be developed for publishing and visualization of accessible tourism data

improve tourism online features for tourists with hearing disabilities when experiencing diverse tourist attractions, in a way that means they can fully participate. On the other hand, Lauría (2016) developed a guidebook with supplementary tourist information in the city of Florence, especially for people with both visual and mobility disabilities. This interactive guidebook, available online and developed in an accessible way, supplies different types of important information regarding accessible conditions. Online multimedia material, therefore, combines different types of information, creating multisensory and bilingual guidebooks, available to every tourist and providing a unique information tool for disabled and non-disabled people.

Technologies in a growth phase

Table 4 outlines the various impacts which ICTs in a development phase can provide. Mobile apps have become important players for more technological tourism and crucial tools for assuring accessible tourism. The evolution that this type of technology has experienced in recent years has turned these devices into almost personalized travel advisors. Mobile apps can have multiple uses for tourism, and this is illustrated by the studies presented. Lozynska et al. (2020) are working on an app that converts speech and text to sign language, fostering connections and helping tourists with hearing disabilities during the trip. Serra et al. (2019) took the different needs in account and developed an app that performs personalized recommendation of points of interest. These examples of development scenarios prove that mobile apps can play an important role in increasing the participation of PwD in tourism activities. The remaining studies did not focus particularly on the development of a mobile app, but also gave important contributions to understanding the impact of this ICT on the development of accessible tourism. Emrouzeh et al. (2017) were able to introduce a process to design a mobile application for tourists with disabilities and Ribeiro et al. (2018a) proposed a framework, which facilitates the gathering of information which is relevant to accessible tourism" and the implementation of common functionalities and access to a large, shared knowledge base.

Even though GPS is mostly incorporated into geographic information systems, there are some differences between these two ICTs. Studies including geographic information systems had different objectives. These systems are intended to work as disseminators of spatial information, providing information to PwD about physical ability and environmental demands. This is the case of Françoso et al. (2013) and Ioannidisa and Vozikisb (2007), who intend to give the most possible information regarding geographic nature questions so that PwD can plan their trips according to their needs.

The main objective of tourism information systems is to work as data centres, storing accessible information that can be made available for many tourists. As Rumetshofer and Wöß (2005) described, these systems should be composed of information about users (user profile), tourism information and information about tourism products. It is also important to allow adaptation and personalization, and this concern is present in the study by Winkler and Wöß, (2006), as they created an intelligent accessibility add-on software, allowing the overall presentation to be organized according to individual needs in a so-called toolbar. The findings of Kołodziejczak (2019) had also major impacts on developing strategies, stating that this type of system should be able to help PwD in all stages of a tourism trip (needs recognition, information search, alternative evaluation, purchase, consumption and post-purchase evaluation). Essentially, these systems have a great impact on accessible tourism because the lack of reliable information has been identified as the main reason constraining PwD, as identified by Zajadacz (2014) in the study carried out.

Table 4. Technological impacts of ICTs in a growth stage

ICT	Description	Author(s)	Impact on Accessible Tourism
Mobile Apps	Software created for mobile platforms such as smartphones	Lozynska et al. (2020); Serra et al. (2019); Agarwal (2019); Ribeiro et al. (2018a); Ribeiro et al. (2018b); Emrouzeh et al. (2017); Neven et al. (2017); Liu et al. (2016); Dekelver et al. (2016); Haworth and Williams (2012);Buhalis and Michopoulou (2011); Baudoin and Venard (2010);	• Platforms can be utilized as information services • Accommodation of a wide range of users and feature • Information sharing • Voice recognition and voice assistant • Reminders of daily activities • Improving orientation and perception processes • Caretakers can provide real-time help, in tourism activities • Enabling programs translating regular language to sign language, giving more opportunities for tourists with disabilities to experience diverse tourist attractions
Geographic Information Systems	System created to analyse and store current spatial or geographic data	Colorado Pástor et al. (2020); Neven et al. (2017); Françoso et al. (2013); Ioannidisa and Vozikisb (2007); Pulugurtha et al. (2006); Sobek and Miller (2006)	• Increase independence in transportation for people with disabilities • Support independent travel on buses or other means of transport • Allow information on both locations and times of arrival/departure • A real-time information, communication and localization environment for improving the mobility of travellers with disabilities • Used to design new systems and services to improve the autonomy and mobility of disabled people in cities.
Tourism Information Systems	Platforms that provide tourism-related communication, information sharing, suggestions based on users' profiles, and provide knowledge transfer among all stakeholders	Kołodziejczak (2019) Zajadacz (2014); Michopoulou and Buhalis (2013); Eichhorn et al. (2008); Winkler and Wöß (2006); Rumetshofer and Wöß (2005); Rumetshofer and Wöß (2004); Geehan and Suen (1993)	• Work as data centres, storing accessible information that can be made available for many tourists • Distribution of information and communication strategies • Satisfy the disabled users' requirements, providing all the necessary information to plan a trip • Provide three essential components of information for accessible tourism conditions: about users (user profile), tourism information (medium profile) and information about tourism objects • Ensure personalization and interoperability • Adapt to the user's individual needs (disabilities) to support appropriate search and presentation features. • Easy access to content in sign language • Provide real-time information for PwD

Technologies in a mature phase

Table 5 sums up the main impacts of ICTs in a more technological mature and stable phase. Studies that include GPS focus more on providing real-time information, communication regarding localization (Gell et al., 2015) or increasing independence of PwD in transportation (Davies et al., 2010).

Despite being considered Transport Low-Technology Solutions, there are ICTs, that still today hold great value for many PwD. The works of both Lamont et al., (2013) and Thomson et al., (2007) are both complementary, as each explain low-technological approaches for transportation, during tourism activities. The first study is essentially focused on people with speech impairments (dyslexia), highlighting the problems accessing and using travel information for this group. The second study in a matter develops a cost-effective system of a monitor and audio supplement, to easy demonstrate dynamic, offering passengers a visual and auditory display of where the bus is in relation to its next stop. These low-technological solutions can be visualized as frontline of other complementary ICTs, especially Geographic Information Systems.

Websites are the most analysed ICT of all those identified. Most studies perform an analysis of the accessibility level of the websites. Several different types of websites were analysed, including websites most used by tourists (Gonçalves et al., 2020); websites belonging to travel agencies (Teixeira et al., 2019;

Table 5. Technological impacts of ICTs in a mature stage

ICT	Description	Author(s)	Impact on Accessible Tourism
GPS	Radio navigation system that allows position to be determined	Colorado Pástor et al. (2020); Françoso et al. (2013); Gell et al. (2015); Baudoin and Venard (2010); (Davies et al. (2010); Panou et al. (2007); Pulugurtha et al. (2006)	• Increases independence in transportation for people with disabilities • Supports independent travel on buses and other transport • GPS on mobile phones allows information on both locations and times of arrival • A real-time information, communication and localization environment for improving the mobility of travellers with disabilities • Used to design new systems and services to improve the autonomy and mobility of disabled people in cities.
Websites	A set of related webpages on the World Wide Web, providing information and tracing back to a common uniform resource locator (URL)	Gonçalves et al. (2020); Perna et al. (2020); Borges et al. (2020); Silveiro et al. (2019); Teixeira et al. 2019) Henriques et al. (2019); Domínguez Vila et al. (2020, 2018, 2017); Domínguez Vila and Brea (2009); Lestari et al. (2019) Luccio and Beltrame (2018); Zhang and Cole (2016); Dattolo et al. 2016); Posada et al. 2014); Wu et al. (2014); Aiello (2012); Lazar et al. (2012); Kim and Lehto (2012); Radosav et al. (2011); Pühretmair and Nussbaum (2010); Pühretmair and Wöß (2008) Williams et al. (2006); Shi (2006) Gutierrez et al. (2005)	• Integrate accessibility information, making it available for a broader target group • Simplified browse pages and simplified selection pages can be applied to tourism webpages, as in the case of tourism objects • Make use of accessibility information stored in a central accessibility database to support potential customers with disabilities with up-to-date, reliable and comparable accessibility information on tourism object(s) • Help disable tourists understand if a tourism object is accessible • Websites can be one of the central communications tools in tourism. • Accessibility of web pages means that PwD can perceive, understand, and browse contents, add their content and use interaction with other users • Take into account that each user has their own profile and needs and therefore integrated adaptation of platforms is crucial • Allow PwD to plan and almost independently decide the trip they want to make • Through the Internet's free flow of information, businesses and attractions can speed up the popularization of accessible tourism concepts and enable more potential customers
Transportation Low-Technology Solutions	Technologies without complex software features (monitor displayer with audio function/ talking signs)	Thomson et al. (2007); Lamont et al. (2013)	• Provide information that helps PwD to decide if they can travel, how to travel and when to travel • Implement auditory systems in the front section of the bus • Allow disabled tourists to feel more confident travelling on buses • Present routes with both audio and visual display • The system would provide relevant information and sufficient time for tourists to disembark • Used for modelling demand for new accessible tourism services

Silveiro et al., 2019); websites containing information on the accessibility of a tourist destination (Perna et al., 2020; Shi, 2006); websites of UNESCO Global Geoparks (Henriques et al., 2019); websites of official/governmental tourism (Domínguez Vila et al., 2020, 2018, 2017; Posada et al., 2014); websites of airlines (Lazar et al., 2012) and lodging websites (Williams et al., 2006). To perform this analysis, automatic analysis tools according the WCAG guidelines (W3C, 2016) were the mostly used. However, due to the fact that they were conducted in different years, different versions of the WCAG guidelines were used. A common feature found across many studies was the low level of web accessibility. This may spell problems for PwD, since websites are a fundamental information source for potential visitors planning their tourism trips (both people with disabilities and people without disabilities). Therefore, they should be developed in an accessible way, and that is exactly what some studies try to recommend.

Luccio and Beltrame (2018) present some guidelines for the development of accessible and also usable tourism websites for tourists with hearing disabilities. Dattolo et al. (2016) studied how a website should be developed to be enjoyable by a user with cognitive disabilities, collecting data such as navigation paths, user profiles, user preferences, and geo-localizations to develop a model. The same rationale was followed by Radosav, et al. (2011), who present guidelines which should be considered in the design of web portals in order to make them accessible to people with disabilities. Finally, Wu et al. (2014) created a website for everyone to post accessible attractions, allowing PwD to share their own experiences, increasing the willingness of potential customers to travel in the process.

The different types of websites also give non-disabled people exposure to accessible tourism, spreading accessibility to a wider audience. In fact, websites are indispensable ICTs for accessible tourism, since similarly to tourism information systems they can work as databases, and help to disseminate information about accessibility in tourism, which holds great value for PwD. The various studies illustrate that websites could promote the accessibility information, making information available to all, since the lack of information represents one of the biggest threats to accessible tourism.

The Impact of ICTs on Consumer Behaviours of Tourists With Disabilities

According to Cohen, et al. (2014), the consumer behaviour model of tourists is a process that can be categorized in three distinct stages: (i) pre-visit (planning, needs recognition, information search, alternative evaluation and purchase), (ii) on-site (purchase and consumption) and (iii) post-visit (post-purchase and evaluation).

The various ICTs analysed have different impacts, under the circumstances of the different stages (Table 6). In fact, most ICTs have a crucial role in all three stages of a tourism trip. For example, mobile apps, as mentioned by Emrouzeh et al. (2017), are related to navigation (wayfinding), social (sharing, collaborating and communicating), mobile marketing, security, emergencies (health monitoring, and weather alerts), transactional (transactions of some sorts for tickets, reservations, and shopping), entertainment (games, movies, e-readers, etc.) and information. These functionalities are connected across the model of tourist consumer behaviour.

Nowadays, with the help of different ICTs, the entire trip can be planned without leaving home. This easier process is especially important for disabled tourists since proper planning can be done, and all information needed can be obtained, especially about specific information and requirements that they need to know about. Virtual reality can help people gain an idea of what a room will look like, or information can be explored in information systems or websites.

Table 6. ICT impacts on disabled tourists' consumer behaviour

ICT	Tourism Trip stages		
	Pre-Visit	On-site	Post-Visit
Virtual Reality	x	x	
Image Recognition Software		x	
Cloud Technologies	x	x	x
Online Multimedia Material	x	x	x
Mobile Apps	x	x	x
Tourism Information Systems	x	x	x
Geographic Information System	x	x	x
GPS		x	
Websites	x	x	x
Transport Low-Technology Solutions		x	

The task of obtaining tourism products on-site can also turn out to be difficult for PwD, and ICTs could provide help and ensure that disabled tourists are not excluded. For example, exhibits can become available via virtual reality, or accessible transportation to get somewhere can be selected with the help of GPS and geographic information systems. On-site information gathering is also crucial for the enjoyment of leisure activities. Overall, information at this point can be seen as a critical factor, and that is why every single identified ICT can contribute to improving accessible tourism activities for PwD.

Finally, the post-trip phase of a journey takes on an important role in assuring accessibility in tourism. Once again, different platforms, information systems, cloud data centres and websites can be used to evaluate accessibility features and promote the sharing of information, so it reaches other PwD. Different mobile apps can allow sharing experiences in social media, which is also a factor in preventing social exclusion of PwD from leisure and tourism activities.

DISCUSSION AND CONCLUSION

A systematic review was performed to understand the impacts of ICTs on the development of accessible tourism, analysing the different technologies that are currently transforming the accessible tourism market. Despite all the potential of technologies to develop accessible tourism, the literature in this area is severely limited, with a total of 65 papers being reviewed, most having been published in more recent years, and especially focused on the USA and Europe. The analysis took the seven segments of the accessible market into consideration (mobility impaired, vision impaired, deaf or hearing impaired, speech impairments, cognitive disabilities, hidden impairments and elderly people), and while most studies converge on all segments, a few concentrate on discovering the technological impacts for a specific segment. Despite a coherent analysis on the topic, some limitations can be pointed out, namely the use of only one research platform, and the fact that the keywords used in research formula, prevent the appearance of some other potential articles. The literature review carried out in this chapter shows that technology may have a considerable role in promoting accessible tourism, as ICTs can provide a shared knowledge base to address every kind of disability. Different disabilities give rise to different information requirements, so specific services need to be provided.

There was huge variety of authors and publishers, demonstrating the high diversity of scientific research areas which are present in the transformation of accessible tourism by ICTs. The research identified 10 ICTs which promote accessible tourism: (i) virtual reality; (ii) voice recognition software; (iii) cloud technologies; (iv) online multimedia material; (v) mobile apps; (vi) tourism information systems; (vii) geographic information systems; (viii) GPS; (ix) websites; and (x) transport low-technology solutions. While some technologies like websites have already reached the peak of their technological innovations, there are still some like virtual reality which are still in a relatively young stage of development. The technological impact is evident and undeniable, especially regarding the sharing of information about accessible features. ICTs also have potential in other areas, like the case of transport in the form of location technology, which can be fundamental to build accessible geographic databases, helping accessible navigation. Mobile apps are other good examples since they can provide proactivity mechanisms and collaborative features. New information technologies focus not just on divulgation of information, but also making sure everything is connected, which in the case of accessible tourism has significant value. Despite the significant importance and influence of ICTs in tourism, they are not the only technologies that can be inserted in the accessible tourism market.

This study reveals that accessible tourism has changed due to the rise of new technologies. Everyday life is changing across many sectors due to technological impacts, and tourism is no exception. Particularly in the case of accessible tourism, this literature review shows that many advantages can be obtained with the help of ICTs to increase the participation of PwD or with other functional limitations. For instance, a great barrier preventing PwD from travelling and enjoying tourism is often due to a lack of Information. The studies about ICTs in tourism also demonstrate that they can provide help in the different stages of a leisure trip. There are technologies more related to the planning stage, like tourism information systems, websites, and geographic information systems, which can help understanding if specific accessibility requirements are included in a tourism offer. Despite this fact, overall, ICTs can have an important role in supporting accessible tourism on-site, as the need for information tend to be a crucial factor for disabled tourists to enjoy tourism and in preventing their social exclusion. Finally, the fact that ICTs also allow disabled people to evaluate and share of content in a post-trip stage can also be a major contribution to accessible tourism, as it can bring improvements in different tourism products and lead them to become more accessible. Simply put, information is the key, and ICTs are the vessels that can help this information flow across different sectors of accessible tourism, promoting the inclusion of PwD in tourism activities.

The identified studies focused on three main function of ICTs: i) information sharing, ii) communication between tourists and other tourism entities and iii) helping tourists in transportation activities, like travelling by bus. These three are undeniably the most critical in generating accessibility. However, there are still a great number of other technological types that can also have important impacts in other areas related to accessible tourism. Technologies like robotics, augmented reality and virtual assistants have incredible advantages that can also be applied to tourism in order to make it more accessible. To complement this study, it would be important to understand how other technological innovations are changing accessible tourism, and what their status is regarding the maturity level of technological innovation.

Another identified gap is the fact that sometimes an ICT which promotes accessibility conditions is not enough, even if PwD cannot use it, due to some types of constraints. For ICTs to be truly accessible for tourism, both provision of accessible content and access conditions need to be ensured. First, accessible conditions must foster, and second, it should be built in an accessible way, so that PwD can use the technology. As seen in many of the ICTs analysed, especially websites, even though they promote information about accessible conditions, they accumulate a large number of failures and mistakes in terms of accessibility, regarding use by PwD. There is a need to create frameworks that can be used by technology developers, to ensure accessibility in usage by PwD. This can be done, especially by inserting PwD in the technological development process, interviewing and understanding requirements or other possible functionalities. Hence, studies to promote the accessible use of technology are a must for the promotion of accessibility conditions and avoidance of exclusion of PwD, regarding technological means.

The use of ICTs in accessible tourism showed that technologies could have a profound impact in this sector, but some disputes may also emerge with a more common use of technologies in leisure activities. The lack of access to the internet and cybersecurity are examples of many challenges that may affect the use of technologies in tourism. The challenges arising from ICTs also extend to other areas of the travel industry, such as employment, new potential business opportunities and adaptability of both travellers (demand point of view) and travel industries (supply point of view). The technological impacts and subsequent digitalization of the travel industry also bring up questions related to sustainable development in tourism and cybersecurity in travel environments. New ICTs create challenges that need to be addressed to make the most of this opportunity provided by a new digital tourism era.

ACKNOWLEDGMENT

This work was developed in the scope of the research project ACTION - POCI-01-0145-FEDER-030376, funded by FEDER, through COMPETE2020 - Programa Operacional Competitividade e Internacionalização (POCI), and by national funds (OE), through FCT/MCTES.

REFERENCES

Agarwal, Y. (2019). Accessibility information in New Delhi for "EasenAccess" Android-based app for persons with disability: An observational study. *Disability and Rehabilitation. Assistive Technology, 14*(7), 645–662. doi:10.1080/17483107.2018.1471743 PMID:29902941

Aiello, L. (2012). Role of meta-search engine in the relationship tourist/e-retailer. In *Proceedings - ICIDT 2012, 8th International Conference on Information Science and Digital Content Technology* (*vol. 3*, pp. 749–756). Academic Press.

Avis, A. H., Card, J. A., & Cole, S. T. (2005). Accessibility and attitudinal barriers encountered by travelers With Physical Disabilities. *Tourism Review International, 8*(3), 239–248. doi:10.3727/154427205774791591

Baudoin, G., & Venard, O. (2010). Information, communication and localization environment for travelers with sensory disabilities in public transports. In *2010 5th International ICST Conference on Communications and Networking in China* (pp. 1-7). IEEE. 10.4108/chinacom.2010.116

Blocksidge, D. (2003). *Dictionary of disability terminology*. http://www.dpa.org.sg/wp-content/uploads/2013/01/Dictionary.pdf

Borges, I., Costa, E., Sousa Pinto, A., & Abreu, A. (2020). Infoaccessibility on the websites of inbound markets of Portugal destination. In Á. Rocha, A. Abreu, J. de Carvalho, D. Liberato, E. González, & P. Liberato (Eds.), *Advances in Tourism, Technology and Smart Systems. Smart Innovation, Systems and Technologies* (Vol. 171, pp. 105–117). Springer. doi:10.1007/978-981-15-2024-2_10

Buhalis, D., & Darcy, S. (2011). *Accessible tourism concepts and issues*. Channel View Publications.

Buhalis, D., & Law, R. (2008). Progress in information technology and tourism management: 20 years on and 10 years after the Internet-The state of eTourism research. *Tourism Management, 29*(4), 609–623. doi:10.1016/j.tourman.2008.01.005

Buhalis, D., & Michopoulou, E. (2011). Information-enabled tourism destination marketing: Addressing the accessibility market. *Current Issues in Tourism, 14*(2), 145–168. doi:10.1080/13683501003653361

ByrneD. M.CorradoC. A. (2017). *ICT services and their prices: What do they tell us about productivity and technology?* (FEDS Working Paper No. 2017-015). https://ssrn.com/abstract=2920451

Chalkia, E., Bekiaris, E., & Madrid, R. I. (2015). Evaluating all-inclusive ICT with developers, end users and stakeholders. In *International Conference on Human Aspects of IT for the Aged Population* (pp. 157-165). Springer. 10.1007/978-3-319-20892-3_16

Cohen, S. A., Prayag, G., & Moital, M. (2014). Consumer behaviour in tourism: Concepts, influences and opportunities. *Current Issues in Tourism, 17*(10), 872–909. doi:10.1080/13683500.2013.850064

Colorado Pástor, B. A., Fois Lugo, M. M., Leyva Vázquez, M., & Hechavarría Hernández, J. R. (2020). Proposal of a technological ergonomic model for people with disabilities in the public transport system in guayaquil. In *International Conference on Applied Human Factors and Ergonomics* (pp. 831-843). Springer. 10.1007/978-3-030-19135-1_81

Dattolo, A., Luccio, F. L., & Pirone, E. (2016). Web accessibility recommendations for the design of tourism websites for people with autism spectrum disorders. *International Journal on Advances in Life Sciences, 8*(3–4), 297–308.

Davies, D. K., Stock, S. E., Holloway, S., & Wehmeyer, M. L. (2010). Evaluating a GPS-based transportation device to support independent bus travel by people with intellectual disability. *Intellectual and Developmental Disabilities, 48*(6), 454–463. doi:10.1352/1934-9556-48.6.454 PMID:21166550

Dekelver, J., Daems, J., Solberg, S., Bosch, N., Van de Perre, L., & De Vliegher, A. (2015). Viamigo: A digital travel assistant for people with intellectual disabilities: Modeling and design using contemporary intelligent technologies as a support for independent traveling of people with intellectual disabilities. In *2015 6th International Conference on Information, Intelligence, Systems and Applications (IISA)* (pp. 1-6). IEEE.

DEO. (2005). *Disability definitions and related links*. https://www.doe.mass.edu/sped/definitions.html

Disabled People's Association (DPA). (2015). *Glossary of disability terminology*. https://www.dpa.org.sg/wp-content/uploads/2015/10/DPA-Disability-Glossary-FINAL.pdf

Domínguez Vila, T., Alén González, E., & Darcy, S. (2017). Website accessibility in the tourism industry: An analysis of official national tourism organization websites around the world. *Disability and Rehabilitation*, *40*(24), 2895–2906. doi:10.1080/09638288.2017.1362709 PMID:28793789

Domínguez Vila, T., Alén González, E., & Darcy, S. (2018). Accessible tourism online resources: A Northern European perspective. *Scandinavian Journal of Hospitality and Tourism*, *19*(2), 140–156. doi:10.1080/15022250.2018.1478325

Domínguez Vila, T., Alén González, E., & Darcy, S. (2020). Accessibility of tourism websites: The level of countries' commitment. *Universal Access in the Information Society*, *19*(2), 331–346. doi:10.100710209-019-00643-4

Domínguez Vila, T., & Brea, J. A. F. (2009). A new challenge: The content and the accessibility on the Spanish tourist web [Un nuevo desafío: El contenido y la accesibilidad al contenido de las web turísticas españolas]. *Revista Galega de Economia*, *18*(1), 1–23.

Eichhorn, V., Miller, G., Michopoulou, E., & Buhalis, D. (2008). Enabling access to tourism through information schemes? *Annals of Tourism Research*, *35*(1), 189–210. doi:10.1016/j.annals.2007.07.005

Emrouzeh, M. P., Dewar, K., Fleet, G., & Bourgeois, Y. (2017). Implementing ICT for tourists with disabilities. In *Proceedings of the 2017 International Conference on E-Education, E-Business and E-Technology* (vol. 6, pp. 50–53). 10.1145/3141151.3141152

Figueiredo, E., Eusébio, C., & Kastenholz, E. (2012). How diverse are tourists with disabilities? A pilot study on accessible leisure tourism experiences in Portugal. *International Journal of Tourism Research*, *14*(6), 531–550. doi:10.1002/jtr.1913

Françoso, M. T., Costa, D. C., Valin, M. M., & Amarante, R. R. (2013). Free software for development of Web GIS in tourism accessibility. *Applied Mechanics and Materials*, *256*, 2953–2956.

Geehan, T., & Suen, L. (1993). User acceptance of advanced traveler information systems for elderly and disabled travellers in Canada. In *Proceedings of VNIS'93-Vehicle Navigation and Information Systems Conference* (pp. 389-392). IEEE. 10.1109/VNIS.1993.585655

Gell, N. M., Rosenberg, D. E., Carlson, J., Kerr, J., & Belza, B. (2015). Built environment attributes related to GPS measured active trips in mid-life and older adults with mobility disabilities. *Disability and Health Journal*, *8*(2), 290–295. doi:10.1016/j.dhjo.2014.12.002 PMID:25637503

Gonçalves, M. J. A., Camarinha, A. P., Abreu, A. J., Teixeira, S., & Da Silva, A. F. (2020). Web accessibility in the tourism sector: An analysis of the most used websites in Portugal. In Á. Rocha, A. Abreu, J. de Carvalho, D. Liberato, E. González, & P. Liberato (Eds.), *Advances in Tourism, Technology and Smart Systems. Smart Innovation, Systems and Technologies* (Vol. 171, pp. 141–150). Springer. doi:10.1007/978-981-15-2024-2_13

Gutierrez, C. F., Loucopoulos, C., & Reinsch, R. W. (2005). Disability-accessibility of airlines' websites for US reservations online. *Journal of Air Transport Management*, *11*(4), 239–247. doi:10.1016/j.jairtraman.2005.01.006

Hamann, L. M. A., Airaldi, L. L., Molinas, M. E. B., Rujana, M., Torre, J., & Gramajo, S. (2015). Smart doorbell: An ICT solution to enhance inclusion of disabled people. In 2015 ITU Kaleidoscope: Trust in the Information Society (K-2015) (pp. 1-7). IEEE.

Haworth, A., & Williams, P. (2012). Using QR codes to aid accessibility in a museum. *Journal of Assistive Technologies*, *6*(4), 285–291. doi:10.1108/17549451211285771

Henriques, M. H., Canales, M. L., García-Frank, A., & Gomez-Heras, M. (2019). Accessible Geoparks in Iberia: A challenge to promote geotourism and education for sustainable development. *Geoheritage*, *11*(2), 471–484. doi:10.100712371-018-0300-5

ICT. (2019). In *Cambridge dictionary*. https://dictionary.cambridge.org/pt/dicionario/ingles/ict

Ioannidisa, C., & Vozikisb, K. T. (2007). *Application of a GIS for the accessibility of archaeological sites by visitors with disability and visitors with reduced mobility*. In *XXI International CIPA Symposium*, Athens, Greece.

Khalil, T. (2000). *Management of technology -The key to competitiveness and wealth creation*. Prentice Hall.

Kim, S. E., & Lehto, X. Y. (2012). The voice of tourists with mobility disabilities: Insights from online customer complaint websites. *International Journal of Contemporary Hospitality Management*, *24*(3), 451–476. doi:10.1108/09596111211217905

Kołodziejczak, A. (2019). Information as a factor of the development of Accessible Tourism for People with Disabilities. *Quaestiones Geographicae*, *38*(2), 67–73. doi:10.2478/quageo-2019-0014

Lamont, D., Kenyon, S., & Lyons, G. (2013). Dyslexia and mobility-related social exclusion: The role of travel information provision. *Journal of Transport Geography*, *26*, 147–157. doi:10.1016/j.jtrangeo.2012.08.013

Lauría, A. (2016). "The Florence Experience": A multimedia and multisensory guidebook for cultural towns inspired by Universal Design approach. *Work (Reading, Mass.)*, *53*(4), 709–727. doi:10.3233/WOR-162256 PMID:26890596

Lazar, J., Jaeger, P. T., Olalere, A., Algarne, M., Augustine, Z., Brown, D'Erasmo, F., Dotson, B., Endiape, G., Govender, L., Hagos, N., Hunt, G., Lawrence, W., Plummer, R., Richardson, A., Saval, J., Schickling, C., Schley, C., Szoka, D., Torrie, R., & Walton, M. (2012). Still up in the air: government regulation of airline websites and continuing price inequality for persons with disabilities online. In *Proceedings of the 13th Annual International Conference on Digital Government Research* (pp. 240-245). ACM. 10.1145/2307729.2307768

Lestari, N. S., Wiastuti, R. D., & Triana, I. (2019). An overview of accessible Tourism information at Taman Mini Indonesia Indah Jakarta. *African Journal of Hospitality, Tourism and Leisure*, *8*(4), 1–11.

Liu, Z., Sokhn, M., Le Calvé, A., & Schegg, R. (2016). City Eye: Accessibility for all. In *Proceedings of the 9th International Conference on Theory and Practice of Electronic Governance* (pp. 350-351). 10.1145/2910019.2910020

Lozynska, O., Savchuk, V., & Pasichnyk, V. (2019). Individual sign translator component of tourist information system. In *International Conference on Computer Science and Information Technology* (pp. 593-601). Springer.

Luccio, F. L., & Beltrame, L. (2018). Accessible tourism for users with hearing loss. In *Proceedings of the 4th EAI International Conference on Smart Objects and Technologies for Social Good* (pp. 243-248). 10.1145/3284869.3284909

Michopoulou, E., & Buhalis, D. (2013). Information provision for challenging markets: The case of the accessibility requiring market in the context of tourism. *Information & Management*, *50*(5), 229–239. doi:10.1016/j.im.2013.04.001

Neven, A., Vanrompay, Y., Declercq, K., Janssens, D., Wets, G., Dekelver, J., Daems, J., & Bellemans, T. (2017). Viamigo: Monitoring tool to support independent travel by persons with intellectual disabilities. *Transportation Research Record: Journal of the Transportation Research Board*, *2650*(1), 25–32. doi:10.3141/2650-04

Panou, M., Bekiaris, E., & Robledo, M. G. (2007). Towards an accessible Europe. In *International Conference on Universal Access in Human-Computer Interaction* (pp. 527-535). Springer.

Perna, G., Varriale, L., & Ferrara, M. (2020). Internet for Supporting and Promoting Accessible Tourism: Evidence from Italy. In Y. Baghdadi & A. Harfouche (Eds.), *ICT for an Inclusive World* (Vol. 35, pp. 149–160). Lecture Notes in Information Systems and Organisation. doi:10.1007/978-3-030-34269-2_12

Petticrew, M., & Roberts, H. (2008). *Systematic reviews in the social sciences: A practical guide*. John Wiley & Sons.

Posada, R. H., Cano, T. C., & Rodríguez, J. A. (2014). Accessibility in eGovernment. In *Proceedings of the XV International Conference on Human Computer Interaction* (pp. 1-2). ACM.

Pühretmair, F., & Nussbaum, G. (2011). Web design, assistive technologies and accessible tourism. In D. Buhalis & S. Darcy (Eds.), *Accessible tourism: Concepts and issues* (pp. 274–285). Channel View Publications.

Pühretmair, F., & Wöß, W. (2008). A flexible concept to establish accessibility information in tourism web-pages. In *International Conference on Computers for Handicapped Persons* (pp. 981-988). Springer. 10.1007/978-3-540-70540-6_147

Pulugurtha, S. S., Hirshorn, B. A., Krishnakumar, V. K., & Stewart, J. E. (2006). Transportation network analysis to access community resources by older populations. In Applications of Advanced Technology in Transportation (pp. 677-682). doi:10.1061/40799(213)108

Radosav, D., Karuovic, D., Markoski, B., & Ivankovic, Z. (2011). Guidelines on accessible web portal design. In *12th International Symposium on Computational Intelligence and Informatics (CINTI)* (pp. 297-302). IEEE.

Ribeiro, F., Silva, A., Barbosa, F., Silva, A. P., & Metrôlho, J. C. (2018b). Mobile applications for accessible tourism: Overview, challenges and a proposed platform. *Information Technology & Tourism*, *19*(1–4), 29–59. doi:10.100740558-018-0110-2

Ribeiro, F., Silva, A., Metrôlho, J. C., Silva, A. P., & Barbosa, F. S. (2018a). A new framework for accessible tourism mobile application development. *International Journal of Mobile Computing and Multimedia Communications*, *9*(2), 31–46. doi:10.4018/IJMCMC.2018040103

Rouse, M. (2014). *ICT (information and communications technology, or technologies)*. https://searchcio.techtarget.com/definition/ICT-information-and-communications-technology-or-technologies

Rumetshofer, H., & Wöß, W. (2004). Tourism information systems promoting barrier-free tourism for people with disabilities. In *International Conference on Computers for Handicapped Persons* (pp. 280–286). 10.1007/978-3-540-27817-7_42

Rumetshofer, H., & Woss, W. (2005). Semantic maps and meta-data enhancing e-accessibility in tourism information systems. In *16th International Workshop on Database and Expert Systems Applications (DEXA'05)* (pp. 881-885). IEEE. 10.1109/DEXA.2005.176

Sedlar, U., Kos, A., Pustisek, M., Bester, J., Pogacnik, M., Mali, L., & Stojmenova Duh, E. (2018). *Tackling the challenges of ICT innovation and talents for Industry 4.0*. https://ltfe.org/reference/tackling-the-challenges-of-ict-innovation-and-talents-for-industry-4-0/

Serra, R., Metrôlho, J., & Ribeiro, F. (2019). App for Inclusive Tourism: A case study. In *2019 14th Iberian Conference on Information Systems and Technologies (CISTI)* (pp. 1-6). IEEE. 10.23919/CISTI.2019.8760623

Shi, Y. (2006). The accessibility of Queensland visitor information centres' websites. *Tourism Management*, *27*(5), 829–841. doi:10.1016/j.tourman.2005.05.012

Siddaway, A. (2014). *What is a systematic literature review and how do I do one?* University of Stirling.

Silveiro, A., Eusébio, C., & Teixeira, L. (2019). Heterogeneity in accessibility of travel agency websites: A study in the central Portugal region. *RISTI - Revista Iberica de Sistemas e Tecnologias de Informação, 35*, 18–34.

Smith, R. W. (1987). Leisure of disabled tourists: Barriers to travel. *Annals of Tourism Research*, *14*(3), 376–389. doi:10.1016/0160-7383(87)90109-5

Sobek, A. D., & Miller, H. J. (2006). U-Access: A web-based system for routing pedestrians of differing abilities. *Journal of Geographical Systems*, *8*(3), 269–287. doi:10.100710109-006-0021-1

Teixeira, L., Eusebio, C., & Silveiro, A. (2019). Website accessibility of portuguese travel agents. In *14th Iberian Conference on Information Systems and Technologies (CISTI)* (pp. 1–6). IEEE.

Thangaraj, M., & Gomathi, K. S. (2019). Design and development of MDOE for virtual tourism management for elderly people. *International Journal of Recent Technology and Engineering*, *8*(2), 3558–3562. doi:10.35940/ijrte.B3067.078219

Thomson, D. J., Gylseth, M., McGarry, R., & Garcia, C. V. (2007). The VVIP system: encouraging the use of public transport in Edinburgh. In CHI'07 Extended Abstracts on Human Factors in Computing Systems (pp. 2091-2096). doi:10.1145/1240866.1240955

W3C. (2016). *Understanding WCAG 2.0*. https://www.w3.org/TR/UNDERSTANDING-WCAG20/

Waschke, S. (2004). *Labeling im barrierefreien tourismus in deutschland–Vergleichende analyse auf basis europäischer beispiele* [Labelling in barrier-free tourism in germany comparative analysis based on european examples]. University of Lüneburg.

Williams, R., Rattray, R., Grimes, A., Darcy, S., Mills, J. E., Han, J. H., & Clay, M. J. (2006). Accessibility of hospitality and tourism websites: A challenge for visually impaired persons. *International Journal of Tourism Research, 5*(1), 28–41.

Winkler, M., & Wöß, W. (2006). Accessibility add-on box enabling barrier-free tourism information systems (TIS). In *International Conference on Computers for Handicapped Persons* (pp. 298-305). Springer. 10.1007/11788713_45

Wu, Y. C. J., Chang, C. L., & Hsieh, Y. J. (2014). Enhancing learning experience of the disabled: An accessible tourism platform. *Journal of Universal Computer Science, 20*(15), 2080–2095.

Zajadacz, A. (2014). Sources of tourist information used by deaf people. Case study: The Polish Deaf community. *Current Issues in Tourism, 17*(5), 434–454. doi:10.1080/13683500.2012.725713

Zajadacz, A., & Szmal, P. (2017). Accessible tourism for deaf people in Poland: The SITur and SITex programs as proposals for accessible urban information. In *International Conference on Universal Access in Human-Computer Interaction* (pp. 348-359). Springer. 10.1007/978-3-319-58700-4_29

Zhang, Y., & Cole, S. T. (2016). Dimensions of lodging guest satisfaction among guests with mobility challenges: A mixed-method analysis of web-based texts. *Tourism Management, 53*, 13–27. doi:10.1016/j.tourman.2015.09.001

ADDITIONAL READING

Alexis, P. (2017). R-Tourism: Introducing the potential impact of robotics and service automation in tourism. *Tourism Management, 29*(3), 458–468.

Buhalis, D., & O'Connor, P. (2005). Information communication technology revolutionizing tourism. *Tourism Recreation Research, 30*(3), 7–16. doi:10.1080/02508281.2005.11081482

Cheung, C. W., Tsang, T. I., & Wong, K. H. (2017). Robot Avatar: A virtual tourism robot for people with disabilities. *International Journal of Computer Theory and Engineering, 9*(3), 229–234. doi:10.7763/IJCTE.2017.V9.1143

Darcy, S., & Dickson, T. J. (2009). A whole-of-life approach to tourism: The case for accessible tourism experiences. *Journal of Hospitality and Tourism Management, 16*(1), 32–44. doi:10.1375/jhtm.16.1.32

Jovicic, D. Z. (2019). From the traditional understanding of tourism destination to the smart tourism destination. *Current Issues in Tourism, 22*(3), 276–282. doi:10.1080/13683500.2017.1313203

Li, Y., Hu, C., Huang, C., & Duan, L. (2017). The concept of smart tourism in the context of tourism information services. *Tourism Management, 58*, 293–300. doi:10.1016/j.tourman.2016.03.014

McKillop, H., & Sills, E. C. (2013). Sustainable archaeological tourism of the underwater Maya project by 3D technology. *Anthropology News*, 12–13.

World Health Organization. (2011). *World report on disability*. https://www.who.int/disabilities/world_report/2011/report.pdf

KEY TERMS AND DEFINITIONS

Accessible Tourism: Tourism for all, that enables people with various access requirements to have access to tourism products, services, and environments, without constraints.

Bibliometric Indicators: Quantitative methods for analyzing scientific publications.

Digital Transformation: Extension of products and services within a digital level.

People With Disabilities: People suffering from any restriction or lack of ability (resulting from an impairment) to perform an activity in the manner or within the range considered normal for a human being.

Technology Life Cycle: Period of time from the emergence of the technological component to its complete decay, when it is replaced by another technology.

Tourists' Consumer Behavior: The way tourists behave according to their attitudes and requirements before, during, and after travelling.

Chapter 2
The Challenges of Accessible Tourism Information Systems for Tourists With Vision Impairment:
Sensory Communications Beyond the Screen

Vicky Richards
Cardiff Metropolitan University, UK

Nic Matthews
Cardiff Metropolitan University, UK

Owen J. Williams
Wales Council of the Blind, UK

Ziad Khan
Digital Accessibility Centre, UK

ABSTRACT

Developments in accessible tourism and the provision of information and communication technologies (ICT), mobile, and assistive technologies have arguably not resulted in equitable opportunities for vision-impaired people. This chapter outlines accessible information needs of vision impaired tourists, drawing upon a small-scale project of nine telephone interviews conducted by Wales Council of the Blind. It considers user experiences in the context of ICT to help vision impaired tourists navigate information systems such as travel apps, social media, and websites, assessing how these technologies meet user needs. Interviews focused on information provision, pre-planning and travel stages of the tourism system, and the challenges for universal design. Designers and tourism providers have roles as facilitators of accessible tourism, enabling vision-impaired tourists to feel included in experiences. This requires collaboration across the tourism ecosystem from digital developers and marketers alongside disabled people as active stakeholders.

DOI: 10.4018/978-1-7998-6428-8.ch002

INTRODUCTION

The tourism industry has experienced continued expansion and diversification to become one of the largest and fastest-growing economic sectors in the world (United Nations World Tourism Organisation, 2020). Despite this economic success, coupled with the social significance of tourism experiences and the legal requirements set out in the UK through the Equality Act (2010), the industry is still working towards equal participation for disabled people. Thus, whilst the industry talks of the value of the 'purple pound' (disabled people), disabled citizens continue to encounter barriers to their participation in tourism experiences. These range from choosing and booking a holiday, to accessing screens while travelling and interacting at visitor attractions. Yet the travel motivations of people with disabilities mirror those of people without disabilities and it has been long recognised that *every* tourist has diverse needs (Yau, McKercher, & Packer, 2004).

This chapter provides an insight into the accessible information needs of vision impaired tourists. It will consider their user assistive technologies (mobile phones, tablets, speech, and magnification) within the context of the development of Information and Communication Technologies (ICT) to help vision impaired tourists navigate the tourism information systems. These include but are not limited to booking sites, information boards, travel apps, social media and websites. The chapter examines how forms of technology and the adaptations of these technologies meet (if at all) the needs of vision impaired users. By presenting their experiences, it brings into focus what needs to be delivered to ensure information systems within tourism are accessible. It will draw upon a small-scale research project by the *Wales Council of the Blind*, that included nine telephone semi-structured interviews, alongside expertise from the Digital Accessibility Centre. The interviews focused on assistive technologies, information provision through the pre-planning and travel stages of the tourism system.

BACKGROUND

Tourism is characterised as a period of fun, free time and of escape, yet it is an activity that requires work and preparation (Richards, Pritchard, & Morgan, 2010). It encompasses pre-planning, negotiating travel and in-destination tourist activities (attractions, transport and hospitality experiences). Buhalis and Darcy (2011) in adapting Leiper's tourism system added accessibility to five areas of amenities, attractions, ancillary services, activities and available tourism packages. Hence, tourism is more than an access issue (Yau et al., 2004). The foundation of any tourism experience is the combination of an accessible destination (Israeli, 2002), appropriate accommodation (Darcy, 2010) and accessible and appropriate information. Darcy and Dickson (2009, p.41) suggest that 'destinations must have knowledge management in place that presents information in a way that allows individuals with access considerations to make informed decisions for their needs'. Here there are very specific issues for people with vision impairment as information is a point of access to the outside world and is evident in all aspects of daily life.

By comparison to other people with disabilities vision impairment creates an additional barrier in that information is often not in an accessible format. For instance, the Wales Council of the Blind (cited in Richards, 2013) reported that 85% of people with sight problems in Wales cannot read their own post. Yet regardless of the advancements in technology, vision impaired people are still relying on a sighted person to read information for them which according to Royal National Institute of the Blind (RNIB) research proves is 'clearly a problem, both in terms of privacy and people's right to live as independently as possible' (RNIB, 2015).

According to the *State of the Nation Eye Health Annual Review* (RNIB, 2017) every day 250 more people start to lose their sight in the UK and one in five of us will live with sight loss in our lifetime (RNIB, 2017). Presently although the prevalence of blindness worldwide has decreased as a result of screening, access to low vison services and treatment (for example through international charities such as Sightsavers), there are 36 million people with some form of sight loss (World Health Organisation, 2020). The WHO has estimated that this could treble due to population growth and an ageing population to 115 million people; 4.1 million of these will be in UK by 2050 (including 1000,000 adults with learning disabilities).

There are many forms of visual impairment and eye conditions from no sight, to various forms of low vision such as glaucoma, cataracts and diabetic retinopathy. This means individuals have different ways of seeing and differential use of any residual vision (Small & Darcy, 2010; Richards et al., 2010). Living with sight loss makes it difficult to read printed information which in turn can result in feelings of isolation, frustration and exclusion; all of which contradict the *Convention on Human Rights* Articles 8 and 9 Access for all and Accessibility, respectively. Accessibility of information is particularly important in the holiday decision-making process. Provision of accessible information is provided for under the Equality Act (2010) in the UK and equivalent legislature worldwide. Incumbent in that provision is an awareness that where information is placed, its format, and the usefulness of its content all influences the user's engagement and experience (Richards, 2013). The significance of the accessibility of information is heightened through an increasing reliance on and assumed access to emerging technologies through websites and Apps. According to research (Byron, Blake, & Bridges, 2013; Flynn & Lord, 2015, cited in RNIB, 2015) the increased use of computers, mobile phones, internet and social media, suggests that visually impaired people may have been affected more by these changes than other disabled people, and those without any impairment.

The development of mobile technologies and assistive technologies for vision impaired people has opened up access to information from using email to browsing the internet and using smartphones and apps. Types of accessibility features include the ability to change the size of the font, the colour of the print or background and for totally blind people they would need to be able to use speech recognition software seamlessly with the interface of the web page. However, emerging technologies in the forms of Apps, information screens and interactive web pages create challenges for the visually impaired browser. In the early days of on-line content, a 'blind user' could access most of the information through text-to speech software (Hackett et al., 2005). However, as design and functionality has advanced to include tables, images, films and text embedded with graphics problems for disabled user groups, particularly visually impaired people have been identified (Curran, Walters, & Robinson, 2007). It has become increasingly challenging to rely on assistive technology and the skills of the web user and unless it is available in alternative formats and compatible with the assistive technologies of users, visually impaired people are unable to access information (Hersh & Johnson, 2008). Furthermore, information is not always accessed at the same speed and spontaneity as non-disabled users might take for granted. These concerns have been expressed in terms of a 'digital divide' (Hersh & Johnson, 2008) where arguably as our society becomes more digitally information driven the gap could increase rather than decrease due to the lack of access to assistive technologies, the lack of training to use such tools and a general deficiency in broadband coverage (WiFi). In fact, the UN Secretary General Ban Ki-Moon stated in 2016, 'Everyone has the right to access tourism services on an equal basis. Even with modern technologies, those with visual, hearing, mobility and cognitive impairments are being left behind in many tourism destinations' (United Nations, 2016).

A natural aspect of holiday planning is gathering information on destinations, attractions and travel options; therefore, tourism materials are crucial to holiday preparations and experiences. They are essential to every potential tourist however as intimated earlier; the planning process is multi-faceted for vision impaired people. The content and quality of information is important, but so is its format, so where information is visual and 'eye catching' it is often inappropriate for vision impaired people (Richards, 2010). In addition, the content of information provided by tourism providers is often inadequate for people with disabilities (Hunter-Jones, 2004).

THE CHALLENGE OF TOURISM INFORMATION SYSTEMS

Over the last decade there has been concerted progress in the advancement of tourism enquiry into the experiences of disabled people in general, and particularly researching the experiences of tourists with a visual impairment (Poria, Reichel, & Brandt, 2011; Richards et al., 2010; Small et al., 2011; Agapito et al., 2014; Small, 2014; Loi & Kong, 2017). There is also a growing recognition and awareness of the accessible tourism landscape. Since the work of Poon (1988) into tourism and information technology, the literature surrounding technology advancements from the traveller perspective has traversed the consciousness of the study of computers in human behaviour (Ayeh, 2015), purchasing behaviours (travel tickets, Bigne et al., 2010) to the influence of User Generated Content (Filho et al., 2018) digital marketing (Cloquet, Palomino, Shaw, Stephen, & Taylor, 2018: Buhalis 2019) and accessibility of websites (Domínguez, González, & Darcy, 2019) .

Defining Disability Within an Accessible Tourism Context

Tourism scholars interested in accessible tourism have approached the marginalisation of disabled people from a social model of disability perspective. The rise of the Disability Rights Movement, the establishment of disability discrimination legislation in many countries, the advent of the *UN Convention on the Rights of Persons with Disabilities* along with the more recent UN Sustainable Development Goals have all contributed to changing perceptions of disability and the increasing removal of barriers to the participation of disabled people in everyday life (Shakespeare, Lezzoni, & Groce, 2009).

Models of disability are important in conceptualising disability, especially when undertaking research *with* people with disabilities and when advocating for a greater collaboration amongst all the stakeholders. Discussions of the medical model and social model of disability have been well documented by disability scholars (for example Barnes & Mercer, 2005; Finkelstein, 2004; Swain & French, 2000, Tregakis, 2003) and tourism scholars, in particular Simon Darcy. The latter exploring an extension of the social model to consider a more sociology view of the body and embodiment construct (Buhalis & Darcy, 2011).

Whilst much of this wider debate is beyond the parameters of this chapter, it is important to examine some of it here to contextualise and challenge traditional views of disability and vision impairment. Traditionally, people with disabilities were seen as 'ill', in need of medical attention and welfare, people who were unable to participate fully in the world and therefore needed to be excluded from it (Barnes & Mercer, 2005). In 1976, the *Union of the Physically Impaired against Segregation* in the UK declared it was society disabling them, hence the social model of disability being defined as 'the restriction imposed on top of our impairments by the way our society is organised – a form of oppression to which all disabled people are subject' (Sheldon, 2004 p. 69). The early 1990s saw a growing movement challenging the social model as it has a tendency to 'absent' the body from its interpretations (Finkelstein, 2004); whereas the sociology of impairment model recognises this relationship between our bodies, selves and environments (Freund, 2001).

According to Buhalis and Darcy (2011) this embodiment is not a homogenous construct as individuals experience different impairments, support levels, (in)dependence, differing use and access to assistive technologies, all of which inform an embodied identity. This perhaps reflects more accurately the idiosyncratic nature of every individual as visually impaired people are 'first and foremost people, who are as idiosyncratic as anyone else' (Duckett & Pratt 2001, p.830) and the inherent complexity of the tourism system and the interplay of the disabled tourist with the journey and touristic activities (Darcy, 2010).

Constraints and Facilitators for Disabled Travellers

Smith (1987, p.376) identified that: 'tourists, especially those with disabilities experience barriers to leisure participation that undermine their sense of freedom and feelings of personal control… barriers that disproportionally affect disabled tourists.' Arguably these sentiments are still relevant. It is the combination of such barriers that influences the amount of satisfaction derived from tourism. This seminal work identifies three main participation barriers for people with disabilities, namely intrinsic, environmental and interactive barriers.

Smith's barriers were extended by Daniels, Drogen-Rodgers and Wiggins (2005) who termed them intrapersonal, interpersonal and structural constraints in order to emphasize the complexity of barriers. It was McGuire (1984, cited in Darcy 2003) who differentiated between constraints and barriers, suggesting that constraints are factors that affect a person's predisposition for participation in an activity, and barriers are those impediments encountered while attempting to participate or whilst participating. Since these early studies, these factors have been investigated by researcher addressing the interplay between constraints, participation, negotiation, tourist behaviour, motivations to travel, disability and the environment (Burnett & aker, 2001; Daniels et al., 2005; Packer, McKercher, & Yau, 2007) all of which confirm Packer et al.'s (2007) assertion that there is indeed a complex interplay between the individual, the tourism context and the environment (Darcy, 2010). Darcy (2003) for example found that the barriers of intrapersonal and interpersonal were largely structural, thus creating a socio-spatial tourism segregation of people with disabilities. Richards et al. (2010) discussed barriers to participation under person-centred, societal and environmental. Accessible information was placed under the environmental barriers although relevant to personal visual function, access to equipment and societal issues of training, policies and communication. Research began to shift the emphasis away from barriers and constraints to highlighting participation and integration manifested in the whole-of-life approach (Darcy & Dickson, 2009), a key theme this this study.

More recently Deville and Kastenholz (2018) have highlighted a further dimension of 'facilitators' for access dependent on the structural, interpersonal and intrapersonal constraints and negotiation of these constraints. For example, support from networks and positive attitudes of others to overcome lack of reliable information and physical barriers. Critical theory in tourism argues that accessible tourism can move beyond the constraints framework to seek transformative outcomes (Pritchard, Morgan and Ateljevic, 2011) for tourists with disabilities.

Technology traverses these constraints. It determines individual access to assistive technologies, access to wireless connections and training to use the technology. Sharpley (2018) argues that traits such as computer anxiety and technophobia will diminish the more technology is integrated into contemporary life. ICT can be transformational and should enhance the tourist experience but there will arguably be additional barriers on top of what are still challenges of accessible information for disabled people, (in particular vision impaired people) in regard to cyber security, lack of trust, fear of making a mistake and the need for a real person (Sharpley, 2018).

TRANSFORMING THE TRAVEL ECOSYTEM THROUGH INFORMATION

Accessible tourism must, by definition, facilitate tourism for all. This starts with ensuring all tourists have access to information. With more of our economic and social lives moving on-line, it is important to understand how that transition has been received and experienced. In a 2017 White Paper on digital transformation, the *World Economic Forum* (WEF) characterised the aviation, travel and tourism industry as being at the 'forefront of digital disruption' (p.4). It identified areas that would bring value for the industry and for society, inclusive of enhancement of security; sector-specific digital enterprises (i.e., tailored use of artificial intelligence and augmented reality experiences); and the enabling of the travel ecosystem (i.e., the advancement of digital platforms). WEF argued this would (could) allow for 'traveller centricity'; optimising the customer experience by understanding their travel habits, preferences and decision-making processes. The challenge for developing accessible tourism is to continue to monitor the extent to which the information systems are responsive to travellers with diverse or complex needs.

Information Systems and the Travel Ecosystem

The impact of technology on information systems across the travel ecosystem is evident and the characterisation of its operation as being at the forefront of innovation and disruption is not unfounded. From platforms that provide meta-searches (i.e., Trivago and Kayak) to on-line travel agencies (i.e., Booking.com and Expedia) through to accommodation providers such as AirBnB and the influence of user generated content (UGC) on social media, the industry is being responsive to consumer expectations for more on-demand services whilst simultaneously harnessing that data to drive future transformation. However, as the industry readies itself for the 'Fourth Industrial Revolution' its' operators must be sensitive to who its customers are (or could be) and be reminded that there are different ways of 'seeing' (Small & Darcy, 2010). Accessible tourism will only become a sustainable reality when the ecosystem works for all. That system is inclusive of industry stakeholders but also others that can address the ICT and other barriers faced by potential tourist. Cassia, Castellani, Rossato and Baccarani (2020) have identified the role that digital ecosystems can play in 'co-creating' tourism experiences.

The digital transformation of travel and tourism is resulting in greater connectivity between service providers and tourists and between tourisms themselves (for example sharing reviews). It also has the *potential* to increase collaboration between operators in the supply chain. The aspiration would be to offer more personalised, 'end-to-end propositions'. This would overhaul the travel experience for tourists, fully tailoring the marketing, booking, travel and transfer and destination experience for the individual, regardless of their needs. Darcy, McKercher and Schweinsberg (2020) term this an 'accessible destination experience' that affords all tourists an 'equality of offering'.

Traveller Centricity

The mediation of the tourism experience through information systems allows for more individual support to be provided to consumers at the point at which they are making their travel decisions. Such data analytics are important to operators, offering behavioural insights, but data are only as good as the ICT systems allow. As new technologies and applications emerge, evidence will be needed on how fit for purpose these systems are. Understanding the tourists' experience of and interaction with such systems is key to enhancing their utility and ensuring that they function for all users. This is challenging for the tourism sector, given the 'idiosyncratic quality of tourism decisions' (Gretzel, 2011, p.760). However, understanding the situational needs of the individual traveller is beneficial in helping the sector use technology in a way that could transform products and services for all (Budd & Ison, 2020, Randle & Dolnicar, 2019).

Designing systems for all starts with understanding the idiosyncratic and multi-faceted nature of the tourist experience. This includes consideration of the decisions made pre-, during- and post-consumption. The 'purple pound' does not represent a homogeneous group of consumers, but there is evidence that disabled tourists face more challenges than most when trying to obtain the information they need in order to make an informed decision about a tourism 'purchase'. Deficiencies in the supply of information have been shown to increase with the degree of disability (Michopoulou & Buhalis, 2013) and this is further complicated for those that require the use of assistive technologies (Drews & Schemer, 2010).

The Potential Power of the Purple Pound

The need for the tourism sector to recognise the purchasing power of disabled tourists and respond is important from a business perspective (the 'purple pound' was estimated to be worth around £80b to the sector, Michopoulou & Buhalis, (2013)). In 2018 the Department for Work and Pensions suggested that the spending power of disabled people and their families was £249b per annum (DWP, 2018). Furthermore, research conducted by VisitEngland, also in 2018, suggested that 430,000 British adults with impairments did not take a domestic holiday as a result of accessibility issues. The industry is missing out on a potential £116.7 million. Tourism for All is a business opportunity and an avenue for competitive advantage amongst destinations (Florim et al., 2019, Michopoulou, Darcy, Ambrose, & Buhalis, 2015). According to Darcy (2011) the business case for access markets is compelling. Factors include demography (a response to an ageing population), corporate social responsibility (a company's contribution to civil society) and a whole of life approach (Darcy & Dickson, 2009), as sooner or later everyone will benefit from universal accessibility in tourism.

The need for the tourism industry to be responsive to the disabled traveller is further heightened by the ease with which pre-, during- and post-travel UGC can be shared through social media. The empowerment of the end user created by UGC and the authenticity of their experience makes them a reliable source of information (Niininen, Buhalis, & March, 2007). Their narratives can fill the gap between the supply of official information and the demand for information. Evidence suggests disabled travellers, and especially those with complex needs, must complete multiple searches in order to plan travel. At best this slows or complicates the planning process, at worst it inhibits travel.

Usability and Universal Design

Given the Convention on the Rights of Persons with Disabilities accessibility standards are starting to be laid out and some centres of excellence exist to test information systems for their accessibility. The Digital Accessibility Centre (DAC) in Wales, for example, tests and accredits content according to the following Web Content accessibility guidelines (2018).

- Website Testing in line with W3C Web content accessibility guidelines (WCAG 2.1) to AA level as recommended by UK and other global governments.
- Mobile testing carried out in line with a combination of W3C WCAG 2.1 recommendations, combined with various Native developer guidelines such as iOS, Android and/or Windows (DAC, 2020)

The transformation of ICT is a technical matter, but the interaction between systems and the end user is important. In line with the World Economic Forum's assertion that the digital transformation can be of value to the tourism industry and society, the experience of engaging with content must be considered. Access alone is not sufficient. Accessible tourism must therefore include an evaluation of the usability of information systems. Usability can, however, be subject to an individual's interpretation of the functionality of a system. A more robust assessment of the 'distance travelled' in achieving accessible tourism is to consider if information systems meet the criteria of universal design. This builds in accessibility and usability for all from the outset. The needs of vision impaired travellers or other groups would be built in rather than be seen as an adaptation or add-on (Eichhorn & Buhalis 2011). Evidence from the current explorative study and the wider literature suggests the digital transformation falls short of this. Yet the growing evidence-base suggests there is scope for ICT systems to be re-engineered and offer all travellers something of value; be that through appropriately designed websites and apps or via emerging technologies such as augmented, mixed and virtual reality which have the potential to inspire all tourists to book, plan and experience travel differently (see Beck, Rainoldi, & Egger, 2019 for a review of the latter). Further examination of what good design looks and feels like is needed, to ensure the travel ecosystem works for all.

INFORMATION AND ASSISTIVE TECHNOLOGY FOR VISION IMPAIRED TRAVELLERS

ICT according to Buhalis and Law (2008) has become an indispensable part of people's lives, bringing radical changes to tourism and embedded in the service delivery process (Rasoulimanesh et al., 2019). Inevitably, technological development, particularly mobile technologies, has advanced rapidly whereby mobile phones/smartphones have become an everyday personal appendage (Chanjaraspong, 2017).

In terms of travel, Bannister (2017) suggests that digital communications enhance passenger travel experience as it helps travellers get to where they need to go safely, efficiently and effectively. The aim is to provide passengers with the right information at the right time to ensure a positive experience. Airport flight information displays and digital signage content encompass for example the Gate Information Display screens (GIDS) (Anonymous, 2002) and interactive touch screens. Metropolitan Airport in Washington D.C. for example expanded its touchscreen directory to become more interactive with its customers. Patrick Hogan, director of public affairs and marketing for the Metropolitan Airports Commission, commented, 'The mobile app and touch-screen directories together map the way to a first-rate airport experience… The devices help travellers find their gate, choose restaurants and shops that fit their taste and budget, and identify locations of the specific merchandise and services' (Anonymous, 2013). Understandably, customers expect information to be where they need it, when they need it. This needs to be presented appropriately especially important in large venues such as airports and train stations which are anxiety inducing environments to minimise apprehension, frustration and stress with finding the correct terminal, finding delay information (Bannister, 2017).

According to AbilityNet (2020), an organisation that supports disabled people of any age, to use technology and provides specialist advice services digital accessibility is not about disability but about universality, reaching as many different people as possible in different environments with diverse capabilities and assistive technologies. The Equality Act (2010) in the UK makes this a legal duty and responsibility for all. Content needs to be easy to read, work on different devices, be easy to navigate, of good design, and include video captions and accessible files. The Clickview Report (Williams & Brownlow, 2019), a study which explored the online shopping experiences of disabled people found though that travel services (e.g. flight, train and bus tickets) was the second highest for inaccessibility after supermarkets despite participants utilising assistive technologies such as screen readers, magnification and dictation software. Effectively as the report states 'in many businesses, the negative online experience of disabled customers suggests disability issues remain niche at best and disabled customers remain peripheral' (Williams & Brownlow, 2019). In terms of competitive advantage, 75% of the respondents would pay more for a product from an accessible site rather than the same product from a website that was less accessible. These discriminatory practices add to the frustration and lack of social inclusion.

It remains the case that access to online and digital information is a challenge. The World Wide Web is complex, characterised as both multimedia and multimodal (text, speech, animations, video, colour and sound effects) where most of this information is obtained via visual and auditory senses (Hersh & Johnson, 2008). For example, access to images is a complex challenge for people with visual impairment or blindness (Mascetti, 2017) as, if the assistive technology finds something it's not compatible with, the information, such as graphs, tables, and photographs become inaccessible to the reader (Hersch & Johnson, 2008).

Not all information communication systems are designed to accommodate the needs of all users (Michopoulau & Buhalis, 2011) hence the need for assistive technologies (see Table 1 for those relevant to vision impaired people). Assistive technology then for vision impaired people should allow an individual to perform a task they would otherwise be unable to do (Williams 2019).

As outlined in the introduction, vision impaired people 'see' in different ways as a result of nuances of the eye condition(s). For a person with glaucoma or retinitis pigmentosa for example they may see clearly centrally but have difficulties with peripheral vision. A person with macular degeneration would have difficulty with central vision (affecting reading in particular) but may have better peripheral vision. This means there will also be different ways of reading and accessing information. Vision impaired individuals are able to surf the Internet using screen reading software, magnification, send and receive electronic mail, access Word processing and spreadsheet software such as Microsoft Excel (Dhalwal & Juyal, 2017). Mobile technologies have similarly become indispensable to vision impaired people after initial concerns around design and usability. Although 'smartphones have been equipped with several accessibility functions such as larger font size, inverted screen colour, and 'speak screen' (text-to-speech

Table 1. Types of assistive technologies for vision impaired people (The Wales Optometry Post Graduate Education Centre Technology report (Williams, 2019)

Types of aids	Examples of assistive technologies
Low Vision Aids for access to printed materials	Video magnification, TV based video magnifiers, portable video magnifiers, desktop video magnifiers, computer-based video magnifiers, auto reader/CCTV system, electronic reading aid (text only)
Accessibility for Computers	Customisation and settings, free software applications, professional software and hardware products, screen magnification, larger monitor & large print keyboard, combined speech and magnification software, screen reading (speech output) software, speech recognition software, Apple accessibility for Mac range of products and the iPad.
Electronic notetakers and Braille devices	Electronic notetaker, Braille embossers and Braille displays/ controllers.
Accessible Smartphones	iPhone, mobile phone software, mobile reading aid (voiceover), GPS software and mobile colour recogniser.
Mobile Apps	Microsoft Soundscape is an exploration app, developed in collaboration with Microsoft, designed to make it easier to explore towns and cities. The app was tested and influenced by Guide Dogs service users and staff with a vision impairment.

applications) to assist vision impaired users access information from the touchscreen these are only 'basic, synoptic nonvisual mechanisms' (Huang, 2018).

THE SMALL-SCALE EXPLORATORY QUALITATIVE STUDY: WALES COUNCIL OF THE BLIND

Research Design

The research was conducted by the *Wales Council of the Blind* (WCB) in consultation with The Digital Accessibility Centre, Neath. Due to Covid-19 restrictions, the planned focus group was replaced by nine semi-structured telephone interviews (see Table 2 for the interview schedule). The aim was to provide an insight into the accessible information needs of vision impaired travellers, their use of assistive technologies with the tourism information communication technologies, the main focus on pre-travel and travelling to and from the holiday destinations.

The participants were recruited by purposive random sampling by contacting members of WCB's regional group in Cardiff. All of the participants were vision impaired people (registered severely sight impaired or registered sight impaired) with varying levels of vision impairment and travel experience. The interviewees were between 22-65 years of age. Five were in paid employment, one was a volunteer and three were unemployed. All are from the South East of Wales (see Table 3 for interviewee details). Each person gave their consent to the interview and to the use of anonymised quotations for dissemination. The interviews were conducted by two WCB volunteers who are vision impaired people themselves. This was in line with the emancipatory disability research paradigm (Stone & Priestley, 1996); framed by the social model of disability and directed by vision impaired people. It also follows the principle of

Table 2. Interview guiding questions

INTERVIEW TOPIC CHECKLIST QUESTIONS
1. Travel experience
2. What technology do they use now, positives and challenges?
3. Experiences with travel information through the travel journey (websites, apps, consoles, social media); Review accessibility of consoles (booking screens in tourism settings, accommodation, travel, interpreted experiences and so on) and what can be done to make them accessible (if it can be achieved).
4. A reflection on where the technology is doing well and where improvements can be made including the skills of providers and designers.
5. What should be basic expectations for accessible information and access?
6. What is the Supply chain – basic itinerary: the ability to have access to online information around a destination
7. What happens when the technology fails, back-ups? Travel contingency plans
8. What would a universal design approach look like for tourism information systems?

research undertaken by Richards et al. (2010) into the tourism experiences of vision impaired people, in regarding the participants as the co-researchers, guiding, empowering, leading and using their lived experiences to inform the process. The researchers found that the interviewees were receptive and keen to share experiences. In embodying the emancipatory approach to the research, these findings will be disseminated to the participants along with the WCB with a view to develop the research further.

Table 3. Composition of Interviewees

Interviewee	Gender	Age	Level of vision impairment Registered severely/sight impaired	Status	Length of interview
A	Female	22	sight impaired	Not employed	30 mins
B	Male	46	sight impaired	employed	25 mins
C	Female	26	Sight impaired	Not employed	25 mins
D	Male	32	Severely sight impaired	volunteer	30 mins
E	Hannah	25	Sight impaired	employed	30 mins
F	Female	65	Severely sight impaired	employed	30 mins
G	Female	34	severely sight impaired	Not employed	35 mins
H	Female	28	Severely sight impaired	employed	30 mins
I	Male	48	severely sight impaired	employed	45 mins

The interviews were transcribed and thematically analysed. Once the transcriptions were complete the research analysis and interpretation framework followed that of Pope and Mays (2000) of immersion in the data, identifying key issues and concepts, indexing, mapping and interpretation. The interview guide in itself assisted in the identification of themes and as a result themes emerged to some extent deductively but also inductively. Each script was thematically analysed line by line to highlight key words, concepts, phrases and sentences, presented then in tables for further analysis and examination.

RESEARCH FINDINGS AND DISCUSSION

Travel Experience

All of the participants have some form of tourism experience, all in and around the UK and internationally. Two participants have and do travel alone, others with friends or with family who are sighted and other vision impaired people. The same considerations will apply as consumers in terms of tourist typology and tourism motivations as with any tourist as set out at the beginning of this chapter.

Travel experience is significant as several of the participants expressed more confidence with travelling in the UK as they were more familiar with the systems used and types of environments (for example train stations and travel agencies) however more challenges were expressed with regard to holidaying abroad due to heightened unfamiliarity of systems and environments and different languages. Participant C 'I use buses and trains to travel in the UK and I am reasonably independent but I do not feel that I can plan journeys using timetables and the Internet. If I fly on my own I always have assistance because of the unfamiliarity and I cannot read screens or know where to go.'

Participant A is most confident travelling in the UK 'I have minimal experience of travelling abroad by myself as it makes me feel uncomfortable because of the inaccessibility', she has though in fact travelled abroad when with a travel company. Participant G also travels abroad but only if she is travelling with her family (her mother and her children). Participant E has regularly travelled abroad where experiences in general have varied as a result of the built environment and its inaccessibility (e.g. Bulgaria) with in contrast travelling with a group of vision impaired friends, deliberately choosing a more accessible environment where there was no need to use public transport. Even here though there were difficulties accessing ticket machines for other activities. All nine of the participants have at some point or other used the 'special' assistance at train stations and airports, another area of the travel system requiring attention in terms of consistency. Interestingly, Participant H stated she would prefer to use the technology if it was fully accessible. Out of the nine participants she is the most confident and has holidayed in the UK and abroad, alone, with family, with sighted friends and with vision impaired friends.

Tourism Information Systems, Assistive Technologies and the Challenges

There are persistent misconceptions about vision impairment so it's important to note that all nine participants have some residual (useful) vision but this varies according to the type of eye condition or combination of conditions. Level of vision will then determine the types of assistive technology and combinations thereof for each person. Vision is also affected by lighting levels and sometimes mood.

Unsurprisingly (although as mentioned there are general misunderstandings by sighted people, an occularcentric view that vision impaired people are unable to use mobile phones) all of the participants use their smartphone and computer for accessing information with one participant also utilising an auto-reader CCTV system, and another a telescope low vision aid (see Table 4). In the main the majority of the participants used their smartphone for accessing bus and train timetables and the internet as Participant G explains: 'It allows me to access information on train times and platforms, flight information, bus timetables, e-tickets, boarding passes and Google maps for navigating new places. It also allows me to access signs and information boards through magnifier apps.'

Table 4. Present use of ICT

Use of Technology	Examples
Type of Equipment	• All use smartphones (I-phone or Samsung) • Computer • Internet • I-pad • Telescope
Platforms	• Facebook • Trainline • Trip Advisor / Soundscape • Google maps • Live GPS (buses)
Assistive Technologies	• Camera Feature: Take a picture & enlarge • Magnifier Apps • Internet – screen readers Zoom text and Jaws • Voiceover on Smartphone

The use of apps is also key to accessing information. These fall into two categories in terms of utilising company apps such as Trainline and TripAdvisor but significantly all of the participants accessed the Google Maps as participant A explains:

When travelling both around the UK and abroad, my iPhone is something I use religiously. It not only allows me to look at departure boards and signs with the camera feature but additionally, there are numerous apps which I can gain easy access to. For example, through apps such as Trainline and Trip Advisor, I can access train times, platform numbers and tourist information regarding attractions and costings. Once I have found the attraction I wish to visit, I then use Google Maps to locate it. (Participant A).

Notably this participant highlights the use of the camera, an in-built accessory that provides an accessibility feature in terms of being able to take a picture of an information board for example and enlarge it on the camera. Participant H notes that her smartphone is her assistive technology 'for things such as Google maps to plan journeys… and to magnify and enlarge different websites.' Participant B provides an overview of accessing information on the smartphone and the internet to access tourism information:

I use the internet and apps on my iPhone to search timetables and when a bus is due (live GPS). Also use the Internet to see when my plane is due to leave and from which gate as I cannot see the terminal screens within the airport and a challenge is that gates aren't always announced as they open and you are relying on hearing your announcement and this mean that you are always on edge and unable to relax. This way I can use my iPhone and feel in control more. I also use the Internet (with zoom built-in magnification software) for booking tickets, accommodation and anything else associated with my trip. This could include looking at food service on a plane ahead of boarding for example. I also use apps to find where my seat is and the nearest toilet so that you can familiarise yourself before you board.

One participant with little useful residual vision relies heavily on speech to text or voiceover apps:

Whilst technology was not used in the initial instance of booking a flight, as this was organised manually by a travel agent, I did however need to access my travel flights through an attached document sent to me by email. At this point my sole reliance was on my smartphone and its voice over assistance to help me understand the contents of the attached document. The voice over assistance on my smartphone would feedback the information on the document. (Participant I).

Challenges

Initial impressions indicate that some of the participants are able to access the required information for their tourism and travel needs in the pre-planning stage in particular with booking tickets, accommodation timetables and some attractions information. However compared to these positive experiences there are some inherent challenges expressed by the participants (see Table 5). Firstly, the participants expressed concern over the accessibility of websites and overall design features in terms of being able to adjust font sizes and colour contrast as Participant E experiences with hotel and attraction websites 'they have different contrast on their websites so I'll often copy and paste information into a word document to make it easier to read.'

Table 5. What are the positives and challenges of using information computer systems?

Positives	Challenges
• Use of the smartphone and Apps ▪ Inbuilt accessibility software ▪ Access train & bus times ▪ Train Platforms ▪ Booking tickets ▪ Google maps in unfamiliar places to plan journeys ▪ Find seat & nearest toilets (in airports) • UK ticket consoles accessible with headphones • E-tickets • Boarding passes • Smartphones magnifier apps • Voice over assistance feeds back the info • Internet to check gate no. & departure • Travel agent confirmations on App	• Difficulty finding the language selector on websites • Automated screens, not fully accessible • Confusing technology • Timing out • Clicking incorrectly • Hotels & attractions websites – different contrasts on websites • Apps not updated regularly; sometimes incorrect information • Information boards can often be in inaccessible places, therefore the magnifier doesn't work • Voice over doesn't pick up visually highlighted text e.g. the warning sign needs text detail • Inconsistency of design and accessibility across countries • Digital platforms are not enough alone - rather speak to a person (at least as back-up / to supplement)

All of the participants cited several times that the booking process is a constant challenge. Visual clutter on websites is a particular barrier as this hinders the accessibility of the other areas of the web page. For instance:

Navigating websites such as Expedia prove quite tricky to get the information that is required without being bombarded with visual clutter. This medium makes choosing destinations more convoluted and causes unnecessary visual confusion when researching flights and accommodation. There is a minefield of information to go through; aspects of it being relevant but most not (Participant A).

Participant F also finds websites cluttered and gives examples 'EasyJet's website is pretty straight forward to use but when you're looking for destinations websites such as TripAdvisor and Expedia can be quite cluttered, a lot of information to go through.' Many of the participants prefer to use apps on their smartphones for information searching and booking, as Participants H describes 'When looking at booking a trip I would usually use apps rather than websites such as Airbnb, booking.com, Trainline or Sky scanner because visually the formats of the apps are easier to navigate than websites.' However, the use of apps is not always a positive experience as according to Participant H:

More often than not a traveller is encouraged to access a service providers booking portal online or by an app. But, more often than not you find when accessing these booking portals a confusing layout and inaccessible functionality. Trying to book a coach transfer using a downloaded app left me unable to complete the enquiry and transaction because of not being able to select an option from the displayed timetable. Therefore, a reliance has to be made on a sighted helper, or a costly phone call to customer services on usually a premium rate telephone number.

Another area to consider is that of public consoles for instance at bus and train stations. An additional aspect of this is individuals do not always wish to be identified as a vision impaired people (not everyone is identifiable by a white cane or guide dog) as Participant B commented 'consoles that have clear concise information with larger text on a contrasting background are easier to use but you have to get so close to the screen that this highlights your sight loss. This can cause greater anxiety in unfamiliar settings and where they are public facing.' Participant H also commented that she would like booking consoles, websites and apps to be more functional with voiceover as 'this would allow me to book things myself and to book things more in advance so I can benefit from cheaper travel.'

Participant D prefers to use the apps as 'the EasyJet app is good as you don't get lost – I find the app easier than using the website as there is less clutter and therefore it's easier to use.' In contrast Participant H observed that 'apps aren't updated regularly, if at all and sometimes can be incorrect.'

There are also difficulties with selecting a different language on tourism sites as participants can be using speech to text and/or magnification software to navigate the website pages as expressed by Participant A 'a significant challenge I face when abroad is having access to English language information on tourism websites. Finding the language selector is often hidden away somewhere that I am unable to find'.

As stated earlier all of the participants travel throughout the UK fairly confidently but do experience more difficulties when travelling abroad. For three of the participants they would rather ask for sighted assistance from family or friends for several reasons. The reasons involve ease of use of websites and confidence. As Participant A expresses:

When going through the motions of booking a future trip I usually look to others for their support. They can navigate the website far easier and book far more quickly than me...this is not only just about expediency but it also involves confidence, or in my case the lack of.

Mobility around unfamiliar places can be a real challenge for vision impaired people, alongside unfamiliar systems, operations and accessibility can create additional concerns for that person as Participant E suggests, 'different countries have different levels of support available for instance booking screens don't always have the same things as the UK to make them accessible like headphones slots and change of contrast.' However, several of the other participants expressed apprehension with the booking screens/consoles in the UK too as according to Participant B 'one challenge is checking in at the airport, the automated screens are difficult to use and not fully accessible.'

All of the participants highlighted the importance of the presence of face to face contact or telephone communications not being overlooked. There are a few reasons for this as several of the participants indicate: 'There is added anxiety of doing things yourself and getting it wrong – this could lead to problems boarding and more anxiety (Participant B); 'I'm terrified of using booking sites using access technologies, unfamiliar and confusing as well as timing me out' (Participant C); 'I use the internet for certain things but mostly I want to speak to a person as it is more accessible and efficient, easier and quicker. It is stress free as if you click on the wrong thing you get lost' (Participant D). Participant C also stated that she 'hates using consoles, I always ask for assistance as I'm anxious of holding the queue up.' Finally, 'technology should not be an excuse to move away from a person-centred approach' (participant B) as it could be argued that assumptions of ICT is that information is accessible: 'There is an initial assumption that because it is on the screen that it is accessible but it is not. Do not assume that because it is electronic that it is accessible – it is not.' Participant I believes that service providers make the assumption that tourists are technology experts and that technology is updated 'the crux of the issue is that staff are trained to simply point a service user in the direction of the app or mobile device based on this assumption.'

A person-centred approach is required also in situations when the technology fails for instance as Participant I describes

When tech fails it tends to be both stressful and anxious particularly when the need for information is of priority. Failure of tech leaves one in a vulnerable situation and in an unfamiliar situation or country this can often cause panic. It's at this moment when emotional stress and anxiety can cause impaired judgement and bad decisions. The inability of accessing digital travel documents could possibly lead to a missed flight or coach travel.

Basic Expectations and Universal Design

So far the participants have explained how they access the usual tourism information systems such as websites, Apps and booking consoles (TripAdvisor, Trainline, Booking.com and Airbnb) with their individual assistive technologies. This section offers the views of the participants in terms of their basic expectations of ICT based on their tourism experiences, soft (human) and hard (technology) expectations. These basic requirements are transferable to the universal design principles (see Table 6).

As stated previously basic expectations of all tourists will be varied and this research highlighted common ones for vision impaired people. Participant I encapsulates what is clearly an equality issue: 'the very least expectation a traveller should reasonably be able to expect is a fair opportunity of accessing information online or via an information console, whether that be to book a holiday or flight, and to also access key information at an airport or terminus.'

Access to a person was a clear expectation and requirement highlighted by every participant emphasising that a major part of the planning and in-transit experience was the presence of a person to help to alleviate the anxiety around these touristic activities. Participant A confirmed that regardless of using the features on the smartphone it is 'having access to a person rather than a booking system or machine that is preferred' and Participant C 'I would rather be able to speak on the phone to a person rather than look things up'. Several of the participants mentioned the importance of staff receiving visual awareness training (Cardiff Airport is a good example of this) and not just customer facing staff, also designers and planners:

Table 6. Pre-planning travel activities; Basic expectations of information; Contingency; Universal design

Pre-Planning Travel	Basic Expectations
• Websites search by sighted others • Decide transport first (websites) • Travel Agents directly • Independent web search • Special travel assistance (train stations & airports) • Social media reviews to decide where to go • Information about locations – accommodation close to transport & attractions • Independently book everything • Local travel agent • Online searches using the smartphone • Online App for National Express	• Access to a person (in person/telephone) • Contrasting coloured signage • Functional positioning of screens (e.g. lower) • Being asked preferred format • Public perceptions of sight loss to be improved • Don't assume that if it's electronic it's accessible • Include disabled people in design • Ability to magnify information on screens • Large text/colour contrast • Contact information • App notifications for transport announcements • Easy to follow steps • Websites tested for accessibility regularly.
When technology fails/contingency strategies:	**Universal Design Approach**
• Ask sighted companions; strangers; tourism providers • Printed tickets • Printed contact details • E-mails • Special transport assistance • Portable charger • Providers to give key content/contact info • Trained staff	• Trained staff / providers / designers / tourism staff • Agreed font size • Voice recognition • Considerate positioning of screens • Tutorials for using Apps • Simplicity • Clear text and colour contrast • Screens, voiceovers (& translate into different languages) • Standardise e.g. booking systems • Employ more people with disabilities/vision impairment in marketing, tourism and digital technologies (paid rather than voluntary)

Members of staff, even those who work in the passport section of airports, should have basic visual awareness training so that no matter which part of the airport someone with sight loss is in, there is adequate and appropriate support for them. (Participant A)

Designers, developers must have exposure to service users with additional needs to better understand implementation of layout, functionality and navigation when accessing digital portals. Even today in the 21st century, designers and developers respond with I did not realise how poor design impacted on users (Participant I)

The technical and practical expectations are also presented in Table 6 above but to summarise these involve such issues of clear signage, clear fonts and adjustable colour contrasts on consoles and web pages including the ability to magnify text and use a voiceover (text to speech). The use of more apps is also desirable for example train and flight announcements when information boards are physically out of reach.

Participant I who tests website accessibility was very clear in stating that all providers such as destination management organisations, tourist boards and on-line travel agencies should have their web sites tested against the guidelines regularly:

What is needed is for service providers to have their websites and information portals tested for accessibility not by a frequent or seasoned traveller, but by the occasional annual holidaymaker. Real-world testing of portals and feedback on environments is paramount to ensure that as many people with access needs can confidently obtain information to benefit their experience.

The participants suggested that collaboration to standardise accessibility should be a goal in achieving fuller inclusion and would more authentically align with universal design principles.

DISCUSSION

The findings have provided some insightful details that highlight the specific issues for vision impaired people and their varying levels of access and connectivity with information communication technology at the pre-planning and in-transit stages of the travel system.

The study has affirmed that expectations of travellers are dependent on accessible information; A traveller, be it first time or frequent traveller, expects a basic expectation of being able to access information and booking portals to enable independent planning of a holiday, event, restaurant booking and so on.

Vision impaired tourists require access as other travellers to the supply chain, and at least having a basic itinerary. This concerns the ability to have access to online information around a destination, the choice of booking a simple flight and accommodation with a travel provider as essential requirements. The next stage is in-transit bookings of transport from home to airport/airport to accommodation, all fundamental to completing a basic itinerary.

Dependency on mobile communication and apps is very high however there remain issues around design and usability. Furthermore, there are pitfalls if technology fails, such as mobile phones losing charge or areas of no signal. Understanding how to help travellers have contingency plans in such situations is key. This can be as simple as important contact information available in terms of accessing key service providers such as airline or accommodation providers quickly, essential to avoid anxiety and distress. Fundamentally being able to locate a clear telephone number or a staff member in person, remembering that the internet should supplement rather than replace other means of communication (Hersh & Johnson, 2008).

Awareness is key in informing tourism service providers of ensuring that accessible information is always present and consistent allowing vision impaired users to navigate using different forms of assistive technology. This requires encouraging a co-produced attitude to design and a universal design approach. Accessibility, as we know, comes in many forms. Information and communication for people with sensory loss is key and tourism information systems must consider this at the earliest stages of planning. One major stumbling block is a lack of understanding about how blind people actually use touch screens for example (Wobbrock et al., 2011) consequently alerting designers to consider user-centred design. Awareness of the issues and requirements remain a key component for what is needed for access and inclusion. Perceptions and awareness of the needs of vision impaired people is still lacking so that staff training of tourism providers (front-line and management) and digital designers particularly digital marketers remains a key area for action. It can be argued that some assumptions are made also about vision impaired people's access to technology, to the assistive technologies equipment, updated software and training to use that equipment.

This collaboration and involvement should be on-going and include reviewing accessibility of websites, consoles (booking screens in tourism settings, accommodation, travel, interpreted experiences and so on) and what could be further developed or enhanced to make them accessible reflecting on where the technology is doing well and where improvements can be made including the skills of providers and designers. See Table 7 for quick tips for digital accessibility (RNIB, 2020).

Table 7. Quick tips for digital accessibility (RNIB, 2020)

Guidelines
• Use clear, formatted headings, to help screen reader users navigate your document or webpage
• Do not use images of text to convey information as they cannot be read by screen readers
• Ensure text can be resized and background and text colours can be modified to suit the reading preferences of users
• Make sure links are written out to describe the document or resource they send the user to
• Make sure information or explainer videos convey the same information in the audio voiceover as the images on screen
• Use image descriptions to share the information given in an image or photograph
• Ensure all downloadable content (Word or PDF) is made accessible.

There are inherent complexities in co-ordinating the complex landscape of the tourism system with advances in digital technology in general and assistive technologies for disabled people. For instance, there are challenges of integrating different information systems within sectors (Beutel et al., 2016) for example with transport. However, access for all calls for standardising of systems, flexible adjustments, for example font size and colour contrast and accessing a person. Such collaboration could include disability champions and disabled staff, include vision impaired people and disabled people in design consultation with destination management organisations and directly with providers with designers and planners (for example designing a new station and positioning of screens and working with access technology suppliers). An example of this is that operated by the eAccessibility network in Europe that has developed a 'cooperative platform for co-ordinating, supporting and improving eAccess in Europe' (ENAT, Muller and Laburda, 2012) and the forthcoming work of the Next Tourism Generation Alliance skills matrix for skills development in areas of digital, green and social skills (NTG, 2020).

CONCLUSION

The fourth industrial revolution that is changing the way people live, work and relate to one another (Shwab, 2016) is fundamentally reshaping tourism experiences (Xiang et al., 2009). This chapter has highlighted there have been considerable developments in the perceptions of accessible tourism, advances in the provision of information and communication technologies, mobile technologies and assistive technologies. Nevertheless, these enhanced opportunities for digital information technology have not completely resulted in equitable opportunities for vision-impaired people. Access to information at the pre-planning stage and during the journey is fragmented and resulting in not only limited information and connectivity but also situations where the affects have significantly influenced the tourism experience at its outset, invoking stress, anxiety, frustration and lack of control. There is a curious discourse here as research in the UK (RNIB, 2017) has shown that sight is the nation's most precious sense by far where 78% of people chose sight as the sense they would least like to lose, ten times more than the next most popular sense, which was smell at only 8%.

Designers and tourism providers have a role to play in being facilitators of accessible tourism enabling vision impaired people and other disabled people to enjoy and feel included in tourism experiences. The business case for accessible tourism is compelling through the increased potential to generate new, repeat and increased income. Disabled people, including vision impaired people are already playing a part in co-creating information through social media platforms and blogs and vlogs. By improving access to information and communication technology marketers and destination management organisations can explore the consumer needs and behaviours of a wider pool of tourists and therefore can ultimately focus more clearly on this heterogenic market range (Darcy & Buhalis, 2011). This requires a co-ordinated approach from the outset and as Eichhorn and Buhalis (2003) have previously argued, it is a change in mind-set. It is a shift from selling inflexible packages to designing tourism offerings that reflect consumer needs and information requirements where 'tourism activity should be devoted to offering services and products that can be engaged by all' (Florim et al., 2019, p. 75).

There are a significant number of people with a broad range of impairments, physical, cognitive, visual and auditory facing barriers, harbouring frustrations who need not be if they are considered appropriately (Williams & Brownlow, 2019). Accessible tourism is therefore an issue for all, beyond the realms of access but shifting to a whole social inclusion approach. Changing the way, we think about impairment and disability is therefore an essential co-requisite to institutional and structural change (Sapey, 2001) to ensure that not only do vision impaired people have access to tourism activities but actually feel included in them, in all the processes of the tourism system chain including digital marketing. Access to information is a crucial facet of that inclusion alongside being enabled to use the emerging technologies (ICT and AT). The concept of Universal design advocates a process whereby access for all is central to design rather than an add-on (Darcy, 2010) it is imperative then that attention is given at every stage of conception and design of new products and features to ensure that the principles of universal and inclusive design are maintained (World Blind Union, 2008). This will require co-operation and collaboration amongst tourism providers across the tourism system, digital developers and designers alongside disabled people as active stakeholders through a co-ordinated approach.

LIMITATIONS AND FUTURE RESEARCH

This was a small-scale study of nine qualitative interviews from a population of one small area in South East Wales facilitated over a two-week period. Although this study gives an insight into the information communication experiences of vision impaired tourists there is scope to do more on a wider scale and through different means. At the time of conducting the research the world was starting to adapt to the social distancing requirements necessitated by Covid-19. Whilst disruptive to the execution of the research it also raised awareness of the salience of research in this area. As travel companies have needed to respond to the changing environment, so customers have wanted to connect with businesses and seek up to the minute guidance. What can be more illustrative of the importance of an accessible travel ecosystem?

Further research could focus on the whole of the tourist's journey and tourism system, usage of assistive technologies and information systems for transport, attractions, hotels, restaurants, theatres, events and so forth by vision impaired people. For example, Asakawa et al. (2018) explored the access of vision impaired people at museums concluding that vision impaired people would feel included with a navigation and audio feedback app for the exhibits.

The interviewees were between the ages of 22-64, which does not take into account young people (prolific smartphone, app and social media users) and the generational shift of consumption habits catalysed by the entrance of the Millennials to the market (Holbrook 2000, cited in Adams & Downey, 2012); or the older generation. The latter are often subject to discriminatory and stereotypical attitudes (Thompson, 2016) but as can been observed by older persons' organisations such as Age UK there is the desire to empower older people in areas of high digital exclusion to get online or to develop their basic digital skills. According to research by Post Office Broadband, 39% of over 65s do not feel confident using a smartphone and two-thirds of people aged over 75 and three out of ten aged 65 to 74 do not use the internet (Age UK, 2020). Poignantly, these figures will include older people with vision impairments. The RNIB *My Voice* research (2015) suggested that one in three blind and partially sighted people felt able to make the most of new technology. However, this was closely linked to age: 86 per cent of 18–29-year olds said that they could make the most of new technology, compared to 25 per cent of people aged 75 and over. There was a significant generational divide in the use of computers, the internet and smartphones, with a particularly low level of technology usage among people aged 75 years and over. The majority of people who do not currently use technology would like to use it if obstacles were removed.

Further research is required to understand the role and work of ICT designers in tourism and their recognition of the access needs of vision impaired people. Designing to include usability, accessibility within a universal design agenda, 'designers must capture the user-centred requirements to ensure the product being designed is based upon target groups' (Smith-Jackson et al., 2003, p.550).

Postscript: The coronavirus pandemic (COVID-19) has brought a surge of online, digital activity but it has also highlighted that access to information is far from equitable for vision impaired people and other disabled people across the UK in 2020. This is particularly salient when the coronavirus disease 2019 (COVID-19) pandemic is likely to disproportionately affect disabled people (Armitage & Nellums, 2020), The Government has consistently shared vital public health information and other updates in a way that excludes people with sensory or complex disabilities (RNIB, 2020), regardless of reminders from RNIB and other organisations. This had led to the third sector campaigning for equal rights of information throughout the pandemic despite the fact that all Government outlets have a legal obligation under the Equality Act (2010) which requires all service providers to make "reasonable adjustments" to support disabled people, including providing information "in an accessible format' (RNIB, 2020). If this is the reality for health in a pandemic then it can be surmised that 'tourism is an idealised commodity that is only accessible to some' (Hall, 2010, p.34), and as Rae (2020) argues what 'technology giveth, technology taketh away.' All it takes is that change of mind-set, awareness and willingness.

ACKNOWLEDGEMENT

Wales Council of the Blind for their guidance and expertise as well as for facilitating the telephone interviews.

This research received no specific grant from any funding agency in the public, commercial or not-for-profit sectors.

REFERENCES

AbilityNet. (2020). *Digital Accessibility*. https://abilitynet.org.uk/accessibility-services/what-is-digital-accessibility

Adams, R., & Downey, C. (2016). Edited Platforms: Anticipating Future Consumption. Journal of Marketing Theory and Practice, 24(2), 224-235.

Agapito, D., Valle, P., & Mendes, J. (2014). The sensory dimension of tourist experiences: Capturing meaningful sensory-informed themes in Southwest Portugal. *Tourism Management*, *42*, 224–237. doi:10.1016/j.tourman.2013.11.011

Age, U. K. (2020). *One Digital*. https://www.ageuk.org.uk/our-impact/programmes/one-digital/

Anonymous. (2002). Delta brings new display screens to San Francisco and Tampa airports. *Airline Industry Information,* 1. Retrieved from http://ezproxy.cardiffmet.ac.uk/login?url=https://search-proquest-com.ezproxy.cardiffmet.ac.uk/docview/210525066?accountid=15588

Anonymous. (2013). New interactive touch-screen directories work with mobile application to put MSP airport information at your fingertips. *Targeted News Service.* Retrieved from http://ezproxy.cardiffmet.ac.uk/login?url=https://search-proquest-com.ezproxy.cardiffmet.ac.uk/docview/1321651369?accountid=15588

Anonymous. (2016). Northern rail to install customer information screens across Cheshire, UK. *Progressive Digital Media Transportation (Incl Airports, Roadways, Railways, Shipping, Automotive & Logistics) News.* Retrieved from http://ezproxy.cardiffmet.ac.uk/login?url=https://search-proquest-com.ezproxy.cardiffmet.ac.uk/docview/1756502983?accountid=15588

Armitage, R., & Nellums, L. B. (2020). The COVID_19 response must be disability inclusive. *The Lancet. Public Health*, *5*(5), 257. doi:10.1016/S2468-2667(20)30076-1 PMID:32224295

Ayeh, J. K. (2015). Travellers' acceptance of consumer-generated media: An integrated model of technology acceptance and source credibility theories. *Computers in Human Behavior*, *48*, 173–180. doi:10.1016/j.chb.2014.12.049

Bannister, D. (2017). Get your digital communications right. *Airport Business, 31*, 10-13. Retrieved from http://ezproxy.cardiffmet.ac.uk/login?url=https://search-proquest-com.ezproxy.cardiffmet.ac.uk/docview/1912912014?accountid=15588

Barnes, C., & Mercer, G. (2005). Disability, work, and welfare. Challenging the social exclusion of disabled people. *Work, Employment and Society*, *19*(3), 527–545. doi:10.1177/0950017005055669

Beck, J., Rainoldi, M., & Egger, R. (2019). Virtual reality in tourism: A state-of-the-art review. *Tourism Review*, *74*(3), 586–612. doi:10.1108/TR-03-2017-0049

Beutel, M. C., Gökay, S., Kluth, W., Krempels, K., Ohler, F.G., Samsel, C., Terwelp, C., & Wiederhold, M. (2016). *Information Integration for Advanced Travel Information Systems*. Academic Press.

Bigné, E., Sanz, S., & Ruiz, C. (2010). Why some internet users don't buy air tickets online. In U. Gretzel, R. Law, & M. Fuchs (Eds.), *Information and Communication Technologies in Tourism 2010* (pp. 209–221). Springer. doi:10.1007/978-3-211-99407-8_18

Budd, L., & Ison, S. (2020). Supporting the needs of special assistance (including PRM) passengers: An international survey of disabled air passenger rights legislation. *Journal of Air Transport Management, 87*, 101851. Advance online publication. doi:10.1016/j.jairtraman.2020.101851

Buhalis, D. (1998). Strategic use of information technologies in the tourism industry. *Tourism Management, 19*(5), 409–421. doi:10.1016/S0261-5177(98)00038-7

Buhalis, D. (2019). Technology in tourism-from information communication technologies to eTourism and smart tourism towards ambient intelligence tourism: A perspective article. *Tourism Review, 75*(1), 267–272. doi:10.1108/TR-06-2019-0258

Buhalis, D., & Darcy, S. (Eds.). (2011). *Accessible tourism: Concepts and issues*. Channel View Publications.

Buhalis, D., & Law, R. (2008). Progress in information technology and tourism management: 20 years on and 10 years after the internet: The state of eTourism research. *Tourism Management, 29*(4), 609–623. doi:10.1016/j.tourman.2008.01.005

Burnett, J. J., & Baker, H. (2001). Assessing the Travel-related Behaviors of the Mobility-Disabled Customer. *Journal of Travel Research, 40*(4), 4–11. doi:10.1177/004728750104000102

Cassia, F., Castellani, P., Rossato, C., & Baccarani, C. (2020). Finding a way towards high-quality, accessible tourism: The role of digital ecosystems. *The TQM Journal*. Advance online publication. doi:10.1108/TQM-03-2020-0062

Chanjaraspong, T. (2016). Acceptance factors for the use of video call via smartphones by blind people. *Kasetsart Journal of Social Sciences*, 1–7. doi:10.1016/j.kjss.2016.02.001

Cloquet, I., Palomino, M., Shaw, G., Stephen, G., & Taylor, T. (2018). Disability, social inclusion and the marketing of tourist attractions. *Journal of Sustainable Tourism, 26*(2), 221–237. doi:10.1080/09669582.2017.1339710

Curran, K., Walters, N., & Robinson, D. (2007). Investigating the problems faced by older adults and people with disabilities in online environments. *Behaviour & Information Technology, 26*(6), 447–453. doi:10.1080/01449290600740868

Daniels, M. S., Drogen-Rodgers, E. B., & Wiggins, B. P. (2005). 'Travel Tales' an interpretive analysis of constraints and negotiations to pleasure travel as experienced by persons with physical disabilities. *Tourism Management, 26*(6), 919–930. doi:10.1016/j.tourman.2004.06.010

Darcy, S. (2010). Inherent Complexity: Disability, accessible tourism and accommodation information preferences. *Tourism Management, 31*(6), 816–826. doi:10.1016/j.tourman.2009.08.010

Darcy, S. (2012). (Dis)Embodied Air Travel Experiences: Disability, Discrimination and the Affect of a Discontinuous Air Travel Chain. *Journal of Hospitality and Tourism Management, 19*(1), E8. doi:10.1017/jht.2012.9

Darcy, S., & Dickson, T. J. (2009). A Whole-of-Life approach to tourism: The case for accessible tourism experiences. *Journal of Hospitality and Tourism Management, 16*(1), 32–44. doi:10.1375/jhtm.16.1.32

Darcy, S., McKercher, B., & Schweinsberg, S. (2020). From tourism and disability to accessible tourism: A perspective article. *Tourism Review, 75*(1), 140–144. doi:10.1108/TR-07-2019-0323

Darcy, S., & Pegg, S. (2010). Towards Strategic Intent: Perceptions of disability service provision amongst hotel accommodation managers. *International Journal of Hospitality Management, 30*(2), 468–476. doi:10.1016/j.ijhm.2010.09.009

Department for Work and Pensions. (2018). *Spending power of disabled people and their families estimated at £249 billion.* https://www.gov.uk/government/news/purple-tuesday-is-the-uks-first-accessible-shopping-day

Deville, E., & Kastenholz, E. (2018). Accessible tourism experiences: The voice of people with visual disabilities. *Journal of Policy Research in Tourism, Leisure & Events, 10*(3), 265–285. doi:10.1080/19407963.2018.1470183

Dhalwal, S. K., & Juyal, S. L. (2017). Assistive technology as a tool of making difference in the life of persons with visual impairment. *Indian Journal of Health and Wellbeing, 8*(4), 319–321.

Disability Accessibility Centre. (2020). *Audit and Accreditation.* http://digitalaccessibilitycentre.org/index.php/audit-and-accreditation

Domínguez, T. V., González, E. A., & Darcy, S. (2019). Accessible tourism online resources: A Northern European perspective. *Scandinavian Journal of Hospitality and Tourism, 19*(2), 140–156. doi:10.1080/15022250.2018.1478325

Drews, W., & Schemer, C. (2010). eTourism for All? Online Travel Planning of Disabled People. *Information and Communication Technologies in Tourism, 2010*, 507–518. doi:10.1007/978-3-211-99407-8_42

Duckett, P. S., & Pratt, R. (2001). The Researched Opinions on Research: Visually impaired people and visual impairment research. *Disability & Society, 16*(6), 815–835. doi:10.1080/09687590120083976

Eichhorn, V., & Buhalis, D. (2011). Accessibility: A Key Objective for the Tourism Industry. In D. Buhalis & S. Darcy (Eds.), *Accessible tourism: Concepts and issues* (pp. 46–62). Channel View Publications.

Equality Act 2010–UK Equality legislation. https://www.legislation.gov.uk/ukpga/2010/15/contents

European Network for Accessible Tourism. (2020). *eAccess+. A Thematic Network for Improving Implementation of eAccessibility throughout Europe.* https://www.accessibletourism.org/?i=enat.en.enat_projects_and_good_practices.1038

Finkelstein, V. (2004). Representing Disability. In J. Swain, S. French, C. Barnes, & C. Thomas (Eds.), *Disabling Barriers – Enabling Environments* (2nd ed., pp. 13–20). Sage.

Florim, de L.M., Viera, D., & Pinho, T. (2019). Digital Marketing Strategies for Accessible Tourism. In J.D. Santos & Ó. Lima Silva (Eds.), *Digital Marketing Strategies for Tourism, Hospitality and Airline Industries*. IGI Global.

Freund, P. (2001). Bodies, Disability and Spaces: The social model and disabling spatial organisations. *Disability & Society, 16*(5), 689–706. doi:10.1080/09687590120070079

Gerino, A., Picinali, L., Bernareggi, C., Alabastro, N., & Mascetti, M. (2015). Towards large scale evaluation of novel sonification techniques for non-visual shape exploration. In *Proceedings of the 17th International ACM SIGACCESS Conference on Computers & Accessibility*. ACM. 10.1145/2700648.2809848

Gretzel, U. (2011). Intelligent Systems in Tourism. *Annals of Tourism Research, 38*(3), 757–779. doi:10.1016/j.annals.2011.04.014

Hackett, S., Parmanto, B., & Zeng, X. (2005). A retrospective look at website accessibility over time. *Behaviour & Information Technology, 24*(6), 407–417. doi:10.1080/01449290500066661

Harrison, E. (2018, Aug 21). Whiteboards used as Gatwick flight information screens fail. *University Wire*. Retrieved from http://ezproxy.cardiffmet.ac.uk/login?url=https://search-proquest-com.ezproxy.cardiffmet.ac.uk/docview/2090168288?accountid=15588

Hersh, M., Johnson, M., & Keating, D. (2008). *Assistive technology for visually impaired and blind people*. Springer. doi:10.1007/978-1-84628-867-8

Huang, H. (2018). Blind users' expectations of touch interfaces: Factors affecting interface accessibility of touchscreen-based smartphones for people with moderate visual impairment. *Universal Access in the Information Society, 17*(2), 291-304. doi:http://dx.doi.org.ezproxy.cardiffmet.ac.uk/10.1007/s10209-017-0550-z

Hunter-Jones, P. (2004). Young People, holiday-taking and cancer – an exploratory analysis. *Tourism Management, 25*(2), 249–258. doi:10.1016/S0261-5177(03)00094-3

Israeli, A. (2002). A Preliminary investigation of the importance of site accessibility factors for disabled tourists. *Journal of Travel Research, 41*(1), 101–104. doi:10.1177/004728750204100114

Loi, K. L., & Kong, W. H. (2017). Tourism for All: Challenges and Issues Faced by People with Vision Impairment. *Tourism Planning & Development, 14*(2), 181–197. doi:10.1080/21568316.2016.1204357

Mascetti, S., Gerino, A., Bernareggi, C., & Picinali, L. (2017). On the Evaluation of Novel Sonification Techniques for Non-Visual Shape Exploration. *ACM Transactions on Accessible Computing, 9*(4), 13. doi:10.1145/3046789

Michopoulou, E., & Buhalis, D. (2013). Information Provision for Challenging Markets: The Case of the Accessibility Requiring Market in the Context of Tourism. *Information & Management, 50*(5), 229–239. doi:10.1016/j.im.2013.04.001

Michopoulou, E., Darcy, S., Ambrose, I., & Buhalis, D. (2015). Accessible tourism futures: The world we dream to live in and the opportunities we hope to have. *Journal of Tourism Futures, 1*(3), 179–188. doi:10.1108/JTF-08-2015-0043

Muller, L., & Laburda, A. (2012). *ENAT Reaching all customers: How do European NTOs compare on online Accessibility*. ENTER 12 IFITT Conference, Helsingborg, Sweden. https://www.accessibletourism.org/resources/enter2012-helsingborg_enat_final_ia--2.pdf

Next Tourism Generation. (2020). *About the Next Tourism Generation Alliance*. https://nexttourismgeneration.eu/about/

Niininen, O., Buhalis, D., & March, R. (2007). Customer Empowerment in Tourism through Consumer Centric Marketing (CCM). *Qualitative Market Research, 10*(3), 265–281. doi:10.1108/13522750710754308

Packer, T. L., McKercher, B., & Yau, M. (2007). Understanding the complex interplay between tourism, disability and environmental contexts. *Disability and Rehabilitation, 29*(4), 281–292. doi:10.1080/09638280600756331 PMID:17364778

Poon, A. (1988). Tourism and information technologies. *Annals of Tourism Research, 15*(4), 531–549. doi:10.1016/0160-7383(88)90048-5

Pope, C., & Mays, N. (2000). Qualitative research in Health Care. In *Analysing Qualitative Data*. https://www.bmj.com/content

Poria, Y., Reichel, A., & Brandt, Y. (2011). Blind People's Tourism Experiences: An exploratory study. In D. Buhalis & S. Darcy (Eds.), *Accessible Tourism: Concepts and Issues*. Channel View Publications.

Pritchard, A., Morgan, N., & Ateljevic, I. (2011). Hopeful Tourism: A New Transformative Perspective. *Annals of Tourism Research, 38*(3), 941–963. doi:10.1016/j.annals.2011.01.004

Rae, J. (2020). As technology giveth, technology taketh away. In The Routledge Handbook of Disability Activism 2020. Taylor and Francis.

Randle, M., & Dolnicar, S. (2019). Enabling people with impairments to use Airbnb. *Annals of Tourism Research, 76*, 278–289. doi:10.1016/j.annals.2019.04.015

Richards, V. (2013). *(Re) Envisioning the Tourism Experiences of vision impaired people* (PhD Thesis). Cardiff Metropolitan University.

Richards, V., Pritchard, A., & Morgan, N. (2010). (Re)Envisioning tourism and vision impairment. *Annals of Tourism Research, 37*(4), 1097–1116. doi:10.1016/j.annals.2010.04.011

Richards, V., Pritchard, A., Morgan, N., & Sedgley, D. (2010). Tourism and Visual Impairment. In S. Cole & N. Morgan (Eds.), *Tourism and Inequality: Problems and Prospects*. CABI. doi:10.1079/9781845936624.0021

Royal National Institute for Blind People. (2015). *My Voice Research: The views and experiences of blind and partially sighted people in the UK*. https://www.rnib.org.uk/sites/default/files/My%20Voice%20 2015%20-%20Full%20report%20-%20Accessible%20PDF_0.pdf

Royal National Institute for Blind People. (2017). *State of the Nation Eye Health Annual Review*. https://www.rnib.org.uk/state-nation-2017

Royal National Institute for Blind People. (2020). *Key Information and Statistics*. https://www.rnib.org.uk/aboutus/Reserach/statistics/Pages/statistics.aspx

Royal National Institute for Blind People. (2020a). *Calling out inaccessible coronavirus information*. https://www.rnib.org.uk/news/campaigning/inaccessible-coronavirus-information

Royal National Institute for Blind People. (2020b). *Quick tips for digital accessibility.* https://www.rnib.org.uk/campaigning/priority-campaigns/accessible-health-information/coronavirus-and-accessible-online-information

Royal National Institute for Blind People. (2020c). *RNIB asks Number 10 for leadership on accessible coronavirus information.* https://www.rnib.org.uk/news/campaigning/letter-to-prime-minister

Sapey, B. (2001). From stigma to the social exclusion of disabled people. In T. Mason, C. Carlisle, C. Watkins, & E. Whitehead (Eds.), *Stigma and social exclusion in healthcare*. Routledge.

Shakespeare, T., Lezzoni, L. I., & Groce, N. E. (2009). Disability and the training of health professionals. *Lancet, 374*(9704), 1815–1816.

Sharpley, R. (2018). *Tourism, Tourists and Society*. Routledge.

Sheldon, A. (2004). Women and Disability. In J. Swain, S. French, C. Barnes, & C. Thomas (Eds.), *Disabling Barriers – Enabling Environments* (2nd ed.). Sage Publications.

Shwab, K. (2016). *The Fourth Industrial Revolution*. World Economic Forum.

Small, J. (2014). Interconnecting mobilities on tour: Tourists with vision impairment partnered with sighted tourists. *Tourism Geographies, An International Journal of Tourism Space. Place and Environment, 17*(1), 76–90.

Small, J., & Darcy, S. (2010). Tourism, Disability and Mobility in Tourism and Inequality. In S. Cole & N. M. Morgan (Eds.), *Tourism and Inequality: Problems and Prospects*. CABI.

Small, J., Darcy, S., & Packer, T. (2011). The embodied tourist experiences of people with vision impairment: Management implications beyond the visual gaze. *Tourism Management*. Advance online publication. doi:10.1016/j.tourman.2011.09.015

Smith, R. W. (1987). Leisure of disabled tourists: Barriers to participation. *Annals of Tourism Research, 14*(3), 376–389.

Smith-Jackson, T., Nussbaum, M., & Mooney, A. (2003). Accessible cell phone design: Development and application of a needs analysis framework. *Disability and Rehabilitation, 25*(10), 549–560.

Stone, E., & Priestley, M. (1996). Parasites, pawns and partners: Disability research and the role of non-disabled researchers. *The British Journal of Sociology, 47*, 699–716.

Swain, J., & French, S. (2000). Towards an Affirmation Model of Disability. *Disability & Society, 15*(4), 569–582.

Thompson, N. (2016). *Anti-discriminatory Practice, Equality, Diversity and Social Justice* (6th ed.). Palgrave.

Tregaskis, C. (2003). Towards inclusive practice: An insider perspective on leisure provision for disabled people. *Managing Leisure, 8*(1), 28–40.

UNESCO. (2019). *ICTs for Persons living with Disabilities.* https://en.unesco.org/partnerships/partnering/icts-persons-living-disabilities

United Nations News. (2016). Accessible tourism will benefit everyone, say senior UN officials on World Day. *UN News*. Accessed May 2020 https://news.un.org/en/story/2016/09/541232-accessible-tourism-will-benefit-everyone-say-senior-un-officials-world-day#.V-vdOPkrKUl

United Nations World Tourism Organisation. (2020). *Why Tourism? Tourism – an economic and social phenomenon*. https://www.unwto.org/why-tourism

VisitEngland. (2018). *Accessibility – non participation*. VisitEngland. https://www.visitbritain.org/sites/default/files/vb-corporate/Documents-Library/documents/England-documents/april_2018_accessibility_non_participation_reportv3.pdf

Wales Council for the Blind. (2008). Statistics in Wales. In *Re) Envisioning the Tourism Experiences of vision impaired people*. Cardiff Metropolitan University.

Web Content Accessibility Guidelines. (2018). http://www.w3.org/WAI/standards-guidelines/wcag

Williams, O. J. (2019). *Low Vision in the 21st century – keeping up with the technology: Wales Optometry Postgraduate Education Centre*. Wales Council of the Blind.

Williams, R., & Brownlow, S. (2019). *The Click-Away Pound Report*. Freeney Williams Ltd.

Wobbrock, J. O., Kane, S. K., & Ladner, R. (2011). Usable Gestures for Blind People: Understanding Preference and Performance. CHI 2011, Vancouver, Canada

World Blind Union. (n.d.). *White Cane Safety Day*. http://www.worldblindunion.org

World Economic Forum. (2017). *Digital Transformation Initiative: Aviation, Travel and Tourism Industry White Paper*. http://reports.weforum.org/digital-transformation

World Health Organisation. (2020). *Blindness and Vision Impairment*. https://www.who.int/news-room/fact-sheets/detail/blindness-and-visual-impairment

Xiang, Z., Gretzel, U., & Fesenmaier, D. (2009). Semantic Representation of Tourism on the Internet. *Journal of Travel Research*, *47*(4), 440–453.

Yau, M. K., McKercher, B., & Packer, T. L. (2004). Traveling with a Disability More than an Access Issue. *Annals of Tourism Research*, *31*(4), 946–960.

ADDITIONAL READING

Finkel, R., Sharp, B., & Sweeney, M. (2018). *Accessibility, inclusion and diversity in critical event studies*. Routledge. doi:10.4324/9781351142243

Gupta, A., Dogra, N., & George, B. (2018). What determines tourist adoption of smartphone apps? An analysis based on the UTAUT-2 framework. *Journal of Hospitality and Tourism Technology*, *9*(1), 50–64. doi:10.1108/JHTT-02-2017-0013

Swain, J., French, S., Barnes, C., & Thomas, C. (Eds.). (2013). *Disabling Barriers – Enabling Environments* (3rd ed.). Sage Publications.

KEY TERMS AND DEFINITIONS

Abilism: As discrimination by the 'able-bodied' of the 'non able-bodied'.

Assistive Technology: An umbrella term for any device or system that allows an individual to perform a task they would otherwise be unable to do.

Design for All: Design for All is the intervention into environments, products and services that aim to ensure that anyone, including future generations, regardless of age, gender, capacities or cultural background, can participate in social, economic, cultural and leisure activities with equal opportunities.

Disablism: Discriminatory, oppressive, or abusive behaviour arising from the belief that disabled people are inferior to others.

Low Vision: Various degrees of vision caused by numerous eye conditions and circumstances.

Mobility: The capacity of facility of movement.

Orientation: The ability to use one's remaining senses to understand one's location in the environment.

Residual Vision: Remaining useful vision.

Servicescape: A man-made built environment in a tourism space.

Vision Impairment: Term for blind, partially sighted, sight loss, visually impaired.

Chapter 3
Travel by People With Physical Disabilities:
Constraints and Influences in the Decision-Making Process

Eugénia Lima Devile
Polytechnic Institute of Coimbra, Portugal

Andreia Antunes Moura
https://orcid.org/0000-0002-1722-3476
Polytecnic Institute of Coimbra, Portugal

ABSTRACT

It is acknowledged that access to tourism by persons with disabilities continues to be shaped by a variety of constraints that are not being clearly or consistently addressed by those in charge of tourism planning and management. Based on these assumptions, this chapter is focused on the identification of structural constraints and negotiation strategies that influence the decision and participation to travel for people with physical disabilities. To achieve these objectives, the chapter used a qualitative methodology, with the analysis and discussion of the information collected through in-depth interviews. Results may ultimately contribute to deepen the knowledge about the factors that influence people with disabilities travel choices, allowing to identify specific recommendations for improving the tourist supply chain and, consequently, for the overall tourist experience quality.

INTRODUCTION

Participation in tourism activities is influenced by a wide variety of factors, whether from the personal and social context of the individual or the environment itself. In the case of people who have physical limitations or special needs, the decision-making process to travel and full tourism participation, in addition to architectural factors, is largely shaped by the capacity of the tourism supply to respond

DOI: 10.4018/978-1-7998-6428-8.ch003

to special and specific needs. It is important to understand how the different barriers or constraints to travel are overcome through adaptive strategies that allow people, even with special needs, to become active travelers. In this context, this chapter objective, beyond identifying travel and tourism constraints, intends to find out how individuals, with physical disabilities, deal with and overcome these difficulties, which inevitably affect the pleasure of the tourism experience.

The quality of lived tourism experiences will have a determining influence on the choice of future trips, and may act as an inhibitor of future participation if the individual evaluates the experience as negative, or, on the contrary, as a facilitating factor, if the overall evaluation of the experience is positive. In the case of a recent disability, the first travel experiences are perceived as a challenge where emotions and feelings are mixed, but as the individual becomes more experienced, he acquires more skills, learning to deal with obstacles in a more efficient way (Packer, Mckercher & Yau, 2007). The experience gained at this level thus allows greater knowledge and confidence, which can stimulate motivations and the desire to travel and, above all, contribute to enriching experiences and experiences that can have a lasting positive impact on people's lives.

Studies published on the decision-making process of tourists with disabilities point to the need for a much longer travel planning phase, where one seeks to minimize the uncertainty associated with the trip and the occurrence of problems through a research process. Accurate and detailed information on the different services involved, including new technologies as the internet, websites and social media, provide participation, learning and connection (Altinay, Saner, Bahçelerli, & Altinay, 2016). According to Neumann & Reuber (2004) the most important factors for choosing a destination are, in decreasing order of importance, accommodation, the existence and availability of adequate means, which allows them to move freely, the availability of excursions and leisure facilities, the existence of cultural activities and the reception and ease of communication. In general, the accessibility conditions of the different services are highly valued when choosing a destination. But there are also diverse needs that require different responses from the tourist destination, namely the availability of different support services, the existence of health equipment and services and the information provided.

The purpose of this chapter is, therefore, to analyze the constraints that negatively affect full participation in tourism by people with physical disability and to identify the adaptation strategies that allow them to overcome the limitations found, namely the role of Information and Communication Technologies (ICT). To do so, we opted for a qualitative methodology, with the analysis and discussion of the information collected through in-depth interviews.

Background

Decision making in tourism has been extensively studied, with many and diverse models found in scientific literature to explain this phenomenon. Most of these models understand this decision making as a process, during which the tourist goes, sequentially, through different phases in which he is making decisions about different aspects of the trip (Decrop, 2006). This is a complex process that starts with the generic decision to travel or not to travel, from which different elements of the tourist product are taken into account. As with other consumers, the decision to travel is influenced by a wide range of variables, among which the social environment of the individual stands out.

Published studies on the decision making process of tourists with disabilities (e.g. Burnett & Baker, 2001; Daniels, Rodgers, & Wiggins, 2005; Darcy, 1998, 2002; Israeli, 2002; McKercher, Yau & Lam, 2003; Murray & Sproats, 1990; Neumann & Reuber, 2004, ; Nyman, Westin & Carson, 2018; Ray &

Ryder, 2003; Shaw & Coles, 2004; Smith, 1987; Yau, McKercher, & Packer, 2004) point to the need for a much longer travel planning phase, where the uncertainty associated with travel and the occurrence of problems are minimized through a rigorous and detailed information search process

Tourists with disabilities use a different decision-making process than other tourists when assessing a tourist site. For most tourists, there is a certain trade-off, where the absence or a lower level of an attribute can be compensated for by an addition of another attribute. However, for visitors with disabilities, accessibility attributes can rarely be compensated for by an addition of another attribute (Israeli, 2002).

More than eliminating barriers, it is necessary to offer a set of effective conditions, which materialize in order to make participation in the trip possible. It should also be noted the influence of previous travel experience, as has been addressed in some of the studies of tourist behaviour with disabilities, such as Blichfeldt & Nicolaisen (2011) Daniels et al. (2005), McKercher et al. (2003), Shaw & Coles (2004) and Yau et al. (2004). This perspective has implicit the travel career of each individual, that is, the experience that is acquired throughout the trips made, allowing the individual to deal with the obstacles in a more positive and knowledgeable way. Specifically, the influences of context are associated with the life cycle of individuals (family structure and age) and their professional occupation. In the case of individuals with disability, in addition to the dynamic influence of these factors, we can add the variable influence that the disability situation can introduce in the process, due to the different perceptions that individuals formulate of it throughout their life cycle.

Participation in tourist activities can thus be a means of boosting these positive changes, which enable the development of tourist skills as individuals become more experienced tourists and thus learn to tackle the different obstacles more effectively.

Such an approach takes us back to the concept of negotiating constraints. This approach holds that constraints are not insurmountable, which is to say that people adopt adaptive strategies to mitigate or circumvent the effects of constraints on their leisure practices, either by modifying their habits or by changing other aspects of their lives in order to overcome them (Daniels et al., 2005; Packer et al., 2007; Yau et al., 2004; Devile & Kastenholz, 2018),

Here, the dynamic and interactive nature of this process is assumed, influenced by a great variety of factors, coming both from the personal context and from the individual's own condition of disability and his social environment. Each of these factors may have a facilitating or constraining effect and will have a variable impact on the various phases of the process. As an example, we can identify communication and information as likely to influence the decision to travel or not at an early stage. In this sense, ICT may have a facilitating role and also an impact on the travel phase itself and/or the evaluation of the experience on how it is evaluated, here too may playing a facilitating or inhibiting role in future participation.

In this context, it is important to highlight the adaptation strategies, reinforcing the role and responsibility of the various stakeholders (tourism sector, health technicians and the disabled themselves and their families) and the need for concerted efforts to increase opportunities for participation in tourism activities. Above all, it constitutes a step forward in assuming that, since the removal of physical barriers is very important, the promotion of accessible tourism activities goes far beyond this, sustained by a change in attitudes, across the different actors in this process.

Negotiation strategies, used by people with disabilities in the context of tourism activity, are a little explored research topic, and little information is available on how constraints interrelate with participation or how different constraints are balanced with adaptive negotiation strategies. The only study that explicitly uses the constraints negotiation construct was the one developed by Daniels et al. (2005). In this way, objective, rigorous information focused on consumer needs constitutes a critical aspect, particularly useful in the decision-making process, so that tourists with disabilities can make conscious and informed decisions, thus avoiding unnecessary risks that can jeopardize all the pleasure associated with a trip (Devile, 2003).

Structural Constraints That Influence the Decision to Travel for People With Physical Disabilities

The leisure constraints literature underlines three types of constraints: structural, intrapersonal and interpersonal. Structural constraints, also called environmental constraints, intervene between the phases of preferences and effective participation (Crawford & Godbey, 1987), and are related with the external context of each individual. Some of the more common are the lack of available time, financial limitations, transportation difficulties, lack of suitable infrastructures, among others. Interpersonal constraints can occur during interactions with an individual's social network, service providers or strangers, or because one lacks a partner with whom to engage in some leisure activity (Crawford & Godbey, 1987). In respect to intrapersonal constraints, the literature emphasizes factors such as person's psychological state, physical functioning or cognitive abilities (Crawford & Godbey, 1987; Smith, 1987), and include factors like social ineffectiveness, stress, anxiety, lack of knowledge and health problems.

The leisure constraints studies achieved a huge theoretical sophistication that have been used on tourism research in order to deepen the knowledge about the factors that influence travel choices and tourism experiences. Specifically, this body of knowledge could be a valid contribution to extend the research about people with disabilities participation in tourism (Daniels et al., 2005; Devile & Kastenholz, 2018). To better understand this market it is important not only to know the barriers and factors that prevent people with special needs from traveling, but above all, to understand how they become active travelers, how they adapt to and negotiate the various constraints that they face (Daniels et al., 2005; Yau et al. 2004, Packer et al., 2007; Devile & Kastenholz, 2018, Nyman et al. 2018).

The above-mentioned constraints influence the travel decision behavior of the population at large and have been extendedly studied, using the leisure constraints construct and others. People with disabilities are affected by the same constraints and, besides that, they face other barriers that can have a specific influence on their tourism decisions and experiences. Within the scope of this chapter, we'll focus only on those factors that have an increased importance in the context of the travel experiences of people with disabilities, leaving aside all others that are common to the population in general.

According to the results obtained from the studies on this topic, although only a few used the leisure constraints construct (e.g. Daniels, Rodgers, & Wiggins, 2005; by Small, Darcy, & Packer, 2012; Devile & Kastenholz, 2018), the literature review revealed a great set of intrapersonal, interpersonal and structural constraints that hamper disabled persons' participation in tourism (e.g.: Daniels, Rodgers and Wiggins, 2005; Devile & Kastenholz, 2018, Kastenholz, Eusébio & Figueiredo, 2015; Nyman et al., 2018; Tao, Goh, Huang & Moyle, 2019; Smith, 1987). Specifically, structural constraints faced by people with physical disability encompass a set of factors related to the lack of accessible tourist services, which can discourage them to participate in tourism. The main issues identified in the published research are associated mainly with the overall accessibility of tourism services – accommodation, attractions, transportation and restaurants, among others.

Accommodation

The aspects associated with the accessibility conditions of the accommodation are the most cited constraints in the literature in the field of accessible tourism (ex: Daniels et al., 2005; Darcy, 2010; Lyu, 2017; Shaw & Coles, 2004). The issues are mainly related to the lack of accessibility of the rooms, including rooms with little space for move with wheelchairs, inadequate bathrooms facilities, among others (Darcy, 2010). Besides that, these rooms are sometimes aesthetically unappealing due to their clinical appearance.

The common areas of the hotel present several situations that are uncomfortable or inaccessible, such as service desks too high, dining spaces, the absence or lack of conditions of the elevators and inaccessible outdoor space, among other situations. Another aspect that stands out in the literature review is the lack of specific, rigorous and detailed information on the conditions related to the accessibility of accommodation (Darcy, 2010; Daniels et al., 2005; Devile & Kastenholz, 2018; O'Neill & Ali Knight, 2000; Yau et al., 2004). Another issue frequently raised by research on this topic is the fact that rooms are often promoted as accessible when in fact they are not (Darcy, 2010).

Several studies have point out that those responsible for accommodation sector do not recognize the advantage of providing accessible rooms (Darcy, Cameron, & Pegg, 2010; Darcy & Pegg, 2011; O'Neill & Ali Knight, 2000). In addition, sometimes these rooms are located in less favorable areas of the hotel and tend to be considered to be lower level rooms. It should be noted that the accessibility requirements are not unrelated to the specificities and severity of the disability. This aspect was analyzed in depth by Darcy (2010), who identified a set of important criteria considered in the selection of accommodation by people with different types of disabilities. According to the author, these criteria vary depending on the type, level of dependence and technical aids used by people with disabilities.

Transportation

In respect to transport, the issues vary according to the means used and involve all elements of the transport chain, hindering both the options to reach the destinations and to move around during the stay (Daniels et al., 2005; Cavinato & Cuckovich, 1992; Lyu, 2017; McKercher et al., 2003). Therefore, despite the progress made in reducing restrictions and improving accessibility, in transport in general, these segments and related infrastructures must be analyzed in a systemic way, including peripheral interface functions.

Regarding air transportation, constraints situations include the likelihood of loss or damage to the passengers' wheelchairs on airplanes and problems with the assembly or loss of the battery of the electric wheelchairs, as well as problems that occur in flight transfers (Packer et al., 2007). Safety and assistance rules, which regulate air transport, can also be considered as constraining factors, especially due to the sense of discrimination that they sometimes indicate (Darcy, 2012). Finally, constraints are also reported on the use of airplane toilets (Darcy, 2012; Poria, Reichel & Brandt, 2010), and also lack of training of professionals about the needs of people with disabilities (Devile & Kastenholz, 2018).

Attractions

Regarding attractions (cultural or natural), the problems pointed out in the literature refer mainly architectural or physical barriers. The issues associated with heritage conservation are often at odds with the necessary changes to promote accessibility, having, in some cases, a very high economic cost (Goodall,

2006; Pearn, 2011). Besides that, but also important, is the case of existing barriers in terms of the enjoyment of intangible heritage.

Another important constraint, transversal of all the tourism services, is the lack of reliable and accurate information, a central issue in order to meet the tourists' differential requirements and allowing them to make informed travel decisions, thus avoiding unpleasant surprises and unnecessary risks (Bieger & Laesser 2001; Cavinato & Cuckovich, 1992; Daniels *et al.*, 2005; Eichhorn, Miller, Michopoulou, & Buhalis, 2008; Packer *et al.*, 2007)

A survey on Portuguese PwD also highlights the relevance of economic/financial constraints (Kastenholz, Eusébio & Figueiredo, 2015). This is due to some additional costs that people with disabilities may have during the trip, such as caregivers travel expenses.

Several studies have shown the persistence of a knowledge gap, from the tourism sector professionals, about the needs of clients with disabilities. This results in a general picture of lack of understanding and awareness, which is often reflected on the quality of the service provided to consumers with disabilities (Devile & kastenholz, 2018; Miller & Kirk, 2002). On the other hand, this lack of knowledge and social sensitivity disability leads to the assumption that people with disabilities are a homogeneous group with similar needs and characteristics (Daruwalla & Darcy, 2005, Richards et al., 2010).

In our study, it was possible to identify different types of structural constraints, as well as diverse negotiation strategies, we tried to organize according to different themes. Some of which have been already mentioned on literature review, but many others are not.

METHODOLOGY

Our empirical work was based on a qualitative methodology, through in-depth interviews. This research intends to contribute to a deeper understanding of the determinants that influence the decision to travel by people with physical disabilities and to identify the adaptation strategies that allow them to overcome the constraints, as well as the role of ITC in this process.

According to Jennings (2005), the in-depth interviews is the most suitable technique to create a space for dialogue, as it favors a natural environment of communication and sharing of ideas and feelings from the interviewees. On the other hand, it also has the advantage of obtaining information on the meaning that the actors attribute to their practices and the events with which they are confronted, allowing a high degree of profundity of the data collected. Conversations with persons in the target group reveal the reflections on their own tourism behavior, uncovering what it means to be a disabled tourist, as well as why and how he/ she makes his travel choices, within their individual context and lifestyle.

As several authors have pointed out (e.g. Kitchin, 2000; Blichfeldt & Nicolaisen, 2011; Richards et al., 2010), to achieve a more in-depth understanding of people with disability' tourist experiences, it is very important to "give them voice". Thus, researchers should seek to explore new ways of understanding accessible tourism, so as to give a central role to people with disabilities in the research process, and thus promote positive social change (Richards et al., 2011).

In order to obtain a holistic understanding of the feelings and perceptions about tourist experiences of people with disabilities, respondents should speak freely of these experiences and express opinions about how their disability influences their travel decisions. For that reason, the questions were drawn according to an unstructured script, marking the contents, but conferring freedom of speech to the interviewees (Quivy & Campenhoudt, 1992).

The identification of the interviewees was based on personal contacts, resulting from our previous work in this field, and also with the support from Salvador Association. This Portuguese disability support organization presented us the potential respondents with the right profile and mediated initial contacts, by sending an email to their associates.

During the selection process of participants, we tried to ensure diversity in terms of travel patterns, type and degree of disability, age, income, and gender. These selection criteria allowed us to identify different factors and to analyze patterns according to different situations and lifestyles. We can therefore consider that a purposeful sampling technique was applied, oriented by the objective of covering a diversity of situations in the studied universe. In total, 12 adults of varying ages (between 30 and 62 years) with physical disability agreed to share their opinions and travel experiences. The disability of five respondents was caused by road accidents or other accidents, two individuals disability resulted from childhood polio, other two resulted from other diseases such as spina bifida, and in one case the cause of disability was due to childbirth problems. As for their family situation, six respondents are married and the remaining single or divorced, of which four live alone, two live with the mother and one with the daughter. Almost all respondents are paraplegic, except for two individuals with quadriplegia, which means that they are not autonomous in carrying out activities of daily living. Regarding their occupation, most respondents are professionally active, apart from two retired individuals.

In general, people were very receptive and available to participate in the research. This facilitated the contact and allowed to create empathy with the interviewees, which was central for creating a facilitating environment for conducting interviews. It became clear that the participants considered their opinion important to eventually introduce changes in a domain of their personal relevance (Quivy & Campenhoudt, 1992).

The interviews were tape-recorded with the participants´ permission and fully transcribed. This process, although time consuming, allows for greater interaction and flexibility in conducting the interview and a more accurate record of the information provided by the respondents (Jennings, 2005; Silverman, 2000). Transcription of 11 hours of interviewing resulted in a total of 295 pages of transcribed text. The transcripts were then sent by email to each respondent in order to obtain further feedback or comments.

The data treatment was based on content analysis identifying major themes and categories in the discourses. This process comprises three phases: a) pre-analysis (organization of the material and definition of procedures); b) exploration (identification of categories and intersections, with more or less narrow amplitudes and connections); c) treatment of results (the data is interpreted and gains meaning). A software supporting qualitative data analysis, namely WebQda was used to create categories, codify, control, filter, search and query the data.

The process of information codification and the production of a system of categories, was based on a mixed method, resulting both from the conceptual framework and the empirical material constituted by the participants discourse, in a dynamic, hybrid deductive and inductive process of confrontation between theory and evidence flowing from the discursive material (Silverman, 2000). As new themes appeared, new fields were added or divided in order to capture the fullness of the discursive material considered relevant to the purpose of our study.

RESULTS AND DISCUSSION

The structural constraints faced by people with physical disabilities during their travel experiences and the negotiation strategies they used to overcome them are summarized in Table 1.

Table 1. Structural constraints and negotiation strategies used by people with physical disabilities

Structural constraints	Meaning	Negotiation Strategies used
Transportation	Poor assistance support /Failures to provide assistance in air and rail transport Fear of damage, loss or exchange of wheelchairs during air transportation Obligation to give advance notice of the need for assistance in rail transport Difficulty moving at stations Lack of lifts or out of use, or difficult to access	Selection of service providers Information sources Notice in advance Travel companion
Service providers lack of knowledge	Lack of knowledge about accessibility Lack of training of professionals, considering people with disabilities as a homogeneous group.	Selection of service providers Information sources
Accommodation	Lack of accessibility of WC bedrooms Location of adapted bedrooms Lack of accessibility of the bedrooms Inexistence of support products & equipment Lack of accessibility in the common areas (ex: main entrance and dining spaces)	Selection of service providers Selection of destinations Information sources Notice in advance Travel companion Adaptation ability
Public space	Obstacles on the sidewalks Uneven pavement Inadequate parking spaces	Selection the places to visit
Rules & regulations	Difficulties in transporting support equipment, imposed by security issues on air transportation	Information sources Rights argumentation
Information availability	Lack of information on accessibility Lack of detail and accuracy of the information on accessibility Lack of personalized information	Information sources
Additional costs	Option for more expensive services to ensure greater accessibility (e.g. taxi, higher category hotels) Need to pay caregivers travel expenses Lower disposable income for travel, resulting from some expenses associated with disability	Selection of service providers
Lack of support services (at destinations)	Lack of professional support services Lack of technical assistance rental services Lack of adapted cars for rent at destinations	Selection of service providers Travel companion
Tourist attractions	Architectural barriers Unfavourable arrangement of seats considered accessible on events	Selection of service providers Selection of places to visit Information sources Notice in advance Adaptation ability
Restaurants	Architectural barriers Equipment design (e.g. height of dining tables) WC accessibility	Adaptation ability

Transportation

The problems encountered at the transport are clearly the most mentioned in the framework of structural constraints, having been highlighted by all interviewees. The type of issues varies depending on the mode of transport, and results, mainly, from the poor assistance provision, especially in air transportation.

The failures in providing assistance in air and rail transport, and the consequent dependence on the professionals, is one of the aspects most emphasized by the interviewees. Although in many countries the assistance service for people with disabilities is mandatory and free of charge in air transport, whether at airports or on aircraft, (except for justified security reasons and provided for by law), the actors' discourse allowed to observe the existence of frequent failures in this service provision:

The problem of assistance ...! Sometimes, when I arrive at Lisbon and all passengers leave the plane and, half an hour later, I am still inside the plane ... and they looking at me [the staff], as if to say "so, you fell asleep, don't you get off the plane?" They even forget that they transport a person with a disability (...) because I often alerted them, in a polite way, I called one of the hostesses and informed that I need assistance and sometimes I heard very unpleasant responses, people did not, no ... they interpreted it as if I had to remind them out to do their work .. (Ent. B7)

It was also very common in the respondent's narratives, the fear related to their wheelchairs, fear of damage, or their loss, or even chair changes, as it was also noted by Packer et al. (2007) and by Darcy (2012):

Another thing is the care they have with wheelchairs. I have already had problems with an airline, they literally broke the remote, I made a complaint, but until today I have not had an answer. It ends up discouraging people to travel" (Ent. B7)

This kind of problems seem to be intensified when there is a need to make stopovers, as was also identified in the studies by Packer et al. (2007). The need to make transfers between flights often requires waiting long hours in uncomfortable situations, as well described in the following quote:

"When I went to Texas, the transfer was 9 hours of waiting, New York, I had to lie down on the floor." (Ent. B11)

In most European countries, rail transportation has an Integrated Mobility Service (IMS), oriented to serve Customers with Special Needs. However, it is often necessary to inform in advance the need for this assistance, requiring advance planning, which is perceived as a constrain factor by some respondents, because it could be difficult to articulate with the time management and options during trips.

In this subcategory, there were also mentioned difficulties in other means of public transport, namely in buses and metropolitan. Even recognizing that there has been a general improvement in this field, the interviewees referred several problems that hinder their mobility, such as lack of accessibility of stations and interfaces and elevators that are out of use or badly located.

The previously mentioned constraints affect people in different ways, depending on whether they travel alone or with a companion. To negotiate these constraints and try to prevent the stress they cause, the main strategy used by the respondents is a rigorous travel planning, undertaking a huge travel information search to verify the service providers conditions to meet their particular needs. This includes to notice in advance, when it is possible, and anticipate and prevent problems.

Service Providers Lack of Knowledge

Lack of knowledge of tourism providers appeared as an important constraint across all respondents' narratives, being reflected in different ways, with diverse impact on their travel experience and satisfaction. Through the interviews content analysis, it was possible to detect a large number of references related to the lack of information and knowledge about disability issues from tourism providers. This could have a negative influence on the different components service of the tourism product, and often results in the emergence of less favourable attitudes towards people with disabilities.

The actors' discourse made it possible to observe, in many cases, that one of the behaviours that reflects the insufficient knowledge of tourism professionals, in relation to clients with disabilities, means avoiding direct contact, rather directing communication to accompanying people without disabilities, as the following interviewee refers:

At lunch [at a restaurant], my wife and I were there and my wife asked what she wanted to eat and the waiter asks: "And what does he want to eat? " and I said: "look, I can talk " and he was embarrassed, the he asked me, but a little bit more embarrassed and always looking more at my wife than at me. (Ent. B12).

Some interviewees recognize that in most cases this kind of attitude reveals lack of training and ignorance on how to deal with guests with disabilities and to help them. Another issue pointed out is the tourism professional's inability to provide specific information on the accessibility of spaces, of great importance for these tourists, as is well documented on the the published research. Also consistent with the literature review, some narratives confirm that regularly this ignorance results on the assumption that people with disabilities are a homogeneous group with similar needs and features (Daruwalla & Darcy & 2005; Devile & Kastenholz, 2018; Richards et al., 2010).

Accommodation

Regarding accommodation, several constraints factors emerged from the contend analysis, especially associated with the lack of accessibility, both of bedrooms and WCs:

My distress, when I look for a hotel, the first thing I will do is enter in the bathroom. And that's what happened to me, for example, last year in Spain. We went to La Manga, with my colleagues, it said '' adapted hotel '', yes sir, we got there, exactly, I didn't enter. (Ent.B5)

There are also some references to the unfavourable location of adapted bedrooms, the lack of accessibility conditions of the common areas of the accommodation (ex: main entrance and dining spaces) and the inexistence of support products & equipment (e.g. bath chairs). As Daniels et al. (2005) point out, a proper accommodation is a critical factor to decision-making process for tourists with disabilities, being crucial to ensure a comfortable stay and to allow their independence from other people. The main negotiation strategies used by our respondents to overcome this kind of constraints is a rigorous travel planning, undertaking a huge travel information search to verify the accessibility conditions to meet their particular needs. This includes to anticipate and prevent problems in advance, especially regarding WC accommodation, in order to select the destinations and accommodations that best meet their needs. Other strategies identified on the narratives are to notice in advance, contacting the service accommodation to confirm all the bedroom details required. Our results also suggest that adaptation ability is an important strategy to handle with the constraints encountered.

Nevertheless, it is likely that the occurrence of some unpleasant situations would have a negative influence on the experience satisfaction. One of the interviewees clearly synthesizes the nature of the impact that situations like this have on the pleasure associated with traveling, making them more dependent from other people:

We stopped being ourselves, the independence that I had... I´m no longer myself. I don't feel like going there anymore, I can't go anymore. I mean, this ... it destroys all the initial joy and expectation and emotion that the person has, right there. Really, the bathroom is one of the worst things, I don't understand why it is like that, I really don't understand. (Ent. B8)

Public Space

In this subcategory, it is possible to conclude that different barriers coexist at the level of public space planning, which prevent or limit the experience in tourist destinations. The aspects highlighted here include the lack of proper parking and poor conditions for circulating resulting mainly from the existence of obstacles on the sidewalks and the uneven pavement in some older historic cities, particularly uncomfortable for wheelchair users. These situations are not new, since they occur in the daily lives of people with physical disabilities and have received attention from the media. What is new here is its negative influence on decision making and choice of tourist destinations.

The discourse analysis allowed us to detect a large number of references related to constraints resulting from the imposition of standards, especially in air transportation, for safety reasons, which we grouped in the subcategory "rules and regulations". As already identified by McKercher et al. (2003), one of the questions raised here concerns the transport of technical or support equipment, necessary to people with physical disabilities:

Especially having to go through the x-rays, disassemble medical machines for fear that they are bombs, they will damage the machines, I have an electro stimulator device in my belly and I'm always afraid when I travel. It is uncomfortable because it is fragile and very expensive material! If a breakdown has to be purchased in England and has to be programmed in Barcelona. Lately it happens more and more. (Ent. B11)

It's a thing ... crazy ... and then we really need some sandals and stuff ... And I took a packet of them, they can't get medication in the country, there has to be a special thing, it was a confusion there ... it's crazy. (Ent. B8)

There are also other unclear internal procedures or regulations, especially in terms of air transport, and, presumably, for security reasons, which raise some indignation on the part of the interviewees, as it was already been identified in Darcy (2012) research. Among others, it is the case of some airlines that define specific seat allocation for people with disabilities and have unclear procedures for their boarding and disembarking process.

Information Availability and ICT Support

The lack of reliable and detailed information or ICT support about accessibility issues emerged as an important constraint across all the narratives, having a huge influence on travel experience:

I have had bad, negative experiences, several situations, like going to a hotel, where it guarantees on the net that the room is accessible and then I get there and it is not. It has steps, small, narrow doors. However, they say they are accessible. (Ent. B1).

This is a very important factor in order to meet tourists with physical disabilities' differential requirements, allowing them to make informed travel decisions, and thus avoiding unpleasant surprises and unnecessary risks (Devile & Kastenholz, 2018; Packer et al., 2007; Daniels et al., 2005; Bieger & Laesser 2001; Eichhorn et al., 2008). The negotiation strategy used to deal with this is contacting directly the tourism providers, by email, or by phone. ICT tools aren't yet the solution on the top of the mind of consumers or tourism providers. However, some of our interviewees observed that most of them are not qualified to provide specific information on accessibility issues, as was already identified before on the theme "Service providers lack of knowledge".

I couldn't get information by phone! I called, therefore, to the responsible company ... that does the space exploration ... I didn't get information. I called again, and it was not possible to get that information! (Ent. B4)

Additional Costs

In this subcategory, we tried to cluster the references related to the increase in travel costs resulting from people with disabilities' specific requirements. Our intention here was not to study the scope of the economic factors, costs per se, which are, for most tourists, an important constraining factor, but rather to analyze the economic factors associated with the disability condition. As expected, economic factors eventually emerged, in the speech of our interviewees, as one of the most significant barriers for travel participation, although only those resulting from disability were included in our analysis. The discourse of respondents made it possible to observe that they tend to choose for more expensive services, as they perceive that higher-level services are the ones that are more likely to have the best conditions of accessibility, both in terms of accommodation and transport and restaurants:

I usually choose an upper-middle-class hotel ... and that has its costs. I can't go to hostels, usually have stairs ... (…). The restaurant is always of the upper middle range ... It also reduces the chances to travel more. We have to choose other conditions. (Ent. B11)

On the other hand, some of our interviewees referred the need to travel with caregivers to help them with their daily routines that also constitutes an additional cost for these interviewees, as reported by the following woman, whose husband has also a physical disability:

This is our biggest obstacle to not traveling more. Sometimes it is more expensive to pay to that person than accommodation, or transportation.... (Ent. B7).

The confluence of these costs with the expenses usually associated with the physical disability condition (for example, in technical aids to ensure greater autonomy and comfort) has a negative effect on the discretionary income for travel. The following descriptions are very illustrative about this situation:

Even our independence has a lot to do with the type of wheelchair that people can afford... it has a lot to do with people's monetary situation. (Ent. B8)

Yeah, because ... it's complicated, that is! a lot, really ... and things are really very expensive, ... when you choose a wheelchair, for example, light alloy ... oh, they are very expensive ... very expensive. (Ent. B6)

It is impossible to have money to travel when you have so many other expenses ... But it is all my effort. I bought a crane to be able to transfer to bed, to the wheelchair, to the bathtub and so on ... In the bathtub I have a stairlift to be able to bathe without help ... I have this device for gymnastics here. (Ent. B9)

Lack of Support Services

The main references included in this category are related to non-existence (or lack of information) of some support services at destinations, available to help PwD in some tasks, especially when they have no autonomy in carrying out some daily life routines. This is the case of three of the participants who are not autonomous and need someone to provide support in some tasks. Depending on the level of dependence and if they travel alone or with companion, there are other types of services needed like technical aids rental services and adapted cars for rent at destinations.

The participants who travel alone or with other people with physical disability negotiate these situations with a rigorous preplanning travel, selecting the service providers and, when it is possible travelling with a companion.

Tourist Attractions

Regarding tourist attractions, besides some transversal factors to the forementioned issues, the main problem is the existence of architectural barriers that prevent the participants to visit some places. A lot of examples were mentioned, mainly cultural heritage sites, but also the lack of accessibility of cultural and tourist events.

It is important to note, that even though the accurate legislation on this, the accessible seats are usually located in uunfavourable sites, which once again influences the experience quality.

IMPROVING THE TOURIST SUPPLY CHAIN FOR PEOPLE WITH DISABILITIES: THE ROLE OF INFORMATION AND COMMUNICATION TECHNOLOGIES

One of the main outputs of our study is the importance of reliable and detailed information about accessibility issues of the tourist supply chain as a negotiation strategy. As we have seen, this is transversal to all components of the tourist supply, emerging from accommodation, transportation, attractions, and could have a huge influence on persons with disabilities tourist experience. In fact, not all the constraints are related to physical accessibility, since there have been undertaken several initiatives, in a large number of countries, in order to improve it. But we found remarkable limitations in the communication and interaction accessibility, or else, between tourist service providers and tourists with disabilities.

In this scope, as has been emphasized by several studies, Information and Communication Technologies (ICT) can be a critical facilitator in providing information to tourists with special needs (Vila et al, 2000; Ribeiro, Silva, Metrôlho, Silva & Barbosa, 2018). As with any consumer, ICT are a powerful tool for accessing information for tourists with disabilities, allowing to access a huge data resources and to provide relevant and accurate information, tailored to their specific needs and preferences.

Furthermore, it enables the exchange of experiences between consumers, the sharing of testimonies and specific recommendations by other consumers with similar needs (Ray & Ryder 2003; Buhalis & Michopoulou, 2011). In addition, it enhances the search for information on consumer-oriented services, according to their individual preferences and requirements, customized to their specific needs across the tourism product (Burnett & Baker 2001; Ray & Ryder 2003, Buhalis & Michopoulou, 2011, Ribeiro et al. 2018). This new digital age in which we live nowadays, stimulated by the phenomenon of the democratization of the Internet under the influence of the World Pandemic caused by COVID-19, shortens the relationship between supply and demand and allows the conception of unique and intimate connectivity links and digital sharing of tips and opinions between tourism providers and tourists with disabilities never seen before (Cassia, Castellani, Rossato and Baccarani, 2020).

The possibility of being able to use systems based on artificial intelligence is another advantage. This allows to identify a profile based on the consumer's previous choices, so that, in this way, the system can identify their preferences, comparing the choices of other tourists with similar profiles, and thus provide more relevant results for the researches carried out (Buhalis & Michopoulou, 2011). It is expected that the continuous development of ICT, with increasing skills in terms of communication, collaboration, mobility and interaction, will improve the communication with these consumers, developing personalized information channels and, consequently, enhancing their participation in social life, in general, and consequently, in tourism activities. A "range of disruptive technologies as: the Internet of Things, the Internet of Everything, fifth generation mobile network (5 G); Radio Frequency Identification (RFID); mobile devices, wearable smartphones and devises; 3 D printing, apps along with APIs, Cryptocurrency and Blockchain, sensor and beacon networks, pervasive computing, gamification as well as enhanced analytical capabilities supported by Artificial Intelligence (AI) and machine learning (ML)… create the infostructure and the smart digital grid that support the seamless interoperability of all stakeholders" (Buhalis, 2019, p. 269).

Nevertheless, for this to be happen, it is crucial to deepen the knowledge about the needs of the tourist with special needs in order to provide an adequate and useful information system, involving all stakeholders of the tourist supply chain.

FUTURE RESEARCH DIRECTIONS

This study has highlighted the lack of the tourism industry's knowledge on the quality of accessible tourism experiences for people with disabilities, drawing attention to the importance of the shared responsibility of stakeholders in the tourism system, in order to promote and implement inclusive tourism. This requires a thorough knowledge about the different dimensions of the disability reality, also considering diversity within this group, and suggests the need for general training and awareness in this field, at different levels (Devile & Kastenholz, 2018).

In this sense, it would be very useful for the deepening of knowledge in this area to extend the study and include people with disabilities who do not have the opportunity to travel. And also extend, or replicate, this study to other types of disabilities, namely the population with hearing impairment, with intellectual disability, and also the senior population.

At the same time, and more broadly, according to Jackson (2000), new directions and research topics are needed to avoid this area of study becoming repetitive and redundant. Therefore, it is suggested that special attention should be given to the following research topics:

1. The negotiation process of leisure constraints and the individual and contextual circumstances that reinforce or limit people's ability to achieve their leisure goals;
2. constraints and transition periods in people's lives - for example, entry into higher education, marriage, birth of the first child, divorce, death of spouse, immigration, etc. To what extent do these periods offer new opportunities for the negotiation of leisure constraints?
3. Turning constraints into opportunities - Why are there very successful people in their leisure lives, not only negotiating constraints, but also seeing constraints in a more positive way as opportunities to develop new interests and leisure goals?
4. Relationship between exclusion and inclusion in tourism policies and practices aimed at people with disabilities;
5. Study of the impact of tourism experiences on people with disabilities.

With this, we advocate a greater involvement of researchers in the field of accessible tourism and within disability issues.

CONCLUSION

Faced with this reality, according to the bibliographical review that we carried out, the implementation of a total accessibility policy in tourism requires a paradigm change in the management of destinations, as well as an organizational change at the level of the whole tourism value system, supported by adequate training of the sector's employees. To this end, it will be essential to recognise the competitive advantages associated with this market and, at the same time, to understand the operational aspects underlying social responsibility practices, understood as pillars of sustainable tourism development.

On the other hand, there is also a lack of research regarding travel motivations, lived and expected experiences and the travel decision process, as well as the constraining and facilitating factors involved in it. It is therefore very important to try to understand these aspects and how they interrelate, how obstacles are perceived and overcome, the importance of the benefits of tourism in daily life and the influence of previous experiences on the travel decision making process.

Although there is some lack of articulation between leisure studies and tourism research, it seems to us that significant advances can be achieved in the area of tourism by applying the leisure constraints approach, which has been the target of interest of numerous scholars over the last three decades, representing today a theoretically consolidated body of knowledge. Specifically, Hinch and Jackson (2000) suggest that tourism researchers could benefit from using leisure inhibitor models to help understand the barriers to participation in tourism,

The need to know consumers' needs, to understand their expectations and experiences is a central aspect for tourist service providers to be able to tailor their offers to actual and potential demand, promoting socially responsible tourism development, because it is more inclusive and friendly.

Throughout this work it has been possible to see that the effective participation of people with disabilities in tourist activities is a dynamic and interactive process in which a number of different factors from their individual context, the social environment and the tourist context interact.

The data from our study, together with the results presented by the literature, reveal that the travel experience, which is being acquired by people with disabilities, allows them to develop tourism skills and thus learn to face the different obstacles more effectively, preventing risks and choosing the services and places that best meet their needs. This same reflection is in line with the study by Blichfeldt and Nicolaisen (2011), which underlines the dynamic nature of disability and the capacity to adapt. Also corroborating Vila, González and Darcy, 2020 studies, which focus information, especially internet and digital tools, as the way forward to accessible tourism. This also explains the fact that the constraints faced by the interviewees are almost always perceived as a learning opportunity, allowing them to make more informed decisions in the future and avoid potentially challenging situations.

Research on constraints influencing the participation of people with disabilities in tourism points to the prevalence of structural constraints, which was confirmed in our study, being more emphasized at the level of transport, the gap in the performance of employees in the tourism sector (insufficient information and knowledge of the reality of disability), the lack of accessibility in accommodation units and public space, and the level of rules and regulations, among others.

Our research has also shown that negative attitudes on the part of tourism professionals seem to have a stronger influence on the tourism experience, often translated into avoiding direct contact with people with disabilities and the option of communicating with the person accompanying them instead.

On the other hand, the decision to travel and the satisfaction obtained with the tourist experience are also the result of the action of the facilitating factors that, contrary to the constraints, promote and encourage participation in tourist activities. At this level, the literature is less profuse. As we have seen, research in this field has clearly focused on the factors that limit, and not so much those that facilitate, the participation of people with disabilities in tourism activities. Our study has allowed, in this field, to identify different mechanisms to mitigate, or overcome, the constraints to participation in tourism activities.

Regarding the trip organization, it was possible to identify a set of negotiation strategies, which together allow effective participation in tourism activities. It is worth mentioning the behaviour of researching information constitutes a privileged negotiation strategy. Hence, the internet, social media and mobile technology as source of information, communication and attraction, allows new solutions, approaches and practice (Moura, Mónico & Mira, 2019), particularly for accessible tourism.

Undoubtedly, disability can have a devastating effect on the lives of people and families it affects. However, it is the widespread representations in society, still marked by the absence of structures for participation and realization of their capabilities, that aggravate the situation of disadvantage and discrimination. Tourism, by its nature, is a socializing activity, allowing it to contribute to the demystification of disability and the reality of the lives of people with disabilities, as well as to the removal of social stereotypes that continue to prevail, despite the evolution felt in recent decades.

Thus, the place given to new technologies and the internet includes the transversality principle in various fields of the tourism industry and is not exhausted with tourism stakeholders. It also assumes great importance for political actors who must ensure the implementation of concrete public policies and the planning of effective actions to enforce them.

REFERENCES

Altinay, Z., Saner, T., Bahçelerli, N. M., & Altinay, F. (2016). The role of social media tools: Accessible tourism for disabled citizens. *Journal of Educational Technology & Society*, *19*(1), 89–99.

Bieger, T., & Laesser, C. (2001). Segmenting Travel on the Sourcing of Information. In *Consumer Psychology of Tourism, Hospitality and Leisure* (pp. 153–167). Cabi Publishing. doi:10.1079/9780851995359.0153

Blichfeldt, B. S., & Nicolaisen, J. (2011). Disabled travel: Not easy, but doable. *Current Issues in Tourism*, *14*(1), 79–102. doi:10.1080/13683500903370159

Buhalis, D. (2019). Technology in tourism-from information communication technologies to eTourism and smart tourism towards ambient intelligence tourism: A perspective article. *Tourism Review*, *75*(1), 267–272. doi:10.1108/TR-06-2019-0258

Burnett, J. J., & Baker, H. B. (2001). Assessing the travel-related behaviors of the mobility-disabled consumer. *Journal of Travel Research*, *40*(1), 4–11. doi:10.1177/004728750104000102

Cassia, F., Castellani, P., Rossato, C. & Baccarani, C. (2020). Finding a way towards high-quality, accessible tourism: the role of digital ecosystems. *The TQM Journal*. doi:10.1108/TQM-03-2020-0062

Cavinato, J. L., & Cuckovich, M. L. (1992). Transportation and Tourism for the Disabled: An Assessment. *Transportation Journal*, *31*(3), 46–53.

Crawford, D., & Godbey, G. (1987). Reconceptualizing barriers to family leisure. *Leisure Sciences*, *9*(2), 119–127. doi:10.1080/01490408709512151

Daniels, M. J., Rodgers, E. B., & Wiggins, B. P. (2005). "Travel Tales": An interpretive analysis of constraints and negotiations to pleasure travel as experienced by persons with disabilities. *Tourism Management*, *26*(6), 919–930. doi:10.1016/j.tourman.2004.06.010

Darcy, S. (2010). Inherent complexity: Disability, accessible tourism and accommodation information preferences. *Tourism Management*, *31*(6), 816–826. doi:.tourman.2009.08.010 doi:10.1016/j

Darcy, S. (2012). (Dis)embodied air travel experiences: Disability, discrimination and the effect of a discontinuous air travel chain. *Journal of Hospitality and Tourism Management*, *19*(1), 91–101. doi:10.1017/jht.2012.9

Darcy, S., Cameron, B., & Pegg, S. (2010). Acessible Tourism and Sustainability: A discussion and case study. *Journal of Sustainable Tourism*, *18*(4), 515–537. doi:10.1080/09669581003690668

Daruwalla, P., & Darcy, S. (2005). Personal and Societal Attitudes to Disability. *Annals of Tourism Research*, *32*(3), 549–570. doi:10.1016/j.annals.2004.10.008

Decrop, A. (2006). *Vacation Decision Making*. Cabi Publishing. doi:10.1079/9781845930400.0000

Devile, E., & Kastenholz, E. (2018). Accessible tourism experiences: The voice of people with visual disabilities. *Journal of Policy Research in Tourism, Leisure & Events*, *10*(3), 265–285. doi:10.1080/19407963.2018.1470183

Eichhorn, V., Miller, G., Michopoulou, E., & Buhalis, D. (2008). Enabling Acess to Tourism through Information Schemes. *Annals of Tourism Research*, *35*(1), 189–210. doi:.annals.2007.07.005 doi:10.1016/j

Figueiredo, E., Eusébio, C., & Kastenholz, E. (2012). How Diverse are Tourists with Disabilities? A Pilot Study on Accessible Leisure Tourism Experiences in Portugal. *International Journal of Tourism Research*, *14*(6), 531–550. doi:10.1002/jtr.1913

Gilbert, C. (1999). Les musées ouvrent leurs portes à tous les publics. *Espaces (Paris)*, *161*, 26–29.

Goodall, B. (2006). Disabled acess and heritage attractions. *Tourism Review International*, *7*, 57–78.

Hinch, T. D., & Jackson, E. L. (2000). Leisure Constraints Research: Its Value as a Framework for Understanding Tourism Seasonability. *Current Issues in Tourism*, *3*(2), 87–106. doi:10.1080/13683500008667868

Israeli, A. A. (2002). A preliminary investigation of the importance of site accessibility factors for disabled touristsl. *Journal of Travel Research*, *41*(1), 101–104. doi:10.1177/004728750204100114

Jackson, E. L. (2000). Will Research on Leisure Constraints Still Be Relevant in the Twenty-first Century? *Journal of Leisure Research*, *32*(1), 62–68. doi:10.1080/00222216.2000.11949887

Jennings, G. R. (2005). Interviewing: a Focus on Qualitative Techniques. In B. W. Ritchie, P. Burns, & C. Palmer (Eds.), *Tourism Research Methods - Integrating Theory with Practice* (pp. 99–117). Cabi Publishing. doi:10.1079/9780851999968.0099

Kastenholz, E., Eusébio, C., & Figueiredo, E. (2015). Contributions of tourism to social inclusion of persons with disability. *Society and Disability*, *30*(8), 1259–1281. doi:10.1080/09687599.2015.1075868

Kitchin, R. (2000). The researched opinions on research: Disabled people and disability research. *Society and Disability*, *15*(1), 25–47. doi:10.1080/09687590025757

Lyu, S. (2017). Which accessible travel products are people with disabilities willing to pay more? A choice experiment. *Tourism Management*, *59*, 404–412. doi:10.1016/j.tourman.2016.09.002

McKercher, B., & Darcy, S. (2018). Re-conceptualizing barriers to travel by people with disabilities. *Tourism Management Perspectives*, *26*, 59–66. doi:10.1016/j.tmp.2018.01.003

McKercher, B., Packer, T., Yau, M., & Lam, P. (2003). Travel agents as facilitators or constraints of travel: Perceptions of people with disabilities. *Tourism Management*, *24*(4), 465–474. doi:10.1016/S0261-5177(02)00107-3

Moura, A., Mónico, L., & Mira, M. R. (2019). Measuring quality regarding destination marketing: Perceptions from local public stakeholders in Portugal. *Tourism & Management Studies*, *15*(1), 44–53. doi:10.18089/tms.2019.150104

Murray, M., & Sproats, J. (1990). The disabled traveller: Tourism and disability in Australia. *Journal of Tourism Studies*, *1*(1), 9–14.

Neumann, P., & Reuber, P. (2004). Economic Impulses of Accessible Tourism for All. Berlin: Federal Ministry of Economics and Technology.

Nyman, E., Westin, K., & Carson, D. (2018). Tourism destination choice sets for families with wheelchair-bound children. *Tourism Recreation Research*, *43*(1), 26–38. doi:10.1080/02508281.2017.1362172

O'Neill, M., & Ali Knight, J. (2000). Disability tourism dollars in Western Australia hotels. *FIU Hospitality Review*, *18*(2), 72–88.

Packer, T. L., Mckercher, B., & Yau, M. K. (2007). Understanding the complex interplay between tourism, disability and environmental contexts. *Disability and Rehabilitation*, *29*(4), 281–292. doi:10.1080/09638280600756331 PMID:17364778

Pearn, M. (2011). Heritage Sites: Attitudinal and Experimental Differences of Disabled and Able-Bodied Visitors. In S. Darcy & D. Buhalis (Eds.), *Acessible Tourism: Concepts and Issues* (pp. 201–213). Channel View Publications.

Poria, Y., Reichel, A., & Brandt, Y. (2010). The flight experience of people with disabilities: An exploratory study. *Journal of Travel Research*, *49*(2), 216–227. doi:10.1177/0047287509336477

Quivy, R., & Campenhoudt, L. (1992). *Manual de Investigação em Ciências Sociais*. Gradiva.

Ray, N. M., & Ryder, M. E. (2003). Ebilities'' tourism: An exploratory discussion of the travel needs and motivations of the mobility-disabled. *Tourism Management*, *24*(1), 57–72. doi:10.1016/S0261-5177(02)00037-7

Ribeiro, F. R., Silva, A., Metrôlho, J. C., Silva, A. P., & Barbosa, F. S. (2018). A New Framework for Accessible Tourism Mobile Application Development. *International Journal of Mobile Computing and Multimedia Communications*, *9*(2), 31–46. doi:10.4018/IJMCMC.2018040103

Richards, V., Pritchard, A., & Morgan, N. (2010). (Re) Envisioning Tourism and Visual Impairment. *Annals of Tourism Research*, *37*(4), 1097–1116. doi:10.1016/j.annals.2010.04.011

Shaw, G., & Coles, T. (2004). Disability, holiday making and the tourism industry in the UK: a preliminary survey. *Tourism Management, 25*(3), 397-403. doi:(03)00139-0 doi:10.1016/S0261-5177

Silverman, D. (2000). *Doing Qualitative Data: A Practical Handbook*. Sage Publications.

Smith, R. W. (1987). Leisure of Disabled Tourists: Barriers to Participation. *Annals of Tourism Research*, *14*(3), 376–389. doi:10.1016/0160-7383(87)90109-5

Tao, B. C., Goh, E., Huang, S., & Moyle, B. (2019). Travel constraint perceptions of people with mobility disability: A study of Sichuan earthquake survivors. *Tourism Recreation Research*, *44*(2), 203–216. doi:10.1080/02508281.2019.1589085

Vila, T., González, E., & Darcy, S. (2020). Accessibility of tourism websites: The level of countries' commitment. *Universal Access in the Information Society*, *19*(2), 331–346. doi:10.100710209-019-00643-4

Yau, M. K., McKercher, B., & Packer, T. L. (2004). Traveling with a Disability: More than an Access Issue. *Annals of Tourism Research*, *31*(4), 946–960. doi:10.1016/j.annals.2004.03.007

Chapter 4
Use of the Internet to Plan Tourism Trips by People With Special Needs

Maria João Carneiro
GOVCOPP, DEGEIT, University of Aveiro, Portugal

Leonor Teixeira
https://orcid.org/0000-0002-7791-1932
IEETA, DEGEIT, University of Aveiro, Portugal

Celeste Eusébio
https://orcid.org/0000-0002-2220-5483
GOVCOPP, DEGEIT, University of Aveiro, Portugal

Elisabeth Kastenholz
GOVCOPP, DEGEIT, University of Aveiro, Portugal

Andreia Antunes Moura
https://orcid.org/0000-0002-1722-3476
Polytechnic Institute of Coimbra, Portugal

ABSTRACT

This chapter aims to examine internet usage to plan a tourism trip by people with special needs (PwSN), particularly addressing differences in the intensity of internet usage according to their age and education level. To achieve these aims, a survey was applied to a sample of Portuguese people with special needs who have already undertaken tourism trips. Various bivariate statistical tests (chi-square, t test, ANOVA, and Kruskal-Wallis) were used to examine differences in internet usage according to respondents' age and education level. Results obtained reveal that the internet is the most important information source used by PwSN in planning their tourism trips. However, respondents report that they are not very satisfied with the online sources used. Moreover, online sources used to plan tourism activities vary according to PwSNs' age and education level.

DOI: 10.4018/978-1-7998-6428-8.ch004

INTRODUCTION

People with special needs (PwSN), including people with disabilities (PwD) and people with other types of needs, represent an important market for tourism destinations. Nevertheless, historically, the population with disabilities has been subject to stereotypical social representations and constructions, based on unfounded beliefs and superstition, which has led to discriminatory attitudes and acts. It is for this reason that, even today, disability has severe consequences for individuals in society (Moura, 2018), leading to their exclusion from many fields of social life, including leisure travel (Kastenholz, Eusébio & Figueiredo, 2015).

Today, any community has the duty to permit and promote access for every person so they can actively participate in society, in a fair and equal way, based on the globalizing approach to human rights organizations (Fontes, Martins, & Hespanha, 2014). Thus, in recent years, human rights protection and the individual value of the person, respect for diversity, the discrimination debate, increasing social awareness, and gradually growing political responsibility are strongly supported by various international organizations. These initiatives have culminated in a broad and vigorous social movement covering more and more countries, leading to innovative and profound societal changes (Moura, 2018).

Gradually, full participation of the individual with special needs in all kinds of social activities is accepted and supported in contemporary societies. In the context of tourism, the emphasis should be on the awareness of governments and tourism industry stakeholders regarding their responsibility to avoid exclusion and not on individuals with disabilities (Packer, McKercher, & Yau, 2007). Therefore, participation in tourism should be understood as a process of interaction in which disability is only one of the factors to be taken into consideration, and not as the determining aspect of the limitations to such participation. Moreover, tourism involvement will enable social inclusion and the enjoyment of benefits that will positively affect health and overall well-being (Kastenholz, Eusébio, Moura, & Figueiredo, 2010). However, travel is still a challenge for many PwSN (including PwD), partly due to the lack of reliable information about tourism destinations (Buhalis, Eichhorn, Michopoulou & Miller, 2006; Darcy, Cameron & Pegg, 2010; Darcy, 2012).

Information and technology are the key to boost accessibility and equal opportunities in tourism planning and enjoyment, playing a crucial role in empowering individuals with disabilities (Altinay, Saner, Bahçelerli, & Altinay, 2016). Thus, Darcy, McKercher and Schweinsberg (2020) stress that it is essential to shift the focus of accessible tourism initiatives from a piecemeal manner, typical in the traditional tourism industry (e.g. transport, accommodation and attractions), towards understanding and developing an integrated and equitable view of the tourism experience as a whole. To do so, the ignition should be precisely the information made available to the tourist in an accessible way. Correspondingly, there is an urgent need for an in-depth investigation of the inclusion and accessibility of information for all, particularly those with disabilities or other difficulties of access (Domínguez Vila, Alén González & Darcy, 2020), and how PwSN use online sources. This is especially true considering the global trend towards democratization of the Internet and other digital tools and the almost doubling of the population with disabilities in the next decade (WTO, 2013). However, there is little research providing a thorough understanding of how PwSN use the Internet to plan and organize their tourism trips. Therefore, the aim of this chapter is to analyse the use of the Internet by PwSN, namely the type of online information sources used to plan trips, the objective with which these sources are used and how this group evaluates these sources. Additionally, this chapter aims to examine whether there are differences in the use of the Internet by PwSN to plan their tourism trips, according to their age and education level.

To achieve these objectives, the present chapter encompasses five main sections. After this introduction, where the relevance and aims of the chapter are explained, there is a literature review about the importance and characteristics of the accessible tourism market, about the use of the Internet as an information source in the scope of tourism and the use of this information source by PwSN. Next, the methodology of the empirical study is presented, and the results of the study are presented and discussed. Finally, the main conclusions and implications of the study are identified.

LITERATURE REVIEW

The Accessible Tourism Market: Relevance and Characteristics

Over time, disability has been associated with rejection, tragedy, imprisonment and even murder. From the horrific traditions of public exposure and abandonment of the bodies of deformed babies in ancient Greece, to the eugenic abuses of Fascist Europe, the culture of segregation towards disability has been perpetuated. Even during the 20th century, the Western world was organized around a well-intentioned and humanist approach, but it neglected equal rights and full social integration, segregating these individuals into special social institutions of health care and social intervention (Moura, 2018).

Fortunately, there have been remarkable changes up to the present day. Individuals with disabilities have been progressively studied and there have been impacting medical, technical-scientific and social developments, which have improved their quality of life and provided opportunities in different sectors of society.

In the field of tourism, politicians, legislators and researchers have developed awareness-raising actions, calling for open-mindedness in relation to disability, recognizing that this concept is no longer likely to be considered from a strictly individual perspective (Kastenholz et al., 2015; Moura, 2018). In fact, Darcy (2010) argues that any architectural changes designed to improve physical accessibility will not bring any benefit to PwD if information and awareness policies are not implemented at a broader social level.

Since 1980, when the World Tourism Organization (WTO) conference presented the Declaration of Manila it is recognized that tourism is a fundamental right of all citizens, addressing global tourism accessibility on a more regular and increasingly in-depth basis. In this context, accessible tourism may be considered as tourism that "enables people with access requirements, including mobility, vision, hearing and cognitive dimensions of access, to function independently and with equity and dignity through the delivery of universally designed tourism products, services and environments. This definition is inclusive of all people, including those travelling with children in prams, people with disabilities and seniors" (Darcy & Dickson, 2009, p. 34). In this sense, it includes every person who suffers restrictions, in a particular tourism environment, in his (her) relational capacity or those who have some kind of special needs. This includes people with disabilities, elderly, pregnant women, families with small children using a stroller, the obese or diabetic, and whoever requires special care temporarily or permanently (WTO, 2013).

At the same time, and in addition to the social relevance of the disability market, it is important to pay attention to its economic significance, since the population with disability cannot be considered a niche market or a minority at all, as it affects not only the individual, but also his/her family and friends.

As we enter the new millennium, over 50 million adults are retired and have free time and financial conditions to travel, but are still limited by some kind of disability, hindering themselves or a loved one

(Gladwell & Bedini, 2004). According to Crawford and Stodolska (2008), 25% of the world's population is affected by disability, either personally or through a relative. Thus, the number of people directly or indirectly affected by disability is clearly significant and growing, given the global demographic propensity towards ageing (UN, 2015). Technological and medical advances that extend and sustain life predict that by 2030 about 60% of the world's elderly population will be over 65 years old and that, as a result, the number of elderly PwD will double by the year 2050 (Ratzka, 2003 in Camisão, 2006) and by 2100 may reach 3.1 billion individuals over 60 with a disability (Gondos & Nárai, 2019).

In 2011, according to the latest World Disability Report, 15% of the world's population was estimated to personally present some kind of permanent disability, and this figure rises to 40% when including those with a temporary disability (World Health Organization, 2011). Recently, the World Bank (2018) found that the population with disabilities already accounts for one fifth of the world's population, i.e. between 110 million and 190 million people.

It therefore appears that a large part of the world's population will or may acquire some form of permanent or temporary disability, which may compromise their independence or autonomy. Tourism destinations must therefore consider the particular social, economic and demographic characteristics of this population (Darcy, 2012), providing appropriate and, whenever necessary, duly adjusted products, activities and services.

In this sense, the potential of the accessible tourism market is unquestionable. To highlight this aspect, the literature also reveals that the frequency and length of holidays taken by the disabled population are factors to be valued (Moura, 2018), while higher expenditure levels, number of persons in a travel group and loyalty are also relevant aspects, making this an appealing target market for destinations (Kastenholz, Eusébio, Figueiredo & Lima, 2012). A study developed by Dwyer and Darcy (2008) in Australia found that, while day trips occur equally, the overnight stay rate at destinations is 21% higher in the population with disabilities and 52% higher in the context of long distance travel. Szcygiel and Piotrowicz (1995) found that 26.4% of all respondents with disabilities were engaged in tourism at least once a year. Darcy (2003) noted that 99.4% of PwD consider themselves perfectly able to travel or very willing to travel. In addition, 74% of PwD surveyed by Darcy (2003) are not satisfied with their level of participation in tourism, due to the barriers they face both when planning the trip and during the trip. Neumann and Reuber (2004) found that German PwD actually travel less frequently than the rest of the population without disabilities, but also found that 48% of these people would travel more intensely if there was more and better accessibility to tourist services, facilities and infrastructure. The same author found that 37% of those surveyed had decided, in the year prior to the survey, not to travel due to lack of accessibility, and that 62% would not even mind paying a higher price if they could fully enjoy accessible tourist products. In the United States, the Open Doors Organization (ODO) conducted a study on the accessibility market in 2015 and found that in that year more than 26 million people with disabilities had travelled and spent approximately US$17.3 billion. At the same time, it is also important to note that 46% of them found some kind of limitation to access during their trip (particularly in terms of accommodation and transport). Another study, developed by the European Commission in 2015, found that the segment of the disability market contributed 352 billion euros to the EU economy between 2012-2013, spending an average €1,000 on a trip. It also found that Europeans with disabilities travel with accompanying persons (1.9 on average) and that China, the US and Brazil are the destinations which are most sought-after by the major emitting markets, namely France, Germany, Italy and the UK.

The various studies analysed (Darcy, 2003; Dwyer & Darcy, 2008; EC, 2015; Neumann & Reuber, 2004; ODO, 2015; Szcygiel & Piotrowicz, 1995) show that disability per se may not be the source of inhibition for tourism participation, and place special emphasis on physical or architectural, financial, social and information obstacles.

Bearing in mind that the basis for the decision to choose a destination, product or tourist service is accessible and reliable information, there is an urgent need to rethink and analyze clear, credible, and simple communication alternatives. According to Packer et al. (2007), tourists with disabilities highlight the personal experiences and/or recommendations of other tourists in the same situation as the most reliable sources of information. However, the Internet and digital tools have received increasing attention from these consumers and it is important to note that a study developed by the European Commission in 2014 found that the Internet is the third most used source of information to support decision-making and the first or most important booking tool for tourist services. This study proved that individuals with disabilities are active users of the Internet, which was considered an easily accessible and user-friendly source of information (EC, 2014).

Internet as an Information Source in Tourism

The development of the Internet in the 1990s marked an era characterized by a great contribution to the practice of communication and information sharing. But while this type of communication is useful in business, in general, in the services and more specifically in the tourism industry, the value of this sharing is exponentially greater (Buhalis & Law, 2008). In this sector the Internet has been widely accepted as a valuable tool for promotion and distribution of tourism products, with impact on tourism-related experience of customers and enhancing consumer experience through co-creation (Buhalis & Sinarta, 2019). The Internet is considered a universal, world-wide accessible technology that includes a set of valuable applications, with the World Wide Web (WWW), or just the Web, being one of the most important services promoted by this universal technology. The usage of this global network to develop online applications, such as electronic commerce (e-commerce) may benefit both consumers and suppliers, allowing the promotion and the consequent purchase of products and services related to the destination without any geographical and time constraint (Öörni, 2004; Werthner & Ricci, 2004). According to these authors, electronic markets substantially benefit from the Internet, since product information can be disseminated at higher speed, at any time and independent of geographical location, with consumable information in the desired quantity and adapted to the clients' demands. Since tourism is a highly intangible sector, and given the globalization of the economy and markets, suppliers need to promote their products to potential customers on a global scale. Thus, tourism is certainly one of the sectors that has most dramatically been transformed by the Internet (World Tourism Organisation Business Council, 1999). As far as Internet usage is concerned, tourism has, in fact, long been seen as a leading sector and even as a driver of business-to-consumer (B2C) e-commerce (Peng & Lai, 2014; Werthner & Klein, 1999).

Almost in the early days of the Internet, Connolly, Olsen, and Moore (1998) stated that Internet-based services enable companies to save time and money, while improving the quality level of services to their customers. Ten years later, Buhalis and Law (2008) presented a study on the impact of information and communication technologies (ICTs) on tourism, highlighting this global web as the main information and communication channel in this sector, with drastic contributions to changes in market conditions experienced by tourist organizations. Today the Internet is a key tool in any organization's strategy, even leading to the reinvention of new business models such as e-commerce, e-services and e-tourism.

The Web, one of the most important services of the Internet, represents a valuable information source, since it promotes accessibility, real-time access and up-to-date information, as well as interactive communications and serves as a distribution channel of information between different geographically distributed actors (Ho, Lin, & Chen, 2012). In this context, the Web has a crucial role in the dissemination of information, providing consumers with details about destinations, accommodation, transport, activities and other kinds of interesting attractions (Domínguez Vila, Alén González, & Darcy, 2017).

In the tourism industry, this technology represents an increasingly important tool to develop business, as tourists and travelers use the Internet for information searches and travel planning (Liu & Zhang, 2014). On the other hand, from the customers' point of view, this channel provides more control in accessing information about touristic products and services, since the tourists can decide what information they want to see and when and where they want to see it. From a supply side perspective, the Internet represents a broad channel of information dissemination without geographical barriers and therefore attracts different kinds of markets. Actually, this channel enables users to access a wide range of services and products around the world, regardless of the supplier's location, reducing the time and effort of searching for information.

In fact, the potential of the Internet as a communication tool, when compared to other mechanisms that can be used by the tourism supply, is unquestionable. While the traditional sources of information, such as newspapers and brochures usually produced by the government agencies or tourism companies, tend to provide local information, the online information available on the Internet informs tourists regardless of where they are (Lee et al., 2019). Some authors indicate that "online information helps people to visit more diverse places, and ultimately reduces the regional tourism gap" (Lee, Chung, & Nam, 2019: p. 1377). Other studies also point out that the Internet, and particularly the communication and information mechanisms powered by it, have an influence on the destination image formation (Llodrà-Riera, Martínez-Ruiz, Jiménez-Zarco, & Izquierdo-Yusta, 2015).

Thus, in the tourism industry, the Internet has proven to be the most popular information source for trip planning, since travelers usually need a considerable amount of information to make a travel plan (Xiang, Magnini, & Fesenmaier 2015).

The tourism business itself has been strongly influenced by the developments of ICT in general and the Internet in particular, both in terms of revolution in tourism objects (products and services) and in the way they are offered to tourists. For example, the transition from a wired network base connected on the Internet to a wireless network base through the use of mobile devices (Gretzel, Sigala, Xiang, & Koo 2015; Koo, Gretzel, Hunter, & Chung, 2015; Li, Hu, Huang, & Duan, 2017) represents one of the results of technological advances in this domain. With this type of technological development, the tourism industry is again entering a new era, known as 'smart tourism' (Gretzel et al., 2015; Li et al., 2017).

The impact of the Internet across the tourism ecosystem is obvious and the role of the emerging ICT as tools in the forefront of innovation and even disruption is the new reality. The study of Cassia, Castellani, Rossato and Baccarani (2020) identifies the role that the emerging tools promoting digital ecosystems can play in 'co-creating' tourism experiences. The digital transformation of tourism boosted by the advent of the 4th industrial revolution is increasing connectivity between supply (tourism service providers) and demand (tourists) and amongst suppliers themselves through information partnerships involving operators in the supply chain and amongst tourists, for example, sharing their reviews.

Despite the increasing importance of the Internet for tourism information searches, it is important to note the influence of some factors in consulting online information sources. Specifically, as far as sociodemographic factors are concerned, younger people and those with higher levels of education are more likely to use online information sources (Ip, Lee, & Law, 2012; Jensen & Hjalager, 2013).

Internet as an Information Source in the Accessible Tourism Market

According to some researchers, accessible tourism provides a great opportunity for the tourism organizations to extend their market, opening future opportunities through new business models (Bowtell, 2015). On the other hand, participation of PwD in tourism activities can promote their well-being, while increasing the level of social integration (Kastenholz et al., 2015; UNWTO, 2016). However, for this group of people to be able to enjoy their trip, they need to have additional information to ensure the essential conditions in the place to which they are travelling. In addition to the information needed by any tourist, PwD require further information related to the accessibility of the destination and all the infrastructures that will be used, as well as of the communication services themselves. Zajadacz (2014: p. 438) states that "travel planning for PwD normally involves a more detailed information enquiry than for travelers without disabilities". The lack of information at different levels has been identified as the main reason why PwD are unable to travel (Stumbo & Pegg, 2005).

Considering these arguments, and in order to improve the mutual benefits between demand and supply of tourism, research has been conducted to understand the accessible tourism market and, consequently, to address the main barriers in it (Bi, Card, & Cole, 2007; Cassia et al., 2020; Lam, Chan, & Peters, 2020; Smith, 1987). Within the vast group of barriers found, communication obstacles appear quite prominently, and the Internet has emerged as one of the main solutions to partially eliminate these barriers, since it removes physical distance, placing tourist products a click away. Tourism is considered an information-intensive phenomenon (Lam et al., 2020) and the Internet can provide several kinds of information at every stage of the travel process, i.e., pre-trip, during the trip, and post-trip. Currently, the Internet is the most universal technology used by the majority of the population worldwide (Internet World Stats, 2020), in the most varied contexts, with the Web being one of the most often used services as a source of information for all people, including PwD (Michopoulou & Buhalis, 2013; Zajadacz, 2014). Therefore, this universal technology has been suggested as vital to help spread information to PwD, improving accessible conditions of tourism practices. Accessible tourism actually represents a market which is heavily dependent on information, both in quantity and quality, emphasizing for the latter the importance of the presentation format and its suitability to the vast range of user profiles. Considering Miller and Kirk's study (2002), as the level of user accessibility requirements increases, less information is available to serve the needs of this public.

Buhalis and Michopoulou (2011) classify the requirements of accessible tourist markets in three categories: (i) physical accessibility related to physical infrastructure, such as the existence of ramps, adapted lifts or adapted bathrooms; (ii) information about accessibility, which concerns information about accessible products and services; and (iii) information accessibility, which addresses the way all information is communicated, ensuring that it reaches everyone, regardless of their condition, with special emphasis on online information. In the latter two categories (information about accessibility and accessible information), the role of the Internet as a facilitating tool in providing these two types of accessibility is highlighted. According to Michopoulou and Buhalis (2013) accessible infrastructure is critical for tourism participation by PwD, but information regarding the accessibility of tourism facilities and destinations represents a fundamental requirement for the disabled market, since PwD cannot ensure that the venues are accessible without information. The need for information depends on the PwD, and this market usually looks for specific information that they need on Web platforms. Cassia et al. (2020) state that the technologies supporting online information can also help in mechanisms of personalization through the specific accessibility requirements, providing tailored responses to the exact needs of a wide range of characteristics that distinguish PwD. If on the one hand this personalization makes it possible to meet the needs of each type of PwD, on the other hand it makes the accessibility of information a complex matter. Actually, each type of PwD has its own specific needs and responding with the same solution to all these specificities is not an easy task. In order to overcome this difficulty, a wide range of studies have explored the issues related to information accessibility (Darcy, 2010; Darcy et al., 2020; Michopoulou, Darcy, Ambrose, & Buhalis, 2020; Silveiro, Eusébio & Teixeira, 2019; Teixeira, Eusébio, & Silveiro, 2019), and some of them examine the communication in terms of Web accessibility. Web accessibility is defined in the literature as the degree to which a website allows access to its content to the largest possible range of users, including people that have visual, auditory, motor or cognitive disabilities or those that experience some kind of environmental constraints (Akgül & Vatansever, 2016).

Indeed, the inventor of the Web, Tim Berners-Lee stated that the power of the Internet/Web lies in its universality and its ability to serve all people, regardless of their condition and disability (Shawn & McGe, 2020). However, the value of its use is related to the ability to efficiently supply the right information at the right time to the right people. Today, several tools and guidelines promoting Web inclusiveness are available to evaluate this Web accessibility, with the Web Content Accessibility Guidelines (WCAG), developed by the World Wide Web Consortium (W3C) (W3C, 2018), being one of the main foundations for assuring that Web contents are accessible. However, while Web accessibility is crucial to remove the communication barriers between supply and demand, the availability of information about accessibility of tourism products and services is also of great value to the accessible tourism market, thus boosting the concept of tourism for all. It is in this context that the present study analyses the use of the Internet as an information source by PwSN.

METHODOLOGY

The methods used to collect the data and the methods used to analyze the results obtained are presented in the next two subsections.

Data Collection Methods

This study was carried out in the scope of a wider funded Portuguese research project ("ACTION – Accessible tourism: co-creation of tourism experiences through a Web-based intelligent system"). The ACTION project intends to develop a Web-based intelligent system to facilitate the co-creation of tourism experiences for the accessible tourism market (PwD, elderly and other people with functional limitations).

To examine the relevance of the Internet as an information source for the Portuguese population with special needs in tourism, the concept of accessible tourism proposed by Darcy and Dickson (2009) was used. Therefore, a survey was directed to Portuguese people of over 17 years of age who belong to the following groups: PwD (mobility, vision, hearing and intellectual) and people with other special needs (seniors, people with food allergies, people with respiratory allergies, pregnant women and people using strollers). It was carried out between July 2019 and January 2020.

A questionnaire was administered to a sample of Portuguese PwSN, selected through a combination of a snowball sampling technique with a convenience approach, given the difficulty of addressing this particular market. Questionnaires designed based on an extensive literature review (e.g. Allan, 2015; Buhalis & Michopoulou, 2011; Buhalis et al., 2006; Ray & Ryder, 2003; Var, Yesiltas, Yayli, & Öztürk, 2011; Zajadacz, 2014), were administered face-to-face, using specialized interviewers, as well as online. The questionnaire included questions related to the information sources used by PwSN to plan tourism trips and the devices used for information searches on the Internet. Moreover, the level of frequency of use of various types of online information sources (websites of tourism suppliers, websites of official tourism organizations, specialized platforms for consulting and booking tourism products, social networks, blogs specialized in tourism, blogs specialized in accessible tourism and platforms specialized in accessible tourism) was examined through a five-point Likert-type scale (1 = never and 5 = very frequently). To analyze the level of frequency of use of online information sources for planning a tourism trip, booking tourism products, obtaining information during a trip, evaluating the tourism products purchased during a trip and sharing their tourism experiences, a similar scale was used (a five-point Likert-type scale). Finally, to assess the level of satisfaction with the online information sources used in terms of information disseminated and the accessibility level of these channels, a five-point Likert-type scale ranging from 1 (very unsatisfied) to 5 (very satisfied) was used.

The questionnaire was translated into Braille to facilitate the participation of people with visual disabilities. Moreover, in order to ensure the validity and reliability of the questionnaire, a pilot study with a group of Portuguese PwSN was carried out in June 2019. A total of 391 PwSN who had previously undertaken tourism trips participated in this survey.

Data Analysis Methods

Univariate and bivariate statistical techniques were used to examine the data using the SPSS software. First, frequencies, mean and standard deviation values were calculated to characterize the sample. Chi-square tests, t-tests, ANOVA and Kruskal-Wallis tests were used to examine whether there are differences in the usage of online information sources and in its evaluation according to the age and the education level of PwSN. The chi-square test was used with qualitative variables, the t-test to compare two groups in terms of quantitative variables and the ANOVA test to compare three groups. When the ANOVA assumptions (normal distribution and homogeneity of variances) were not observed, the corresponding non-parametric test (Kruskal-Wallis test) was used. In all bivariate tests, a 5% level of significance was used.

RESULTS AND DISCUSSION

Sample Profile

Sociodemographic data reveal a predominance of female respondents (63.7%), while there is a balance between single people (51.3%) and those with other marital status (Table 1). The sample is diversified regarding age, education level and status in terms of an economic activity, with the majority being employed (36.6%) or having another situation besides being employed or retired (e.g. being a student or being unemployed) (46.7%). Regarding education level, it is interesting to note that the predominant groups are those at the extremes, i.e. those having less than secondary education (36.0%) and those with higher education (39.4%), which may result from the age differences among respondents, but also from differences in abilities and opportunities in the scope of this group of respondents. Despite the average age being 42 years there is a wide variety in terms of age, since the standard deviation is quite high (16.7).

People with different types of social needs were included in the sample. Nevertheless, there is a prevalence of people (e.g. seniors) with other special needs than disabilities (30.2%), and with motor and mental disabilities (24.3% and 20.0%, respectively) (Table 1). The degree of disability is quite high (75.8%), although it varies widely across the sample (standard deviation = 20.3%). Although more than three quarters of respondents (78.5%) state that they do not need help to move, this data must be analyzed with care, since respondents may need other types of help.

Table 1. Sociodemographic profile of the sample

Profile		N	%	Mean	Std. Deviation
Gender					
	Male	142	36.3		
	Female	249	63.7		
Marital status					
	Single	200	51.3		
	Other	190	48.7		
Education level					
	Less than secondary education	139	36.0		
	Secondary education	95	24.6		
	Higher education	152	39.4		
Economic activity status					
	Employed	140	36.6		
	Retired	64	16.7		
	Other	179	46.7		
Age		386		41.9	16.664
Type of special needs					
	Hearing	18	4.6		
	Visual	35	9.0		
	Motor	95	24.3		
	Mental	86	22.0		
	Multiple	39	10.0		
	Other	118	30.2		
Percentage of disability		132		75.8	20.331
Need of help to move					
	Yes	84	21.5		
	No	307	78.5		

Relevance of the Internet as an Information Source to People With Special Needs

If it is true that the majority of the respondents report that they search for information on their own when planning trips (53.6%), many also refer that they are helped by their relatives (37.6%), social organizations (21.0%) and friends (14.0%) in this task at the planning stage (Table 2). The fact that there are not more PwSN searching information by themselves may be partially associated with some failures in complying with requirements for providing information to this group, as already remarked by Buhalis and Michopoulou (2011) – to provide information about accessibility and to ensure that information is accessible. Chi-square tests show that the help of social organizations is more frequent in the group of people aged up to 42 years and for those who do not have higher education. The cooperation of other organizations or people not listed in the questionnaire is more reported by people over the age of 42, possibly because these organizations, although not being considered social organizations, are especially designed to cater to the needs of specific groups, such as seniors. The help of relatives is especially important among those with less than secondary education. On the other hand, people with higher education reveal a great degree of autonomy, with 82.9% reporting having sought information on their own. This finding may result from higher education providing greater knowledge on how to search for information to plan a trip, but may also mean that these people had a degree of disability which simultaneously let them progress into higher education and enabled them to access information more easily than people with greater needs. The data previously presented also highlight the important role of social organizations and relatives in helping PwSN in trip planning, specifically in searching for information, suggesting the importance of keeping social organizations and families of PwSN informed about tourism supply, including tourism trips or accommodation which are accessible for PwSN.

The information sources most used by respondents are the Internet (76.8%), followed by recommendation of relatives and friends (49.1%), previous experience (46.1%) and social networks (31.0%) (Table 2). These findings provide empirical evidence of the vital role of the Internet for the accessible market in line with Xiang et al. (2015) who consider that Internet has been revealed as the most popular information source for trip planning. It is worthy of notice that, even when these respondents look for information autonomously, they consider the opinion of others (e.g. family, friends) and search for information provided by organizations such as travel agencies (18.1%), other tourism organizations (18.8%) and social organizations (8.9%). This fact reinforces the need of tourism suppliers to invest in promoting their products among these organizations and persons. Information search seems, in the case of the accessible market, to be a collaborative process shared with others that usually support them: relatives, friends, and also tourism organizations. Older people are more likely to contact other tourism organizations than younger people, eventually because they have more knowledge about the existence of these organizations or because they are not as well supported in their searches for information as younger people, for example by social organizations that can help them plan trips. People with secondary and higher education are more likely to consider the suggestion of friends and relatives. However, this data has to be analyzed with care, since relatives also help plan the trip for 45.3% of those with less than secondary education. Thus, this information only reinforces the crucial role that relatives have for PwSN, independently of their level of education, either by directly helping to plan the trip or by giving them some advice.

Table 2. Information sources used, by age and education level

Information sources	Age[a]						Education Level[a]						
	Total		Less than or equal to 42 years	Older than 42 years	Chi-square test		Total		Less than secondary education	Secondary education	Higher education	Chi-square test	
	N	%	(N = 223, 47.8%)	(N = 163, 42.2%)	Value	p-value	N	%	(N = 139, 36.0%)	(N = 95, 24.6%)	(N = 152, 39.4%)	Value	p-value
Who searches information to plan tourism trips*													
Myself	207	53.6	53.8	53.4	0.007	0.932	207	53.6	23.7	50.5	**82.9**	102.647	**0.000**
Relatives	145	37.6	36.8	38.7	0.142	0.707	143	37.0	**45.3**	37.9	28.9	8.388	**0.015**
Friends	54	14.0	13.9	14.1	0.003	0.953	53	13.7	12.9	14.7	13.8	0.154	0.926
Social organizations	81	21.0	**26.9**	12.9	11.167	**0.001**	83	21.5	**39.6**	**23.2**	3.9	57.784	**0.000**
Other	31	8.0	5.4	**11.7**	5.021	**0.025**	31	8.0	8.6	11.6	5.3	3.264	0.196
Information sources used*													
Internet	208	76.8	80.3	71.9	2.565	0.109	207	76.7	66.7	78.8	79.6	4.055	0.132
Recommendations of relatives and friends	133	49.1	31.0	18.1	2.925	0.087	132	48.9	35.1	**43.9**	**56.5**	8.366	**0.015**
Previous experience	125	46.1	46.5	45.6	0.021	0.886	124	45.9	40.4	36.4	52.4	5.610	0.078
Social networks (e.g. Facebook, Twitter, Instagram, LinkedIn)	84	31.0	34.4	26.3	2.015	0.156	83	30.7	28.1	27.3	33.3	10.280	0.598
Other tourism organizations	51	18.8	13.4	**26.3**	7.239	**0.007**	50	18.5	15.8	19.7	19.0	0.369	0.831
Travel agencies	49	18.1	15.3	21.9	1.968	0.161	48	17.8	12.3	22.7	17.7	2.285	0.319
Travel books/guides	43	15.9	12.7	20.2	2.736	0.098	43	15.9	8.8	12.1	20.4	5.098	0.078
Social organizations	24	8.9	10.2	7.0	0.824	0.364	24	8.9	15.8	6.1	7.5	4.362	0.113
Magazines and newspapers	21	7.7	5.7	10.5	2.123	0.145	20	7.4	5.3	6.1	8.8	b)	
Television and radio	18	6.6	4.5	9.6	2.869	0.090	18	6.7	7	4.5	7.5	b)	

Values in bold correspond to the highest values when statistically significant differences exist.
a Percentage in column
b) The assumptions of Chi-square test were not observed
* Only the values corresponding to people who said 'yes' are presented

The devices most often used to collect information to plan trips are the computer/laptop (used by 72.6%), followed by the smartphone/mobile phone (62.8%), with only about a fifth (19.2%) using a tablet (Table 3). This data shows that the accessible tourist market tends to use more than one kind of device to have access to tourist information. It also shows the relevance that mobile technologies have in this scope, highlighting the importance of making tourist information for PwSN available on this kind of devices, ensuring the provision of accessible information, as suggested by Buhalis and Michopoulou (2011). People aged 42 or less and those with secondary and higher education tend to use the smartphone/mobile phone more than the other groups, which suggests that younger people are more predisposed to use these devices, but also that some knowledge is required to search for tourism information when using them.

The most frequently used online sources are specialized platforms for consulting and booking tourism products (e.g. Booking.com, TripAdvisor, Airbnb) (3.78 on a scale from 1 (never) to 5 (very frequently)), followed by tourism supplier websites (e.g. hotels, travel agencies, restaurants) (3.51), social networks (e.g. Facebook, Instagram, Twitter, Snapchat, LinkedIn) (2.80) and official tourism organization websites (e.g. Turismo de Portugal) (2.57) (Table 4). These findings especially show the relevance of commercializing tourism products for the accessible market through online platforms that not only permit information to be consulted, but also to book products, while tourism suppliers (e.g. those who work in accommodation, restaurants, and even the intermediaries) should also use platforms with information designed for and accessible to this market. Specialized platforms in accessible tourism (e.g. Tour4All) are rarely used (1.62) by respondents, which should be further analyzed to uncover the reasons underlying this low level of use. Nevertheless, there are also several significant differences in the frequency with which respondents look for information in these online sources, according to age and education level. While older people tend to consult specialized platforms in accessible tourism (e.g. Tour4all) more frequently than younger people, the opposite happens with social networks. These findings probably reveal different interests in terms of information and opinions sought, as well as regarding social media usage among people with different ages. As far as education is concerned, people with higher education are most likely to consult several of these online sources, namely websites of official tourism organizations, specialized platforms for consulting and booking tourism products, blogs specialized in tourism and websites of tourism suppliers, with this last source being also frequently consulted by those with secondary education. This may happen for several reasons, such as those with lower levels of education relying more on sporadic searches, those with the highest levels of education having more patience and interest to do longer searches or searching for information more often due to having a higher propensity to travel, typically associated with higher income. This should be analyzed in more detail in the future.

Table 3. Devices used for searching information on the Internet, by age and education level

Devices used for Internet search, including social networks*	Age[a]						Education Level[a]						
	Total		Less than or equal to 42 years	Older than 42 years	Chi-square test		Total		Less than secondary education	Secondary education	Higher education	Chi-square test	
	N	%	(N = 223, 47.8%)	(N = 163, 42.2%)	Value	p-value	N	%	(N = 139, 36.0%)	(N = 95, 24.6%)	(N = 152, 39.4%)	Value	p-value
Computer/Laptop	193	72.6	73.9	70.8	0.306	0.580	193	72.8	62.3	72.7	6.7	4.103	0.129
Smartphone/Mobile Phone	167	62.8	**69.9**	53.1	7.886	**0.005**	166	62.6	43.4	**68.2**	**67.1**	10.507	**0.005**
Tablet	51	19.2	19.6	18.6	0.044	0.834	50	18.9	9.4	25.8	19.2	5.137	0.077

Values in bold correspond to the highest values when statistically significant differences exist.
a Percentage in column
* Only the values corresponding to people who said 'yes' are presented

Online information sources are most frequently used by this market to get information to plan trips (4.33) and to book tourism products (4.05), and are also often used to get information during the trip (3.75), but less used to evaluate the tourism products (3.19) or share the experience with others (3.01) (Table 5). However, data suggests that the propensity to use online sources for the previously mentioned purposes varies by age and education level. Younger people are revealed as more frequently using these sources to obtain information during the trip, probably because they are more accustomed to using these sources during the day, and they most often use the sources to book products and share their experience, possibly because they are more used for doing these tasks in other domains of their lives. Those with higher education engage in most of these activities more often than people from other groups (except for sharing the experience and other tasks not specified in the questionnaire), which suggests that these

Table 4. Frequency of consulting different online sources, by age and education level

Types of online information sources*	Age						Education Level								
	Total		Less than or equal to 42 years	Older than 42 years	T- test		Total		Less than secondary education	Secondary education	Higher education	ANOVA		Kruskal-Wallis test	
	N	Mean	(N = 136, 60.0%)	(N = 89, 40.0%)	Value	p-value	N	Mean	(N = 139, 36.0%)	(N = 95, 24.6%)	(N = 152, 39.4%)	Value	p-value	Value	p-value
Tourism supplier websites (e.g. hotels, travel agencies, restaurants)	225	3.51	3.55	3.44	0.581	0.562	224	3.50	2.81[1]	**3.49**[2]	**3.74**[2]	7.082	**0.001**		
Official tourism organization websites (e.g. *Turismo de Portugal*)	222	2.57	2.54	2.62	-0.387	0.699	221	2.57	2.50[1,2]	2.16[1]	**2.77**[2]	4.089	**0.018**		
Specialized platforms for consulting and booking tourism products (e.g. Booking.com, TripAdvisor, Airbnb)	224	3.78	3.89	3.60	1.380	0.169	223	3.77	2.52	3.46	**4.33**			42.854	**0.000**
Social networks (e.g. Facebook, Instagram, Twitter, Snapchat, LinkedIn)	223	2.80	**2.99**	2.51	2.371	**0.019**	222	2.79	3.15	2.61	2.75	1.627	0.199		
Blogs specialized in tourism	219	2.26	2.37	2.10	1.502	0.135	219	2.26	1.88[1]	1.96[1,2]	**2.52**[2]	5.914	**0.003**		
Blogs specialized in accessible tourism	220	2.03	2.10	2.01	-0.369	0.712	220	2.03	1.88	1.89	2.14	1.174	0.311		
Specialized platforms in accessible tourism (e.g. Tour4All)	220	1.62	1.50	**1.81**	-2.017	**0.046**	220	1.62	1.60	1.55	1.66	0.228	0.796		
Other	195	1.31	1.32	1.29	0.201	0.841	195	1.31	1.48	1.24	1.28	0.913	0.403		

Bold values represent the highest mean values when the differences between the segments' means are statistically significant.
Post hoc test: Scheffe test (1Homogenous subset 1; 2Homogenous subset 2)
* 5-point Likert-type scale where 1 = never and 5 = very frequently

tasks may require some knowledge that not all respondents hold and that this group may be using the Internet more regularly, in general.

Evaluation of the Internet as an Information Source for People With Special Needs

In general, respondents are not very satisfied with the websites they used to obtain information, neither regarding the information provided (answering 3.51 on a scale from 1 = very unsatisfied to 5 = very satisfied) nor concerning the accessibility levels of the websites (3.50) (Table 6). This suggests that the accessibility of websites must be improved. This probably happens because, as argued by Zajadacz (2014), while the accessible market has more information needs than the general public, websites rarely cater to these special needs. Findings thus show that websites fail to fulfil one of the main requirements of accessible tourism pointed out by Buhalis and Michopoulou (2011), namely information about accessibility and accessible information.

However, the results also reveal that people with higher education are more satisfied with the level of accessibility than other people, probably because they are better prepared to find credible and useful information on the Internet and deal with problems resulting from limitations concerning information accessibility.

Table 5. Functions performed by online information sources, by age and education level

Functions of online information sources*	Age						Education Level								
	Total		Less than or equal to 42 years	Older than 42 years	T- test		Total		Less than secondary education	Secondary education	Higher education	ANOVA		Kruskal-Wallis test	
	N	Mean	(N = 136, 60.0%)	(N = 89, 40.0%)	Value	p-value	N	Mean	(N = 139, 36.0%)	(N = 95, 24.6%)	(N = 152, 39.4%)	Value	p-value	Value	p-value
Get information to plan a tourism trip	224	4.33	4.43	4.19	1.524	0.130	223	4.33	4.20	3.95	**4.55**			8.721	**0.013**
Booking tourism products (e.g. accommodation, transports)	224	4.05	**4.23**	3.78	2.307	**0.016**	223	4.05	3.54	3.79	**4.34**			10.133	**0.006**
Get information during a tourism trip	224	3.75	**3.94**	3.44	2.649	**0.009**	223	3.74	3.05	3.38	**4.14**			24.118	**0.000**
Evaluate tourism products and the tourism trip carried out	223	3.19	3.26	3.09	0.800	0.425	222	3.18	2.63^{1}	$2.89^{1,2}$	**3.50^{2}**	6.827	**0.001**		
Share a tourism experience	224	3.01	**3.17**	2.76	2.037	**0.043**	223	3.00	3.15	2.88	3.01	0.402	0.670		
Other	185	1.23	1.16	1.35	-1.473	0.143	185	1.23	1.36	1.02	1.28	4.750	0.093		

Bold values represent the highest mean values when the differences between the segments' means are statistically significant.
Post hoc test: Scheffe test (1Homogenous subset 1; 2Homogenous subset 2)
* 5-point Likert-type scale where 1 = never and 5 = very frequently

CONCLUSIONS AND IMPLICATIONS

PwSN are an important and growing group of travelers who demand and deserve better adapted, inclusive and accessible tourism provision (Kastenholz et al., 2015; Moura, 2018; WTO, 2013) and require, amongst other factors, improved information accessibility. Such information on accessibility in tourism supply should be both accessible and credible (Buhalis & Michopoulou, 2011), while it should also be able to address the specific information requirements of specific groups (Cassia et al., 2020).

As a matter of fact, the PwSN market, travelling with higher perceived (and real) risks, particularly need to plan well in advance to avoid discomfort and problems while at the destination, with pre-trip information search being an essential strategy to cope with barriers to travel and ensure a positive travel experience (Lima Devile, Kastenholz, & Santiago, 2012). Indeed, information and technology may greatly enhance accessibility in tourism planning and enjoyment (Altinay et al., 2016; Buhalis et al., 2006; Darcy et al., 2010; Darcy, 2012).

The findings reported here, obtained from a diversified sample of Portuguese PwSN, provide empirical evidence of the vital role of the Internet amongst PwSN when planning a trip, as also found by Xiang et al. (2015). Apart from searching alone, this group additionally relies heavily on family, social organizations and friends in the planning phase. However, age and education level stand out as moderating variables.

The group with higher education in the sample particularly shows more autonomy in their information search, searching in more online sources than other groups, and also shows greater satisfaction with their online information search; they probably use the internet more regularly, in general, and are better prepared to find relevant information. Younger respondents also show a relatively high propensity to use online sources. This fact suggests an increasing importance of the Internet in the future amongst PwSN, as new generations will be more acquainted with it.

Globally, information sources are most often used for trip planning and booking, while information search during the trip is also referred to, mostly by the more educated and the younger groups. The Internet is used less to evaluate tourism products, with the more highly educated doing so, or to share

Table 6. Satisfaction with websites, by age and education level

Satisfaction*	Age						Education Level						
	Total		Less than or equal to 42 years	Older than 42 years	T- test		Total		Less than secondary education	Secondary education	Higher education	Kruskal-Wallis test	
	N	Mean	(N = 136, 60.0%)	(N = 89, 40.0%)	Value	p-value	N	Mean	(N = 139, 36.0%)	(N = 95, 24.6%)	(N = 152, 39.4%)	Value	p-value
Satisfaction with the information provided on the websites	225	3.51	3.60	3.36	1.631	0.104	224	3.50	3.24	3.37	3.65	5.315	0.070
Satisfaction with the accessibility level of the websites used	225	3.50	3.58	3.38	1.292	0.198	224	3.50	3.07	3.39	**3.70**	7.823	**0.020**

Bold values represent the highest mean values when the differences between the segments' means are statistically significant.
Post hoc test: Scheffe test (1Homogenous subset 1; 2Homogenous subset 2)
* 5-point Likert-type scale where 1 = very unsatisfied and 5 = very satisfied

experience with others, with younger respondents more active in this domain. Even though, these last mentioned activities must be considered as very important to tourism businesses and destinations, since evaluations, comments and experiences shared by friends and relatives (online or other) are, indeed, highly important information sources, as also confirmed in this study, and largely influence others in travel decision making. The older and those with a lower education level depend more on social organizations and particular tourism organizations they may be familiar with, probably due to their limited familiarity with the Internet. Overall, satisfaction with information provided online and with accessibility of this information is just above middle of the scale, suggesting a need to improve.

In brief, the results suggest the need for businesses and destinations to provide and invest more in useful and easily accessible information in their offerings on the Internet, which should also be targeted to specific needs and groups. They should be aware of online recommendations and criticism, as well as shared experiences, significantly influencing others, and try to improve their service provision as well as the correctness of information to avoid dissatisfaction.

Still, there also seems to be a need to address social and tourism organizations in order to capacitate them in providing relevant and updated information to PwSN who still rely heavily on them to find the most appropriate travel solution. The trip planning process for PwSN seems, indeed, to be a collaborative process shared with organizations that usually support them, with family and friends as well as with some tourism organizations.

Overall, they should know about and monitor the role of diverse information sources, where the Internet is apparently playing an increasing role, to better understand the diverse groups' needs, so as to a) attract more of these potential clients, b) improve their destination experience and c) enhance their market position within a growing and significant tourist market.

ACKNOWLEDGMENT

This work was also financially supported by the project POCI-01-0145-FEDER-030376, funded by FEDER, through COMPETE2020 – Programa Operacional Competitividade e Internacionalização (POCI), and by national funds (OE), through FCT/MCTES.

References

Akgül, Y., & Vatansever, K. (2016). Web Accessibility Evaluation of Government Websites for People with Disabilities in Turkey. *Journal of Advanced Management Science*, *4*(3), 201–210. doi:10.12720/joams.4.3.201-210

Allan, M. (2015). Accessible tourism in Jordan: Travel constraints and motivations. *European Journal of Tourism Research*, *10*, 109–119.

Altinay, Z., Saner, T., Bahçelerli, N. M., & Altinay, F. (2016). The role of social media tools: Accessible tourism for disabled citizens. *Journal of Educational Technology & Society*, *19*(1), 89–99.

Bi, Y., Card, J. A., & Cole, S. T. (2007). Accessibility and attitudinal barriers encountered by Chinese travellers with physical disabilities. *International Journal of Tourism Research*, *9*(3), 205–216. doi:10.1002/jtr.603

Bowtell, J. (2015). Assessing the value and market attractiveness of the accessible tourism industry in Europe: A focus on major travel and leisure companies. *Journal of Tourism Futures*, *1*(3), 203–222. doi:10.1108/JTF-03-2015-0012

Buhalis, D., Eichhorn, V., Michopoulou, E., & Miller, G. (2006). *Accessibility market and stakeholder analysis*. OSSATE, University of Surrey. Retrieved from: https://www.accessibletourism.org/resources/ossate_market_analysis_public_final.pdf

Buhalis, D., & Law, R. (2008). Progress in information technology and tourism management: 20 years on and 10 years after the Internet - The state of eTourism research. *Tourism Management*, *29*(4), 609–623. doi:10.1016/j.tourman.2008.01.005

Buhalis, D., & Michopoulou, E. (2011). Information-enabled tourism destination marketing: Addressing the accessibility market. *Current Issues in Tourism*, *14*(2), 145–168. doi:10.1080/13683501003653361

Buhalis, D., & Sinarta, Y. (2019). Real-time co-creation and nowness service: Lessons from tourism and hospitality. *Journal of Travel & Tourism Marketing*, *36*(5), 563–582. doi:10.1080/10548408.2019.1592059

Camisão, V. (2006). Turismo e Acessibilidade. In W. M. Guia, A. Pereira, T. Brizolla, M. F. Krahl, M. D. Pinto, & A. C. Santos (Eds.), *Turismo Social: Diálogos do turismo - Uma viagem de inclusão* (pp. 320–356). Ministério do Turismo e Instituto Brasileiro de Administração Municipal.

Cassia, F., Castellani, P., Rossato, C., & Baccarani, C. (2020). (in press). Finding a way towards high-quality, accessible tourism: The role of digital ecosystems. *The TQM Journal*. Advance online publication. doi:10.1108/TQM-03-2020-0062

Connolly, D. J., Olsen, M. D., & Moore, R. G. (1998). The internet as a distribution channel. *The Cornell Hotel and Restaurant Administration Quarterly*, *39*(4), 42–54. doi:10.1177/001088049803900408

Crawford, J. L., & Stodolska, M. (2008). Constraints experienced by elite athletes with disabilities in Kenya, with implications for the development of a new hierarchical model of constraints at the societal level. *Journal of Leisure Research*, *40*(1), 128–155. doi:10.1080/00222216.2008.11950136

Darcy, S. (2003). *Riding the Wave of Tourism and Hospitality Research*. Paper presented at the CAUTHE Conference - Disabling Journeys: The tourism patterns of people with impairments in Australia, Lismore: Southern Cross University.

Darcy, S. (2010). Inherent complexity: Disability, accessible tourism and accommodation information preferences. *Tourism Management*, *31*(6), 816–826. doi:10.1016/j.tourman.2009.08.010

Darcy, S. (2012). (Dis)Embodied air travel experiences: Disability, discrimination and the affect of a discontinuous air travel chain. *Journal of Hospitality and Tourism Management*, *19*(1), 1–11. doi:10.1017/jht.2012.9

Darcy, S., Cameron, B., & Pegg, S. (2010). Accessible tourism and sustainability: A discussion and case study. *Journal of Sustainable Tourism*, *18*(4), 515–537. doi:10.1080/09669581003690668

Darcy, S., & Dickson, T. J. (2009). A whole-of-life approach to tourism: The case for accessible tourism experiences. *Journal of Hospitality and Tourism Management*, *16*(1), 32–44. doi:10.1375/jhtm.16.1.32

Darcy, S., McKercher, B., & Schweinsberg, S. (2020). From tourism and disability to accessible tourism: A perspective article. *Tourism Review*, *75*(1), 140–144. doi:10.1108/TR-07-2019-0323

Domínguez Vila, T., Alén González, E., & Darcy, S. (2017). Website accessibility in the tourism industry: An analysis of official national tourism organization websites around the world. *Disability and Rehabilitation*, *40*(24), 2895–2906. doi:10.1080/09638288.2017.1362709 PMID:28793789

Domínguez Vila, T., Alén González, E., & Darcy, S. (2020). Accessibility of tourism websites: The level of countries' commitment. *Universal Access in the Information Society*, *19*(2), 331–346. doi:10.100710209-019-00643-4

Dwyer, L., & Darcy, S. (2008). Economic contribution of disability to tourism in Australia. In S. Darcy, B. Cameron, L., Dwyer, T., Taylor, E., Wong, & A. Thomson (Eds.), Technical report 90040: Visitor accessibility in urban centres (pp. 15–21). Gold Coast, Australia: Sustainable Tourism Cooperative Research Centre.

European Commission. (2014). *Economic Impact and Travel Patterns of Accessible Tourism in Europe – Final Report*. Retrieved from: http://ec.europa.eu/DocsRoom/documents/7221/attachments/1/translations/en/renditions/native

European Commission. (2015). *Economic impact and travel patterns of accessible tourism in Europe Final report*. Retrieved From: https://op.europa.eu/en/publication-detail/-/publication/d04fbb3d-8d15-4fe3-b554-844bde75c171/language-en/format-PDF/sourcesearch#

Fontes, F., Martins, B. S., & Hespanha, P. (2014). The emancipation of disability studies in Portugal. *Disability & Society*, *29*(6), 849–862.

Gladwell, N., & Bedini, L. (2004). In search of lost leisure: The impact of caregiving on leisure travel. *Tourism Management*, *25*(6), 685–693. doi:10.1016/j.tourman.2003.09.003

Gondos, B., & Nárai, M. (2019). The opportunities in accessible tourism. *The USV Annals of Economics and Public Administration*, *19*(1), 48–57.

Gretzel, U., Sigala, M., Xiang, Z., & Koo, C. (2015). Smart tourism: Foundations and developments. *Electronic Markets*, *25*(3), 179–188. doi:10.100712525-015-0196-8

Ho, C.-I., Lin, M.-H., & Chen, H.-M. (2012). Web users' behavioural patterns of tourism information search: From online to offline. *Tourism Management*, *33*(6), 1468–1482. doi:10.1016/j.tourman.2012.01.016

Internet World Stats. (2020). http://www.Internetworldstats.com/stats.htm

Ip, C., Lee, H., & Law, R. (2012). Profiling the Users of Travel Websites for Planning and Online Experience Sharing. *Journal of Hospitality & Tourism Research (Washington, D.C.)*, *36*(3), 418–426. doi:10.1177/1096348010388663

Jensen, J. M., & Hjalager, A.-M. (2013). The role of demographics and travel motivation in travellers' use of the internet before, during, and after a trip. *International Journal of Tourism Policy*, *5*(1-2), 34–58. doi:10.1504/IJTP.2013.054051

Kastenholz, E., Eusébio, C., & Figueiredo, E. (2015). Contributions of tourism to social inclusion of persons with disability. *Disability & Society*, *30*(8), 1259–1281. doi:10.1080/09687599.2015.1075868

Kastenholz, E., Eusébio, C., Figueiredo, E., & Lima, J. (2012). Accessibility as Competitive Advantage of a Tourism Destination: The Case of Lousã. In K. F. Hyde, C. Ryan, & A. Woodside (Eds.), Field Guide for Case Study Research in Tourism, Hospitality, and Leisure (Advances in Culture, Tourism and Hospitality Research, Volume 6) (pp. 369-385). Bingley: Emerald Publishers.

Kastenholz, E., Eusébio, M. C., Moura, A., & Figueiredo, E. (2010). Animação turística acessível: Do sonho à realidade. In N. Abranja, A. Marques, F. Coelhoso, & I. Carneiro (Eds.), *Turismo acessível: Estudos e experiências* (pp. 171–192). Edições Pedago.

Koo, C., Gretzel, U., Hunter, W. C., & Chung, N. (2015). The Role of IT in Tourism. *Asia Pacific Journal of Information System*, *25*(1), 99–104. doi:10.14329/apjis.2015.25.1.099

Lam, K. L., Chan, C.-S., & Peters, M. (2020). Understanding technological contributions to accessible tourism from the perspective of destination design for visually impaired visitors in Hong Kong. *Journal of Destination Marketing & Management*, *17*, 1–9. doi:10.1016/j.jdmm.2020.100434

Lee, H., Chung, N., & Nam, Y. (2019). Do online information sources really make tourists visit more diverse places? Based on the social networking analysis. *Information Processing & Management*, *56*(4), 1376–1390. doi:10.1016/j.ipm.2018.01.005

Li, Y., Hu, C., Huang, C., & Duan, L. (2017). The concept of smart tourism in the context of tourism information services. *Tourism Management*, *58*, 293–300. doi:10.1016/j.tourman.2016.03.014

Lima Devile, E., Kastenholz, E., & Santiago, R., (2012). Inibidores, facilitadores e estratégias de negociação associadas às experiências turísticas das pessoas com incapacidade. *RDT/Journal of Tourism and Development, 17/18*, 1417-1430.

Liu, J. N. K., & Zhang, E. Y. (2014). An investigation of factors affecting customer selection of online hotel booking channels. *International Journal of Hospitality Management, 39*, 71–83. doi:10.1016/j.ijhm.2014.01.011

Llodrà-Riera, I., Martínez-Ruiz, M. P., Jiménez-Zarco, A. I., & Izquierdo-Yusta, A. (2015). A multidimensional analysis of the information sources construct and its relevance for destination image formation. *Tourism Management, 48*, 319–328. doi:10.1016/j.tourman.2014.11.012

Michopoulou, E., & Buhalis, D. (2013). Information provision for challenging markets: The case of the accessibility requiring market in the context of tourism. *Information & Management, 50*(5), 229–239. doi:10.1016/j.im.2013.04.001

Michopoulou, E., Darcy, S., Ambrose, I., & Buhalis, D. (2020). Accessible tourism futures: The world we dream to live in and the opportunities we hope to have. *Journal of Tourism Futures, 1*(3), 179–188. doi:10.1108/JTF-08-2015-0043

Miller, G., & Kirk, E. (2002). The disability discrimination act: Time for a stick? *Journal of Sustainable Tourism, 10*(1), 82–88. doi:10.1080/09669580208667154

Moura, A. (2018). *Benefícios do Turismo para Lidar com o Stress da Incapacidade: Turismo Acessível, Stress-Coping e Incapacidade Motora e Sensorial.* Editorial Académica Española - OmniScriptum Publishing Group.

Neumann, P., & Reuber, P. (2004). *Estímulos económicos do turismo acessível para todos.* Munique e Berlim: Ministério Federal da Economia e do Trabalho Alemão (BMWA) – versão portuguesa, não publicada, disponibilizada pelo Secretariado Nacional de Reabilitação e Integração da Pessoa com Deficiência.

Öörni, A. (2004). Consumer objectives and the amount of search in electronic travel and tourism markets. *Journal of Travel & Tourism Marketing, 17*(2/3), 3–14. doi:10.1300/J073v17n02_01

Open Doors Organization. (2015). *Marketing Outlook Forum 2015, New ODO Market Study Finds Positive Trends in Accessible Travel.* Retrieved From: https://opendoorsnfp.org/wpcontent/uploads/2016/05/ODOMarketingForecast2015.pdf

Packer, T. L., McKercher, B., & Yau, M. K. (2007). Understanding the complex inter-play between tourism, disability and environmental contexts. *Disability and Rehabilitation, 29*(4), 281–292. doi:10.1080/09638280600756331 PMID:17364778

Peng, L., & Lai, L. (2014). A service innovation evaluation framework for tourism ecommerce in China based on BP neural network. *Electronic Markets, 24*(1), 37–46. doi:10.100712525-013-0148-0

Ray, M., & Ryder, E. (2003). 'Ebilities´ tourism: An exploratory discussion of the travel needs and motivations of the mobility-disabled. *Tourism Management, 23*(1), 57–72. doi:10.1016/S0261-5177(02)00037-7

Shawn, L. H., & McGe, L. (2020). *Accessibility, World Wide Web Consortium.* Available: https://www.w3.org/standards/webdesign/accessibility

Silveiro, A., Eusébio, C., & Teixeira, L. (2019). Heterogeneity in accessibility of travel agency websites: A study in the central Portugal region. *Revista Ibérica de Sistemas y Tecnologías de la Información*, *35*, 18–34. doi:10.17013/risti.35.18-34

Smith, R. W. (1987). Leisure of disable tourists: Barriers to participation. *Annals of Tourism Research*, *14*(3), 376–389. doi:10.1016/0160-7383(87)90109-5

Stumbo, N. J., & Pegg, S. (2005). Travelers and tourists with disabilities: A matter of priorities and loyalties. *Tourism Review International*, *8*(3), 195–209. doi:10.3727/154427205774791537

Szcygiel, A., & Piotrowicz, M. (1995). Leisure activities of people with special needs, their selfconcept and social interactions. In A. Slezynski & M. Petrrynski (Eds.), *Progress in Tourism of People with Special Needs* (pp. 280–289). AGAT.

Teixeira, L., Eusébio, C., & Silveiro, A. (2019). Website Accessibility of Portuguese Travel Agents. 14th Conferência Ibérica de Sistemas y Tecnologías de Información, 1–6.

United Nations. (2015). *World population ageing 2015*. Retrieved from: https://www.un.org/en/development/desa/population/publications/pdf/ageing/WPA2015_Report.pdf

UNWTO. (2016). *Manual on accessible tourism for all: Principles, tools and best practices*. Retrieved from: https://www.e-unwto.org/doi/pdf/10.18111/9789284418077

Var, T., Yesiltas, M., Yayli, A., & Öztürk, Y. (2011). A study on the travel patterns of physically disabled people. *Asia Pacific Journal of Tourism Research*, *16*(6), 599–618. doi:10.1080/10941665.2011.610143

W3C. (2018). *Web Content Accessibility Guidelines (WCAG) 2.1*. Retrieved from: https://www.w3.org/TR/2018/REC-WCAG21-20180605

Werthner, H., & Klein, S. (1999). *Information technology and tourism: A challenging relationship. Viena*. Springer-Verlag. doi:10.1007/978-3-7091-6363-4

Werthner, H., & Ricci, F. (2004). E-commerce and tourism. *Communications of the ACM*, *47*(12), 101–105. doi:10.1145/1035134.1035141

World Bank (2018). *Disability Inclusion and Accountability Framework*. Retrieved from: file:///D:/Docs%20Andreia%202019_2020/2.%20ARTIGOS/1.%20ARTIGOS%202019_2020/ICOTTS_2020/126977-WP-PUBLIC-DisabilityInclusionAccountabilitydigital.pdf

World Health Organization. (2011). *Relatório Mundial sobre a Deficiência*. Retrieved from: https://apps.who.int/iris/bitstream/handle/10665/70670/WHO_NMH_VIP_11.01_por.pdf;jsessionid=5BBD98488DC0CF3AD13E7C780D460A2F?sequence=9

World Tourism Organization. (2013). *Recommendations on Accessible Tourism*. Madrid: World Tourism Organization. Retrieved from: http://business.turismodeportugal.pt/SiteCollectionDocuments/all-for-all/recommendations-accesible-tourism.pdf

World Tourism Organization Business Council. (1999). *Marketing Tourism Destinations Online: Strategies for the Information Age*. Madrid: World Tourism Organization. https://www.e-unwto.org/doi/book/10.18111/9789284403288

Xiang, Z., Magnini, V. P., & Fesenmaier, D. R. (2015). Information technology and consumer behavior in travel and tourism: Insights from travel planning using the internet. *Journal of Retailing and Consumer Services*, *22*, 244–249. doi:10.1016/j.jretconser.2014.08.005

Zajadacz, A. (2014). Sources of tourist information used by Deaf people. Case study: The Polish Deaf community. *Current Issues in Tourism*, *17*(5), 434–454. doi:10.1080/13683500.2012.725713

KEY TERMS AND DEFINITIONS

Accessible Tourism: Tourism that enables people with access requirements, including mobility, vision, hearing, and cognitive dimensions of access, to function independently and with equity and dignity through the delivery of universally designed tourism products, services and environments. This definition is inclusive of all people including those travelling with children in prams, people with disabilities and seniors.

Information Search: Active acquisition of information in which people may consult several information sources such as friends and relatives, Internet, travel agencies, television, radio, magazines, and newspapers.

Online Information Sources: Systems that make information available to users via the web.

People With Disabilities: People suffering from any restriction or lack of ability (resulting from an impairment) to perform an activity in the manner or within the range considered normal for a human being.

People With Special Needs: People with disabilities (e.g., visual, hearing, motor disabilities) and people with other special needs such as seniors, pregnant women, people using strollers and people with respiratory allergies.

Tourism Trip: Travel made to a main destination outside the usual environment, which lasts less than a year, and that is done for any purpose (e.g., leisure and recreation, business, health, education) other than being employed by a resident entity in the country or place visited.

Trip Planning: Process occurring in the pre-trip stage, where people make decisions and take care of all the arrangements necessary to travel and participate in tourism, including searching for information about tourism services.

Chapter 5 Smart Cities and Accessible Tourism: A Systematic Review

Ana Dias
University of Aveiro, Portugal

Alexandra Queirós
University of Aveiro, Portugal

Gonçalo Santinha
https://orcid.org/0000-0002-4732-5959
University of Aveiro, Portugal

Carlos Rodrigues
https://orcid.org/0000-0001-6403-6959
University of Aveiro, Portugal

Mário Rodrigues
University of Aveiro, Portugal

Nelson P. Rocha
https://orcid.org/0000-0003-3801-7249
University of Aveiro, Portugal

ABSTRACT

Promoting accessibility in tourism can impact on other areas, including sustainable mobility, social inclusion, and territorial marketing, since it implies taking the needs of all people into account to access tourism products and services, including those with permanent or temporary disabilities. Smart cities may change the way people experience their surroundings and their ability to provide contextual services is a key aspect to make cities more accessible, comprehensible, and enjoyable. The systematic review reported by the present chapter aimed to identify relevant research studies supported by smart cities infrastructures with an impact on accessible tourism. The literature search and the analysis of the retrieved articles were performed according to the preferred reporting items for systematic reviews and meta-analyses (PRISMA) guidelines. The results identify the most relevant achievements related to accessible tourism in the context of smart cities, including the types of smart services being developed and their maturity level.

DOI: 10.4018/978-1-7998-6428-8.ch005

INTRODUCTION

Smart cities promote the integration of traditional urban infrastructures, existing knowledge and experiences, by combining aspects of intelligence and sustainability and using Information and Communication Technologies (ICT) to optimize urban services and, therefore, to improve the quality of life of the citizens (AlDairi, 2017; Azevedo Guedes et al., 2018; Santinha et al., 2019). Several authors link the implementation of smart cities to a set of major areas, namely, economy, mobility, environment, governance, social and human capital, and living conditions of the citizens, which includes health conditions, cultural and education facilities, housing quality, and touristic attractiveness, among other issues (see *inter alia* Giffinger & Gudrun, 2010; Lazaroiu & Roscia, 2012; Vanolo, 2014).

As such, the optimization of tourism services is also considered in the scope of the smart cities' paradigm. Indeed, smart cities are portrayed as digital ecosystems where economic, tourist, recreational and logistic aspects should all be considered together (Colace, 2016). Moreover, as cities become progressively complex, ICT bring smartness into organisations and societies, towards, for example, more competitive tourism destinations (Encalada, Boavida-Portugal, Ferreira & Rocha, 2017).

With the expansion of smart cities, the achievement of tourist information and services has become more fitting, promoting the development of intelligent tourism and self-service travel. Enriched information access together with new tourists' profiles, enthusiastic with online content and ready to share information, help to increase knowledge of tourist behaviour regarding their spatial and temporal distribution in the destinations. Consequently, smart tourism exposes individuals as information makers, filtering the existing options for tracking their position.

This context is seen as an opportunity for decision-makers to envisage new ways of planning and managing towards a sustainable and smarter future (Encalada et al., 2017). The current trend of smart tourism is becoming very significant for urban renovation due to integration of ICT in tourism. Under these circumstances, smart tourism is a critical piece and a hands-on venture of the smart cities' strategies (Gautam, Asami, Batajoo & Fujisaki, 2016). Smart tourism destinations are, thus, characterised by a ubiquitous presence of new technologies able to improve the tourist experiences. The word "experience", under the concept of "experience economy" (Pine & Gilmore, 1998), has been widely associated with product names, destination locations and digital media. Many factors contributed to the wide acceptance of this perspective through which the economic scenario is viewed. The logic of the experience provided a new basis for looking for new value creation activities. Moreover, the idea that consumers value experiences more than goods and services, was affirmed by Pine and Gilmore (2013). Therefore, increasing the knowledge of touristic experiences is crucial for supporting decision makers and planners in producing informed decisions seeking to increase the destinations' competitiveness (D'aniello, Gaeta & Reformat, 2017) considering the interests and needs of the tourists (Lijing, Yanrong & Jianhua, 2015).

Smart cities may change the way people experience their surroundings and their ability to provide contextual services is paramount to make cities more accessible, comprehensible and enjoyable. Therefore, it should be emphasized the growing importance of smart cities as tourism destinations through innovation, communication, and interactivity with visitors, not only to enhance tourists' experiences but also to improve the accessibility of the available products and services for impaired people.

Tourism is part of the modern lifestyle, affirming the potential of accessible recreational tourism for improving the quality of life, self-confidence and self-esteem of the population with disabilities, as well as an opportunity for active social inclusion of a population group that remains isolated and faces barriers in daily life, often difficult to overcome. This fact is even more challenging in a tourist context, despite the growing efforts in the area of social inclusion, particularly present in the political and institutional discourse. However, barriers to participation in leisure and tourism persist, although this population group shows needs and desires similar to non-disabled population and reveals even more intense experiences and benefits derived from participation (Kastenholz, Eusébio & Figueiredo, 2015).

In fact, travel planning and experience is not an easy task, especially for impaired people. Planning simple trips and deciding where, when, how to travel, how to move in particular areas and how to choose proper means of transportation can be extremely difficult for persons with disabilities or reduced mobility. Hence, providing organized and timely relevant touristic information to people with any kind of disability, temporary or permanent, can have a major impact in their experiences. Moreover, promoting accessibility in tourism can impact on other areas, including sustainable mobility, social inclusion and territorial marketing, since it implies taking the needs of all people into account to access tourism products and services, including those with permanent or temporary disabilities (Liu, Sokhn, Le Calvé & Schegg, 2016; Rodrigues, Sousa, Martins & Rosa, 2018).

Accessible tourism is an evolving field of research and, as in other areas of tourism, it is a multidisciplinary area, influenced by other areas of knowledge such as disability, psychology, geography, economics, public policies, and marketing (Buhalis & Darcy, 2010). Collaborative practices among stakeholders were identified by Michopoulou, Darcy, Ambrose and Buhalis (2015) as the main factor for the development of accessible tourism solutions. A collaborative approach is seen as necessary to recognize the complementary nature of different paradigms and to reshape and transform the future of the accessible tourism industry. The future of accessible tourism is also presented as influenced by policy initiatives on accessibility and sustainability. However, it is also argued that the future of accessible tourism is to create its own stimuli and contribute with unique learning about the diversity of tourism markets that will shape tourism concepts and global policy initiatives.

The improving financial impact that tourism may have, coupled with the rapid evolution of ICT, have made tourism not only more accessible, but also more competitive with an online "try-before-you-buy" tourism experience now becoming a reality, what is particularly relevant for impaired people. By promoting the integration of traditional urban infrastructures and ICT, including Internet of Things (IoT) enabled sensors, the use of the smart city concept allows cities to succeed socially and economically as well as to provide high quality and sustainable urban services (da Costa Liberato, Alén-González & de Azevedo Liberato, 2018; Louw, 2017; Perez-Jimenez, Rabadan, Rufo, Solana & Luna-Rivera, 2015; Pierdicca et al., 2016; Puma, 2019).

Although there are a growing number of reviews related to smart cities (e.g., Azevedo Guedes et al., 2018; Chauhan, Agarwal, Irshad, 2016; Kar, 2016; Purnomo & Prabowo, 2016; Rocha et al., 2019), there is still a gap concerning the linkage between smart cities and accessible tourism. Since systematic evidence is required to inform smart cities' stakeholders and researchers about state-of-the-art solutions, the systematic review here reported aimed to identify relevant research studies supported by smart cities infrastructures to optimise touristic experiences and with an impact on accessible tourism. Considering this objective, the authors performed a systematic literature review according to the Preferred Reporting Items for Systematic Reviews and Meta-Analyses (PRISMA) guidelines (Moher, Liberati, Tetzlaff & Altman, 2009). The results identify the most relevant achievements related to accessible tourism in the context of smart cities, and the types of smart services being developed as well as their maturity level, scope and target audience.

The remaining sections are organized as follows. The next sections outline the research questions and the research methodology following the PRISMA guidelines, address the topics of each research question raised, present the synthesis of the results and discuss the findings and implications of the present research.

METHODS

For the specific objective of the systematic review reported by this article, the following research questions were considered:

- RQ1: What are the smart tourism services being developed?
- RQ2: What is the smart tourism services application scope?
- RQ3: What are the smart tourism services target audiences?
- RQ4: What are the reported smart tourism services maturity levels?
- RQ5: Do the proposed solutions address accessibility issues? Are they somehow related to accessible tourism?

Boolean queries were prepared to include all the articles published before 31.12.2019 that have in their titles, abstract or keywords at least one of the following expressions: 'Smart City', 'Smart Cities' 'Smartcity', 'Smartcities', 'Smart-city' and 'Smart-cities'. The resources considered to be searched were two general databases, Web of Science and Scopus, and one specific technological database, IEEE Xplore. The literature search was performed in March 2020.

As inclusion criteria, the authors aimed to include all the articles that report the development of information services to optimise the touristic experience with explicit use of smart cities' infrastructures.

Considering the exclusion criteria, the authors aimed to exclude all the articles not published in English, without abstracts or without access to full text. Furthermore, the authors also aimed to exclude all the articles that were not scientific reports, report overviews or reviews or that were not relevant for the specific aim of this study.

The analysis of the retrieved articles was performed according to the following steps:

- First, the authors removed the duplicates and the articles without abstract.
- Then, the authors assessed the titles and abstracts of the retrieved articles using semi-automatic techniques to verify if the abstracts were related to smart cities and tourism. Those that are not scientific reports, are overviews or reviews or are clearly outside the scope of the present systematic review, were removed.
- Afterwards, the abstracts of the remaining articles were assessed and only those related to improving tourist experience were selected.
- Finally, authors assessed the full text of the retrieved articles and classified them. This classification was performed by using a synthesis process based on the method proposed by Ghapanchi and Aurum (2011) (i.e., terms and definitions used in the included articles were identified to create a primary list of application domains, which was afterwards refined by further analyses).

In all these steps the articles were analysed by at least two authors. Any disagreement was discussed and resolved by consensus.

RESULTS

Figure 1 presents the PRISMA flowchart of this systematic review.

Figure 1. Flowchart of the systematic review

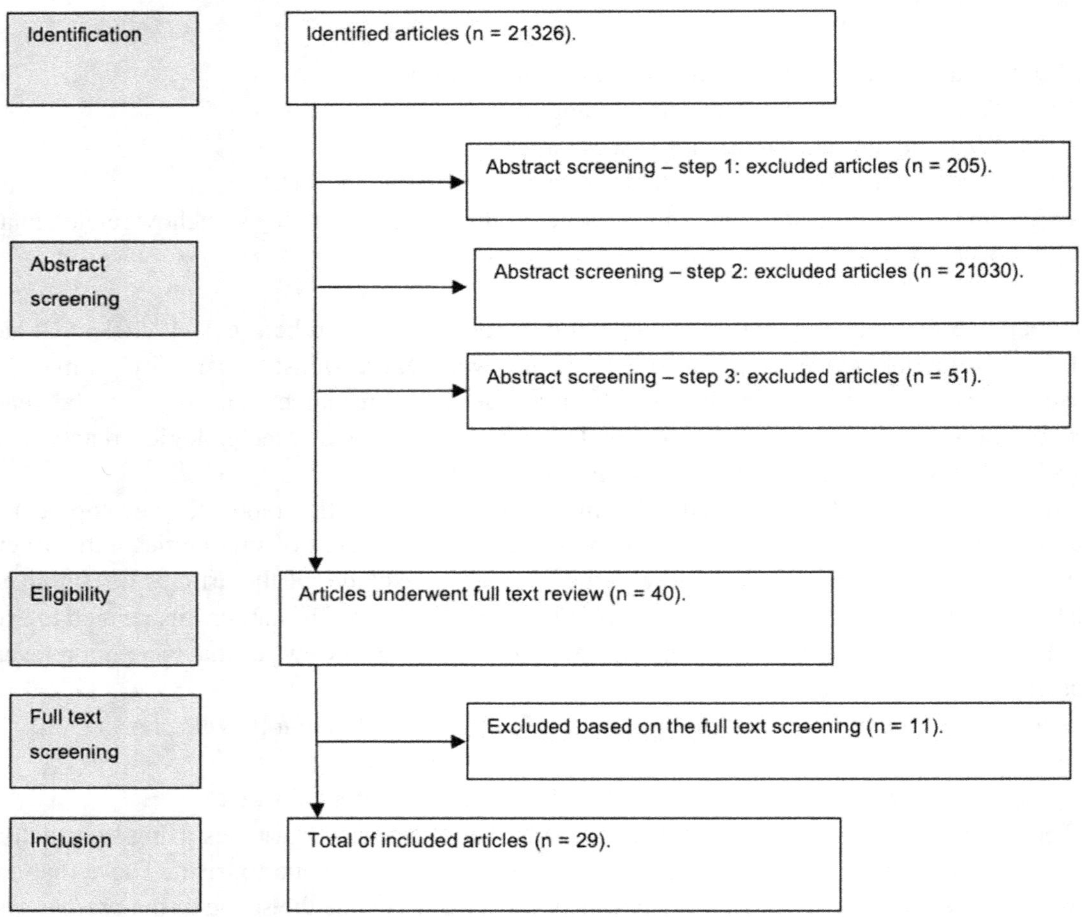

A total of 21326 articles were retrieved from the initial search on Web of Science, Scopus and IEEE Xplore.

In the first screening step, 202 articles were removed because they were duplicated and other three articles were removed because they did not have abstracts.

Then, 21030 articles were removed because they were not scientific reports (51 articles), were overviews or reviews (175 articles), or were clearly outside the scope of the present systematic review (20804 articles).

Afterwards, the abstracts of the remaining 91 articles were analysed, and it was concluded that 40 articles were related to tourist experience improvement.

Considering the 40 remaining articles, it was not possible to access the full texts of nine articles and two were still considered out of scope. As such, 11 references were excluded from this analysis.

Considering the 29 included articles, 23 were published in conference proceedings, while six articles were published in scientific journals (Table 1).

After the respective full-text analyses, the included 29 articles were divided into and organized according to the following two groups: Mobile Apps (five articles) and Personalized/Integrated systems (24 articles).

Table 1. Included articles according to the publication year and publication type

Year	Articles published in Scientific Journals	Article published in Conference Proceedings
2011		(Luberg, Tammet & Järv, 2011)
2013		(Giovannella, Iosue, Moggio, Rinaldi & Schiattarella, 2013)
2014	(Garau, 2014)	(Barbu, Kuhn, Spirescu, Lamorte & Scanu, 2015) (Barile, Calandra, Caso, Dauria, Di Mauro, Cutugno & Rossi, 2014)
2015	(Pereira, Sousa, Barata, Oliveira & Monsieur, 2015)	(Cacho et al., 2015) (Mighali et al., 2015)
2016	(García-Crespo, González-Carrasco, López-Cuadrado, Villanueva & González, 2016) (Cervantes, Gutiérrez, Gutiérrez & Sánchez, 2016)	(Benfares, El Bouzekri El Idrissi & Amine, 2016) (Kaur & Maheshwari, 2016) (Liu et al., 2016) (Yamamoto & Ikeda, 2016)
2017		(Bhattacharya, Painho, Mishra & Gupta, 2017) (Dalli & Bri, 2017) (Figueredo, Cacho, Thome, Cacho, Lopes & Araujo, 2017) (Marjury, Karen, Diana & Gabriel, 2017) (Quarati et al., 2017) (Spiel, Werner, Hödl, Ehrenstrasser & Fitzpatrick, 2017)
2018		(Abdul Rahman, Zaaba & Hussain, 2018) (Nowak Da Costa & Bielski, 2018) (Rodrigues et al., 2018)
2019	(Choi, Ryu & Kim, 2019) (Rosa & Lopes, 2019)	(Colace, de Santo, Lombardi & Santaniello, 2019) (Herzog, Sikander& Worndl, 2019) (Nurhasan, Pradibta, Suryadi & Alfinda, 2019) (Racz & Zilizi, 2019)

Type of Solutions

Mobile Apps

Within the five articles selected that describe mobile applications related to mobile tourism services, in the scope of smart cities (Kaur & Maheshwari, 2016; Marjury et al., 2017; Spiel et al., 2017; Nurhasan et al., 2019; Racz & Zilizi, 2019), Marjury et al. (2017) propose an offline mobile application for places identification with augmented reality, which main goal is to work without internet connection. Also, in the domain of augmented reality, Nurhasan et al. (2019) propose an application to provide tourism information, to determine the current location and destination point. To get this, the application must be able to scan perfectly in the object marker field. The reference object is a photo of a tourist location. The scanning process is carried out by the users, and the user can see information from the tourist objects they visit; two technologies are used, namely Geographic Information Systems (GIS) and augmented reality. The use of GIS technology aims to provide information that is geographically related to a tourism object, while augmented reality is used to bring up 3D objects along with information relating to the tourism object. This application is presented as a smart tourist guide application for the introduction of Malang city tourism potential using hybrid technology.

In the field of virtual reality, Racz and Zilizi (2019) offer an application which implements a virtual tour on a Mediterranean tourist destination, to reveal a new opportunity for Mediterranean islands. Kaur & Maheshwari (2016) also present a smart tourist application for Dubai visitors, enabling them to create their itineraries based on their interests, budgets and duration of visit, so that tourists can see all the nearest spots on the map according to their current location; they can take a virtual tour of any point of interest.

Spiel et al. (2017) propose a mobile application which digitally enhances a public urban space by providing timely relevant and spatially rich information about the immediate neighbourhood of a particular way-finding pillar.

Personalized/Integrated systems

Twenty-four articles pertain studies reporting integrated systems (smart tourism systems) that aim to aggregate touristic data from multiple sources to provide integrated and/or personalized services, in the context of smart cities.

Most of the studies presented here describe solutions to support tourists in planning their visits, exploring places and moving around efficiently, signifying integrated solutions (semantically integrated knowledge base and a set of services), complementing other services (Barbu et al., 2015).

On the field of local attractions' management and heritage tourism, three articles should be highlighted. One concerns a solution to provide information about local attractions to visitors but also to collect the movement of tourists to facilitate efficient allocation of tourist resources (Cacho et al., 2015). Other regard services selection of automation with the user profile in the field of material and immaterial heritage tourism (Benfares, El Bouzekri El Idrissi & Amine, 2016) and a bottom-up conceptual model of a platform based on user-generated data to offer paths based on user-specific requirements; paths based on cultural information, local folklore and natural trails (Garau, 2014).

The offer of personalized experiences and some type of interactivity with the user were also identified and, in this study, five articles stand out, one corresponding to one collaborative system that allows players to carry out games and challenges that improve their tourism experience; a cloud based platform, combining augmented reality features, gamification and geolocation in a smart city environment (García-Crespo et al., 2016). In another article (Giovannella et al., 2013), a prototype containing eight applications is presented: an art game (arkanoid based), an interactive glass case of a museum, the lighting control of the glass case, the exploration of a festivals' program and that of an event, an interactive restaurant window, the exploration of a 3D reconstruction of an ancient fountain and a sign book. This service is aimed at enhancing and supporting active learning experiences to be developed during touristic explorations of a territory. In Mighali et al. (2015), a smart system, an IoT-aware architecture, able to improve the cultural experience of the user, by involving the most important recent innovations in the ICT field, is proposed. The purpose is to contribute to the provision of interactive and personalized cultural experience, where IoT can give a significant contribution, coupled with wearable and mobile devices and other components included in the proposed system architecture. The fundamental concept of the system should be the location-awareness, that is, all the provided services have to act according to the users' position and their movements. Colace et al. (2019) describe a methodology to design a chatbot based on a Context-Aware System able to recommend contents and services according to contexts and users' profile. The chatbot can be understood as modern tourist guide that allows the provision of services or narrative content, integrated and adjusted to the needs of users. Due to the lack of existing applications in the current market that help the tourists in managing their planner travel journey in Malaysia, Abdul Rahman, Zaaba and Hussain (2018) proposed an android application that operate as virtual tourist guide and additional features such as recommendation places to go.

The use of social data to personalize recommendations in smart tourism is also reported in some of the included articles of this systematic review. From the set of twenty-four references, six articles stand out. A profiling mechanism and a recommendation system, for single users and groups of tourists, that filters the available choices to display and simplify the users' decision-making process by obtaining their preferences from the social network Facebook is proposed by Barile et al. (2014). Yamamoto and Ikeda (2016) present a recommendation media GIS specially tailored to recommend tourist spots in urban areas, a system which integrates a Web-GIS, a Social Networking System and a recommendation system. Another solution reported (Figueredo et al., 2017) is able to detect tourist implicit preferences based on photos, obtained from personal social media. Using deep learning techniques, it is possible to classify locations usually visited by a specific tourist and, finally, the fuzzy classification determines better tourism class to a potential tourist (i.e., tourists are classified in categories).

One additional article (Cervantes et al., 2016) reports a solution that makes productive use of semantically-related social data, in order to help users (tourists and citizens in general) to discover cultural heritage, points of interest and available services in a smart city; extracted knowledge that can be used to personalize recommendations in a smart tourism application.

Choi et al. (2019) suggest a travel recommender system for automating word-of-mouth (WOM) effects and providing personalized travel-planning services to tourists; proposes a travel recommender system that helps a tourist build his/her personalized travel plan based on collaborative filtering and constraint satisfaction filtering.

In Luberg et al. (2011), a personalised recommendation system is based on the personal preference profile. Based on the created profile, the suggestion engine finds interesting objects for the given user. For each found object a score of "interestingness" is found.

In this review and from this set of references under analysis, four articles were identified as particularly focused on accessibility issues. In Liu et al. (2016), a linked-data, technology-based solution, in the domains of tourism services and e-governance, to help impaired people planning their trips, is presented.

Regarding a specific project, the "Accessibility for All in Tourism" (ACCES4ALL) two articles were identified (Rodrigues et al., 2018; Rosa & Lopes, 2019). One of the project aims is presented in Rodrigues et al. (2018), a study about an Internet of Everything (IoE) eco-friendly, accessible and smart bus stop that integrates all the features from the existing bus stops, as well as integrating intelligent features that allow the bus stop to adapt itself to each different user or group of users; greatly contributing to the improvement of the passenger experience, providing concentrated community services information. In Rosa and Lopes (2019), some results of ACCES4ALL project are presented, namely the perceptions of elderly tourists of bus stop environments located in their own countries, to assess the relevance of some attributes to their mobility. Its main objective is to develop a pilot study: an accessible, smart and sustainable bus stop to be located at Faro International Airport. The focus of Nowak Da Costa and Bielski (2018), is the design of new accessibility maps. As existing structures and/or historical buildings in cities, built prior to mobility requirements, must be retrofitted, a significant number of potential interesting locations are not accessible to persons with reduced mobility. So, the focus of this research is to make detailed mobility related information available and improve the communication of accessibility information to users. The purpose is to evaluate the accessibility potential of public places and spaces to help persons with reduced mobility planning their outings by taking advantage of smart data. This research will continue on this topic through the alternative design and implementation of accessibility map prototypes.

Other issues are reported in the remaining articles. In Quarati et al. (2017), a service able to exploit and integrate weather sensor data provided by free contributions of citizen scientists Personal Weather Stations (citizen enthusiasts in hydro meteorology) communities grouped in Weather networks, is presented. Bhattacharya et al. (2017) describe an integrated traffic situation assessment and alert communication system, a geographical information-based decision-making related to traffic destinations and routes. A solution to integrate route attractiveness attributes into a tourist trip recommendation system for walking tourists is presented in Herzog et al., (2019).

In the field of Open Data, an Application Programming Interface (API) for accessing cities touristic data is presented to enable applications to seamlessly use data from several cities, increasing their potential market while reducing the development costs (Pereira et al., 2015).

Another particular issue is tackled in Dalli and Bri (2017), where an electronic ticketing system to integrate the information of events that could interest tourists, is presented. With this system, tourists can use their smart phone to buy electronic tickets of the destinations or to events and travel agencies and co-operators can set the information of products and services via electronic ticket package management server.

Application Scope

Regarding application scope (Figure 2), within the five articles that describe mobile applications related to mobile tourism services, in the context of smart cities (Kaur & Maheshwari, 2016; Marjury et al., 2017; Spiel et al., 2017; Nurhasan et al., 2019; Racz & Zilizi, 2019), three of them (Kaur & Maheshwari, 2016; Marjury et al., 2017; Spiel et al., 2017) consider that the proposed solution applies specifically to the urban context, to urban tourism. The two remaining references (Nurhasan et al., 2019; Racz & Zilizi, 2019) particularize other issues, specifically e-tourism applied to marketing, entertaining and education (Racz & Zilizi, 2019) and promotional and educational purposes (Nurhasan et al., 2019).

Inside the group of references categorised as "Personalized/Integrated Systems", five articles are focused on heritage and cultural tourism (Barile et al., 2014; Benfares et al., 2016; Colace et al., 2019; Garau, 2014; Mighali et al., 2015).

Additional articles focus complementary fields to tourism arena, namely urban infrastructures management and services related to transportation, safety and water consumption (Cacho et al., 2015); the personal entertainment, professional training and simulation (e.g., training simulation for evacuation plans); the mobility assistance (García-Crespo, et al., 2016); and learning (Giovannella et al., 2013).

Six articles (Barbu et al., 2015; Figueredo et al, 2017; Herzog et al., 2019; Pereira et al., 2015; Quarati et al., 2017; Yamamoto & Ikeda, 2016) make it very explicit that their focus is on the city and the remaining six articles (Abdul et al., 2018; Bhattacharya et al., 2017; Cervantes et al., 2016; Choi et al., 2019; Dalli & Bri, 2017; Luberg et al., 2011) present tourism in general as their application scope.

Four references (Liu et al., 2016, Nowak Da Costa & Bielski, 2018; Rodrigues et al., 2018; Rosa & Lopes, 2019) propose accessible tourism as their central field of action.

Figure 2. Distribution of the articles in terms of application scope

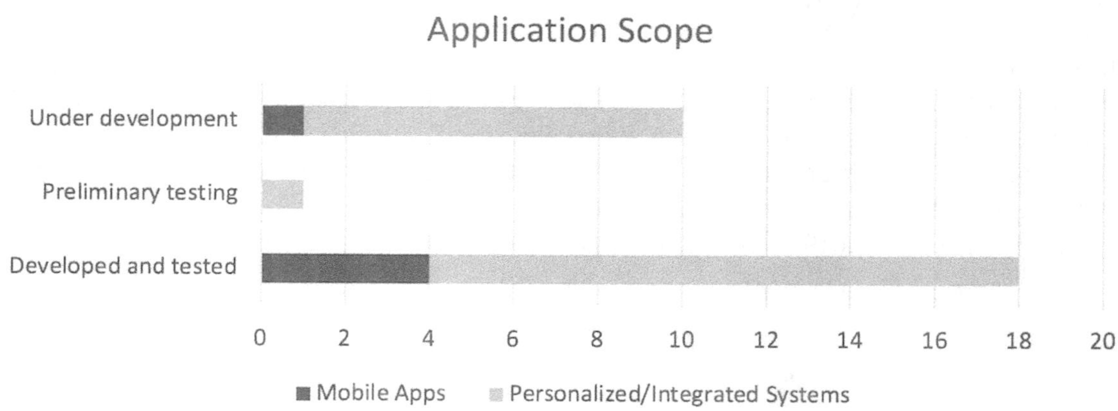

Target Audience

In what regards the target audience, within the first group of references, comprising five articles classified as "Mobile Apps", four articles (Kaur & Maheshwari, 2016; Nurhasan et al., 2019; Racz & Zilizi, 2019; Spiel et al., 2017) mainly target tourists and one (Marjury et al., 2017) focuses on organizations providing services to visitors. Still, and in relation to the first four references, two articles also mention other audiences, in particular: locals (Spiel et al., 2017) and public and private tourism promoters (Nurhasan et al., 2019).

Within the twenty-four articles that describe "Personalized/Integrated Systems", only six (Bhattacharya et al., 2017; Cervantes et al., 2016; Choi et al., 2019; Colace et al., 2019; Giovannella et al., 2013; Yamamoto & Ikeda, 2016;) refer to a less varied audience, in this case, tourists and users / citizens.

In each of the remaining eighteen articles, a diversity of audiences are mentioned, beside tourists: researchers and developers (Barbu et al., 2015; Barile et al., 2014; Figueredo et al., 2017;); local authorities, public services and policy makers (Abdul et al., 2018; Benfares et al., 2016; Cacho et al., 2015; Garau, 2014; Herzog et al., 2019; Luberg et al., 2011; Mighali et al., 2015; Pereira et al., 2015); private services in tourism industry (Dalli & Bri, 2017; García-Crespo et al., 2016; Figueredo et al, 2017; Herzog et al., 2019; Liu et al., 2016; Pereira et al., 2015; Rodrigues et al., 2018;); managers of public or private weather forecast networks (Quarati et al., 2017); people with specific needs, in this case senior tourists (Rodrigues et al., 2018; Rosa & Lopes, 2019), people with impaired mobility (Nowak Da Costa & Bielski, 2018) and people with other impairments (Liu et al., 2016).

Maturity Level

In terms of maturity level, different development stages were identified (Figure 3). Within the "mobile apps" group, four applications were developed and tested (Marjury et al., 2017; Nurhasan et al., 2019; Spiel et al., 2017; Racz & Zilizi, 2019) while in Kaur and Maheshwari (2016), although it is referred that the app was developed there is no explicit information on its implementation or assessment.

Inside the larger group of references, classified as "Personalized/Integrated Systems", nine (Abdul et al., 2018; Benfares et al., 2016; Dalli & Bri, 2017; Garau, 2014; Liu et al., 2016; Mighali et al, 2015; Nowak Da Costa & Bielski, 2018; Rodrigues et al., 2018; Rosa & Lopes, 2019) of the twenty- four solutions proposed, are on the concept/design phase, under development and not tested or assessed; one article is presenting the system architecture overview and describing preliminary testing (Luberg et al., 2011) and the remaining fourteen articles report solutions with a higher level of development and all of them have been tested. Within this last group, six articles specifically related to prototypes (Barile et al., 2014; Bhattacharya et al., 2017; Choi et al., 2019; Colace et al., 2019; Giovannella et al., 2013; Quarati et al., 2017), five refer to proofs of concept (Barbu et al., 2015; Cervantes et al., 2016; García-Crespo, et al., 2016; Herzog et al., 2019; Yamamoto & Ikeda, 2016) and one (Pereira et al., 2015), an open data

Figure 3. Distribution of the articles in terms of maturity level

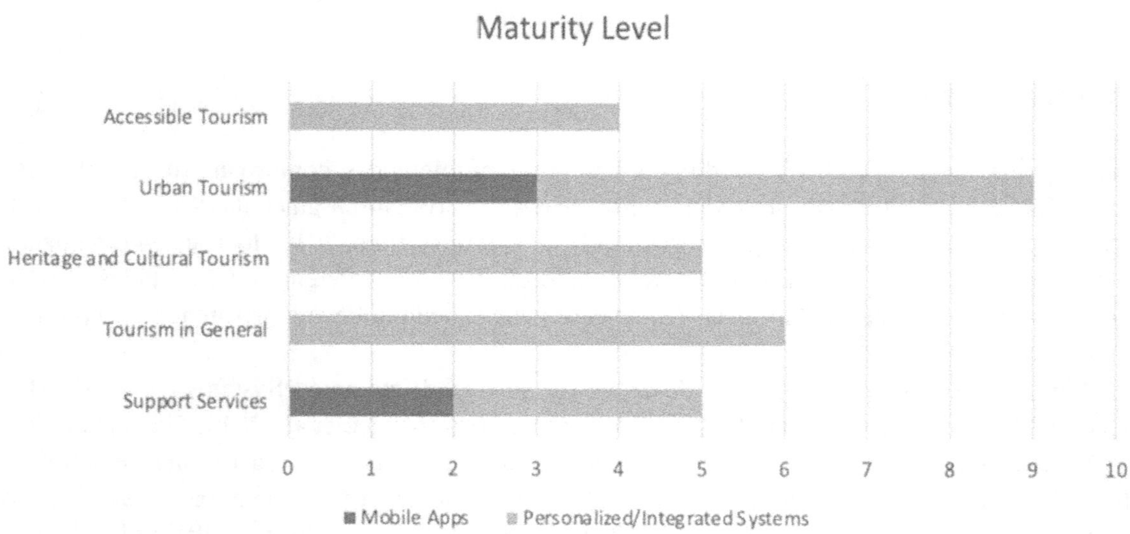

model, is currently available for five cities and several companies have developed mobile applications that use this solution, but with an uncertain future because of its reliance on European funding.

Accessible Tourism

This review was conducted with the main purpose of knowing if the solutions developed within the scope of smart cities and tourism, address accessibility issues and if, in some way, they are related to accessible tourism. What resulted from this review was that, within the twenty-nine references included, considering apps and integrated systems, only four articles (Liu et al., 2016; Nowak Da Costa & Bielski, 2018; Rodrigues et al., 2018; Rosa & Lopes, 2019) report solutions tackling these particular issues and targeting persons with specific needs. In the great majority of the remaining twenty-five references, there is no allusion to that issue or to that audience.

It is now important to add more detail concerning the four articles focused on accessible tourism issues.

In Liu et al. (2016), the purpose is to design and implement a visualized travel system by using linked data for impaired people. For this project, they intend to include points of view from people with different levels of disabilities within one platform. This linked data-based innovation system should make the decision process easier for impaired people and also make their transportation, visits, and leisure activities less difficult.

Rodrigues et al. (2018) consider the needs of the elderly and other people with reduced mobility (e.g., wheelchair people, blind, deaf and tourists with trolleys). It is reflected that a bus stop should have some instructions, not displayed on the screen, which should work for most blind users. These instructions should be in Braille, but not simple instructions. They have built-in sensors that detect when those instructions are being read. This way, the system can detect that a blind person is there and can help/interact with the user by using the proper way. Also, when visually impaired (or blinds) users are detected (e.g., when the users' mobile phone Bluetooth is detected) two pilot tones will be used. One to guide the visually impaired people to the SBS screen, not only to show visual information, but also to interact with the SBS by using sound and voice commands. The second pilot tone can guide the user to the bus door.

Rosa and Lopes (2019) report a collaborative design process, the assessment of the environments of bus stops being made through the perspectives and the lens of older customers' experience. The project is particularly focused on elderly people and on accessible tourism.

Finally, impaired people correspond to the target users in Nowak Da Costa and Bielski (2018), being the focus of this research the design of new accessibility maps, with the ultimate goal to get citizens, local government and ministerial projects involved to help maximise the accessibility of public spaces and a part of the smart city design.

DISCUSSION AND CONCLUSION

Concerning the first research question (i.e., what are the smart tourism services being developed?), the results show that, in the context of smart cities' implementation, there is a concern related to the optimization of services directed at improving tourists' experiences. The included articles describe solutions to support tourists in planning their visits, exploring places and moving around efficiently. These solutions were implemented using different technologies, including augmented reality and virtual reality. Referring to the concept of "experience economy", which supports that consumers value experiences more than goods and services (Pine & Gilmore, 2013), a topic of great interest also in the area of accessible tourism, the results reveal that this perspective has been applied in smart cities.

With regard to the second research question (i.e., what is the smart tourism services application scope?), there is a focus on the urban context with various areas of application, namely: heritage and cultural tourism; e-tourism applied to marketing, entertaining, education and promotion. Additional articles defined other complementary fields besides tourism arena, namely urban infrastructures' management and services related to transportation, safety and water consumption, personal entertainment, professional training and simulation and sustainable mobility. With regard to accessible tourism, within the references included, only four articles identify it as their field of action.

Regarding the third research question (i.e., what are the smart tourism services target audiences?), a large majority of articles refer to several other audiences in addition to tourists, such as: researchers and developers; public services and policy makers, as well as private services in tourism industry. Within the scope of accessible tourism, inside the group of the included references, a minority (four articles) specifically refer that they target senior tourists, people with impaired mobility and people with other impairments.

Referring now to the fourth research question (i.e., what are the reported smart tourism services maturity levels?), there is a set of reported solutions already endowed with a certain degree of development, as evidenced by the reference to the performance of some tests and evaluations. However, solutions with a higher degree of maturity in terms of implementation and assessment, allowing for their use in a larger scale and real context, have not been identified. Therefore, the outlook is still very technology oriented.

In what respects accessible tourism (i.e., the fifth research question), most of the included articles did not mention this topic. Three of the included articles only make superficial allusions. Spiel et al. (2017) talk about a workshop, in which the recruited local expert group, a heterogeneous group of nine with diverse backgrounds, included a hearing-impaired and an active senior. They did not explicitly invite seniors with impaired mobility, but they wanted that representatives of a neighbourhood centre to be present, in order to bring a diversity of perspectives. In turn, Garau (2014), discusses the problem of the digital divide and the risk that the data collected could be biased and reflect the desires of a small proportion of the community (not considering, for instance, the ageing population). To overcome this weakness, the author suggests the need for evaluating the effective level of utilization of the solution, user types and/or its usability (Garau, 2014). Moreover, Yamamoto and Ikeda (2016) proposed that the people who are more than eighteen years old, both in and outside the region, would be users of the system. In spite of mentioning a concern in developing clear and easy to understand questions in building the user profiles, it is referred that "(...) the system is for use with ordinary people" (Yamamoto & Ikeda, 2016), stating the exclusion of non-ordinary users, not being clear who they could be.

Finally, four of the included articles are focused on accessible tourism (Liu et al., 2016; Nowak Da Costa & Bielski, 2018; Rodrigues et al., 2018; Rosa & Lopes, 2019). They report a travel guidance application to provide information for impaired people (Liu et al., 2016), an accessible bus stop (Rodrigues et al., 2018), an accessibility assessment of the environments of bus stops (Rosa & Lopes, 2019) and the design of accessible maps (Nowak Da Costa & Bielski, 2018). All these four solutions were under development and were not tested or evaluated, which is essential to translate and disseminate the solutions being developed.

With these results it is possible to affirm that the potential of smart cities has been insufficiently harnessed to promote accessible tourism as well as the necessary raise of participation and social inclusion mentioned in Kastenholz, Eusébio and Figueiredo (2015). The results obtained reveal a still significant divergence from the real needs of the accessible tourism market, as well as a deficit in the capacity to exploit the opportunities it offers.

Regarding the future and, being recognized the accessible tourism area as multidisciplinary and, therefore, influenced by other areas, including public policies (Buhalis & Darcy, 2010), it would be important to discuss the role of smart cities in the construction of the future of tourism, and in particular of accessible tourism. This is an aspect that is somehow also reinforced in Michopoulou, Darcy, Ambrose and Buhalis (2015), when it is argued that the future of accessible tourism is influenced by policy initiatives on accessibility and sustainability, with an emphasis on collaborative practices, particularly meaningful in the concept of smart cities.

Despite some limitations about both the chosen keywords and the databases that were used in the selection of the articles, this systematic review followed a rigorous method for retrieving, selecting and analysing the articles, so that its results are reliable.

The results show that the number of studies to take advantage of the smart cities' paradigm to make tourism more accessible (i.e., four articles) is insignificant and not very representative within the total of studies related to smart cities. Moreover, these studies do not represent any added value in terms of translation from research to concrete solutions with impact in daily life of the citizens since their proposals are in an early development stage and were not assessed.

According to the results of this systematic review, it is possible to conclude that the articles reporting the development of information services using smart cities to support accessible tourism, similarly to other results related to the development of smart cities' applications to support impaired people (Rocha et al., 2020), present limited value creation. Therefore, it seems that smart cities create plenty of room to develop accessible tourism research.

REFERENCES

Abdul Rahman, A. A., Zaaba, Z. F., & Hussain, A. (2018). Usable tourism application: MyTouristPlanner (MTP). *2018 International Conference on Promising Electronic Technologies, ICPET 2018*, 31-36. 10.1109/ICPET.2018.00012

AlDairi, A., & Tawalbeh, L. (2017). Cyber security attacks on smart cities and associated mobile technologies. *Procedia Computer Science, 109*, 1086–1091. doi:10.1016/j.procs.2017.05.391

Azevedo Guedes, A. L., Carvalho Alvarenga, J., Dos Santos Sgarbi Goulart, M., Rodriguez, R., Vicente, M., & Pereira Soares, C. A. (2018). Smart cities: The main drivers for increasing the intelligence of cities. *Sustainability, 10*(9), 3121. doi:10.3390u10093121

Barbu, C., Kuhn, K., Spirescu, A., Lamorte, L., & Scanu, C. (2015). 3cixty: An innovative way to explore cities. *2015 IEEE 1st International Smart Cities Conference, ISC2 2015*. 10.1109/ISC2.2015.7366178

Barile, F., Calandra, D. M., Caso, A., Dauria, D., Di Mauro, D., Cutugno, F., & Rossi, S. (2014). ICT solutions for the OR.C.HE.S.T.R.A. project: From personalized selection to enhanced fruition of cultural heritage data. *10th International Conference on Signal-Image Technology and Internet-Based Systems, SITIS 2014*, 501-507. 10.1109/SITIS.2014.12

Benfares, C., El Bouzekri El Idrissi, Y., & Amine, A. (2016). Smart city: Recommendation of personalized services in patrimony tourism. *Colloquium in Information Science and Technology, CIST*, 835-840. 10.1109/CIST.2016.7805003

Bhattacharya, D., Painho, M., Mishra, S., & Gupta, A. (2017). Mobile traffic alert and tourist route guidance system design using geospatial data. *International Archives of the Photogrammetry, Remote Sensing and Spatial Information Sciences - ISPRS Archives, 42*(4W3), 11-18. 10.5194/isprs-archives-XLII-4-W3-11-2017

Buhalis, D., & Darcy, S. (2010). *Accessible tourism: Concepts and issues. Accessible tourism: Concepts and issues*. Channel View Publications. doi:10.21832/9781845411626

Cacho, A., Estaregue, D., Figueredo, M., Lucas, J., Aurelio, M., Farias, H., . . . Alves, C. (2015). A smart destination initiative: The case of a 2014 FIFA world cup host city. *2015 IEEE 1st International Smart Cities Conference, ISC2 2015*. 10.1109/ISC2.2015.7366223

Cervantes, O., Gutiérrez, E., Gutiérrez, F., & Sánchez, J. A. (2016). SOCIAL METRICS APPLIED to SMART TOURISM. *ISPRS Annals of the Photogrammetry, Remote Sensing and Spatial Information Sciences, 4*(4W1), 117-124. 10.5194/isprs-annals-IV-4-W1-117-2016

Chauhan, S.; Agarwal, N.; Kar, A. K. (2016). Addressing big data challenges in smart cities: a systematic literature review. *Info, 18*(4), 73-90.

Choi, I. Y., Ryu, Y. U., & Kim, J. K. (2019). A recommender system based on personal constraints for smart tourism city. *Asia Pacific Journal of Tourism Research*, 1–14. Advance online publication. doi: 10.1080/10941665.2019.1592765

Colace, F., de Santo, M., Lombardi, M., & Santaniello, D. (2019). Chars: A cultural heritage adaptive recommender system. *TESCA 2019- 1st ACM International Workshop on Technology Enablers and Innovative Applications for Smart Cities and Communities, Co-Located with the 6th ACM International Conference on Systems for Energy-Efficient Buildings, Cities, and Transportation, ACM BuildSys 2019*, 58-61. 10.1145/3364544.3364830

Colace, F., Santo, M. D., Lemma, S., Lombardi, M., Rossi, A., Santoriello, A., ... Vigorito, M. (2016). How to describe cultural heritage resources in the web 2.0 era? *11th International Conference on Signal-Image Technology and Internet-Based Systems, SITIS 2015*, 809-815. 10.1109/SITIS.2015.50

D'aniello, G., Gaeta, M., & Reformat, M. Z. (2017). Collective perception in smart tourism destinations with rough sets. *2017 3rd IEEE International Conference on Cybernetics, CYBCONF 2017*. 10.1109/CYBConf.2017.7985765

da Costa Liberato, P. M., Alén-González, E., & de Azevedo Liberato, D. F. V. (2018). Digital technology in a smart tourist destination: The case of Porto. *Journal of Urban Technology, 25*(1), 75–97. doi: 10.1080/10630732.2017.1413228

Dalli, A., & Bri, S. (2017). Design of electronic ticket system for smart tourism. *12th International Conference on Signal Image Technology and Internet-Based Systems, SITIS 2016*, 490-492. 10.1109/SITIS.2016.82

Encalada, L., Boavida-Portugal, I., Ferreira, C. C., & Rocha, J. (2017). Identifying tourist places of interest based on digital imprints: Towards a sustainable smart city. *Sustainability (Switzerland), 9*(12). doi:.. doi:10.3390u9122317

Figueredo, M., Cacho, N., Thome, A., Cacho, A., Lopes, F., & Araujo, M. (2017). Using social media photos to identify tourism preferences in smart tourism destination. *2017 IEEE International Conference on Big Data, Big Data 2017,* 4068-4073. 10.1109/BigData.2017.8258423

Garau, C. (2014). Smart paths for advanced management of cultural heritage. *Regional Studies, Regional Science, 1*(1), 286–293. doi:10.1080/21681376.2014.973439

García-Crespo, Á., González-Carrasco, I., López-Cuadrado, J. L., Villanueva, D., & González, Á. (2016). CESARSC: Framework for creating cultural entertainment systems with augmented reality in smart cities. *Computer Science and Information Systems, 13*(2), 395–425. doi:10.2298/CSIS150620006G

Gautam, B. P., Asami, H., Batajoo, A., & Fujisaki, T. (2016). Regional revival through IoT enabled smart tourism process framework (STPF): A proposal. *2016 Joint 8th International Conference on Soft Computing and Intelligent Systems and 2016 17th International Symposium on Advanced Intelligent Systems, SCIS-ISIS 2016,* 743-748. 10.1109/SCIS-ISIS.2016.0162

Ghapanchi, A. H., & Aurum, A. (2011). Antecedents to IT Personnel's Intentions to Leave: A Systematic Literature Review. *Journal of Systems and Software, 84*(2), 238–249. doi:10.1016/j.jss.2010.09.022

Giffinger, R., & Gudrun, H. (2010). Smart cities ranking: an effective instrument for the positioning of the cities? *ACE: Architecture, City and Environment, 4*(12), 7-26.

Giovannella, C., Iosue, A., Moggio, F., Rinaldi, E., & Schiattarella, M. (2013). User experience of kinect based applications for smart city scenarios integrating tourism and learning. *2013 IEEE 13th International Conference on Advanced Learning Technologies, ICALT 2013,* 459-460. 10.1109/ICALT.2013.142

Herzog, D., Sikander, S., & Worndl, W. (2019). Integrating route attractiveness attributes into tourist trip recommendations. *The Web Conference 2019 - Companion of the World Wide Web Conference, WWW 2019*, 96-101. 10.1145/3308560.3317052

Irshad, M. (2016). A systematic review of information security frameworks in the internet of things (iot). In *IEEE 18th International Conference on High Performance Computing and Communications; IEEE 14th International Conference on Smart City; IEEE 2nd International Conference on Data Science and Systems (HPCC/SmartCity/DSS)* (pp. 1270-1275). IEEE.

Kastenholz, E., Eusébio, C., & Figueiredo, E. (2015). Contributions of tourism to social inclusion of persons with disability. *Disability & Society, 30*(8), 1259–1281. doi:10.1080/09687599.2015.1075868

Kaur, M. J., & Maheshwari, P. (2016). Smart tourist for Dubai city. *2016 2nd International Conference on Next Generation Computing Technologies, NGCT 2016,* 30-34. 10.1109/NGCT.2016.7877385

Lazaroiu, G. C., & Roscia, M. (2012). Definition methodology for the smart cities model. *Energy, 47*(1), 326–332. doi:10.1016/j.energy.2012.09.028

Lijing, Z., Yanrong, P., & Jianhua, H. (2015). The development strategy for the tourism in Hebei under the background of smart city based on data mining. *7th International Conference on Intelligent Computation Technology and Automation, ICICTA 2014*, 991-994. 10.1109/ICICTA.2014.238

Liu, Z., Sokhn, M., Le Calvé, A., & Schegg, R. (2016). City eye: Accessibility for all. *ACM International Conference Proceeding Series, 01-03-March-2016* 350-351. 10.1145/2910019.2910020

Louw, C. (2017). Good, better, best practise - a comparative analysis of official destination websites. *African Journal of Hospitality, Tourism and Leisure*, *6*(3).

Luberg, A., Tammet, T., & Järv, P. (2011). Smart City: A Rule-based Tourist Recommendation System. In R. Law, M. Fuchs, & F. Ricci (Eds.), *Information and Communication Technologies in Tourism 2011*. Springer. doi:10.1007/978-3-7091-0503-0_5

Marjury, D. H., Karen, B. C., Diana, M., & Gabriel, L. F. (2017). Offline mobile application for places identification with augmented reality. *2017 4th International Conference on eDemocracy and eGovernment, ICEDEG 2017,* 261-264. . doi:10.1109/ICEDEG.2017.7962546

Michopoulou, E., Darcy, S., Ambrose, I., & Buhalis, D. (2015). Accessible tourism futures: The world we dream to live in and the opportunities we hope to have. *Journal of Tourism Futures*, *1*(3), 179–188. doi:10.1108/JTF-08-2015-0043

Mighali, V., Del Fiore, G., Patrono, L., Mainetti, L., Alletto, S., Serra, G., & Cucchiara, R. (2015). Innovative IoT-aware services for a smart museum. *WWW 2015 Companion 24th International Conference on World Wide Web*, 547-550. 10.1145/2740908.2744711

Moher, D., Liberati, A., Tetzlaff, J., & Altman, D. G. (2009). Preferred reporting items for systematic reviews and meta-analyses: The PRISMA statement. *Annals of Internal Medicine*, *151*(4), 264–269. doi:10.7326/0003-4819-151-4-200908180-00135 PMID:19622511

Nowak Da Costa, J., & Bielski, C. (2018). Towards «tourism for all» - improving maps for persons with reduced mobility. *International Archives of the Photogrammetry, Remote Sensing and Spatial Information Sciences - ISPRS Archives, 42*(4) 547-550. 10.5194/isprs-archives-XLII-4-475-2018

Nurhasan, U., Pradibta, H., Suryadi, S. B., & Alfinda, A. (2019). Smart tourist guide application for the introduction of Malang city tourism potential using hybrid technology. *Journal of Physics: Conference Series*, *1402*(6) 10.1088/1742-6596/1402/6/066051

Pereira, R. L., Sousa, P. C., Barata, R., Oliveira, A., & Monsieur, G. (2015). CitySDK tourism API - building value around open data. *Journal of Internet Services and Applications*, *6*(1), 1–13. doi:10.118613174-015-0039-z

Perez-Jimenez, R., Rabadan, J., Rufo, J., Solana, E., & Luna-Rivera, J. M. (2015). Visible light communications technologies for smart tourism destinations. *2015 IEEE 1st International Smart Cities Conference, ISC2 2015*. 10.1109/ISC2.2015.7366193

Pierdicca, R., Frontoni, E., Zingaretti, P., Malinverni, E. S., Galli, A., Marcheggiani, E., & Costa, C. S. (2016). *Cyberarchaeology: Improved way findings for archaeological parks through mobile augmented reality*. . doi:10.1007/978-3-319-40651-0_14

Pine, B. J., & Gilmore, J. H. (1998). Welcome to the experience economy. *Harvard Business Review*, *76*(4), 97–105. PMID:10181589

Pine, B. J., & Gilmore, J. H. (2013). *The experience economy: Past, present and future. Handbook on the experience economy*. Edward Elgar Publishing., doi:10.4337/9781781004227.00007

Puma, P. (2019). *The Cortona heritage project: Digital applications in education to architectural and archaeological heritage* . doi:10.1007/978-3-030-12240-9_71

Purnomo, F., & Prabowo, H. (2016). Smart city indicators: A systematic literature review. *Journal of Telecommunication. Electronic and Computer Engineering*, *8*(3), 161–164.

Quarati, A., Clematis, A., Roverelli, L., Zereik, G., D'Agostino, D., Mosca, G., & Masnata, M. (2017). Integrating heterogeneous weather-sensors data into a smart-city app. *2017 International Conference on High Performance Computing and Simulation, HPCS 2017*, 152-159. 10.1109/HPCS.2017.33

Racz, A., & Zilizi, G. (2019). Virtual reality aided tourism. *2019 Smart Cities Symposium Prague, SCSP 2019*. 10.1109/SCSP.2019.8805727

Rocha, N., Dias, A., Santinha, G., Rodrigues, M., Queirós, A., & Rodrigues, A. (2020). A Systematic Literature Review of Smart Cities' Information Services to Support the Mobility of Impaired People. *International Conference on Health and Social Care Information Systems and Technologies.*

Rocha, N. P., Dias, A., Santinha, G., Rodrigues, M., Queirós, A., & Rodrigues, C. (2019). Smart Cities and Public Health: A Systematic Review. *Procedia Computer Science*, *164*, 516–523. doi:10.1016/j.procs.2019.12.214

Rodrigues, J. M. F., Sousa, N., Martins, M., & Rosa, M. (2018). IoE accessible bus stop: An initial concept. *ACM International Conference Proceeding Series*, 137-143. 10.1145/3218585.3218659

Rosa, M. P., & Lopes, J. D. C. (2019). Senior tourists' perceptions of bus stop environments used in collaborative design. *WIT Transactions on the Built Environment*, 121-132. 10.2495/UT190111

Santinha, G., Dias, A., Rodrigues, M., Queirós, A., Rodrigues, C., & Rocha, N. P. (2019). How Do Smart Cities Impact on Sustainable Urban Growth and on Opportunities for Entrepreneurship? Evidence from Portugal: The Case of Águeda. In *New Paths of Entrepreneurship Development* (pp. 31–53). Springer. doi:10.1007/978-3-319-96032-6_3

Spiel, K. K., Werner, K., Hödl, O., Ehrenstrasser, L., & Fitzpatrick, G. (2017). Creating community fountains by (re-)designing the digital layer of way-finding pillars. *19th International Conference on Human-Computer Interaction with Mobile Devices and Services, MobileHCI 2017*, 10.1145/3098279.3122135

Vanolo, A. (2014). Smartmentality: The smart city as disciplinary strategy. *Urban Studies (Edinburgh, Scotland)*, *51*(5), 883–898. doi:10.1177/0042098013494427

Yamamoto, K., & Ikeda, T. (2016). Social recommendation GIS for urban tourist spots. *18th IEEE International Conference on High Performance Computing and Communications, 14th IEEE International Conference on Smart City and 2nd IEEE International Conference on Data Science and Systems, HPCC/SmartCity/DSS 2016*, 50-57. 10.1109/HPCC-SmartCity-DSS.2016.0018

KEY TERMS AND DEFINITIONS

Accessible Tourism: Allows all people to independently access, with equity and dignity, tourism products, and/or services.

Smart Cities: Cities capable of collecting and analysing vast quantities of data, through low-cost sensors, data-actuated devices and wireless communication networks. These data are collected and analysed to promote automate and intelligent processes (e.g., advanced data analytic), to improve the quality of the available services and to promote the emergence of new ones.

Smart Tourism: A more fitting achievement of tourist information and services, promoting the development of intelligent tourism and self-service travel. Enriched information access together with new tourists' profiles, to increase knowledge of tourist behaviour regarding their spatial and temporal distribution in the destinations. Smart tourism exposes individuals as information makers, filtering the existing options for tracking their position.

Chapter 6

Smart and Accessible Cities?

Not Always – The Case for Accessible Tourism Initiatives in Buenos Aries and Sydney

Ana Clara Rucci
https://orcid.org/0000-0002-5094-8460
Universidad Nacional de La Plata, Argentina, and Universidad de Alicante, Spain

Natalia Porto
Universidad Nacional de La Plata, Argentina

Simon Darcy
UTS Business School, University of Technology Sydney, Australia

Leandro Becka
Universidad Nacional de La Plata, Argentina

ABSTRACT

This chapter aims to discuss disability, aging, and accessibility and their relationship with technology trends, taking into account a world full of different kind of constraints, taking Buenos Aires (Argentina) and Sydney (Australia) as study cases. Moreover, the authors also study how a hypothetically smart and accessible city could set the basis for making it touristic for all. Particularly, this chapter focuses on how cities that are being designed and planned under the smart city/destination model do not always consider people with disabilities and seniors in their models. So, these models that leave out more than the 30% of the world's population are not only inappropriate but also ineffective and inefficient, and consequently challenge the underlining sustainability of such projects.

DOI: 10.4018/978-1-7998-6428-8.ch006

INTRODUCTION

Disability, ageing, and accessibility have a close relationship. On one side, it is estimated that 15% of the population worldwide live with one or more disabling conditions (WHO, 2011) and, on the other side, more than 46 percent of seniors – those aged 65 years and over—have disabilities at an increasing rate. Looking ahead, the global trends in ageing populations and the higher risk of disability in older people are likely to lead to further increases in the population affected by disability (UN, 2015). Indeed, for the first time in history, by 2050, the number of senior citizens -those over 60 years old- will be higher than the number of children in the 0-14 aged group (European Commission, 2014). With this reversing of the "population pyramid" (UN, 2019), it is envisaged that the ageing of the world population is leading to significant social transformation with major implications for all sectors of society (Patterson & Balderas, 2020). These areas include labour and financial markets, tourism, leisure, housing, transportation, and social protection, among others (Sudbury & Simcock, 2009; UN, 2015). In this scenario, accessibility is defined as the conditions which enable PWD, seniors, pregnant women, among others, to use and enjoy places, products, and services through the principles of universal design. Darcy & Dickson (2009) calculate that 31% of the population benefit from accessibility provisions. Nevertheless, accessibility becomes an active enabler of mobility, daily activities such as self-care and communication, if technological factors (digital advances and innovation) are present (Sert, 2019; Darcy, 2010).

Problems with wayfinding and mobility to move from one place to another or the lack of information about the right direction to take are some of the difficulties that PWD (PwD) experience. In this scenario, the Internet of Things (IoT) plays a role as an enabler technology, offering PwD the assistance and support they need to achieve a good quality of life and allows them to participate in the social, cultural and economic life (Nitti et al., 2018; Buhalis & Michopoulou, 2011; Michopoulou & Buhalis, 2013). Within the UN Convention on the Rights of Persons with Disability's Article 9 places the built environment transport and information communication technologies (ICT) in the same article. ICTs like the Internet and social media are ubiquitous with modern travel and have democratized the travel process for many people: PwD traveling is much easier now than ever before. Accessing public spaces is such an important issue that the promotion and development of ICT services for PwD are considered one of the axes of smart city models. Although ICTs are valuable tools to make a city smart, it does not create a smart city by itself, but it is the integration of ICTs and the fulfilment of basic needs and services as a whole that creates the foundation for the development of smart cities (Sánchez et al., 2019). It is important to point out that the accessible and technological features of a smart city benefit not only residents but also tourists. An accessible city for residents is a tourist city for visitors with disabilities.

This chapter aims to discuss disability, ageing, and accessibility, and their relationship with technology trends, taking into account a world full of different kind of constraints, taking Sydney (Australia) and Autonomous City of Buenos Aires (CABA, for its Spanish acronym, Argentina) as study cases, considering that both cities have interesting backgrounds in accessible tourism (Porto et al., 2019) and are in different stages of smartness development (ESI Smarter Cities 2025, 2018). Moreover, we also study how a hypothetically smart and accessible city could set the basis for providing tourism for all. In particular, this chapter focus on how cities that are being designed and planned under the smart city/destination model, do not always consider people with access needs – considering seniors with access needs, PwD or reduced mobility or communication, pregnant women, people with temporary limitations, among others– (Buhalis & Michopolou, 2011) in their models. So, these models are not only inappropriate but also ineffective and inefficient, and consequently challenging the underlining sustainability of such projects (Darcy & Dickson, 20 09; Darcy, McKercher & Schweinsberg, 2020).

The chapter is structured as follows. A conceptual framework in which the relationship between accessible tourism, technology, and smart tourism destinations is described, focusing on the existence of smart tourism models that include accessibility in them. Then, study cases are presented, describing their accessibility and smart destinations characteristics. The collection of data has been done mainly through the collection of secondary data from each of the case study cities together with the management information systems, corporate plans and planning documents (Yin, 2004; Veal & Darcy, 2014). Finally, there is a discussion on the topics under study.

ACCESSIBLE TOURISM, TECHNOLOGY, AND SMART SCENARIOS

Challenges of Becoming a Smart City

Before the advent of the COVID-19, the global economy was experiencing a transformation that has few precedents in history, in terms of its size, speed, and range of change. Quite simply, changing the ways people produce, consume, work, and live (Schwab, 2017). Forecasts suggest that urban spaces will become increasingly dense and will face numerous problems related to the management of scarce resources, the provision of public services, the management of information, urban mobility and traffic, as well as energy efficiency and, in general, sustainability. In this context, cities will continue to transform into increasingly complex ecosystems (Prego, 2018; Sun et al., 2016). Considering estimations from the World Tourism Organization (UNWTO, 2019) for 2050, almost 20% of the world population (which will also continue to increase) will be traveling and generating additional demand for services and infrastructure in the different cities of the world. Therefore, as cities grow, they face significant challenges such as the growing social and economic inequalities, the climate emergency and the problem of decent work, which are some of the topics included in humanity through the 17 Sustainable Development Goals (SDGs) of the 2030 agenda of the United Nations (UN, 2015). Some of the most challenging issues are the development of sustainable cities and communities, where tourism and technological advances promoted by ICTs appear as strategic tools for its achievement, including Artificial Intelligence (AI), Internet of Things (IoT), Big Data analysis, 3D printing, and smart sensors, among others (Sánchez et al., 2019).

Given this situation, cities must maximize their efficiency without neglecting people's welfare, which is facilitated through the smart city scheme. Although there is no formal and widely accepted definition of Smart City, the concept refers to criteria linked to better use of public resources, the improvement of the quality of services offered to citizens, while reducing public administration operating costs (Zanella et al., 2014). As Figure 1 shows, the areas of intervention of smart cities are related to the economy, mobility (efficient traffic management, optimization of public transport routes, waste management, sustainable infrastructures, reduction of the emission of harmful gases, promotion of the use of public and private electric vehicles (eMobility), natural environment (systems that allow energy savings, efficient water consumption, recycling, among others), citizenship and quality of life (telemedicine, telecare, data management and patient records, alerts to emergency services automatically when an elderly or disabled person has fallen or deviated from their route), tourism applications to facilitate tourist visits, leisure and consumption guides (shopping, restaurants) adapted to the individual interests of each individual

Figure 1. Dimensions of a Smart City

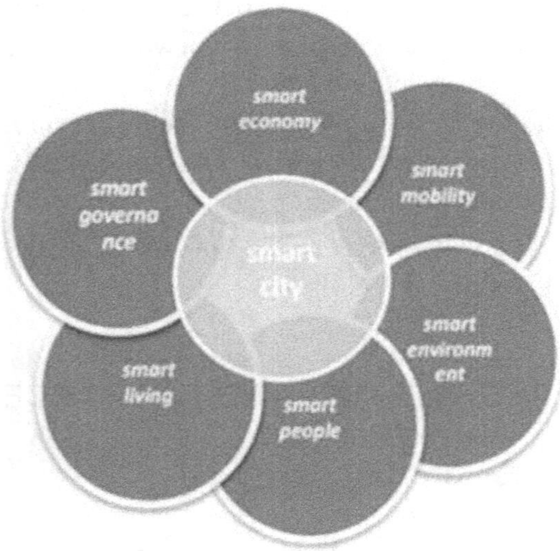

and governance (system electronic administration, online payment platforms, iCloud environments, broadband for mobile phones and free public wi-fi) (Winkowska et al., 2019; Giffinger et al., 2007).

Transforming into a smart city helps better and more automatic management of urban infrastructure and services, which should result in a reduction in public spending and the upgrading of the quality of the services provided, the improvement of information for citizens, and in decision making (COIT, 2018). The coexistence of numerous heterogeneous problems associated with the irreversible process of urban agglomeration makes the city an ideal platform to develop new technologies. The objective is to face local challenges in terms of smart, sustainable development, strengthening the relationship between the urban environment and ICTs. Smartness, as a distinctive feature of cities, should be perceived as the sum of critical advances in urban infrastructure, resources, and public services (Allwinkle & Cruickshank, 2011).

Among the difficulties that arise when implementing the concept of smart cities, it should be mentioned that problems such as the deployment of smart technologies in cities with complex social problems can exacerbate social inequalities. In particular, most investments in the development of the concept of smart cities focus on creating new facilities instead of modernizing old ones (Ravetz, 2017; Naphade et al., 2011; Krukowska, 2018; Proseedmag, 2017; Sikora-Fernandez, 2017; Bashynska & Dyskina, 2018). Although technology is a crucial tool to make a city smart, it does not create a smart city by itself, but it is the integration of ICTs and the fulfilment of basic needs and services as a whole that creates the basis for the development of smart cities (Sánchez et al., 2019). Also, every time a city plans to be smart, it must tackle specific problems that arise from city planning. Sánchez et al. (2019) study a wide range of challenges, divided by areas of influence, and, in turn, establish a degree of priority assigned to each challenge. This research establishes that in the first place a city must meet the needs of its inhabitants and provide basic services as a prerequisite to be later considered a smart city and only, as a second instance, should ICTs be introduced. Moreover, accessibility is one of the assets to be included. However, it is essential to take into consideration the development of each city and the possibilities of them to adapt in terms of technology and accessibility.

Accessibility, Technology, and Tourism

In tourism, accessibility is an asset that people with access needs assess and value mostly when technological issues are included. Although the IoT is considered as an enabling technology - offering PwD the assistance and support they need to achieve a good quality of life (Nitti et al., 2018)- or that thanks to the rise the ICTs, the Internet, and social media, PwD are finding that traveling is much easier now than ever before (Puhretmair & Nussbaum, 2011), 23% of Americans with disabilities never go online compared with 8% of the general US population (G3ict & World Enabler, 2017) or senior still prefer brochures rather than searching on the Internet (Littrell et al., 2004). That means that technology has two faces that need to be analysed: the enabler, and the inequity created of the digital divide.

On the one hand, the specific right of persons with disabilities within the *Convention of Rights of PwD* (UN, 2006) through articles 9 (Accessibility), 20 (Personal Mobility) and 30 (Participation in Cultural Life) need to be integrated into all aspects of a smart city's travel trip chain. Making a good travel plan is more complicated for PwD than other people. For them, even simple trips, such as traveling to local shops or visiting a museum or an exhibition, can be complicated (Liu et al., 2016). A wide variety of travel planning needs to be made days, weeks, or months in advance depending on the complexity of the trip whether it is a single destination trips that require planning for where, when, and how to travel, as well as how to get around and how to choose a suitable means of transportation and accommodation (Darcy, 2011).

On the other hand, the IoT, the technological factor besides contributing PwD to travel, is considered to be part of the solution to the challenge of marrying continued growth with a more 1e and responsible tourism sector (UNWTO, 2018). The evolution of new technologies in the last years has modified how tourists search and share information (Del Chiappa & Baggio, 2015; Wang et al., 2013) and the fast development of ICTs has changed the behavior of tourists. The vast increase of distribution channels (Buhalis & Amaranggana, 2013) and communication platforms (Gretzel et al., 2015) have provided tourists with new ways of obtaining information about the destinations and the organizations of their trips (Gretzel, 2011; Wang et al., 2013; Del Chiappa & Baggio, 2015). Especially in tourism, innovation is a critical factor for explaining the growth in tourism and its adaptation to the new needs of the demand (Hjalager, 2002; Shanker, 2008; Baggio & Cooper, 2010). In this context, destinations are adjusting to these new technologies, not only in order to satisfy new traveler's needs (Gretzel et al., 2015) and to help tourists to anticipate their needs or to make real-time decisions (Boes et al, 2016) but also to develop smart tourism destinations applying smartness over the whole touristic value chain to correct current problems related to sustainability, wealth distribution, accessibility, overtourism, among others (Ramón-Rodríguez & Such-Devesa, 2018; Perles-Ribes et al., 2017). However, some authors (Lamsfus et al., 2015) affirm that a destination is not smart because it makes intensive use of technology, but because it uses this technology to understand and facilitate human mobility.

In this context, smart tourism destinations have emerged as new models of governance and tourism development that consider innovation and technology as means to achieve the sustainability of the activity and the improvement of citizens` quality of life. Accessibility is among some of the axes that support the models of smart tourism destinations. It is understood as the conditions with which an environment, product or service must be available, so that it is used and enjoyed in conditions of equality, dignity and comfort by all people, including those with disability (IMSERSO, 2003).

Models of Accessibility Into Smart Cities and Destinations

Depending on the way the smartness is measured, different models of smart cities and destinations arise, and the most important smart city indices were analyzed (Table 1).[1] *Cities in Motion* (IESE, 2019) is the one with the greatest geographical coverage, analyzing 174 cities (79 of them capitals) from 80 countries. *Smarter Cities 2025* (ESI, 2018) is a report produced by a coalition of consulting firms, leaders and providers of smart city services, platforms, and products. A smart city maturity score was awarded to 136 cities throughout the world, being able to be categorized as "Beginners", "Transitioning", or "Leaders", and according to these categories, it states different application premises for each situation state. The *Smart City Index* (IMD, 2019) ranks the performance of the 102 smartest cities in the world based on the analysis, first from a structural perspective and then from a technological perspective, of the set of selected dimensions.

Table 1. Twenty smartest cities by index

Ranking / Index	Cities in Motion	Smart City Index	Smarter Cities 2025*
1st	London	Singapore	Boston
2nd	New York	Zurich	Chicago
3rd	Amsterdam	Oslo	Copenhagen
4th	Paris	Geneva	Edinburgh
5th	Reykjavík	Copenhagen	Gothenburg
6th	Tokyo	Auckland	London
7th	Singapore	Taipei City	Oxford
8th	Copenhagen	Helsinki	Paris
9th	Berlin	Bilbao	Rome
10th	Vienna	Dusseldorf	San Francisco
11th	Hong Kong	Amsterdam	Shanghai
12th	Seoul	San Francisco	Sydney
13th	Stockholm	Vancouver	Tel Aviv
14th	Oslo	Sydney	Tokyo
15th	Zurich	Toronto	Vienna
16th	Los Angeles	Montreal	Yinchuan
17th	Chicago	Vienna	Zurich
18th	Toronto	Bologna	Singapore
19th	Sydney	Prague	Amsterdam
20th	Melbourne	London	Toronto

Source: Elaborated by authors based on IESE (2019); IMD (2019); ESI (2018)

Note: * As there are no ranking positions in the index, the first 17 cities are included into "leaders" category alphabetically ordered. Singapore, Amsterdam, and Toronto belong to the "transitional" category.

Even though different indices were presented, none of them take in consideration accessibility as one of the measurement variables, and they barely mention tourism. This in itself indicates both the ablest normative understanding of cities that dismisses the UN CRPD and its national signatory obligations together with excluding all major world economic indicators. The touristic factor is a small section of the "International Projection" topic in *Cities in Motion* (IESE, 2019). This document points out that maintaining global projection involves improving the City's brand and its international recognition through strategic tourism plans, attracting foreign investment and representation abroad. The three indexes mentioned consider the issue "accessible" related to the access to public information, or others to governance concerns but not to its disabled citizenry for international visitors.

When considering the incorporation of accessibility and tourism into smart scenarios, we found two models applied to different countries in the last years (Table 2). The first one is the *Smart Cities for all,* created byG3ICT and World Enabler (2017). It is one of the first projects that analyses the relationship between accessibility and smart cities and considers the smart cities capabilities (infrastructure, transportation, citizen engagement, public safety) to make them more accessible for PwD. The project started in 2017 considering the inclusion of PwD but nowadays it also includes solutions for elderly people. The toolkit is identified by global experts as solutions to the barrier in the digital inclusion of persons with disabilities and older persons in Smart Cities. Those tools are open to be downloaded for everybody and can be used for different organizations related to Smart Cities, including government managers, policy makers, IT professionals, disability advocates, procurement officials, technology suppliers, and developers who design smart city apps and solutions. Also, they show the results of the surveys made to experts around the world, which can help in different processes of decision taking of the toolkit users.

The second one, the *Smart Tourism Destination* (STD), is modelled by the State Society for the Management of Innovation and Technology Tourism of the Ministry of Industry, Energy, and Tourism of Spain (Segittur, 2015). It is based on UNE 178.501 standard (Management System for Intelligent Tourist Destinations). The STD model is based on concepts around innovation, technology, sustainability,

Table 2. Accessibility in Smart City and Smart Tourism Destinations' models

	Accessibility into smart tourism destination models	
Model	Smart Cities for All (SC4A) (G3ICT, 2019)	Smart Tourism Destination (Segittur, 2015)
Data collection	400 surveys to international experts, interviews, and secondary data sources	National data taken from the municipal registers that each city provides
Methodology	Seven guidelines: Inclusive Innovation Playbook, Communicating Digital, Inclusion in Cities, Solutions for Digital Inclusion in Cities, Adopting an ICT Accessibility Policy, Implementing, Accessibility Standards, and Global Strategy for Digital Inclusion	Application of Smart Tourism Destination Technical norms: UNE 178501 & UNE 178502.
Number of dimensions	One: Accessibility	Four axes: technology, innovation, accessibility, and sustainability.
Accessibility variable	All the documents are based and aimed towards application of accessibility on cities	One of the 4 main axes on which smart tourism destinations are based
Cases	Europe, Australia, Central & South America Asia, Middle East; Africa.	27 cities within Spain

Source: Elaborated by authors based on G3ict & World Enabler, 2019; Segittur, 2015.

and accessibility which takes into consideration "the adaptation to the needs of people with any type of disability, as well as children, pregnant women, the elderly, and the sick".

Review of good practices or initiatives on accessibility in tourism and intelligence of the world

Best practices or initiatives in accessibility in tourism have been selected to show exemplary and innovative solutions in sustainable and accessible tourism development as well as innovative and inclusive digital tools to enhance the visitor's experience (Table 3). *European Capital of Smart Tourism* is a model taken by the European Commission since 2018, which rewards European cities and gives them the opportunity to share their practices as smart tourism destinations in one of four categories: accessibility, digitalization, sustainability, creativity and cultural heritage. Each city that wants to be taken in consideration will be asked to write four essays to present its best practices and measures implemented under each of the categories previously mentioned. A group of experts compared and selected two winners as "European Capital of Smart Tourism" and other four cities were recognized with European Smart Tourism Awards for their outstanding achievements in the individual categories of the competition.

Table 3. Smart accessible initiatives

Initiative	Accessibility	City / Destination
Best Practices	Physical accessibility as a new normal	Málaga (Spain)
	Barrier free digital accessibility of the city's offers	Helsinki (Finland)
	The smart city cards	Copenhagen (Denmark), Ljubljana (Slovenia), Lyon, (France), Tallinn, (Estonia)
	Visitors with disabilities	Brighton (United Kingdom), Málaga (Spain), Valencia (Spain), Tallinn (Estonia), Lyon (France)
	Personal city helpers'	Helsinki (Finland), Lyon (France), Nice (France)
	A collaborative platform for accessibility in tourist destinations – 'Tur4all programme'	Valencia (Spain)
	Using innovative signage in historic sites	Lyon (France)
	Annual Accessibility Guide	Nantes (France)
	Best practice in UNWTO accessible Tourism manuals	Many cities highlighted
Awards	Category: Accessibility (2019)	Málaga (Spain)
	Category: Accessibility (2020)	Breda (Netherlands)
	Melbourne	Sydney (Australia)

Source: Elaborated by authors based on EC, 2020

One of the main contributions to identify constraints for PWD when doing tourism activities in Europe, was the study named "Mapping and Performance Check of the Supply of Accessible Tourism Services in Europe" (EC, 2015). That was an important precedent in the accessible tourism legacy of the region because it allowed working on the improvements in that field. Considering the region of

Europe as an example that other regions often imitate, the chapter enquiries into other regions such us Latin America or Oceania.

Different from Europe, the region of Latin America is characterized by a high level of inequality (Abramo et al., 2020). The overage rate of disability indicates that 12% of total population with some form of disabilities; 5.4% in the Caribbean and 12.4% in the rest of Latin America (ECLAC, 2013). At the same time, the economy of those countries has a high dependency on tourism representing 26% of total GDP in the Caribbean and 10% in Latin America, respectively (ECLAC, 2020). Both disability and the tourism sector are two important issues for the region that mean to be studied in depth.

In this scenario, accessible tourism has been studied and developed by countries for the last decades (Rucci, 2018), however, smartness has incipient participation in the region. In this line, the Inter-American Development Bank (IDB) has been preparing action plans for municipalities in Latin America and the Caribbean since 2011 in order to support the development of urban sustainability strategies for the transformation into Smart Cities. The IDB's Emerging and Sustainable City (ICES) initiative has a cross-cutting approach that is based on three pillars: urban sustainability, environmental and climate change sustainability, and fiscal sustainability and governance. ICES initiative works in cities with populations between 100,000 and 2 million inhabitants. This non reimbursable technical assistance program provides direct support to central and local governments in the development and execution of urban sustainability plans. In 2017, the initiative had 77 cities in the region with which it is currently working. More recently, the following stand out: Santiago de Chile with practices to promote electric transport or smart meters; Buzios (Brazil) with the use of LED street lighting technologies; Bogotá (Colombia) with the pioneering implementation of the mass public transport system; Montevideo (Uruguay) with the 'Bus Rapid Transit' that has become the largest free software exporter in LATAM (Bouskela et al., 2016).

Autonomous City of Buenos Aires (CABA, for its acronym in Spanish) is doing its first steps in Smart Tourism Destination through the implementation of the Tourism Intelligence System (Esper, 2020). In Latin America, it is considered as the smartest city (Ospina, 2018).

CASE STUDIES

In this section, the study is focused at city level taking Sydney (Australia) and CABA (Argentina) as study cases, considering that both cities have interesting backgrounds in accessible tourism (Porto et al., 2019) and are in different stages of smartness development (ESI Smarter Cities 2025, 2018).

This is not the first time that Australia and Argentina have been studied from the accessible tourism perspective, but it is the first time that the analysis is based on the local level. At country level, through a nationally comparative tourism accessibility measure (Tourism Accessibility Index, Porto et al., 2019), the results indicate that Australia has the highest accessibility value of the index for tourists with disabilities in the period of analysis (1990 to 2015). Argentina ranked second in 1990 but lost its position after 2001, as a consequence of the major social-economic crisis in that year. Both countries have a strong political will in working on the inclusion of people with access needs in tourism. However, in the smart city field, while Australia has been committed to that challenge for a long time, Argentina is starting to get involved in the problem. Although Argentina and Australia have been studied individually from an accessible tourism perspective before, it is interesting to compare their advances with regard to the application of smartness to accessibility, considering their territorial particularities.

General characteristics related to demographic and tourism variables of Sydney and CABA as well as data from the corresponding countries, are presented in Table 4.

Table 4. Background indicators of the case studies, 2018-2019

Indicator	Australia	Sydney	Argentina	CABA
Total population (in millions)	25.20	5.30	44.56	3.07
Population with disabilities (% total population)	17.70	16.30	10.20	8.00
Territorial expansion (km^2)	7.60 million	12,368	2.70 million	203
Gross Domestic Product (USD per capita)	57,374.69	59,200.60	11,683.95	20,036.00
International tourist arrivals (in millions)	8.70	4.10	6.9	2.40
Average stay (in nights)	32	20	12	9
International tourist per 100 inhabitants	34.50	77.20	15.48	126.00
International tourism receipts (in millions USD)	30,858	6,925	5,563	1,914
Position of the study cases in the surveyed indices				
Cities in Motion (IESE, 2019)		19th		77th
Smarter Cities 2025 (ESI, 2018)		Leader		Transitioning
Smart City Index (IDM, 2019)		14th		87th

Source: Elaborated by authors based on Buenos Aires Tourism Observatory (2020); World Bank (2020); INDEC (2019, 2020); ENDI (2018); ABS 2018; TRA 2019

Sydney (Australia)

Australia has a long tradition in issues regarding people with disabilities, accessibility and, in particular, accessible tourism. The Commonwealth *Disability Discrimination Act* (DDA), 1992 [Comm], and the New South Wales government's *Disability Inclusion Act*, 2014 [NSW] are considered the settlements In particular, the new state-based provision requires all state and local governments to have an inclusion plan, hence, taking responsibility for disability and inclusion across all business units of the organization. In particular, the City of Sydney (Figure 2) represents a paradigmatic approach, according to its background, history and innovative related initiatives.

The *Sustainable Sydney 2030* (ex- Community strategic plan 2017-2021, and now under revision 2050)[2] refers to a strategic plan developed for the community "including people young and old, and people with disability" (City of Sydney, 2011). The plan considers sustainability as its main aim, but accessibility is also included and it expressed the community's vision and the city's commitment to the sustainable development of Sydney to 2030 and beyond (City of Sydney, 2017b).

Yet, the City of Sydney has not always had a model approach to disability and inclusion. From the awarding of the Olympics in 1993 and in the lead up to the Sydney 2000 Olympic and Paralympic games, the City of Sydney was challenged in the Federal Court for practices regarded as discriminatory under the DDA by the Physical Disability Council of NSW for four practices: sidewalk curb cuts that did not comply to Standards Australia AS1428.1-4; curb heights lower than the standard that created a greater gradient for disembarking low floor buses; the systematic installation of streetscape furniture on the

Figure 2. The City of Sydney Boundaries

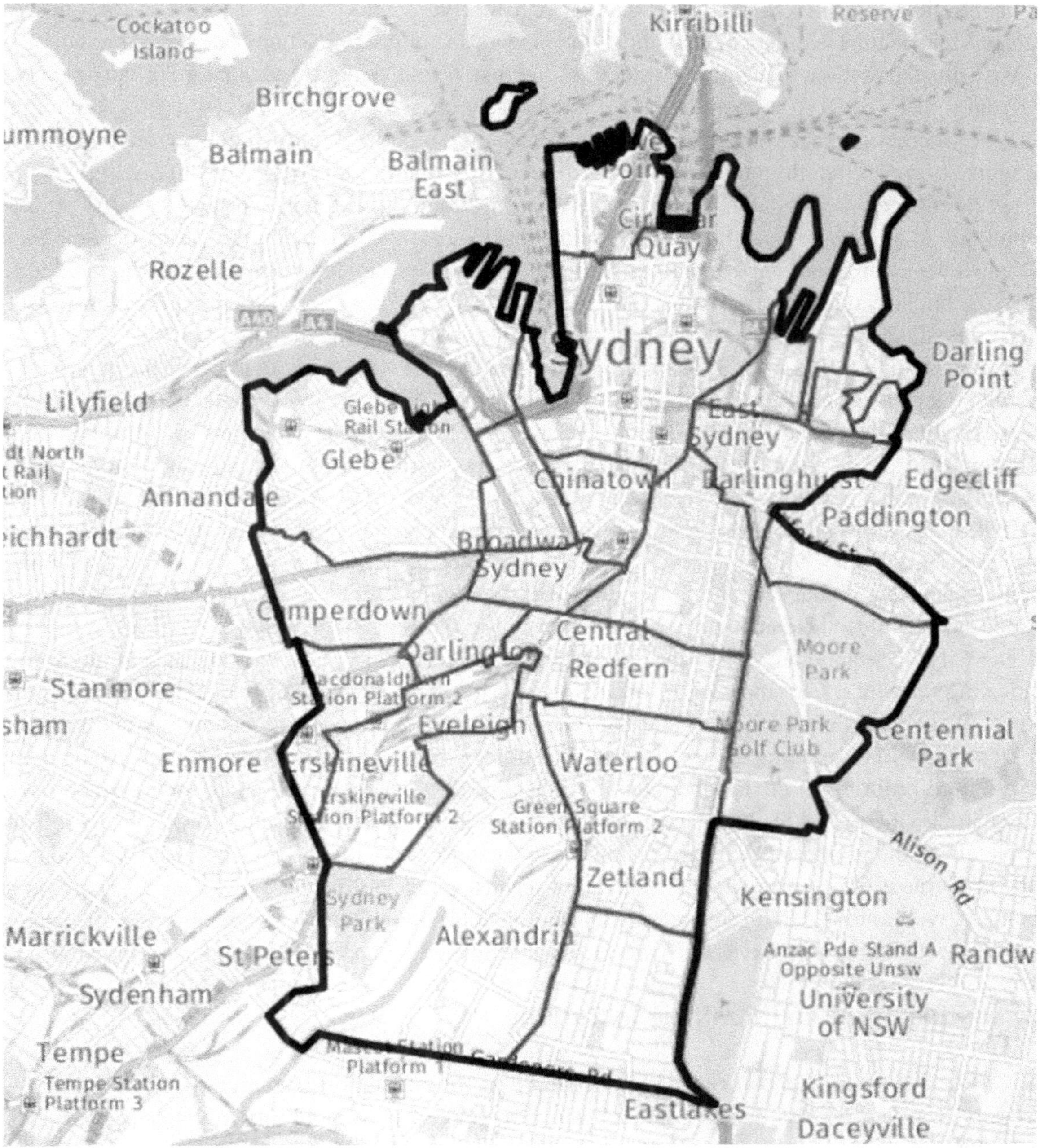

streets of Sydney; and the removal of an over street bridge linking buildings together that provided access for wheelchair users to a car parking station for drivers with disability (Cashman & Darcy, 2008; Darcy, 2001, 2003). Rather than an open dialogue with PwD, Sydney took an adversarial stance where the challenge to these actions went through legal processes. After Sydney 2000 had been completed, the City of Sydney's disposition mellowed, they employed a disability access officer in 2001, undertook the first access plan in 2002 established the City of Sydney Access Committee in 2003, and promoted a series of events to improve the accessibility of the City (City of Sydney, 2002). More recently Sydney has been praised for its inclusive practice on major developments and cultural practice (Darcy & Almond, 2016).

With this foundation, the Digital Strategy set "inclusion" as the highest reason for digitalization (City of Sydney, 2016 p.10). It is mentioned that amongst the greatest beneficiaries of digital technologies are PWD increasing access to information and enhancing mobility and the quality of daily life. They recognize however that digitalization must also recognize that those people who are not digitally connected or engaged are less able to experience the educational, workforce, and civic benefits of digital technology. While recognizing the 'digital divide' that many PWD find themselves due to a combination of their lower socioeconomic status, partly due to much lower employment rates and the need for assistive technology that comes at a cost (Alper et al., 2015; Darcy et al., 2019; Goggin & Newell, 2005; Yerbury et al., 2019), it does not suggest any strategies to alleviate the divide.

The Inclusion Plan (City of Sydney, 2017a) outlined a series of actions to improve the quality of life of PwD in the city. It includes creating opportunities for social connection and engagement and improving the quality of personal, social, and business interaction, that enhance people's social and emotional well-being. In this way, the inclusion plan synchronizes with both the tech start-up community, the resident's well-being, and workforce documents (City of Sydney, 2012, 2015), while tourism and the Digital Strategy are the most important considerations to integrate. With this development, most of the ingredients are in place for developing a more sophisticated economic development strategy that builds upon Sydney as a global city and at the same time recognizing the importance for local initiatives to upskill individuals to take advantage of the opportunities present (City of Sydney, 2019a).

The City of Sydney has been putting into place this series of policy documents building towards embracing the Smart City movement. In a series of City Talks, the City of Sydney promoted its place as part of the world's global cities and the importance of benefiting from technology and smart city developments. In developing its own Smart City Strategic Framework, it sought to have five outcomes: a city future-proofing its environment and bolstering resilience; connected and empowered communities; shaping vibrant and livable places; a global economic competitiveness and attracting and retaining global talent; and providing efficient, customer-centric services (City of Sydney, 2020a). Central to this is the importance of the visitor economy and how it can leverage off the vibrancy of the host community, a competitive economic environment, making sure that its environment is sustainable and resilient, and liveable through efficient services. However, is there a place for accessibility and accessible tourism?

Accessible Tourism Initiatives in Sydney

The City of Sydney is the most crucial player concerning the integration of smart city considerations with accessible tourism. This is most prevalent through the three key accessibility initiatives on its website that draw upon parking, accessible public transport, and accessibility map. Each initiative is based on a geographic information system approach to identify the three considerations together with accessible public toilets. This aligns with what some have called the three Ps of accessibility: parking; accessible public toilets; and public transport (Darcy, 2003). For example, Figure 3 identifies accessible on-street and parking stations that are integrated through Google Street View that makes planning a day to the city easier for those who are unable to use public transport and have to rely on a private modified vehicle (Darcy & Burke, 2018).

Figure 3. Accessible Parking map with pop-up details and integrated with Google Street View

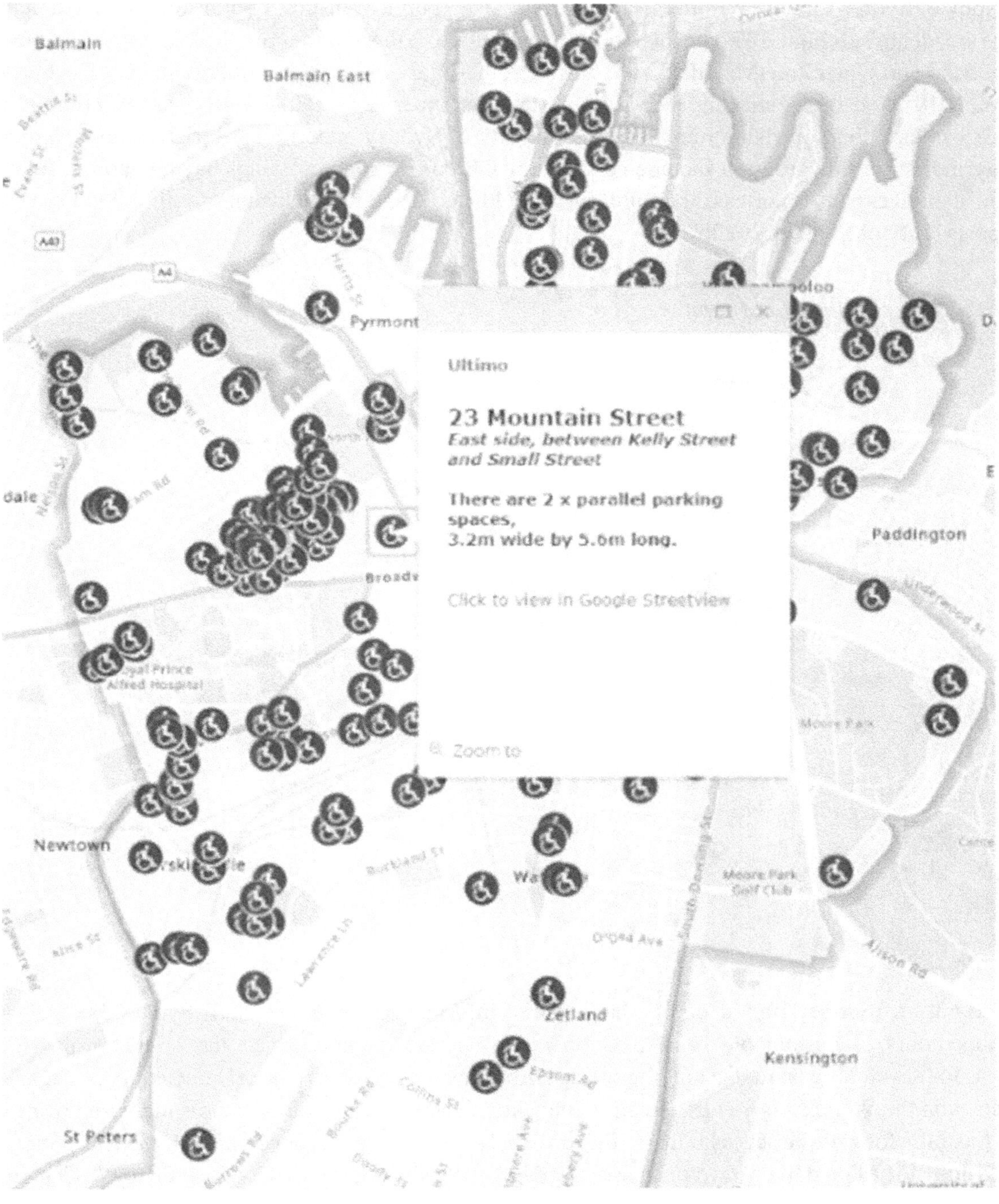

The Accessibility Map provides a visible wayfaring system for negotiating the sometimes-steep topography of Sydney streets. Within the accessibility map, public transport options, parking, and public toilets are identified (Figure 4). The interactive map brings together data sources from the National Public Toilet Map and OpenStreetMap data, also links to other maps and apps including the Sydney For All (NSW, 2019) website identifying accessible destination experiences. However, maps are no good for people who are blind or vision impaired, and the City of Sydney undertook a series of improvements of wayfinding for this group. It included all Sydney CBD pedestrian crossings having audible signals, the substantial expansion of tactile ground surface indicators and the inclusion of rail on Sydney street signposts (City of Sydney, 2020b).

Figure 4. Accessibility Map Features

The City of Sydney provided very little tourist information for PWD, until Sydney For All that developed out of a Sustainable Tourism Cooperative Research Centre project that sought to provide a process for developing an understanding of accessible destination experience (Darcy et al., 2008a). The project won the World Leisure Organization innovation prize amongst other awards including from Vision Australia for website accessibility as the tourism NSW website was not compliant to W3C protocols and Sydney for All had to be developed as a standalone website (Darcy, 2013, 2014; Darcy & Cameron, 2009a, 2009b, 2013; Darcy et al., 2008b). One of the missing links of that website was the accommodation module. Sydney, through its community grants, funded a project to operationalize research on the information collection, marketing, and distribution of accessible hotel accommodation (Darcy, 2010, 2011; Darcy & Cameron, 2009/2010). This pilot project of 15 hotels for accessible rooms in the Sydney CBD is still available on the Easy Access Australia website.

While suitable for its time, Sydney for All is now outdated and has not been continuously updated by Destination NSW. One such opportunity has been to nurture the tech start-up community, which the Ultimo Precinct, in particular, has become part of a tech start-up hub (City of Sydney, 2019b). We are now seeing the development of start-ups, explicitly focusing on disability tech solutions, and some of these are focusing on accessibility and accessible tourism. We will now briefly review two of these start-ups and their focus.

BindiMaps is a disability tech solution for people who are blind or vision impaired wayfinding indoors. Historically, GPS systems do not work indoors, and BindiMaps provides a solution through hardware and software. Through the use of bluetooth beacons, sensors within the smartphone, and installed guidance system synchronised with audio and other technologies, BindiMaps provides an indoor navigation system for people with vision impairments or who are blind. The added advantage of the system is that it offers all those advantages for people who do not have a vision impairment and is, in essence, a universal

Figure 5. Example of Wheeleasy Manly Visitor Information Centre listing

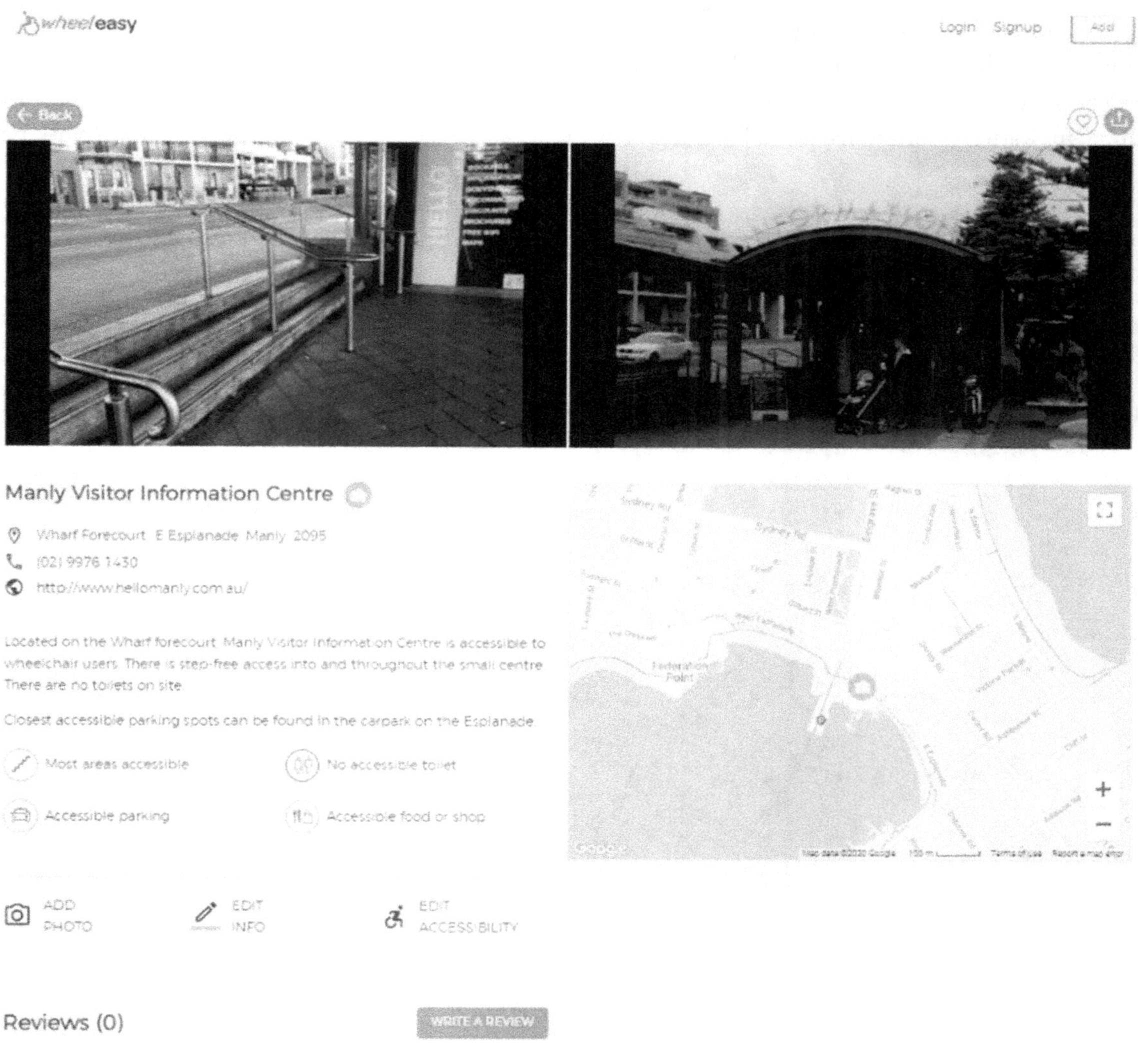

navigation tool for indoors. Not surprisingly one of the first organizations to install the system was Guide Dogs NSW/ACT and operates in another 20 centres across the country including tourist attractions.

WheelEasy is a website and app that has taken off from where Sydney for All started and really developed on steroids. The underlying philosophy is through detailed information provision and access icons, providing enough information on which to allow consumers to make an informed choice as to whether the listing will be accessible to them. While Sydney for All did this across disability access types (mobility, vision, hearing and cognitive access), WheelEasy focuses predominantly on mobility access. Certainly, in today's age of social media, the site provides opportunities for users to add photos, edit information and provide further information about the relative accessibility. It also has a review feature. While the app has started in Central Sydney it is expanding geographically. The site has many signifiers of Max Burt's previous career in advertising with a high-quality look, crisp feel and plenty of white space. Figure 5 provides an example of a listing and associated information.

Autonomous City of Buenos Aires (CABA, Argentina)

The topic of accessibility in Argentina has been included in national and provincial laws and in certain municipal regulations, seeking compliance with the rights of PwD in different areas, which impact directly and indirectly into the accessibility chain of tourism destinations. Although Argentina is the only country in South America that has a National Law of Accessible Tourism (Law 25.643, 2002) until now,

Figure 6. CABA Communes Map

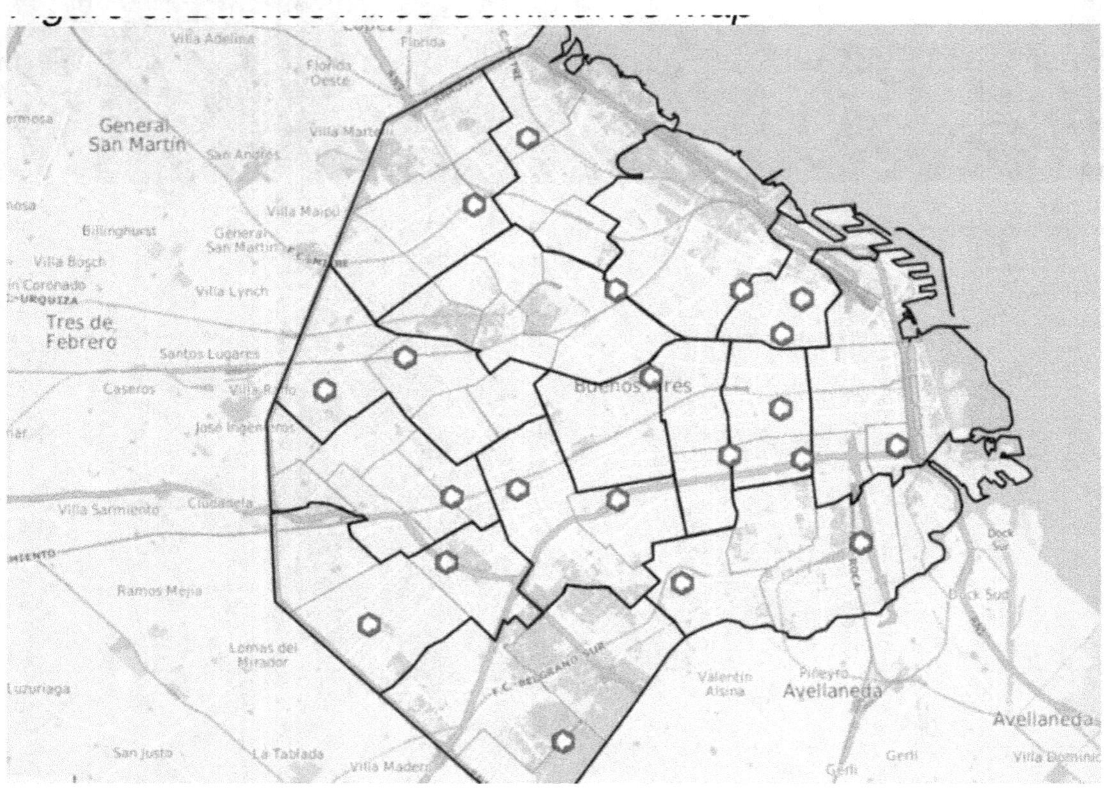

it is pointed out that the Argentinian legislation for accessible tourism was never truly regulated (Rucci, 2018; Casanova Ferro, 2011).

At national level, a goal in the tourism accessible sector was the creation of the Accessibility Guidelines for Tourist Accommodations in 2008 and the Accessibility Guidelines for Tourism Services in 2011. They not only allow the standardization of the accessibility criteria of the tourism sector services but also to set common expectations to tourists and tour operators when the place or service has an Accessibility Distinction from these Guidelines.

While Argentina has been working on accessibility for more than a decade, the experience and involvement in the smartness field is recent. A great step was taken by the country through the creation of the Argentine Network of Smart Tourism Destinations (ANSTD) (Instituto Ciudades del Futuro, 2020), taking as basis the Spanish model (Segittur, 2015) and adapting it to the contextual and geographical characteristics of the country.[3] Although the Smart Tourism Destination model is actually in an incipient stage of development in Argentina, the case of CABA is a leading example at the country level (Figure 6).

The city stands out in terms of legislation with Law No. 962 (2003) "Physical accessibility for all", Law No. 1080 (2003) "Reservation of parking spaces", Law No. 66 (2005) "Menu in Braille System", Law No. 2510 (2007) "Permission to access transportation and public places with an assistance dog", among others.

Figure 7. Accessibility references - Accessible Tourism Guide

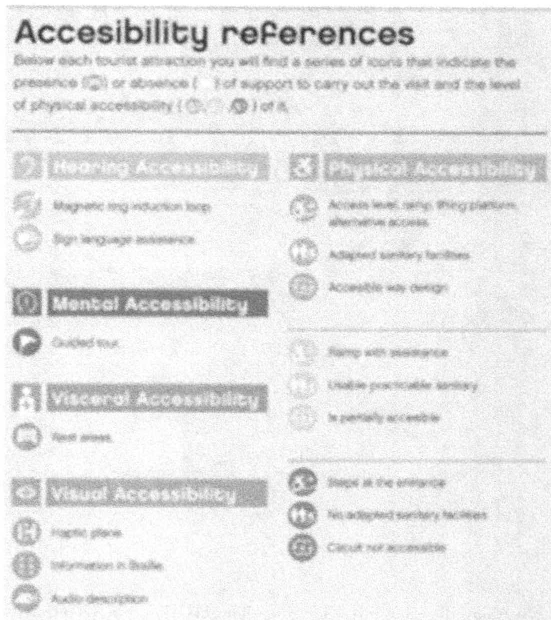

Accessible Tourism Initiatives in Buenos Aires

The Buenos Aires City Tourist Board has been working on tourism accessibility policies from 2016 being some of its goals that CABA has the highest number of accessibility distinctions in the tourism sector of the country.

Regarding accessibility in the provision of information, although the Buenos Aires City Tourist Board website is not accessible, CABA is on the tenth edition of the Accessible Tourism Guide (Figure 7) in which information about accessibility conditions of tourist resources, services and facilities are provided in a clear and simple format. Easy reading, easy to understand with maps and photographs incorporated to facilitate the location and identification of tourist places of interest. Also, the Guide is in several languages (Spanish, English and Portuguese), in audio description and in Argentinian Sign Language (ALS). Another online initiative led by the Tourism Board is an online map of tourist establishments in the city which shows the different accessibility conditions they meet (Figure 8)

Figure 8. Interactive Map of Accessible Places

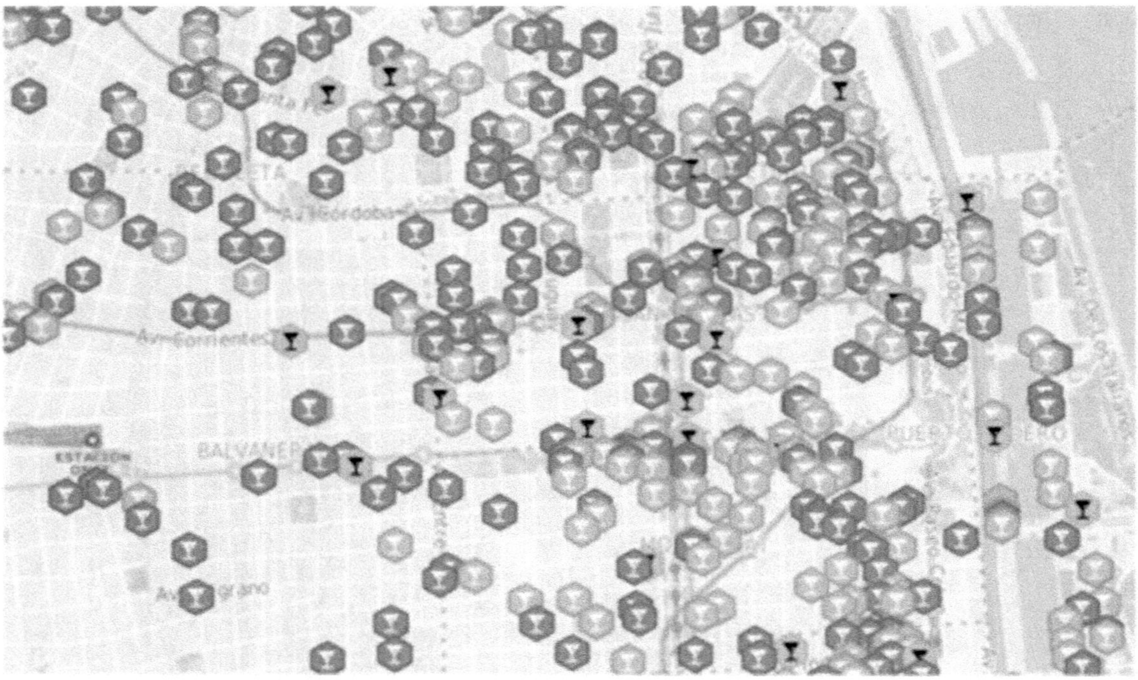

As for transportation and terminals, the city has accessibility conditions at the train stations and eight corridors of the Metrobus are 100% buses low floor (Rucci & Ravlic, 2020). The two city-tour buses are also accessible too. One of the initiatives led by the Tourism Board is an online map of tourist establishments in the city which shows the different accessibility conditions they meet (Figure 9)

One of the most outstanding buildings in 2019 for its accessibility was "*Jardin Japones*" (Japanese garden), which made a new haptic map for people with visual disability, low vision and the elderly (Figure 9). Its restaurant was also recognized by Accessibility Guidelines, given that it has a Braille and macro-type menu, adapted tables, accessible toilets and an elevator (L. Ghezzi, personal communication, 6 June 2020).

Figure 9. Haptic Map 3D

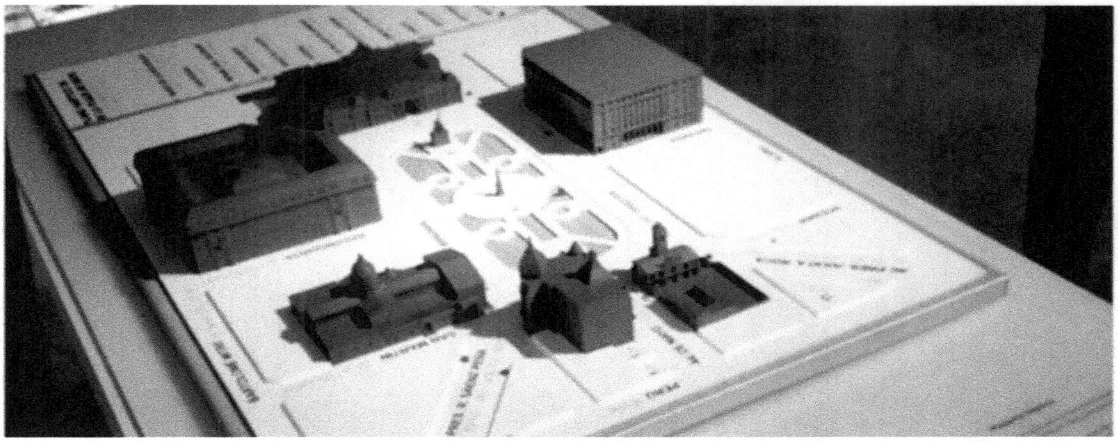

Two main challenges are faced by the City. The first one is related to transport in all of its forms: for example, only 50% of the subway stations have accessibility conditions; lack of ramps to go down in wheelchairs of the cruises in the port; absence of accessible taxis to connect the different arrival and departure terminals with accommodation, attractions and tourist services. The second challenge is the relationship between accessibility and technological aspects. Since 2019, Buenos Aires City Tourism Board in interaction with the Tourism Observatory of CABA started to survey tourists with disabilities at tourist assistance centres and cruise terminal stations to collect data about tourists with disabilities and their satisfaction in the city (L. Ghezzi, personal communication, 6 June, 2020).

Regarding smartness, CABA created the Tourism Observatory in 2014. The aim of this institution is to: i) measure the economic impact of tourism activity, job creation and investments, projections, and positioning with respect to the national and international market; ii) obtain information for the generation of new products (for example, tourists who attend congresses, or tourists from cruise ships); and, iii) obtain information through a Tourism Intelligence System (Figure 10) with Big Data, about air connectivity, hotels, gastronomy, tourist mobility, most visited tourist attractions, use of credit cards, among others, is obtained. The Tourism Intelligence System uses innovative tools to enhance the experience of tourists throughout the study of tourist behaviour and generating information to other areas of government to the private sector (companies and entrepreneurs) and academia.

As for mobile applications that facilitate tourism information for PwD, there is one called "Accesapp" that has the most relevance in the field, both in the City and nationally. Accesapp is a mobile platform that provides accessibility information in more than 500 tourism places and services and related services in Argentina (Figure 11). Each provider of tourism services can access it through a membership (monthly or annual) meeting certain accessibility criteria and commitment to inclusion. Users can comment and rate the place, service or establishments according to their experience and level of accessibility perceived. That information helps other people about accessibility conditions of the tourism sector.

Figure 10. Capture of the Tourism Intelligence System website

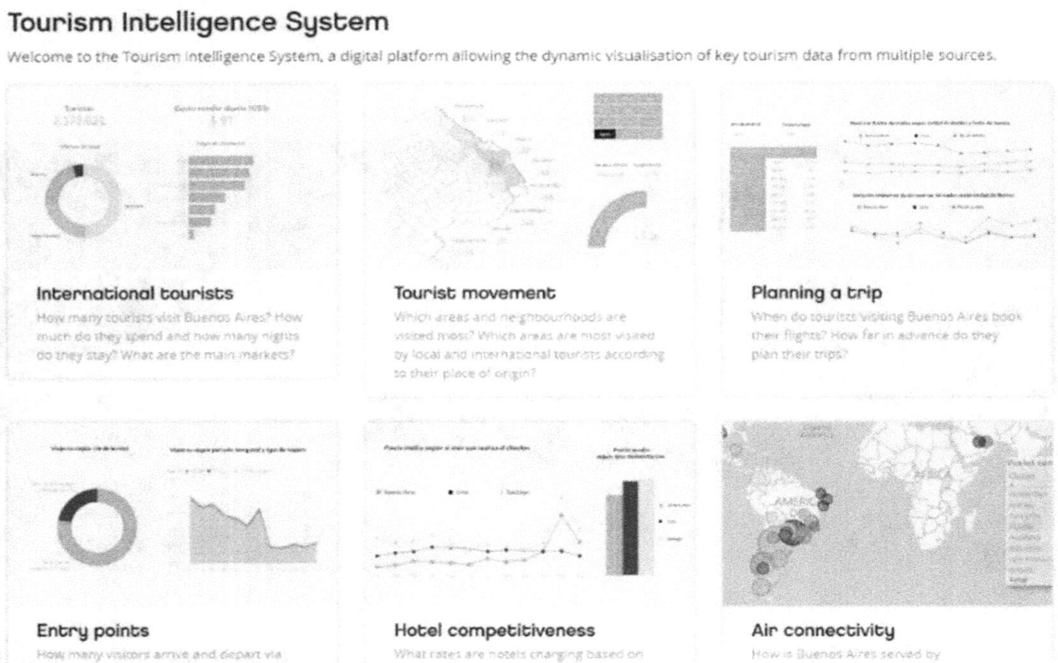

Figure 11. Captures of Accesapp Menus

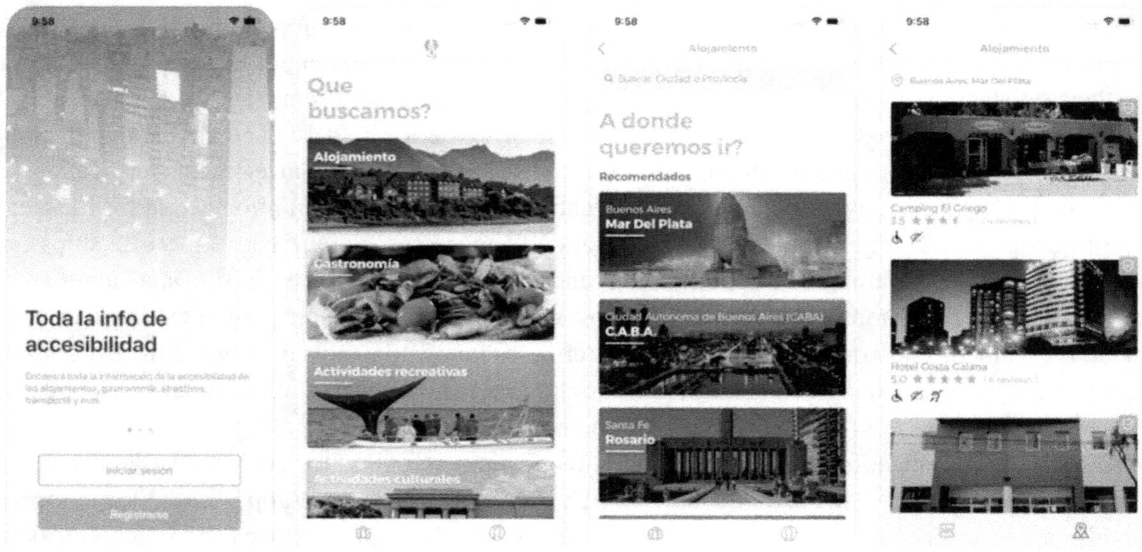

Table 5. Main differences between the study cases

	Sydney (Australia)	CABA (Argentina)
National level	Solid background on PWD rights, accessibility issues and accessible tourism at national level	Incipient background at national level Argentine Network on Smart Tourism Destinations
Local level	Independent initiatives at the local level, complementing the national perspective: Sustainable Sydney 2030; Digital Strategy	Initiatives at the local level based on national perspective: Accessibility Guidelines for Tourist Accommodations (2008) and for Tourist Services (2011)
Planning	Specific initiatives: Inclusion Plan, City Talks, Sydney for All	Non-specific initiatives: Buenos Aires Tourism Board, Tourism Observatory
	Working on an own Smart City Strategic Framework	Tourism Intelligence System (2019)
Accessible Tourism Initiatives	Accessible Parking map Accessible Map	Interactive map of accessible places
	BindiMaps	Jardin Japones's haptic map
	WheelEasy	Accessapp
Future actions	Consolidate strategies taking into account PWD, accessibility, digital and smartness issues	Design a specific plan or programs for PWD and accessible tourism
	Update Sydney for all	Create a similar platform for CABA
	Replicate and incorporate more sophisticated start-ups technologies	Incorporate start-up technologies

Source: Elaborated by authors.

DISCUSSION

The new paradigm in the world of modern tourism recognizes accessibility and technological issues as the pillars over which a successful tourism destination must emerge and consolidate. Disability, ageing, accessibility, technological issues and smart conditions are critical and fundamental concepts that academia should study both theoretically and empirically to provide policy makers with useful insights to promote smart accessible tourism destinations. This chapter provides a review of smart tourism models developed around the world (Cities in Motion, Smart City Index and Smarter Cities 2025) and goes a step further in identifying the inclusion of accessibility on them (Smart Cities for all, Smart Tourism Destination). This analysis gives some insights to develop and discuss the particularities of the selected case studies.

This chapter aims to discuss disability and accessibility, and their relationship with technology trends, taking into account a world full of different kind of constraints, taking Sydney (Australia) and Autonomous City of Buenos Aires (Argentina) as case studies, considering that both cities are well-recognized touristic cities all over the world, have interesting backgrounds in accessible tourism (Porto et al., 2019) and are in different stages of smartness development (ESI Smarter Cities 2025, 2018). Regarding accessibility and smartness, Sydney has been involved in the problematic for a long time while CABA is in an emergent stage of development, learning from more experienced and traditional destinations.

From an international point of view, Australia and Argentina have as accessible tourism as smartness backgrounds at different levels of development. While tourism is considered a great indicator in terms of arrivals and income in both cases, the results of Tourism Accessibility Index (TAI) indicate that Australia has a higher value for tourists with disabilities than Argentina (Porto et al., 2019).

While Sydney ranks within the 20 smartest cities in the world, CABA is in the 80th place. However, one promising feature is that CABA is positioned in first place in regional terms, that is, in Latin America. Both cities, each of them into their regional contexts, are working on the development and implementation of smartness, being aware of the contribution of tourism and the smartness to the 17 Sustainable Development Goals (SDGs) of the 2030 agenda of the United Nations (UN, 2015) and the problems that smartness, as an innovative and challenging issue, probably will face. Each of these cities, with a clear position and role, Sydney as a leader and CABA as a transitioning, are good examples of the way in which a destination should implement critical dares to create a strategy of innovative and intelligent tourism for all.

From a theoretical point of view, both cases studied are in line with the idea of a model for a smart tourism destination. While Sydney has been working for years on the inclusion of an integral plan of accessibility with different approaches, CABA has made some progress since 2019, following the Spanish model (Segittur, 2015). Concerning accessibility, Sydney has been working hard to overcome the existing barriers in the city (Darcy, 2001, 2003) specially related to the preparation of the city to hold Olympic and Paralympic Games during the late 90's. Afterwards, the challenge of digitalization emerged and, moreover, the challenge of the couple accessibility-digitalization emerged, when Sydney discovered that the Digital Strategy have to take into account that PWD are one of the main beneficiaries of this strategy, through increasing access to information, enhancing mobility and improving the quality of daily life" (City of Sydney, 2016). The materialization of Sydney going through accessibility is the policy of the three Ps of accessibility: parking; accessible public toilets; and public transport.

Looking into CABA, accessibility is not in policymakers` minds until late 2000, more than ten years after Sydney. A National Law of Accessible Tourism (year 2002) is the background and the point of departure of a promising stage of development on the field. However, while CABA has improved a lot in terms of digitalization, there is still a lack of relationship between intelligence and accessibility policies. Although CABA is working on the access to information and the incorporation of technology in decision-making for PwD, it does not have a tourism website or mobile apps accessible for PWD, what is considered as "digital divide".

"A City for All: Inclusion (Disability) Action Plan 2017-21" is another issue to be highlighted in the accessibility commitment of Sydney because it focuses on creating a liveable community more than just increasing physical accessibility. Through that Plan, Sydney understood the importance of adopting a city under the universal design for all its citizens, including PWD and seniors, and its collateral benefits for tourists. In line with this purpose, during the period 2016-2018, more than 20 thousand ramps were created in CABA, which means that 7 out of 10 corners of the city are accessible. This shows the stage that CABA is working on, focusing on physical accessibility instead of in an integrated policy.

Our case studies are a clear example of a leader and a transitioning on the game of studying, developing, implementing, and evaluating strategies and policies of accessible tourism smart destinations. A leader city that has to consolidate and take care of the achievements of more than 30 years of continuous work. A transitioning city that has to learn and take advantage of successful experiences worldwide.

Both cities are making their way in the smartness path of development and are conscious of the accessibility feature of tourism in their planning strategies. Sydney is an example of a city that has not only worked on accessibility policies but also on its technological implementation, considering PWD as the primary beneficiaries. Buenos Aires has a lot to learn from Sydney, mainly to not being smart without being accessible.

FUTURE RESEARCH

This research enhances the current understating of the technical challenges involved in creating more inclusive, accessible, innovative, and smart tourism destinations. This allows the identification and implementation of different initiatives to give powerful insights to researchers, stakeholders, travel agents and tour operators, policy-makers and Destination Management Organizations (DMO) to adapt to this promising market niche, learning to deal with the daring features of the modern, independent, and more demanding tourism. Also, this research is a first approach to the comparison of case studies from different countries and continents, considering globally recognised examples of smart accessibility, in which policy-makers and the tourism industry are still working on their implementation and improvement. Some cities learn from their past experience while others may capture the insights from the other. There are great opportunities for future research. The determinants of the effectiveness of smart accessibility initiatives and a focused policy based on the needs and motivations of PWD as tourists are on the agenda.

REFERENCES

G3ict & World Enabled. (2017). *Smart Cities for All: Results of the 2016 Global Expert Survey.* Retrieved from: https://smartcities4all.org/wp-content/uploads/2017/06/Smart-Cities-for-All-A-Vision-for-an-Inclusive-Accessible-Urban-Futur...-min.pdf

Abramo, L., Cecchini, S., & Ullmann, H. (2020). Addressing health inequalities in Latin America: The role of social protection. *Ciencia & Saude Coletiva*, *25*(5), 1587–1598. doi:10.1590/1413-81232020255.32802019 PMID:32402029

ABS. (2018). Disability, Ageing and Carers, Australia: Summary of Findings. Australian Bureau of Statistics.

Accessible Tourism Guide. (2019). Retrieved from https://www.buenosaires.gob.ar/sites/gcaba/files/c0_en_opt.pdf

Allwinkle, S., & Cruickshank, P. (2011). Creating smarter cities: An overview. *Journal of Urban Technology*, *18*(2), 1–16. doi:10.1080/10630732.2011.601103

Alper, M., Ellcessor, E., Ellis, K., & Goggin, G. (2015). Reimagining the Good Life with Disability: Communication, New Technology, and Humane Connections. In H. H. Wang (Ed.), Communication and the "Good Life" (pp. 197-212). New York: International Communication Association Theme Book Series, vol 2 - Peter Lang.

Aydemir, B., & Kılıç, S. N. (2017). Dünyada ve Türkiye'de Üçüncü Yaş Turizmi. *Yüzüncü Yıl Üniversitesi Sosyal Bilimler Enstitüsü Dergisi*, *1*(3), 1–11.

Baggio, R., & Cooper, C. (2010). Knowledge transfer in a tourism destination: The effects of a network structure. *Service Industries Journal*, *30*(10), 1757–1771. doi:10.1080/02642060903580649

Boes, K., Gretzel, U., Zhong, L., Koo, C., Buhalis, D., & Inversini, A. (2016). Smart tourism destinations: Ecosystems for tourism destination competitiveness. *International Journal of Tourism Cities*, *2*(2), 108–124. doi:10.1108/IJTC-12-2015-0032

Bouskela, M., Casseb, M., Bassi, S., De Luca, C., & Facchina, M. (2016). *The Road toward Smart Cities: Migrating from Traditional City Management to the Smart City*. Retrieved from: https://publications.iadb.org/publications/english/document/The-Road-toward-Smart-Cities-Migrating-from-Traditional-City-Management-to-the-Smart-City.pdf

Buenos Aires City. (2019). *Mapa Interactivo de lugares accesibles*. Retrieved from: www.buenosaires.gob.ar/copidis/accesibilidad/mapa-interactivo-de-lugares-accesibles

Buenos Aires City. (2020). *Communes Map*. Retrieved from: https://mapa.buenosaires.gob.ar/mapas/?lat=-34.620818&lng=-58.366615&zl=11&modo=transporte&map=comunas

Buhalis, D., & Amaranggana, A. (2013). Smart Tourism Destinations. In Z. Xiang & I. Tussyadiah (Eds.), *Information and Communication Technologies in Tourism 2014* (pp. 553–564). Springer. doi:10.1007/978-3-319-03973-2_40

Buhalis, D., & Darcy, S. (Eds.). (2011). *Accessible tourism: Concepts and issues*. Channel View Publications.

Casanova Ferro, G. (2011). *Turismo: El camino de la Discapacidad a la Accesibilidad*. Retrieved from: https://www.academia.edu/594410/Turismo_El_camino_de_la_discapacidad_a_la_accesibilidad

City of Sydney. (2002). *Disability Action Plan 2002-2005*. Retrieved from https://www.cityofsydney.nsw.gov.au/AboutSydney/CBDDisabledAccess/Default.asp

City of Sydney. (2012). *City of Sydney workforce strategy 2012-2015*. Retrieved from: https://www.cityofsydney.nsw.gov.au/surveys-case-studies-reports/community-indicators-report

City of Sydney. (2015). *City of Sydney Wellbeing survey 2015*. Retrieved from: https://www.cityofsydney.nsw.gov.au/surveys-case-studies-reports/community-indicators-report

City of Sydney. (2016). *Digital Strategy* Retrieved from: https://www.cityofsydney.nsw.gov.au/__data/assets/pdf_file/0005/288167/Digital-Strategy.pdf

City of Sydney. (2017a). *A city for all: inclusion (disability) action plan 2017-21*. Retrieved from Sydney:https://www.cityofsydney.nsw.gov.au/__data/assets/pdf_file/0003/287121/Inclusion-Disability-Action-Plan-2017-2021_Accessible.pdf

City of Sydney. (2017b). *Sustainable Sydney 2030: community strategic plan 2017-2021*. Retrieved from Sydney: https://www.cityofsydney.nsw.gov.au/__data/assets/pdf_file/0011/288173/Adopted-Sustainable-Sydney-2030_Accessible-Version.pdf

City of Sydney. (2019a). *Economic development strategy - Sydney's economy: global City, local action*. Author.

City of Sydney. (2019b). *Technology start-ups action plan*. Retrieved from: https://www.cityofsydney.nsw.gov.au/business-economy/tech-startups

City of Sydney. (2019c). *Accessibility Map*. Retrieved from: https://www.cityofsydney.nsw.gov.au/explore/getting-around/accessibility/accessibility-map

City of Sydney. (2020). *Sydney maps*. Retrieved from: https://www.cityofsydney.nsw.gov.au/learn/maps

City of Sydney. (2020a). *Smart City Strategic Framework.* Retrieved from Sydney: https://www.cityofsydney.nsw.gov.au/council/your-say/smart-city-strategic-framework

City of Sydney. (2020b). *Wayfinding signage (for blind and vision impaired).* Retrieved from https://www.cityofsydney.nsw.gov.au/vision/sustainable-sydney-2030/transport-and-access/liveable-green-network/wayfinding-signage

City of Sydney. (2020c). *Accessible Parking map*. Retrieved from: https://cityofsydney.maps.arcgis.com/apps/Minimalist/index.html?appid=c6d11d61dc994bce8715335e924f5151

Comisión Europea (2014). *Economic impact and travel patterns of accessible tourism in Europe – Final report.* Service Contract SI2. ACPROCE052481700 – European Commission, DG Enterprise and Industry.

Darcy, S. (2001). A games for everyone?: Planning for disability and access at the Sydney 2000 Games. *Disability Studies Quarterly*, *21*(3), 70–84. doi:10.18061/dsq.v21i4.319

Darcy, S. (2003). The politics of disability and access: The Sydney 2000 Games experience. *Disability & Society*, *18*(6), 737–757. doi:10.1080/0968759032000119497

Darcy, S. (2010). Inherent complexity: Disability, accessible tourism and accommodation information preferences. *Tourism Management*, *31*(6), 816–826. doi:10.1016/j.tourman.2009.08.010

Darcy, S. (2011). Developing Sustainable Approaches to Accessible Accommodation Information Provision: A Foundation for Strategic Knowledge Management. *Tourism Recreation Research*, *36*(2), 141–157. doi:10.1080/02508281.2011.11081315

Darcy, S. (2013). Sydney for All: developing an equality of experience for tourists with disabilities. In I. Cooper, J. Smith, & K. Moss (Eds.), *The World Leisure International Innovation Prize Winning Projects 2006-2012* (pp. 97–103). World Leisure Organisation.

Darcy, S. (2014). Sydney for All. In UNWTO (Ed.), Manual on Accessible Tourism for All: Principles, Tools and Good Practices (pp. 50-66). Madrid: UNWTO, the Spanish ONCE Foundation and the European Network on Accessible Tourism (ENAT).

Darcy, S., & Almond, B. (2016). Case study: Sydney's Cultural Ribbon. In F. Aragall, P. Neumann, & S. Sagramola (Eds.), *Design for All in Tourism Destinations - European concept for accessibility* (pp. 76–81). European Concept for Accessibility.

Darcy, S., & Burke, P. F. (2018). On the road again: The barriers and benefits of automobility for PWD. *Transportation Research Part A, Policy and Practice*, *107*(Supplement C), 229–245. doi:10.1016/j.tra.2017.11.002

Darcy, S., & Cameron, B. (2009/2010). *Competitive Grant Application $30,000 - Urban accessibility and hotel accommodation.* Academic Press.

Darcy, S., & Cameron, B. (2009a). *Uniquest Award - Highly Commended Sydney for All Web portal tourism innovation.* https://www.uniquest.com.au/index.php?sectionID=863

Darcy, S., & Cameron, B. (2009b). *Vision Australia - Making a Difference award for Sydney for All Web portal.* http://www.visionaustralia.org/info.aspx?page=2267#TourismNSW

Darcy, S., & Cameron, B. (2013). Sydney for All - Collaborative Marketing of Accessible Attractions within the Sydney Destination Management Precinct. In I. Ambrose & A. Barragán Iturriaga (Eds.), Exemplars in Global Accessible Tourism Management. Geneva: UN World Tourism Organisation in conjunction with the European Network For Accessible Tourism and the ONCE Foundation for Cooperation and Social Inclusion For PWD.

Darcy, S., Cameron, B., Dwyer, L., & Taylor, T. (2008b). *Research Note: Understanding accessible spaces and places*. www.sydneyforall.com

Darcy, S., Cameron, B., Dwyer, L., Wong, E., & Thomson, A. (2008a). *Visitor Accessibility in Urban Centres (Project # 90064)*. Retrieved from Gold Coast: https://www.researchgate.net/publication/228716596_Visitor_Accessibility_in_Urban_Centres

Darcy, S., & Dickson, T. (2009). A Whole-of-Life Approach to Tourism: The Case for Accessible Tourism Experiences. *Journal of Hospitality and Tourism Management*, *16*(1), 32–44. doi:10.1375/jhtm.16.1.32

Darcy, S., Yerbury, H., & Maxwell, H. (2019). Disability citizenship and digital capital: The case of engagement with a social enterprise telco. *Information Communication and Society*, *22*(4), 538–553. doi:10.1080/1369118X.2018.1548632

Del Chiappa, G., & Baggio, R. (2015). Knowledge transfer in smart tourism destinations: Analyzing the effects of a network structure. *Journal of Destination Marketing & Management*, *4*(3), 145–150. doi:10.1016/j.jdmm.2015.02.001

ECLAC. (2020). *Recovery measures for the tourism sector in Latin America and the Caribbean present an opportunity to promote sustainability and resilience*. Retrieved from: https://repositorio.cepal.org/bitstream/handle/11362/45767/4/S2000440_en.pdf

Economic Commission for Latin America and the Caribbean (ECLAC). (2013). *Social Panorama of Latin America 2012*. Retrieved from: https://www.cepal.org/en/publications/1248-social-panorama-latin-america-2012

Ente de Turismo de la Ciudad de Buenos Aires. (2018). *Noticia: Buenos Aires presentó sus avances como destino inteligente*. Retrieved from: https://turismo.buenosaires.gob.ar/es/article/noticia-buenos-aires-present%C3%B3-sus-avances-como-destino-inteligente

ESI. (2018). *Smarter Cities 2025*. Retrieved from https://econsultsolutions.com/esi-thoughtlab/smarter-cities-2025/

Esper, F. (2020). Enfoques, modelos y buenas prácticas internacionales en DTI: los casos de Buenos Aires y la Red Argentina. In *Sistemas de inteligencia y modelos de trabajo en red para una gestión turística inteligente en Argentina*. INVAT-TUR, Webinar.

European Commision (EU). (2015). *Mapping and Performance Check of the Supply of Accessible Tourism Services*. Retrieved from: https://op.europa.eu/en/publication-detail/-/publication/f0948779-1303-481a-a2d8-5aa043035b81

European Union. (2020). Guide for applicants. *Competition For The European Capital Of Smart Tourism 2020*.

Fleischer, A., & Pizam, A. (2002). Tourism constraints among Israeli seniors. *Annals of Tourism Research, 29*(1), 106–123. doi:10.1016/S0160-7383(01)00026-3

Fundación, O. N. C. E. (2017). *Observatorio de Accesibilidad Universal del turismo en España.* Retrieved from: http://www.drtturismo.com/sites/default/files/observatorio_de_turismo_110917_4.pdf

GfK. (2015). Economic Impact on Travel Patterns of Accessible Tourism in Europe – Final Report. GfK, University of Surrey, The Neumann Consulting and PRO Solutions, Surrey.

Giffinger, R., Fertner, C., Kramar, H., Meijers, E., & Pichler-Milanovic, N. (2007). *Smart cities: Ranking of European medium-sized cities*. Centre of Regional Science. Retrieved from: http://www.smart-cities.eu/download/smart_cities_final_report.pdf

Goggin, G., & Newell, C. (2005). Disability. In *Australia: Exposing a Social Apartheid.* University of New South Wales Press.

Gretzel, U., Koo, C., Sigala, M., & Xiang, Z. (2015). Special issue on smart tourism: Convergence of information technologies, experiences and theories. *Electronic Markets, 25*(3), 175–177. doi:10.100712525-015-0194-x

Gretzel, U., Sigala, M., Xiang, Z., & Koo, C. (2015). Smart tourism: Foundations and developments. *Electronic Markets, 25*(3), 179–188. doi:10.100712525-015-0196-8

Hjalager, A. M. (2002). Repairing innovation defectiveness in tourism. *Tourism Management, 23*(5), 465–474. doi:10.1016/S0261-5177(02)00013-4

Huber, D., Milne, S., & Hyde, K. F. (2018). Constraints and facilitators for senior tourism. *Tourism Management Perspectives, 27*, 55–67. doi:10.1016/j.tmp.2018.04.003

IESE. (2019). *Índice IESE. Cities in Motion.* Universidad de Navarra. Retrieved from: https://media.iese.edu/research/pdfs/ST-0509.pdf

IMD. (2019). *Smart City Index 2019.* Retrieved 04/05/2019 from: www.imd.org/research-knowledge/reports/imd-smart-city-index-2019

INDEC. (2019). *Informes técnicos Turismo Internacional.* Retrieved from: https://www.indec.gob.ar/indec/web/Nivel4-Tema-3-13-55

INDEC. (2020). *Proyecciones y estimaciones poblacionales.* Retrieved from: https://www.indec.gob.ar/indec/web/Nivel3-Tema-2-24

Instituto Ciudades del Futuro. (2020). *Red Argentina de Destinos turísticos inteligentes*. Retrieved from: https://www.google.com/search?q=red+argentina+dti&oq=red+argentina+dti&aqs=chrome.0.69i59j0l4j69i60l3.1410j0j7&sourceid=chrome&ie=UTF-8

Instituto Nacional De Estadísticas y Censos (INDEC). (2018). *Estudio Nacional sobre el Perfil de las Personas con Discapacidad.* Retrieved from: https://www.indec.gob.ar/ftp/cuadros/poblacion/estudio_discapacidad_12_18.pdf

Jang, S. S., Bai, B., Hu, C., & Wu, C. (2009). Affect, travel motivation, and travel intention: A senior market. *Journal of Hospitality & Tourism Research (Washington, D.C.)*, *33*(1), 51–73. doi:10.1177/1096348008329666

Kiss, G., Jansen, H., Castaldo, V. L., & Orsi, L. (2015). The 2050 City. *Procedia Engineering*, *118*, 326–355. doi:10.1016/j.proeng.2015.08.434

La Rosa, G. (2020). Enfoques, modelos y buenas prácticas internacionales en DTI: los casos de Buenos Aires y la Red Argentina. In Sistemas de inteligencia y modelos de trabajo en red para una gestión turística inteligente en Argentina. INVAT-TUR, Webinar.

Lamsfus, C., Martin, D., Alzua-Sorzabal, A., & Torres-Manzanera, E. (2015). Smart Tourism Destinations: An Extended Conception of Smart Cities Focusing on Human Mobility. In L. Tussyadiah & A. Inversini (Eds.), *Information and Communication Technologies in Tourism 2015* (pp. 363–375). Springer. doi:10.1007/978-3-319-14343-9_27

Law N° 25.643 (2002). Accessible Tourism Law.

Littrell, M. A., Paige, R. C., & Song, K. (2004). Senior travellers: Tourism activities and shopping behaviours. *Journal of Vacation Marketing*, *10*(4), 348–362. doi:10.1177/135676670401000406

Liu, Z., Sokhn, M., Le Calvé, A., & Schegg, R. (2016). City Eye: Accessibility for All. In *ICEGOV '15-16 Proceedings of the 9th International Conference on Theory and Practice of Electronic Governance* (pp. 350-351). ACM.

Neumann, P., & Reuber, P. (2004). Economic Impulses of Accessible Tourism for All. Study commissioned by the Federal Ministry of Economics and Technology and Federal Ministry of Economics and Labour, Berlin.

Nikitina, O., & Vorontsova, G. (2015). Aging population and tourism: Socially determined model of consumer behavior in the "senior tourism" segment. *Procedia: Social and Behavioral Sciences*, *214*, 845–851. doi:10.1016/j.sbspro.2015.11.736

Nitti, M., Giusto, D., Zanda, S., Di Francesco, M., Casari, C., Clemente, M. L., Lai, C., Milesi, C., & Popescu, V. Using IoT for Accessible Tourism. In A. R. Garcia Ramirez & M. Gitirana Gomes Ferreira (Eds.), *Smart Cities, Assistive Technologies in Smart Cities* (pp. 31–50). IntechOpen.

NSW. (2019). *Sydney for All*. Retrieved from: https://sydneyforall.com/

Ospina, J. P. (2018). *Ciudades inteligentes en América Latina*. Banco Interamericano de Desarrollo (BID). Retrieved from: https://conexionintal.iadb.org/2018/11/27/267_e_ideas6/

Patterson, I., & Balderas, A. (2020). Continuing and Emerging Trends of Senior Tourism: A Review of the Literature. *Population Ageing*, *13*(3), 385–399. doi:10.100712062-018-9228-4

Perles Ribes, J. F., Ramón Rodriguez, A. B., Vera Rebollo, J. F., & Ivars Baidal, J. (2017). The End of Growth in Residential Tourism Destinations: Steady State or Sustainable Development? The Case of Calpe. *Current Issues in Tourism*. Retrieved from: https://www.tandfonline.com/loi/rcit20

Porto, N., Rucci, A. C., Darcy, S., Garbero, N., & Almond, B. (2019). Critical elements in accessible tourism: Principal component analysis from Oceania and South America. *Tourism Management, 75*, 169–185. doi:10.1016/j.tourman.2019.04.012

Puhretmair, F., & Nussbaum, G. (2011). Web Design, Assistive Technologies and Accessible Tourism. In D. Buhalis & S. Darcy (Eds.), *Accessible Tourism: Concepts and Issues* (pp. 274–285). Channel View Publications.

Ramón-Rodríguez, A. B., & Such-Devesa, M. J. (2018). Is hotel efficiency necessary for tourism destination competitiveness? An integrated approach. *Tourism Economics, 24*(1), 3–26. doi:10.5367/te.2016.0555

Rucci, A. C. (2018). *Discapacidad en el MERCOSUR: accesibilidad turística como una propuesta de desarrollo regional* (Master thesis). Universidad Nacional de La Plata, Argentina.

Rucci, A. C., & Ravlic, I. (2020 in press). La accesibilidad en turismo desde 2005: una mirada desde la LNT. In E. Amadasi & J. L. López Ibáñez (Eds.), El turismo en la Argentina desde 2005: una mirada desde la Ley Nacional de Turismo. Academic Press.

Sánchez-Corcuera, R., Nuñez-Marcos, A., Sesma-Solance, J., Bilbao-Jayo, A., Mulero, R., Zulaika, U., & Almeida, A. (2019). Smart cities survey: Technologies, application domains and challenges for the cities of the future. *International Journal of Distributed Sensor Networks, 15*(6), 1–36. doi:10.1177/1550147719853984

Schwab, K. (2017). *The Fourth Industrial Revolution*. Crown Business.

Sert, A. N. (2019). Senior Tourism in the aging world. In V. Krystev, R. Efe & E. Atasoy (Eds.), Theory and Practice in Social Sciences (pp. 488-498). Sofía, Bulgaria: St. Kliment Ohridski University Press.

Shanker, D. (2008). ICT and Tourism: Challenges and Opportunities. In *International Conference "Tourism in India-Challenges Ahead"*. Indian Institute of Management Kozhikode.

Shoemaker, S. (2000). Segmenting the mature market: 10 years later. *Journal of Travel Research, 39*(1), 11–26. doi:10.1177/004728750003900103

Sniadek, J. (2006). Age of seniors–a challenge for tourism and leisure industry. *Studies in the Physical Culture and Tourism, 13*, 103–105.

Sociedad Estatal para la Gestión de la Innovación y las Tecnologías Turísticas (SEGITTUR). (2015). *Informe destinos turísticos: construyendo el futuro*. Madrid: Ministerio de Industria, Energía y Turismo.

Sun, Y., Song, H., Jara, A. J., & Bie, R. (2016). Internet of Things and Big Data Analytics for Smart and Connected Communities. *IEEE Access: Practical Innovations, Open Solutions, 4*, 766–773. doi:10.1109/ACCESS.2016.2529723

TRA - Tourism Research Australia. (2019). *Latest International Visitor Survey - Results Year ending December 2019*. Retrieved from: https://www.tra.gov.au/International/international-tourism-results

UN. (2019). *Perspectivas de la población mundial 2019*. Retrieved from: https://population.un.org/wpp/

UNE 178501 and UNE 178502 (2016). *Destinos Turísticos Inteligentes*. AENOR, Spain.

United Nations. (2006). *Convención sobre los Derechos de las Personas con Discapacidad y Protocolo Facultativo*. Nueva York: UN.

United Nations, Department of Economic and Social Affairs, Population Division. (2015). *World Population Prospects: The 2015 Revision, Key Findings and Advance Tables*. Working Paper No. ESA/P/WP.241.

United Nations, Department of Economic and Social Affairs, Population Division. (2019). *World Population Prospects 2019: Highlights*. Working Paper No. ST/ESA/SER.A/423.

UNWTO. (2014). *Manual sobre Turismo Accesible para todos. Alianzas público-privadas y buenas prácticas sobre Turismo Accesible*. UNWTO.

UNWTO. (2018). *Official Celebration of World Tourism Day 2018 - Tourism and digital transformation*. Retrieved from: http://media.unwto.org/es/event/celebracion-oficial-dia-mundial-del-turismo-2018-turismo-y-la-transformacion-digital

Wang, D., Li, R., & Li, Y. (2013). China's "smart tourism destination initiative: A state of the service-dominant logic. *Journal of Destination Marketing & Management*, *2*(2), 59–61. doi:10.1016/j.jdmm.2013.05.004

WHO. (2015). *World report on ageing and health 2015*. Retrieved from: https://www.who.int/ageing/events/world-report-2015-launch/en/

Winkowska, J., Szpilko, D., & Pejić, S. (2019). Smart city concept in the light of the literature review. *Engineering Management in Production and Services*, *11*(2), 70–86. doi:10.2478/emj-2019-0012

World Bank. (2020). *Argentina Data*. Retrieved from: https://data.worldbank.org/country/argentina

World Health Organization (WHO). (2011). *Informe Mundial sobre la Discapacidad*. Ginebra: WHO.

Yerbury, H., Darcy, S., & Burridg, N. (2019). Accessing learning resources: experiences of students with disability. *Proceedings of the Tenth International Conference on Conceptions of Library and Information Science, Ljubljana, Slovenia, June 16-19, 2019. Information Research, 24*(4), paper colis 1929. Retrieved from http://InformationR.net/ir/24-4/colis/colis1929.html

Zanella, A., Bui, N., Castellani, A., Vangelista, L., & Zorzi, M. (2014). Internet of things for smart cities. *IEEE Internet of Things Journal*, *1*(1), 22–32. doi:10.1109/JIOT.2014.2306328

ADDITIONAL READING

Darcy, S., McKercher, B., & Schweinsberg, S. (2020). From tourism and disability to accessible tourism: A perspective article. *Tourism Review*, *75*(1), 140–144. doi:10.1108/TR-07-2019-0323

Dominguez Vila, T., Darcy, S., & Gonzalez Alen, E. (2015). Competing for the disability tourism market - A comparative exploration of the factors of accessible tourism competitiveness in Spain and Australia. *Tourism Management*, *47*, 261–272. doi:10.1016/j.tourman.2014.10.008

Fraser, D. (2017). "Perceptions of risks and barriers to participation in tourism for the disabled" (PhD Thesis, University of Plymouth, UK), available at: http://hdl.handle.net/10026.1/8808

Gassiot Melian, A. (2016). "Accessible tourism: an integrated model of the behaviour of tourists with disabilities in a destination" (PhD thesis, Universitat de Girona, Spain), available at: https://www.tdx.cat/handle/10803/398249 (accessed 15 July 2020)

KEY TERMS AND DEFINITIONS

Accessibility: Conditions universally designed that enable everybody to fully participate in activities.

Assistive technology: Technology used for the assistance of people with disabilities, in order to make them more independent.

Constraints: Different kinds of barriers that people with disabilities find when doing diary activities.

People with access needs or requirements: Some authors includes to this group to people with permanent disabilities, those with temporary disabilities such us obesity, pregnant woman, among others, and people older than 65 years old. Those that requires accessibility conditions.

Smart accessible destination: Those destinations based on technological issues that includes everyone -including people with access needs.

ENDNOTES

1. The different indices are built from the selection of certain key dimensions with rigorous statistical supports to define which cities are identified as the most innovative, globalized and environmentally sustainable. The different rankings and measurements reveal the subjectivity with which they can determine the "intelligence" of cities, depending on the methodologies used. Smarter Cities 2025 index (ESI, 2018) is the only one of those indices which doesn't rank the cities numerically. Instead, it calculates a maturity score of a smart city for all the cities analysed, based on interviews with government leaders. It considers 10 pillars (governance, economy, infrastructure, talent, funding, mobility, environment, public safety, public health, payment systems). Then, it divides the score into four equally weighted components (level of smart city investments, use of data analytics, application of smart technologies, and self-rating on stage of smart city maturity). Each city reaches an overall score combining the normalized scores for each pillar.
2. This document has been adapted over the years. From this foundation, the more detailed "A City for All: Inclusion (Disability) Action Plan 2017-21" (City of Sydney, 2017a) is the 4th disability action plan undertaken by the City of Sydney Council.
3. La Rosa (2020) some references were made regarding connectivity problems, lack of data generation to carry out analytics, problems in the digitization of the tourist offer, accessibility, etc.

Chapter 7
Accessibility to Spa Experiences

Eleni Michopoulou
https://orcid.org/0000-0002-1857-4462
University of Derby, UK

Sarah J. Hilton
University of Derby, UK

ABSTRACT

This chapter aims to highlight and raise awareness of the previously unknown barriers currently faced by wheelchair using consumers in the spa industry and the implications of these barriers for consumer and industry alike. Existing research on accessibility within this specific environment is extremely limited (if any). This study shows that access to accurate information is a key issue, a key barrier to participation and not only for those who have not visited a spa before. Gaining information pre visit in tourism is increasingly done online and there is the opportunity to use technologies and especially websites and social media platforms to help provide this information. The chapter also illustrates the potential for health and greater mental and social wellbeing the spa industry and the wider wellness tourism industry have for wheelchair users and how they could mutually benefit each other, as well as further promoting the case for barrier free accessible tourism and leisure opportunities.

INTRODUCTION AND CONTEXT OF STUDY

Despite a higher social conscience dictating global policy on disability, intended to promote acceptance and equality (United Nations, 2006a; 2006b), disabled people face widespread discrimination and a diverse range of disabling barriers to participation in everyday life (Gray, Gould, & Bickenbach, 2003; Harpur, Connolly, & Blanck, 2017; World Health Organisation, 2017). With calls from industry, academia and disabled people themselves for better understanding of the experiences and specific needs of this growing demographic of consumer (Jackson, 2018; Tarasoff, 2017; Williams, 2017), the importance of undertaking research on this subject is becoming increasingly apparent; particularly in a world facing the implications of increased life expectancies (Kasnauskiene & Michnevic, 2017) and ageing populations (Buhalis & Darcy, 2011).

DOI: 10.4018/978-1-7998-6428-8.ch007

The disabled population is projected to steadily increase, with disability and inclusion considered a growing social concern globally (Michopoulou, Darcy, Ambrose, & Buhalis, 2015). Yet much of the traditional thinking and policy making around disability has led to the inadequate physical and social environment that exists today. Modern thinking now views disability as a social construct (Masala & Petretto, 2008) caused and/or exacerbated by a disabling environment (Bogart, Rottenstein, Lund, & Bouchard, 2017; Haegele & Hodge, 2016) and not by placing the disabled person at fault (Masala & Petretto, 2008). Social action and responsibility are needed to change this disabling environment into an enabling environment, ensuring full and equal participation in life as those who are able bodied, minimising the impact of impairment and enabling those who are disabled to enjoy a full and meaningful life as possible (Jackson, 2018).

Tourism and its sub sectors are a key area in which a disabled person can escape daily life and enjoy experiences with loved ones in new and exciting locations, giving a sense of freedom and helping to minimise the effects of mobility impairment (Kaganek et al., 2017; Kastenholz, Eusebio, & Figueiredo, 2015). However, disabling barriers often limit and adversely affect these experiences (Kaganek et al., 2017; Smith, 1987; Yau, McKercher, & Packer, 2004).

Disability and inclusion are becoming progressively more important factors as part of the supply and demand of the tourism industry, with the disabled being recognised as an important consumer demographic and considered vital for tourism economic growth (Ambrose, Darcy, & Buhalis, 2012). Many of the most lucrative travel markets are from developed countries who are experiencing ageing populations with sizeable disposable incomes, yet who are more prone to disability as they age further and will therefore have increasingly more complex and specific needs as a tourism consumer (Darcy & Dickson, 2009; Global Wellness Institute, 2018b). This is a particularly pressing concern for operators of wellness tourism (Crismariu, 2017; Morris, Mueller, & Jones, 2010) of which the spa industry is a key component of, as ageing baby boomers are its key consumer. Wellness tourism, and spa in particular can bring numerous health and wellbeing benefits for wheelchair users, by the diverse therapies and unique experiences they offer (Gomez et al., 2013; Ortega et al., 2017; Suarez et al., 2011).

These industries and institutions will need to be ready to receive future consumers with access needs, and there is an urgent need to implement accessible tourism practices for better and more equal inclusion. This can be achieved by firstly gaining a greater understanding of their complex needs and consumer experiences, in order to improve the current offering. Hence, building on the above discussions, this study investigates the barriers to and within the spa industry currently encountered by wheelchair consumers, and considers the ability of the industry to service this important and growing tourist market segment. In particular, the objectives of the study are to: (1) understand the barriers to participation wheelchair tourists who have not visited a spa may face; (2) identify the barriers currently encountered by wheelchair tourists during the spa customer journey; and (3) consider how these barriers affect their participation in the spa experience.

THEORETICAL BACKGROUND

The study was focused on identifying disabling barriers, and therefore it adopts the social model stance on disability, where disability manifests as a result of shortcomings and barriers occurring in a disabling society, not as a result of a person's medical condition or impairment (Haegele & Hodge, 2016). Removal of these barriers therefore lessens the extent of a person's disability. As the study also focused on the impacts of these barriers to social participation, it acknowledges elements of the ICF model (Johnson-Migalski and Drout, 2018) which better highlights the multifaceted nature of disability beyond impairment. It considers external barriers, and also includes personal factors such as how differences in impairment can affect an individual's activity and participation in everyday life situations beyond employment.

Barriers are obstacles which compromise a person's independence and can negatively affect their ability to participate in physical and community activities, limiting freedom and personal control (Harris, Yang, & Sanford, 2015; Kaganek et al., 2017). Participation, especially for those disabled people with a spinal cord injury (SCI) has been found to have a direct and powerful influence on their quality of life, so it can be argued that these barriers not only affect the participation of an experience in the short term, but can gradually have an accumulative effect on a person's life overall (Chang, Wang, Jang, & Wang, 2012).

Across academic literature there are various descriptions for barrier types, however this study focuses on environmental (in this case meaning physical), informational and social barriers. This is because of how they can affect wheelchair users (people with other impairments such as visual impairments will face their own specific barriers):

Environmental – These are arguably the most disabling barriers for wheelchair users as they physically prevent full participation and enjoyment by restricting access to areas and surroundings. In a world primarily catering for the able bodied person, "environmental" describes architectural and design issues in the physical world, such as access restricting infrastructure including narrow doorways, steps and steep ramps, lack of appropriate facilities such as disabled toilets and changing rooms, and can also describe hotel bedrooms with unsuitable bed and shower heights. This category also includes natural obstacles such as rocky paths, long grass or exposed tree roots that make it more difficult to manoeuvre a wheelchair over. Combatting environmental barriers can lead to painful or humiliating experiences (McKercher and Darcy, 2018) if the wheelchair user must leave their chair, or just total lack of participation if they are unable to.

Informational – These barriers refer to missing or inaccurate access information on websites. Wheelchair users rely on accurate information to inform their purchase decision or intention to visit (Kolodziejczak, 2019); the Internet is a great source of information, however it is often perceived as inaccurate or unreliable (Buhalis & Michopoulou, 2011; Michopoulou & Buhalis, 2013). This barrier category can also describe inaccurate information received from members of staff at an organisation, either in person, or via email or telephone, and can also describe lack of signage with which to locate accessible facilities.

Social – This describes poor social attitudes such as rudeness or contempt from the general public or those in public service, such as receptionists or waiting staff.

Studies of Barriers Encountered By Wheelchair Users

Research has started to identify barriers encountered by wheelchair users in various industry sectors around the world, most often using qualitative interviews or focus group research in order to fully examine the participants' experiences in detail. It must be noted that experiences of barriers can vary between different countries and cultures, and are of course subjective to each individual wheelchair user based on his/her capabilities, type of wheelchair, severity of impairment and expectations of an experience. Personal values and morals may also heavily influence reactions to social barriers.

Table 1 below, shows a non-exhaustive list of studies, which have identified examples of existing environmental (physical), informational and social barriers across various other industries and sectors:

Table 1. Environmental, informational and social barriers found in existing literature

Industry/sector	Barrier		
	Environmental	**Informational**	**Social**
Transport			
Aviation (Chang & Chen, 2012; Davies & Christie, 2017; Poria. Reichel, & Brandt, 2010; Yau et al., 2004)	• Narrow plane aisles mean wheelchair users may have to leave their chair, increasing their vulnerability. • Inaccessible toilets leading to use of catheters or starving oneself, in order to prevent the need to use the toilet, can lead to humiliating experiences. • Uncomfortable/ • Unsuitable cabin seats.	• Poor information on each individual airlines wheelchair policy, both online, on websites and via the telephone. • Lack of accurate information when delays and flight changes occur.	• Staff unaware of how to lift the wheelchair user from chair to airline seat. • Crew members communicating to wheelchair users as if they also have a cognitive disability e.g. deliberately slowly and loudly. • Crew staff responding to accompanying persons regarding information requests, not to the wheelchair user directly.
Healthcare			
Hospitals/healthcare setting (Stillman, Williams, Bertocci, Smalley, & Frost, 2017; Tarasoff, 2017)	• Inaccessible washrooms, examination tables and beds.	• Lack of information on the medical implications of pregnancy and medication on a wheelchair user.	• Poor attitudes such as mocking other nurses when they are assisting a wheelchair user. • Lack of knowledge on lifting a wheelchair user. • Misunderstandings over disability related needs.
Dentistry (Rachid-Kandvani, Nicolau, & Bedos, 2015)	• Heavy doors. • Narrow doorframes. • Cluttered reception areas making manoeuvring a wheelchair difficult. • Inaccessible or unsuitable toilets. • Very high reception desks. • Uncomfortable dental chair.	• Finding accurate accessibility information when searching for a suitable dental practice, online or by phone.	• Staff refusing treatment to wheelchair users via telephone or in practice. • Able-bodied people being served first at reception desks. • Dental professionals unwilling or unknowledgeable on how to transfer the wheelchair user onto the dental couch
Leisure, recreation and tourism			
Hotels (Poria, Reichel, & Brandt, 2011)	• Furniture within hotel rooms makes manoeuvring a wheelchair difficult. • Narrow and heavy hotel room doors. • Thick carpets are difficult to roll over • The heights of shelves, basins and switches for air conditioning etc are too high.	• Poor signage of accessible features such as a lift. • Hotels inaccurately described as accessible on websites when in reality it is not.	• Staff sometimes too helpful or over protective. • Staff unsure how to lift the wheelchair user from the wheelchair to a restaurant chair. • Staff assuming that a wheelchair user has a cognitive disability – speaking loudly or slowly or just to the person accompanying him/her.
Leisure/tourist attractions (Jackson & Searle, 1985; United Nations, 2006c)	• Parking problems • Uneven paths • Lack of entrance ramps • Inaccessible toilets, cafes, gift shops, information centres	• Lack of accurate and detailed information on accessibility on websites and provisions for wheelchair users	• Staff sometimes too helpful or over protective. • Staff unsure what was expected of them or how to help.
Libraries (Leong & Higgins, 2010)	• Steps at entrances and lack or ramps to enter the building.	• Poor signage for entrance and lift locations.	• Staff unsure what was expected of them or how to help.

continued on following page

Table 1. Continued

Industry/sector	Barrier		
	Environmental	Informational	Social
Casinos (Wan, 2013)	• Gaming tables too high, need assistance placing chis and reading other peoples' cards. Tall information desk and cashpoint. • Lack of space to manoeuvre • Lack of accessible toilets and lifts • Wheelchair users have to use the alternative entrances as the main entrance is inaccessible	• Lack of signage to the accessible entrance and the toilets.	• Wheelchair users made to feel like they are not the preferred customers of casinos, were told they are not a priority. • Staff lack training in how to speak to people with disabilities, not just wheelchair users, talked to them like they were stupid. • Poor attitudes from other customers blaming them for slow play.

This table highlights the presence of multiple barriers in a variety of daily life scenarios and activities. Common themes show physical inaccessibility being caused by architectural physical barriers primarily in the form of narrow doorways and steps. The most common information barrier is lack of accessibility information when using technologies such as websites when searching for a suitable premises or venue. Unsolicited staff conduct and lack of knowledge are the most commonly mentioned social barriers. These barriers on their own are enough to adversely affect the experience for the wheelchair user; therefore it can be assumed that when these barriers are experienced in combinations, the experience is even more deeply affected.

An extra barrier to participation to consider is the intrinsic barrier, directly associated with the individuals' own physical, psychological or cognitive condition, and their personal capabilities with it, as opposed to those barriers as previously discussed which are externally imposed upon the individual (Kastenholz et al., 2015; Smith, 1987). Examples of intrinsic barriers can include health conditions, such as chronic pain, where an individual is in so much pain due to their medical condition, that participation is impossible. Physical and psychological dependency is also an intrinsic barrier, particularly for those with a progressive neurological disorder, or a severe SCI resulting in quadriplegia. In these cases, an individual is completely dependent upon another person as their primary care giver in order to be able to participate, if that person is unavailable, the disabled person cannot therefore independently participate. Intrinsic barriers can go beyond the context they are experienced in and can adversely affect intentions to participate in future activities, sometimes resulting in no future participation at all (Kastenholz et al., 2015; Yau et al., 2004). However, these are again subjective and different wheelchair users will have different thresholds as to where these intrinsic barriers start to affect their participation.

Barriers Faced By Wheelchair Users In The Context Of Tourism Activities

Tourism provides an escape from daily life and its problems, and it is widely accepted that the desire to travel is the same for persons with or without a disability (Yau et al., 2004). Wheelchair users do travel, despite the challenges involved, however they often face the same problems but in new and unfamiliar situations and locations (Kaganek et al., 2017; Yau et al, 2004) with barriers not only restricting their options for activities on their trip, but critically undermining the key components and objectives of tourism and leisure – freedom and escapism (Smith, 1987). Due to the medical model of disability dictating policy for so long, the able-bodied tourist and their capabilities have been prioritised; urban planning and countryside recreation have centred around the active seeing and mobile body (Aitchison, 2003). For disabled people overall, barriers are one of the many reasons why participation rates in and qualities of experience of tourism are lower than that of the general population (McKercher & Darcy, 2018).

As previously shown in table 1, air travel, a popular and sometimes the only way of reaching destinations, is fraught with multiple barriers and can be a hugely unpleasant and complicated ordeal for wheelchair users. Such negative experiences can lead to the triggering of the intrinsic barrier, where a person feels they are not able to travel due to the stress and effort required (Yau et al., 2004). Access to historic sights while on a trip frequently involve negotiating many - sometimes uneven - steps, and are enjoyed from a vantage point at which to observe them, and therefore those with mobility impairments are immediately at a disadvantage.

Domestic opportunities to socialise and meet new people are often fraught with arguments and stress (Olsen, 2018). A trip to the cinema with friends can also involve unpleasant experiences; often the only place for a wheelchair is at the very front of the auditorium, particularly when stadium style seating is in place (Hammerle, 2005). This results in a painful viewing angle, straining eyes and necks in order to see properly, and splits groups of friends up as those who are able bodied can sit further back at a more comfortable distance (Hammerle, 2005). This does not provide an equally enjoyable experience for wheelchair users, and is not an experience they are likely to want to repeat; another example of a social outing that wheelchair users would be more inclined to decline in future (Hattan, 2004)

Booking concert tickets can be a lengthy and expensive process for those in wheelchairs; disabled access tickets are often more expensive and more complicated to buy than those with no access needs. They are often in very short supply and must be booked well in advance, which means that the option to join friends at short notice is extremely limited (Olsen, 2018). If tickets are successfully pre-booked, areas for wheelchair access are often placed in areas with a poor view of the stage, or in areas where people can stand in front of them. This leads to the disabled person giving up on attending these types of social outings, due to the stress and financial burden, often triggering intrinsic barriers. According to a survey conducted by disability led charity 'Attitude is Everything' (2018), more than 80% of deaf and disabled music fans have experienced problems booking tickets to live music events, with one in ten considering legal action over the difficulties they experience when making their ticket purchase. A lack of comprehensive online information (about the venue), and restrictions on purchases such as disabled tickets only available via a phone call, and difficulties acquiring an accompanying carers ticket, were some of the issues raised by survey respondents.

These barriers directly affect participation in activities that should bring joy and relief to a person; instead, they add further difficulty and stress.

The Case For Accessible Tourism

In several points of the UN Convention on the Rights of Persons with Disabilities (CRPD) Article 30, it was declared that UN states must take appropriate measures to ensure that those with disabilities enjoy access to places for tourism services, access to tourism venues and access to tourism activities (United Nations, 2006c), encouraging worldwide awareness and the widespread implementation of accessible tourism. The word accessibility when used in a disability context implies that disabled people can, without assistance, reach, enter or pass to and from, and make use of facilities without being made to feel that one is the object of charity (Nyanjom, Boxall, & Slaven, 2018). Accessible tourism therefore describes this inclusive access for disabled people, to tourism products services and environments, including those with permanent or temporary mobility, vision, hearing or cognitive impairments (Darcy & Dickson, 2009), or to put it briefly "tourism for all" (Michopoulou et al., 2015). Tourism New South Wales has added to this definition, by stating that accessible tourism should make it easy for all people to *enjoy* tourism experiences (Darcy & Dickson, 2009). The addition of the word enjoy is key, as one of the key aspects of tourism is fun and recreation; if the experience is not enjoyable then the whole purpose of the excursion is futile.

As well as those with disabilities, other sections of society – estimated as being roughly 31% of a developed countries population - also benefit from accessible tourism (Darcy & Dickson, 2009), including those who are obese, older people and families with young children/toddlers who may also be pushing a buggy. Because of its implications, accessible tourism has been described as a vehicle with which to promote individual and social well-being, not just for socially marginalised groups and their families (disabled people), but for society as a whole (Devile & Kastenholz, 2017).

Accessible tourism is an evolving area, for both industry practice and academia (Buhalis & Darcy, 2011; Devile & Kastenholz, 2017). As more research is conducted into the needs and experiences of those who have access requirements, a greater understanding is achieved with which to lay the foundations in order to direct policy and industry, with the aim to create more inclusive practices within tourism. Removing barriers to achieve accessible tourism for disabled people, unlocks those stimulating and empowering activities as previously discussed, leading to increased participation, improved health and a better quality of life. As well as the individual and social benefits, there are economic advantages of accessible tourism too.

As well as gaining momentum as a pressing human rights issue, accessible tourism is also gaining recognition for its important contribution to the economics of the tourism industry (Ambrose et al., 2012). The Purple Pound describes the spending power of disabled households in the UK, and its contribution to accessible tourism in England alone is worth over £12.1 billion (Visit England, 2015; Williams, 2017). Overall, the value of the Purple Pound in the UK is worth £250 billion (Williams, 2017) making the disabled a consumer group with huge spending potential. Hence, the spending power of the disabled as a consumer group is sizeable, and should be considered by accessible businesses and industries.

Wellness Tourism and Spa

Wellness tourism is the area in which the vast tourism and wellness industries overlap. It primarily describes the branch of tourism catering for both health and tourism motivations, where one of the main objectives of taking leisure time away from home is to maintain or improve the state of one's personal health and wellbeing (Clift & Page, 1996; Global Wellness Institute, 2018b). It is one of the fastest growing tourism subsectors (Global Wellness Institute, 2018b), offering unique social leisure experiences each with their own specific health benefits as the name suggests (Michopoulou, 2017), and therefore could be a particularly appealing area of tourism for wheelchair users to explore.

At the turn of the new millennium, it was predicted to be a significant and powerful subsector of the tourism industry (Weiermair & Mathies, 2004) and this prediction was realised; wellness tourism is thriving and growing more than twice as fast as general tourism itself and was worth approximately $639.4 billion in 2017 (Global Wellness Institute, 2018a). A key component of the wellness tourism industry is the $118.8bn spa industry (Global Wellness Institute, 2018a).

Wellness tourism is fuelled by increasingly active, health conscious and wealthy generations, particularly the baby boomers, coupled with younger generations who are becoming more stressed and seeking an escape from fast paced lifestyles and a poor work/life balance (Global Wellness Institute, 2018a; Hudson, Thal, Cardeas, & Meng, 2017; Kim, Chiange, & Tang, 2016; Mintel, 2015; Voigt, Brown, & Howat, 2011). Baby boomers are the most powerful demographic within tourism overall, due to their large numbers and considerable disposable income (Borges Tiago, Couto, Tiago, & Dias Faria, 2016; Hung & Petrick, 2009; Tsiotsou & Hudson, 2010) and are also considered the key wellness tourist responsible for driving the demand of its products and services as they seek out active and fulfilling experiences to preserve and prolong their youth in their retirement years; spa destinations and resorts being an ideal place of interest in which to achieve this (Darcy & Dickson, 2009; Hung & Lu, 2016; Ozkuk et al., 2018; Voigt & Pforr, 2014).

They are also the demographic of tourist that are the most likely to have disabilities due to accidents or injuries, and experience changes in mental, physical or sensory abilities due to their age (Crismariu, 2017; Morris, Mueller & Jones, 2010). Strokes, diabetes and both osteo and rheumatoid arthritis, give rise to a greater chance of mobility impairment likelihood of needing a wheelchair (Crismariu, 2017; Smith, Sakakibara, & Miller. 2016).

Wellness tourism destinations and facilities will be facing the increased probability of hosting customers who are mobility impaired or wheelchair dependent in the coming years, heightening the need for the adoption of accessible tourism practices within this sector. Making a wellness tourism facility such as a spa accessible and wheelchair friendly has benefits for both parties.

The Potential Benefits of Spa For Wheelchair Users

As well as being well placed to meet the needs of the health conscious baby boomers, spa visits have also been described as ideal for people living with lifelong disabilities and many of these institutions market themselves on the rejuvenating benefits to be gained from their facilities and treatments (Smith & Puczko, 2017) particularly those in Eastern Europe which take a more clinical spa approach (Rawlinson & Heap, 2017). Table 2 below displays selected spa therapies and their benefits for symptoms of those who may be wheelchair dependent either due to age, injury, or congenital condition:

As seen in table 2, it becomes clear that spa therapies offer a range of physical, physiological and psychological benefits. Additionally, the often peaceful and relaxing environment and surroundings of spas, especially those by the coast can calm anxiety, lower blood pressure and promote better sleep (Charlier & Chaineaux, 2009; Global Wellness Summit, 2019). Finally, resort and hotel spas often offer other leisure activities, that may appeal to more active accompanying family members or carers, such as tennis, yoga, golf or cycling, making them an ideal location that offers something for everyone to enjoy.

However, there is little existing knowledge on the barriers that may influence wheelchair users' participation in spa experiences. If a wheelchair user does visit a spa, it is unknown which barriers they encounter and how this affects their customer experience. There is therefore a pressing need to identify and understand both the barriers to participation currently existing that prevent wheelchair users from being able to take part in a spa experience, as well as those that are encountered by a wheelchair user within the spa customer journey. This will allow industry to better accommodate this demographic of spa consumer better, and thus improving their experience, creating value for both parties as well for society at large.

Table 2. Potential benefits of spa therapies for wheelchair users

Spa therapy	Health benefit
Massage	• Lowers heart rate and blood pressure, reducing anxiety and mental fatigue (Beck, 2010) • Powerful sedative, deeply relaxing – promotes sleep (Johnson, 2011; Koog, Jin, Yoon, & Min, 2010) • Pain relief (Keeratitanont, Jensen, Chatchawan, & Auvichayapat, 2015; Peamruetai, et al., 2016; Sritoomma, Moyle, Cooke, & O'Dwyer, 2012;)
Thalassotherapy (therapeutic use of seawater/sea air/ sea products/ coastal climate)	• Promotes sleep (Charlier & Chaineaux, 2009; Pereira, 2018) • Effective for dry or inflamed skin disorders, pain relief, anti-depressant and boosts the metabolism and circulation (Pereira, 2018)
Pelotherapy (therapeutic use and application of thermal muds/ mud baths/ rasul chambers)	• Anti-inflammatory and analgesic (Gomez et al., 2013; Ortega et al., 2017; Suarez et al., 2011). • Stimulates circulation (Gutenbrunner, Bender, Cantista, & Karagulle, 2010)
Hydrotherapy/ balneotherapy (therapeutic use of water/mineral water)	• Pain relief for muscles and joints (Karagulle & Karagulle, 2015; Ozkuk et al., 2018; Ucok et al., 2008; Zhang, Roxburgh, Huang, Parsons, & Davies, 2014) • Reduces anxiety (Ozkuk et al., 2018) • Beneficial for joint, spinal and mobility conditions (Gass & Gass, 2001; Karagulle & Karagulle, 2015; Ucok et al., 2008; Varga et al., 2008; Zhang et al., 2014)
Aromatherapy (therapeutic use of essential oils)	• Deep relaxation (Koog et al., 2010) • Promotes sleep (Domingos & Braga, 2014) • Reduces anxiety (Domingos & Braga, 2014)

METHODOLOGY

This exploratory study adopted a qualitative approach and was conducted using a mono-method cross sectional design, allowing collection of information rich data from multiple participants at a single snapshot in time in the UK (Bryman, 2016; Marshall & Rossman, 2011). Sample inclusion criteria included anyone who was dependent on a wheelchair (no facilitators), and filtering criterion was whether they had visited a spa or not.

Questions to participants were informed based on literature, and particularly influenced by similar studies by authors such as Rachid-Kandvani et al. (2015) as the customer journey a dental patient takes resembles that of which a spa guest would take (i.e. both environments require transfer to a treatment couch, and both can be classed as vulnerable situations). The questions covered the entire spa customer journey, from the initial search for suitable premises and first contact with its staff, physical arrival at the organisation, through to any service recovery on departure. Whether the participant had visited a spa or not determined which questions they would be asked; those who had not visited were questioned as to why this was, and those who had visited a spa were questioned in more detail about their experiences. Questions were asked via online and telephone interviews, and where this was not practical for the participant, they were emailed interview questions to fill out at their own convenience, as well as any follow up questions as required by the researcher. Data was analysed using thematic analysis, which allowed key and sub themes to be identified.

FINDINGS AND DISCUSSION

Sample Profile

A total of nine participants were recruited for the study, three were known to the researcher, four were recruited via snowball sampling, one via an appeal via self-selection via email to a disability organisation, and one through social media. Participants varied in the severity of their mobility impairment, ranging from those who were able to walk short distances, to those who were completely wheelchair dependent. The reason for mobility impairment and subsequent wheelchair use also varied from advanced neurological disease, to old age. Four of the participants had not visited a spa, and five had visited. The sample profile is shown in Table 3:

After the data was analysed using thematic analysis, it was possible to rank the barrier types in order of significance for each group of participants, shown in Table 4

Table 3. Sample profile

Participant	Gender	Age	Has/has not visited a spa	Reason for wheelchair use	Mobility capabilities
P1	F	35	Has	Myalgic Encephalomyelitis (M.E.) Postural orthostatic tachycardia syndrome (POTS)	Can only walk short distances
P2	F	24	Has	Myalgic Encephalomyelitis (M.E.)	Can only walk short distances
P3	M	41	Has	Cerebral palsy	Can only walk short distances
P4	F	60	Has not	Multiple Sclerosis	Cannot walk
P5	F	34	Has	Spinal Muscular Atrophy	Cannot walk
P6	F	38	Has not	Myalgic Encephalomyelitis (M.E.)	Can only walk short distances
P7	F	42	Has	Spinal cord injury	Cannot walk
P8	F	26	Has not	Myalgic Encephalomyelitis (M.E.) Postural orthostatic tachycardia syndrome (POTS) Suffers from seizures	Can only walk short distances
P9	F	86	Has not	Old age and arthritis	Can barely walk

This data shows that barriers hold different levels of significance for different consumer groups. However, the key finding for this study was that information barriers were the most critical barriers for both groups of wheelchair users when accessing the spa industry.

Table 4. Rank of significance of each barrier type for different spa consumers

Rank of significance	Non-spa visitors	Spa-visitors
1	Information	Information
2	Intrinsic	Environmental
3	Environmental	Social
4	Social	Intrinsic

Understanding The Internal And External Barriers To Participation Wheelchair Tourists Who Have Not Visited A Spa May Face

Informational barriers

These were the most influential barriers to spa participation. As stated in the literature review, wheelchair users rely on accurate information to inform their purchase decision or intention to visit (Kolodziejczak, 2019) and the findings of this study further reinforced this. The study participants all had varying degrees of mobility and severity of their health condition, meaning that all have diverse but specific needs, which they must ensure will be met. The very fact that they cannot just turn up at a spa and hope for the best means a visit requires careful researching of spas in order to gain the relevant information needed in order for them to decide whether to visit it or not.

The research *process to find a suitable spa can be very lengthy and time consuming,* with issues described by participants as being similar to those experienced when trying to buy concert tickets (Attitude is Everything, 2018; Olsen, 2018). P6 wished to visit a spa - but has had to make several telephone calls in a so far fruitless search to find a suitable one.

This research process is often made harder due *to lack of information*. It is often necessary for wheelchair users to view the spas website or contact the spa itself, in order to determine whether their individual needs will be satisfied, either by asking questions or viewing the available information provided. Here, the study has found that the lack of information from both websites and from spa staff about their own workplace have been a barrier to participation. Vague answers are not enough for someone who is already vulnerable, when wanting to enter an environment that has the potential to make them even more vulnerable. If persons do not feel comfortable talking on the telephone regarding their needs with a disability, websites are therefore the next best information option and are available even when the spa itself is closed. However, participants expressed frustration again at the lack of available information online and a general distrust regarding their accuracy.

These findings are consistent with information barriers found in other industries and sectors; lengthy complicated online searches for suitable premises are also found when searching for an accessible dental practice, often resulting in a wheelchair user giving up the search (Michopoulou & Buhalis, 2013; Rachid-Kandvani et al., 2015). Lack of information on websites and from staff again corresponds with the recent studies on booking concert tickets (Attitude is Everything, 2018; Olsen, 2018) but also reinforces earlier findings from studies on aviation (Chang & Chen, 2012; Davies & Christie, 2017; Poria et al., 2010) hotels (Poria et al., 2011) and when visiting other leisure and tourist attractions (United Nations, 2006c) suggesting it is a widespread and well established problem.

Intrinsic barriers

For non-spa visitors, the intrinsic barrier is also highly influential, where wheelchair users own health conditions and associated limitations affect their intention and ability to visit a spa. Participants agreed that some *medical condition makes it very difficult to visit.* For those participants who had more severe mobility impairment such as P4 who has advanced MS, there were concerns they would need special assistance; a worry echoed by P9, an elderly participant who is barely able to walk or stand up unassisted. P8 with specific needs as a result of their complex and fluctuating medical condition was concerned about these not being met, or understood. Finally, three out of the four non-visiting participants felt that visiting a spa would be too much of a struggle overall.

These findings show that intrinsic barriers act as a key barrier to spa participation after information barriers. Some wheelchair users have highly specific individual needs resulting from medical conditions, and require specialist equipment or trained personnel. If these are not available, spa can be considered another environment that is best avoided due to the stress and effort required, and the potential for problems to occur (Yau et al., 2004).

Environmental (physical) barriers

Environmental barriers were not always physically experienced; however they were anticipated. Participants expressed concern over *potential inaccessibility of the sites*, and the possibility of architectural obstacles preventing access to the building, and lack of specialist equipment needed, such as hoists. Environmental barriers such as internal steps prevent full and equal participation to spa experiences compared to those who are able bodied, also a negative factor on the intention to visit. In the case of P4, who wishes to visit a spa with friends, the steps or architectural features she was informed about would exclude her from some areas that her friends would enjoy, another example of barriers causing social isolation and exclusion for wheelchair users (Craig, Tran, & Middleton, 2009). This, when combined with all the other barriers was summed up by P8 when she said: *"I would not risk going and spending a lot of money to not feel like I could relax and enjoy the experience the same way able-bodied people can."*

Environmental barriers are found in many other sectors, some identified in table one such as hotels (Poria et al., 2011) dentistry (Rachid-Kandvani et al., 2015) and libraries (Leong & Higgins, 2010) and throughout daily life. It is therefore deemed easier to avoid entering an environment for an activity that can be considered recreational (i.e. not an essential daily activity) where there are more potential obstacles to negotiate.

Social Barriers

Social barriers were mentioned by only one of the four non-visiting participants, who explained that she was made to feel that her business was not welcome as a wheelchair user when making initial enquiries with the spa. Negative first impressions of an organisation and its service such as this can be classed as a barrier to participation, particularly in a non-essential service environment such as this, where the consumers can decide to go elsewhere if they receive poor service. What is encouraging is that the other three participants did not mention this, perhaps indicating that this barrier is not as prevalent as others. Although not a medical environment, spa can still be considered a caring environment and should be welcoming to all.

In comparison with the literature on existing disability studies, this experience correlates most closely with those service attitudes expressed in casinos (Wan, 2013) a leisure environment that may lack the higher levels of service training and professionalism from its staff members compared to sectors such as those within healthcare.

Barriers Encountered By Spa Visitors Using Wheelchairs

This consumer group were most affected by information and environmental barriers, initially struggling to find a spa to visit, and once there, finding that architectural features limited access, leading to a lesser experience than able-bodied people enjoyed; despite paying the same price.

Informational barriers

Information barriers were identified as affecting spa visiting wheelchair tourists in similar ways to those who didn't visit, with lack of information from both websites and staff members resulting in a *lengthy and difficult search*. However, information barriers were also experienced once the wheelchair user was in the spa too, with staff not knowing the implications of having certain treatments that had been booked.

Four out of the five spa-visiting participants mentioned a lack of information on websites, as well as staff being unaware of the accessibility of their own workplace. *Lack of information* from staff for these participants was not limited to accessibility, but also included lack of information about the suitability and implications of having certain treatments, which has resulted in one of the participants being refused a pre booked treatment when they arrived at the spa, causing disappointment, distress and embarrassment. Once arrived at the spa, lack of information can also apply to signage of accessible features, such as the entrance, or where the lifts are located.

For those wheelchair tourists visiting a spa, these findings indicate that information barriers are more widespread and are not only limited to the search process. These findings first align with the findings from non-spa users regarding the lengthy search for a spa and gathering information, as well as with the studies in other industries as previously stated in table 1, such as when accessing dental practices (Rachid-Kandvani et al., 2015) and the aviation industry as a consumer (Chang & Chen, 2012; Davies & Christie, 2017). Additionally the lack of knowledge from staff regarding treatment implications echoes the findings of studies of barriers within healthcare environments (Stillman et al., 2017; Tarasoff, 2017). Poor signage to accessible entrances and lifts were also identified in studies in hotels (Poria et al., 2011), libraries (Leong & Higgins, 2010) and casinos (Wan, 2013).

Environmental (physical) barriers

For spa visiting participants, these barriers were inevitably experienced more profoundly than those who had not visited, and were experienced throughout the customer journey through the spa, from the entrance, through to the changing rooms, thermal facilities, treatment rooms and finally the relaxation rooms. Barriers preventing access to the spa building itself were mentioned, correlating with studies on libraries (Leong & Higgins, 2010) and dentistry (Rachid-Kandvani et al., 2015). P1 and P2 mentioned having to use an alternative "tradesman's entrance" due to *inaccessibility* of the entrance, and this has also been experienced by wheelchair users when visiting casinos (Wan, 2013).

The lack of suitable accessible changing and toilet facilities in spas is a common occurrence again within casinos (Wan, 2013) as well as having prevalence in aviation (Chang & Chen, 2012; Davies & Christie, 2017; Poria et al., 2010; Yau et al., 2004). hospitals (Stillman et al., 2017; Tarasoff, 2017), dental practices (Rachid-Kandvani et al., 2015) and other leisure attractions, with comments that they were either used for other purposes such as being used as a storeroom/cleaning cupboard (P3) or lacking in specialised equipment that would make them accessible for those who require hoists (P5).

Some spa thermal facilities such as saunas and the pool itself were deemed not accessible again due to architectural features such as steps or again, lack of lifting equipment. Narrow doorways and small and cluttered rooms prevent access and manoeuvrability for wheelchairs, as has been identified in dental practices, hotels and casinos (Poria et al., 2011; Rachid-Kandvani et al., 2015; Wan, 2013).

Regarding treatment rooms, narrow doorways either result in total inaccessibility (and means the wheelchair user cannot have a treatment) or lead to the wheelchair user having to leave their chair, leaving them more vulnerable and not at ease, defeating the object of the experience and mirroring similar negative emotions identified in the studies on aviation, where plane aisles are too narrow to fit a standard wheelchair (Chang & Chen, 2012; Davies & Christie, 2017; Poria et al., 2010; Yau et al., 2004). These barriers also occur within the relaxation room, where the beds and beanbags are also inaccessible, whilst furniture arrangements and small dimensions can make manoeuvring a wheelchair very difficult, also expressed in the study of dental practices (Rachid-Kandvani et al., 2015) and healthcare settings (Stillman et al., 2017; Tarasoff, 2017).

Spa environmental barriers at best cause inconvenience for wheelchair users when they require use of a separate entrance, and at their worst cause total lack of access to a facility or room, resulting in a lesser experience than an able bodied person. In the case of P3, difficulties getting onto a treatment bed were the direct cause of their reluctance to book treatments.

Social barriers

This type of barriers is the most subjective; participants react and interpret behaviour in different ways and it affects individuals to different degrees. However, there is a negative theme in the participant's experiences that staff have been visibly uncomfortable at serving them, or their body language has made them feel like a burden or an effort, adversely affecting the experience.

Participants discussed *unsolicited staff conduct* and disclosed that they had faced negative attitudes from staff both at reception and in the treatment rooms both by body language and in treatments they have received. This negative conduct has affected the welcome experience at the spa, as well as the experience and atmosphere within the treatment room (P1). P5 had a negative treatment experience and directly stated that she felt it was because of their disability. Negative attitudes from staff have also been experienced in casinos (Wan, 2013) where staff spoke to disabled customers as if they were stupid. Not knowing how to lift or move a wheelchair user, or not feeling confident about performing a treatment on them, can also be blamed on a lack of training, which has been experienced within healthcare settings and aviation, when transferring customers to an airline seat, or medical/dentistry table (Chang & Chen, 2012; Davies & Christie, 2017; Poria et al., 2010, 2011; Rachid-Kandvani et al., 2015; Yau et al., 2004).

Intrinsic barriers

Intrinsic barriers have made the least impact for these participants and have not adversely affected the intention to visit, as *medical condition should not cause limited participation*. However, the presence of the intrinsic barriers was identified within this group of participants, as it limited the level of participation that the wheelchair user is able to achieve. For example, P1 was happy to go to a spa, but due to their medical condition, would not visit alone. Other participants do not feel safe using certain facilities at a spa due to their medical condition (P2 and P3)

CONCLUSION

Following the discussions above it becomes clear that within the spa industry, the able bodied person has been prioritised, mirroring the wider social and built environment (Aitchison, 2003). These findings can help to further strengthen the argument of McKercher and Darcy (2018) who claim that for disabled people overall, barriers are one of the many reasons why participation rates in and qualities of experience of tourism are lower than that of the general population.

This study expanded the existing body of academic knowledge by adding spa to the list of industries where the barriers facing wheelchair users have been investigated. It has identified previously unknown barriers which are experienced and by whom, and how they affect the participation of a wheelchair tourist; it allows a greater understanding of the subject of accessibility, uniquely within a spa context. Findings confirm that the spa industry is also marred with barriers, and has similar issues regarding accessibility with other industries and sectors, particularly those in healthcare and dentistry.

With regards to the ability of the spa industry to service the important and growing market segment of wheelchair tourists it is clear that the able bodied have been prioritised, much like the rest of society and the built environment that follows the medical model of disability. Primary data has shown that wheelchair tourists wish to and do visit spas, however barriers are present throughout their customer journey and adversely affect their participation. Therefore at present, even though the spa industry is serving these consumers, there are improvements to be made.

Hence, the following recommendations are proposed to help industry become more inclusive and accessible. There is a clear need for accurate, reliable and easy to access information, both regarding spa accessibility and the implications of treatments. This was the key barrier for both sets of participants, however it is arguably the most straightforward to remedy, using ICT tools such as websites and social media, and improving their content with precise and up to date accessibility information that takes into account different impairments, not just mobility. Websites should be simple to navigate and information easy to find. Accessibility information should cover all areas of the spa, including any additional areas such as where food is consumed. An example of required information specific to wheelchair users would be how wide the spa treatment room doors are, and if the spa has any lifts.

These technologies have the potential to act as a 24 hour virtual "spa reception" providing a valuable source of instant information - the importance of them should not be underestimated. Photos and even walkthrough videos can add to accessibility information to help add further clarity for a potential visitor, better allowing them to make an informed decision. Access information could also be detailed on a pdf, which could be attached to an email sent in reply to enquiries from wheelchair users.

Training staff on the accessibility of their own workplace will further add to this, perhaps keeping a manual by the telephone that they can refer to, with door dimensions or step locations and heights clearly stated. The language that staff use when serving not just wheelchair users, but all disabled people can also be improved, as was highlighted by the presence of the social barrier. This can be achieved via specialist training programmes.

Finally, the prevalence of environmental (physical) barriers throughout spas shows that for wheelchair users, they are sometimes not the relaxing and stress relieving places they market themselves as. It is appreciated that spas are often located in old and converted buildings, where drastically improving access is not possible, however simple measures such as acquiring portable ramps will allow a wheelchair to negotiate single steps, and ensuring that treatment rooms have adjustable couches is also important, as this eases the process of transfer from wheelchair to treatment couch.

Spas of the future should be designed with a variety of users in mind, not just wheelchair users but those with other impairments too. Every aspect of the customer journey should be considered, from the entrance through to the changing rooms, thermal facilities and treatment rooms. This includes the dimensions of rooms, width of doors and corridors, as well the installation or provision of specialist equipment such as hoists. This equipment may not be aesthetically pleasing, however portable hoists are available and can easily be stored in a cupboard (PN-Paraplegia News, 2005).

These recommendations are intended to improve the service that the spa industry provides for wheelchair tourists, in order to increase participation and elevate the spa experience. Better experiences may result in positive word of mouth recommendations, encouraging more wheelchair users to visit. Additionally, as a result of these recommendations, non-spa visitors may feel better reassured that they can participate in spa, lessening the limiting effects of their own intrinsic barriers. By becoming more accessible, spas can take both a social and economic advantage over their competitors, increasing their sustainability.

Limitations Of Study And Future Research Directions

As with all research this study also has inherent limitations. The number of participants was limited, and a bigger sample should be used in the future to corroborate the findings. Also, future research should look into more detail into the requirements generated by different levels of motor dis/ability. For example, a set of stairs may be a challenge for someone who is able to walk, albeit with difficulty, and uses a wheelchair for convenience and speed, however it would be practically impossible for someone who is totally paralysed to negotiate them. With this in mind, it is recommended that further studies concentrate on those with more similar levels of impairment, in order to further investigate specific needs. This research concentrated on mobility impairment, however other sensory or cognitive impairments should also be investigated, as these people may face different barriers than the ones identified for wheelchair users here, or the same barriers but with different implications.

REFERENCES

Aitchison, C. (2003). From leisure and disability to disability leisure: Developing data, definitions and discourses. *Disability & Society*, *18*(7), 955–969. doi:10.1080/0968759032000127353

Ambrose, I., Darcy, S., & Buhalis, D. (2012). *Best Practice in Accessible Tourism: Inclusion, Disability, Ageing Population and Tourism*. Channel View Publications. doi:10.21832/9781845412548-029

Attitude is Everything. (2018). *State of Access Report 2018*. Available at: http://www.attitudeiseverything.org.uk/uploads/general/State_of_Access_Report_2018.pdf

Beck, M. (2010). *Theory and Practice of Therapeutic Massage* (5th ed.). CENGAGE Learning.

Bogart, K. R., Rottenstein, A., Lund, E. M., & Bouchard, L. (2017). Who self-identifies as disabled? An examination of impairment and contextual predictors. *Rehabilitation Psychology*, *62*(4), 553–562. doi:10.1037/rep0000132 PMID:28581320

Borges Tiago, M., Couto, J., Tiago, F., & Dias Faria, S. (2016). Baby Boomers turning grey: European profiles. *Tourism Management*, *54*, 13–22. doi:10.1016/j.tourman.2015.10.017

Bryman, A. (2016). *Social Research Methods* (5th ed.). Oxford University Press.

Buhalis, D., & Darcy, S. (2011). *Accessible Tourism*. Channel View Publications.

Buhalis, D., & Michopoulou, E. (2011). Information-Enabled Tourism Destination Marketing: Addressing the Accessibility Market. *Current Issues in Tourism*, *14*(2), 145–168. doi:10.1080/13683501003653361

Chang, F., Wang, Y., Jang, Y., & Wang, C. (2012). Factors Associated With Quality of Life Among People With Spinal Cord Injury: Application of the International Classification of Functioning, Disability and Health Model. *Archives of Physical Medicine and Rehabilitation*, *93*(12), 2264–2271. doi:10.1016/j.apmr.2012.06.008 PMID:22728700

Chang, Y., & Chen, C. (2012). Meeting the needs of disabled air passengers: Factors that facilitate help from airline and airports. *Tourism Management*, *33*(3), 529–536. doi:10.1016/j.tourman.2011.06.002

Charlier, R., & Chaineux, M. (2009). The healing sea: A sustainable coastal ocean resource: Thalassotherapy. *Journal of Coastal Research*, *25*(4), 838–856. doi:10.2112/08A-0008.1

Clift, S., & Page, S. (1996). *Health And The International Tourist* (1st ed.). Routledge.

Craig, A., Tran, Y., & Middleton, J. (2009). Psychological morbidity and spinal chord injury: A systematic review. *Spinal Cord*, *47*(2), 108–114. doi:10.1038c.2008.115 PMID:18779835

Crismariu, O. (2017). Accessible Tourism – An Expression of Social Responsibility. *Cactus*, *2*, 33.

Darcy, S., & Dickson, T. (2009). A whole-of-life approach to tourism: The case for accessible tourism experiences. *Journal of Hospitality and Tourism Management*, *16*(1), 32–44. doi:10.1375/jhtm.16.1.32

Davies, A., & Christie, N. (2017). An exploratory study of the experiences of wheelchair users as aircraft passenger – implications for policy and practice. *IATSS Research*, *4*(2), 89–93. doi:10.1016/j.iatssr.2017.05.003

Devile, E., & Kastenholz, E. (2017). Accessible tourism experiences: The voice of the people with visual disabilities. *Journal of Policy Research in Tourism, Leisure & Events*, *10*(3), 265–285. doi:10.1080/19407963.2018.1470183

Domingos, T., & Braga, E. (2014). Meaning of aromatherapy massage in mental health. *Acta Paulista de Enfermagem*, *27*(6), 579–584. doi:10.1590/1982-0194201400094

Gass, E., & Gass, G. (2001). Thermoregulatory responses to repeated warm water immersion in subjects who are paraplegic. *Spinal Cord*, *39*(3), 149–155. doi:10.1038j.sc.3101117 PMID:11326325

Global Wellness Institute. (2018a). *Global Wellness Economy Monitor – October 2018*. Available at: https://globalwellnessinstitute.org/wp-content/uploads/2019/04/GWIWellnessEconomyMonitor2018_042019.pdf

Global Wellness Institute. (2018b). *Global Wellness Tourism Economy – November 2018*. Available at: https://globalwellnessinstitute.org/industry-research/global-wellness-tourism-economy/

Global Wellness Summit. (2019). *2019 Wellness Trends*. Available at: https://www.globalwellnesssummit.com/2019-global-wellness-trends/download-report/

Gomez, C., Carretero, M., Maraver, F., Cantista, P., Armijo, F., Legido, J., ... Delgado, R. (2013). Peloids and Pelotherapy: Historical Evolution, classification and glossary. *Applied Clay Science*, *75-76*, 28–38. doi:10.1016/j.clay.2013.02.008

Gray, D., Gould, M., & Bickenbach, J. (2003). Environmental barriers and disability. *Journal of Architectural and Planning Research*, *20*(1), 29–37.

Gutenbrunner, C., Bender, T., Cantista, P., & Karagulle, Z. (2010). A proposal for a worldwide definition on health resort medicine, balneotherapy, medical hydrology and climatology. *International Journal of Biometeorology*, *54*(5), 495–507. doi:10.100700484-010-0321-5 PMID:20532921

Haegele, J., & Hodge, S. (2016). Disability Discourse: Over view and Critiques of the Medical and Social Models. *Quest*, *68*(2), 193-206.

Hammerle, J. (2005). The rise of stadium seating in movie theatres and the disabled's fight for a comparable seat. *De Paul Law Review*, *2*, 589.

Harpur, P., Connolly, U., & Blanck, P. (2017). Socially Constructed Hierarchies of Impairments: The Case of Australian and Irish Workers' Access to Compensation for Injuries. *Journal of Occupational Research*, *27*(4), 507–519. doi:10.100710926-017-9745-7 PMID:29181808

Harris, F., Yang, H., & Sanford, J. (2015). Physical Environmental Barriers to Community Mobility in Older and Younger Wheelchair Users. *Topics in Geriatric Rehabilitation*, *31*(1), 42–51. doi:10.1097/TGR.0000000000000043

Hattan, A. (2004). Stadium-Style Seating Movie Theatres: Does the Cornerstone of the Theatre Industry's Recent Transformation Violate the Americans with Disabilities Act? *Seton Hall Journal of Sports & Entertainment Law*, *14*(1), 267–303.

Hudson, S., Thal, K., Cardeas, D., & Meng, F. (2017). Wellness tourism: Stress alleviation or indulging healthful habits? *International Journal of Culture, Tourism and Hospitality Research*, *11*(1), 35–52. doi:10.1108/IJCTHR-09-2015-0111

Hung, K., & Lu, J. (2016). Active living in later life: An overview of aging studies in hospitality and tourism journals. *International Journal of Hospitality Management*, *53*, 133–144. doi:10.1016/j.ijhm.2015.10.008

Hung, K., & Petrick, J. (2009). How Do We Get Baby Boomers And Future Seniors On Motorcoaches? *Tourism Analysis*, *14*(5), 665–675. doi:10.3727/108354209X12597959359338

Jackson, E., & Searle, M. (1985). Recreation Non-Participation And Barriers To Participation: Concepts, And Models. *Loisir et Societe. Society and Leisure*, *8*(2), 693–707. doi:10.1080/07053436.1985.10715236

Jackson, M. (2018). Models of Disability and Human Rights: Informing the Improvement of Built Environment Accessibility for People with Disability at Neighborhood Scale? *Laws*, *7*(1), 10. doi:10.3390/laws7010010

Johnson, J. (2011). *Deep Tissue Massage*. Human Kinetics.

Johnson-Migalski, L., & Drout, M. O. (2018). Using the International Classification of Functioning, Disability, and Health (ICF) in Adlerian approaches. *Journal of Individual Psychology*, *74*(1), 38–54. doi:10.1353/jip.2018.0003

Kaganek, K., Ambrozy, T., Mucha, D., Jurczak, A., Bornikowska, A., Ostrowski, A., Janiszewska, R., & Mucha, T. (2017). Barriers To Participation In Tourism In The Disabled. *Sport Tourism*, *24*(2), 121–129. doi:10.1515/pjst-2017-0013

Karagulle, M., & Karagulle, M. (2015). Effectiveness of balneotherapy and spa therapy for the treatment of chronic low back pain: A review on latest evidence. *Clinical Rheumatology*, *34*(2), 207–214. doi:10.100710067-014-2845-2 PMID:25535198

Kasnauskiene, G., & Michnevic, K. (2017). Contribution of increased life expectancy to economic growth: Evidence from CEE countries. *International Journal of Economic Sciences*, *2*, 82.

Kastenholz, E., Eusebio, C., & Figueiredo, E. (2015). Contributions of tourism to social inclusion of persons with disability. *Disability & Society*, *30*(8), 1259–1281. doi:10.1080/09687599.2015.1075868

Keeratitanont, K., Jensen, M., Chatchawan, U., & Auvichayapat, P. (2015). The efficacy of traditional Thai massage for the treatment of chronic pain: A systematic review. *Complementary Therapies in Clinical Practice*, *22*(1), 26–32. doi:10.1016/j.ctcp.2015.01.006 PMID:25682523

Kim, E., Chiange, L., & Tang, L. (2016). Investigating wellness tourists' motivation, engagement and loyalty: In search of the missing link. *Journal of Travel & Tourism Marketing*, *34*(7), 867–879. doi:10.1080/10548408.2016.1261756

Kolodziejczack, A. (2019). Information As A Factor Of The Development Of Accessible Tourism For People With Disabilities. *Quaestiones Geographicae*, *2*(2), 67–73. doi:10.2478/quageo-2019-0014

Koog, Y., Jin, S., Yoon, K., & Min, B. (2010). Interventions for hemiplegic shoulder pain: Systematic review of randomised controlled trials. *Disability and Rehabilitation*, *32*(4), 282–291. doi:10.3109/09638280903127685 PMID:20055567

Leong, I., & Higgins, S. (2010). Public Library Services For Wheelchair-Bound Young People in Singapore. *Public Library Quarterly*, *29*(3), 210–229. doi:10.1080/01616846.2010.502033

Liu, P., Zhang, H., Zhang, J., Sun, Y., & Qiu, M. (2019). Spatial-temporal response patterns of tourist flow under impulse pre-trip information search: From online to arrival. *Tourism Management*, *73*, 105–114. doi:10.1016/j.tourman.2019.01.021

Marshall, C., & Rossman, G. (2011). *Designing Qualitative Research* (5th ed.). SAGE Publications Inc.

Masala, C., & Petretto, D. (2008). From disablement to enablement: Conceptual models of disability in the 20th Century. *Disability and Rehabilitation*, *30*(17), 1233–1244. doi:10.1080/09638280701602418 PMID:18821191

McKercher, B., & Darcy, S. (2018). Re-conceptualizing barriers to travel by people with disabilities. *Tourism Management Perspectives*, *26*, 59–66. doi:10.1016/j.tmp.2018.01.003

Michopoulou, E. (2017). Marketing for the Spa Industry. In S. Rawlinson & T. Heap (Eds.), *International Spa Management*. Goodfellow Publishers.

Michopoulou, E., & Buhalis, D. (2013). Information Provision for Challenging Markets: The Case of the Accessibility Requiring Market in the Context of Tourism. *Information & Management*, *50*(5), 229–239. doi:10.1016/j.im.2013.04.001

Michopoulou, E., Darcy, S., Ambrose, I., & Buhalis, D. (2015). Accessible tourism futures: The world we dream to live in and the opportunities we hope to have. *Journal of Tourism Futures*, *1*(3), 179–188. doi:10.1108/JTF-08-2015-0043

Mintel. (2015). *Complementary and Alternative Medicine – UK – December 2015.* Available at: https://academic.mintel.com/display/716196/

Morris, J., Mueller, J., & Jones, M. (2010). Tomorrows Elders with Disabilities: What the Wireless Industry Needs to Know. *Journal of Engineering Design*, *21*(2/3), 131–146. doi:10.1080/09544820903303431

Nyanjom, J., Boxall, K., & Slaven, J. (2018). Towards inclusive tourism? Stakeholder collaboration in the development of accessible tourism. *Tourism Geographies*, *20*(4), 675–697. doi:10.1080/14616688.2018.1477828

Olsen, J. (2018). Socially disabled: The fight disabled people face against loneliness and stress. *Disability & Society*, *33*(7), 1160–1164. doi:10.1080/09687599.2018.1459228

Ortega, E., Galvez, I., Hinchado, M., Guerro, J., Martin-Cordero, L., & Torres-Piles, S. (2017). Anti-inflammatory effect as a mechanism of effectiveness underlying the clinical benefits of Pelotherapy in osteoarthritis patients: Regulation of the altered inflammatory and stress feedback response. *International Journal of Biometeorology*, *61*(10), 1777–1785. doi:10.100700484-017-1361-x PMID:28456879

Ozkuk, K., Uysal, B., Ates, Z., Okmen, B., Refia, S., & Gamze, D. (2018). The effects of inpatient versus outpatient spa therapy on pain, anxiety, and quality of life in elderly patients with generalized osteoarthritis: A pilot study. *International Journal of Biometeorology*, *62*(10), 1823–1832. doi:10.100700484-018-1584-5 PMID:30022244

Peamruetai, B., Khunnara, N., Angkana, A., Sukree, K., Suton, J., Premsuda, J., ... Pravit, A. (2016). The Effect of the Posture of the "Hermit Doing Body Contortion" on Relief of Shoulder and Scapular Pain Caused by Chronic Myofascial Pain Syndrome: A Randomised, Parallel Group, Controlled Trial. *Siriraj Medical Journal*, *68*(6), 350–357.

Pereira, L. (2018). Seaweeds as Source of Bioactive Substances and Skin Care Therapy – Cosmeceuticals, Algotherapy and Thalassotherapy. *Cosmetics*, *5*(4), 68. doi:10.3390/cosmetics5040068

PN-Paraplegia News. (2005). Pool Party: Why Swelter in Summers Heat? Whether for therapy or just cooling off, residential swimming pools and spas can be accessed by just about everyone. *PN- Paraplegia News, 5*, 55.

Poria, R., Reichel, A., & Brandt, Y. (2010). The Flight Experiences of People with Disabilities: An Exploratory Study. *Journal of Travel Research*, *49*(2), 216–227. doi:10.1177/0047287509336477

Poria, Y., Reichel, A., & Brandt, Y. (2011). Dimensions of hotel experience of people with disabilities: An exploratory study. *International Journal of Contemporary Hospitality Management*, *23*(5), 571–591. doi:10.1108/09596111111143340

Rachid-Kandvani, F., Nicolau, B., & Bedos, C. (2015). Access to Dental Services for People Using a Wheelchair. *American Journal of Public Health*, *105*(11), 2312–2317. doi:10.2105/AJPH.2015.302686 PMID:26378836

Rawlinson, S., & Heap, T. (2017). *International Spa Management: Principles and practice*. Goodfellow Publishers Ltd.

Smith, E., Sakakibara, B., & Miller, W. (2016). A review of factors influencing participation in social and community activities for wheelchair users. *Disability and Rehabilitation. Assistive Technology*, *11*(5), 361–374. doi:10.3109/17483107.2014.989420 PMID:25472004

Smith, M., & Puczko, L. (2017). *The Routledge Handbook of Health Tourism*. Routledge.

Smith, R. (1987). Leisure of disable tourists. *Annals of Tourism Research*, *14*(3), 376–389. doi:10.1016/0160-7383(87)90109-5

Sritoomma, N., Moyle, W., Cooke, M., & O'Dwyer, S. (2012). The effectiveness of Swedish massage and traditional Thai massage in treating chronic low back pain: A review of the literature. *Complementary Therapies in Clinical Practice*, *18*(4), 227–234. doi:10.1016/j.ctcp.2012.07.001 PMID:23059437

Stillman, M., Williams, S., Bertocci, G., Smalley, C., & Frost, K. (2017). Healthcare utilization and associated barriers experienced by wheelchair users: A pilot study. *Disability and Health Journal*, *10*(4), 502–508. doi:10.1016/j.dhjo.2017.02.003 PMID:28245968

Suarez, M., Gonzalez, P., Dominguez, R., Bravo, A., Melian, C., Perez, M., & Fagundo, J. (2011). Identification of organic compounds in San Diego de los Banos Peliod (Pinar del Rio, Cuba). *Journal of Alternative and Complementary Medicine (New York, N.Y.)*, *17*(2), 155–165. doi:10.1089/acm.2009.0587 PMID:21309710

Tarasoff, L. (2017). Sexual and Reproductive Health of Women with Disability: "We don't know. We've never had anybody like you before. Barriers to perinatal care for women with physical disabilities. *Disability and Health Journal, 10*(3), 426–433. doi:10.1016/j.dhjo.2017.03.017 PMID:28404229

Tsiotsou, R., & Hudson, S. (2010). Wooing zoomers: Marketing to the mature traveller. *Marketing Intelligence & Planning, 28*(4), 444–461. doi:10.1108/02634501011053568

Ucok, K., Mollaoglu, H., Demirel, R., Ucok, K., Mollaoglu, H., Kavuncu, V., ... Evcik, D. (2008). Effects of balneotherapy with exercise in patients with low back pain. *Journal of Back and Musculoskeletal Rehabilitation, 21*(4), 263–272. doi:10.3233/BMR-2008-21407

United Nations. (2006a). *Convention on the Rights of Persons with Disabilities – Articles.* Convention on the Rights of Persons with Disabilities (CRDP). Available at: https://www.un.org/development/desa/disabilities/convention-on-the-rights-of-persons-with-disabilities/convention-on-the-rights-of-persons-with-disabilities-2.html

United Nations. (2006b). *Article 19 – Living independently and being included in the community.* Convention on the Rights of Persons with Disabilities (CRDP). Available at: https://www.un.org/development/desa/disabilities/convention-on-the-rights-of-persons-with-disabilities/article-19-living-independently-and-being-included-in-the-community.html

United Nations. (2006c). Article 30 – Participation in cultural life, recreation, leisure and sport. *Convention on the Rights of Persons with Disabilities (CRDP).* Available at: https://www.un.org/development/desa/disabilities/convention-on-the-rights-of-persons-with-disabilities/article-30-participation-in-cultural-life-recreation-leisure-and-sport.html

Varga, J., Gaal, J., Varga, J., Szekanecz, Z., Kurko, J., Ficzere, A., ... Bender, T. (2008). Balneotherapy in elderly patients: Effect on pain from degenerative knee and skin conditions and on quality of life. *The Israel Medical Association Journal, 10*(5), 365–369. PMID:18605361

Visit England. (2015). *The Purple Pound. Volume and Value of Accessible Tourism in England.* Available at: https://www.visitbritain.org/sites/default/files/vb-corporate/Documents-Library/ve_purplepound2015.pdf

Voigt, C., Brown, G., & Howat, G. (2011). Wellness tourists: In search of transformation. *Tourism Review, 66*(1), 16–30. doi:10.1108/16605371111127206

Voigt, C., & Pforr, C. (2014). *Wellness Tourism – A destination perspective.* Routledge.

Wan, Y. (2013). Barriers for people with disabilities in visiting casinos. *International Journal of Contemporary Hospitality Management, 25*(5), 660–682. doi:10.1108/IJCHM-Jul-2012-0112

Weiermair, K., & Mathies, C. (2004). *The Tourism and Leisure Industry: Shaping the Future.* The Haworth Press, Inc.

Williams, V. (2017). The 'Purple Pound' and Breaking down Barriers in the Workplace in the UK. *International Journal of Therapy and Rehabilitation, 24*(3), 97. doi:10.12968/ijtr.2017.24.3.97

World Health Organisation. (2017). *10 facts on disability.* Available at: https://www.who.int/features/factfiles/disability/en/

Yau, M., McKercher, B., & Packer, L. (2004). Traveling with a disability. *Annals of Tourism Research*, *31*(4), 946–960. doi:10.1016/j.annals.2004.03.007

Zhang, Y., Roxburgh, R., Huang, L., Parsons, J., & Davies, T. (2014). The effect of hydrotherapy treatment on gait characteristics of hereditary spastic paraperesis patients. *Gait & Posture*, *39*(4), 1074–1079. doi:10.1016/j.gaitpost.2014.01.010 PMID:24556467

KEY TERMS AND DEFINITIONS

Accessibility: The practice of making products and services, spaces, information, and processes usable by as many people as possible.

Barrier: A circumstance or obstacle that prevents movement or access.

Disability: A physical or mental condition that limits a person's movements, senses, or activities.

Leisure: Time when one is not working or occupied; free time.

Spa: A commercial establishment offering health and beauty treatment through such means as steam baths, exercise equipment, and massage.

Wellbeing: The state of being comfortable, healthy, or happy.

Wellness: The state of being in good health, especially as an actively pursued goal.

Chapter 8
An Empirical Investigation of the Accessibility of Official European Tourism Websites

Katja Kous
Faculty of Electrical Engineering and Computer Science, University of Maribor, Slovenia

Gregor Polančič
https://orcid.org/0000-0002-4746-1010
Faculty of Electrical Engineering and Computer Science, University of Maribor, Slovenia

ABSTRACT

This chapter presents an investigation into the accessibility of official tourism websites of 26 EU member states. It is divided into two main parts. The first part is focused on the theoretical background of web accessibility, where the definition of accessibility, accessibility laws, accessibility recommendations, accessibility automatic tools, and related research are presented. The second part includes an empirical investigation with the main focus being on evaluating the compliance of tourism websites with web content accessibility guidelines 2.1 (WCAG 2.1) and the identification of the most commonly violated WCAG 2.1 success criteria. The results, which are based on automatic tool multiguideline accessibility usability validation environment (MAUVE), show that all investigated websites included at least one error. The most frequently identified errors were related to success criteria (1) reflow; (2) name, role, value; (3) link purpose (in context); and (4) purpose identification. Based on the acquired knowledge, the recommendations for improving web accessibility are presented.

INTRODUCTION

The media on the Internet have become the most common channels for acquiring information (Domínguez Vila, Alén González, & Darcy, 2018) for all people, including people with disabilities, who represent about 15% of the world's population (World Health Organization, 2018). In the European Union (hereinafter referred to as EU), the percentage of people with disabilities is slightly lower and includes approximately

DOI: 10.4018/978-1-7998-6428-8.ch008

10% of the population (European Network for Accessible Tourism, 2008). Across EU member states, the percentage of people with long-term disabilities varies between 12% and 41% (Eurostat, 2018). The lowest proportions of self-perceived long-standing disabilities were reported in Malta (12%) and Sweden (13%), whereas the highest were reported in Latvia (41%), Slovenia (36%), and Estonia (35%) (Eurostat, 2018). In 2020, the EU is expected to have 120 million people with disabilities (European Commission, 2018; Rajšp, Kous, Kuhar, Šumak, & Šorgo, 2019). Additionally, the population in Europe is aging. When the elderly (aged 65 or above) are added, the number of people who live with some form of disability additionally increases.

In light of this, accessibility is presented as a significant segment of website quality characteristics in terms of ensuring equal opportunities in accessing online information. All websites should be accessible for all users, regardless of their characteristics or needs. This means that websites, as well as tools and technologies, should be designed and developed so that all people (including people with visual, auditory, physical, speech, cognitive, and neurological disabilities) can use them effectively. Many recommendations, guidelines, standards, and concepts have been published to ensure web accessibility (e.g., WCAG, ISO/IEC 40500:2019, EN 301 549, Section 508, Design for all, etc.) (Kous, Kuhar, Pušnik, & Šumak, 2019; Kous & Polančič, 2019). Although many guidelines and standards describe development principles and how to design an accessible website, websites still demonstrate a relatively low degree of accessibility. The existing research among the most commonly researched domains (e.g. education, and government) (Nunez, Moquillaza, & Paz, 2019) reports that many websites include a large number of accessibility-related problems and, usually, the investigated websites are inconsistent with the guidelines presented in the related standards (Acosta, Acosta-Vargas, & Lujan-Mora, 2018; Ahmi & Mohamad, 2015; Al-Khalifa, Baazeem, & Alamer, 2017; Espadinha, Pereira, Da Silva, & Lopes, 2011; Ismail & Kuppusamy, 2016). Our previous research, which was performed in governmental (Kous & Pavlinek, 2017) and educational domains (Kous et al., 2019), also reported similar results. Between 2013 and 2017, we investigated the web accessibility of 189 Slovenian municipalities' websites, where only one website demonstrating full compliance with WCAG 2.0 in 2013 and 2014 (Kous & Pavlinek, 2017). However, in more recent years, all websites included at least one case of noncompliance with the same guidelines (Kous & Pavlinek, 2017). The second study was focused on a web accessibility investigation of the faculties' websites at the University of Maribor between 2018 and 2019 (Kous et al., 2019). The results, based on an automatic evaluation, indicated that none of the investigated web pages satisfied all WCAG 2.0 recommendations (Kous et al., 2019). Nevertheless, when comparing 2018 to 2019, the number of violated success criteria decreased among 59% of web pages. For 12% of web pages, the number of violated success criteria remained unchanged, and the number of violated success criteria increased in 29% of web pages (Kous et al., 2019). To summarize, although awareness of web accessibility on a theoretical level is very high, when it comes to the practical level, web accessibility is still a challenge and largely depends on web designers and implementers. Therefore, many websites are still inaccessible for certain groups of the population, especially for people with disabilities (Kous et al., 2019) and the elderly.

The websites have also become one of the significant sources of getting information in the tourism domain, since users get information about the destination, including details about activities, accommodation, transport and other attractions that they may wish to access (Domínguez Vila et al., 2018; Teixeira, Eusébio, & Silveiro, 2019). This has also contributed to the use of mobile devices that provide real-time access to information, which is also essential for the domain. Therefore, official tourism websites are important not only in persuading people to choose a specific destination and to find the searched-for information in real-time but also to present a source of information to plan a trip and help to create and promote the long-term development and marketing of a destination (Domínguez Vila et al., 2018). A surprising bit of information, mentioned by the European Network for Accessible Tourism (Eurostat, 2018), is that 70% of people with officially registered disabilities are physically and financially able to travel. To ensure the accessibility of tourist destinations, products, and services to all people, including people with disabilities and the elderly, the term "accessible tourism" was created a decade ago. In 2009, Darcy and Dickson defined it thusly: "Accessible tourism enables people with access requirements, including mobility, vision, hearing and cognitive dimensions of access, to function independently and with equity and dignity through the delivery of universally designed tourism products, services and environments. This definition is inclusive of all people, including those traveling with children in prams, people with disabilities and seniors." (Darcy & Dickson, 2009, p. 34) . According to European Network for Accessible Tourism (ENAT, 2018), accessible tourism refers to "tourism that caters to the needs of a full range of consumers including persons with disabilities, older persons and cross-generational families. It entails the removal of attitudinal and institutional barriers in society and encompasses accessibility in the physical environment, in transportation, information and communications and other facilities and services. It encompasses publicly and privately-owned tourist locations". Nevertheless, people with disabilities encounter challenges when they travel and one of the main barriers is related to the search for and access to information (Vila, González, Elisa, & Darcy, 2019).

In light of the above, web accessibility is also essential in the tourism domain. Based on this, we were motivated to achieve an insight into web accessibility in the tourism domain and to verify whether the official tourism websites of EU member states are compliant with WCAG 2.1.

The chapter is organized into seven sections. The next section includes the presentation of the theoretical background of web accessibility. In the beginning, web accessibility definitions are presented. Based on a review of different web accessibility guidelines, concepts, and standards, a brief description, with a focus on the detailed presentation of WCAG 2.1, is given. The following section is focused on evaluation tools for automated accessibility assessment. Therefore, an overview of the evaluation tools is given, including the tools' classification based on different criteria (e.g. type, license, supported format). The third section describes similar research obtained by literature review in the tourism domain. The fourth section presents a detailed description of research, including the research objective, investigated websites, selected evaluation tool and execution of the research. In the fifth section, the results with a discussion are presented. After that, the recommendations for the improvement of web accessibility are described. The chapter is finished with conclusions.

WEB ACCESSIBILITY

According to existing literature, there are numerous definitions for the term accessibility. ISO 9241-171 uses a more general description and relates accessibility to the term usability. It defines accessibility as the "usability of a product, service, environment or facility by people with the widest range of capabilities" (ISO 9241-171, 2008). ISO 30071-1 uses a detailed and extended definition and defines accessibility as the "extent to which products, systems, services, environments and facilities can be used by people from a population with the widest range of user needs, characteristics and capabilities to achieve identified goals in identified contexts of use" (ISO/IEC 30071-1:2019, 2019). This extended definition is also presented and used by other ISO standards, such as (ISO/TR 22411:2008, 2008), (ISO 26800:2011, 2011), (ISO

Table 1. The list of laws in countries around Europe

Country	Name (date enacted)	Type	Scope	Web Only
Finland	Act on Electronic Services and Communication in the Public Sector (2003)	Accessibility law	Government	no
France	Law N° 2005-102 Article 47 (2005)	Accessibility law	Public sector	no
France	Order of 29 April 2015 on the general accessibility framework for public administrations (2015)	Accessibility law	Public sector	yes
France	Law N° 2016-1321 Article 106 (2016)	Digital Governance law	Public sector, Private sector	no
Germany	Act on Equal Opportunities for Disabled Persons of 2002 (2002)	Non-discrimination law	Public sector, Private sector	no
Germany	Federal Ordinance on Barrier-Free Information Technology (2011)	Accessibility policy	Government	yes
Ireland	The Disability Act (2005)	Accessibility law	Public sector	no
Ireland	Equal Status Acts 2000 to 2004 (2004)	Non-discrimination law	Public sector, Private sector	no
Ireland	Employment Equality Acts 1998 and 2004 (2004)	Non-discrimination law	Public sector, Private sector	no
Italy	Law of 9 January 2004, n. 4 "Provisions to support the access of disabled people to I.T. tools" (Stanca Law) (2004)	Accessibility law	Public sector, government	no
Sweden	Discrimination Act (2008:567)	Non-discrimination law	Public sector, Private sector	no
United Kingdom	Equality Act 2010 (2010)	Non-discrimination law	Public sector, Private sector	no

Source: W3C WAI (2018d).

9241-112:2017, 2017). W3C specifies the most commonly used definition stating that *"websites, tools, and technologies are designed and developed so that people with disabilities can use them"* (W3C WAI, 2018a). More specifically, people (including people with visual, auditory, physical, speech, cognitive, and neurological disabilities) can (W3C WAI, 2018a): (1) perceive, understand, navigate, and interact with the Web; and (2) contribute to the Web.

Accessibility Laws and Policies

Accessibility is one of the general principles of the Convention on the rights of persons with disabilities, which is supported by several legal laws and acts defined on both international and national levels. In the EU, three web accessibility laws and policies have been adopted to ensure equal access for people with disabilities (EUR-Lex, 2018), namely: (1) the United Nations Convention on the Rights of Persons with Disabilities; (2) the Web and Mobile Accessibility Directive, and (3) the European Standard EN 301 549 - Accessibility requirements suitable for public procurement of ICT products and services (Kous, Kuhar, Pavlinek, Heričko, & Pušnik, 2020). Before the EU adopted uniform accessibility laws, many countries within the EU published specific accessibility laws and policies as presented in Table 1.

Most of the laws presented in Table 1 are defined as the accessibility law (42%) or non-discrimination law (42%) and only two laws are defined differently - one law (8%) as the Digital Governance law and one (8%) as the Accessibility policy. Two of the presented laws (16%) are applied for the government, three (25%) for the public sector, one (8%) for both the public sector and government and six (50%) for the public as well as the private sector. Most of the presented laws (84%) describe the general accessibility law and only two laws (16%) are specified only for web accessibility.

To summarize, the adopted laws indicate that awareness of the importance of accessibility is on a high-level political sphere at the international as well as national levels. Therefore, the existing laws on a legal level ensure equal opportunities for all people and eliminate accessibility challenges for them, especially for people with disabilities (Kous et al., 2020). Some of the existing laws are based on accessibility recommendations and guidelines. Therefore, they are related to practical knowledge and include practical guidelines for ensuring web accessibility. The most known recommendations are presented in the following subsection.

Relevant Recommendations for web Accessibility

Many recommendations, guidelines, standards, and concepts have been published to ensure knowledge about web accessibility and accessible websites. The common denominator of these documents is to provide instructions to designers and developers on how they can create and develop a website that is useful for a multitude of people with different abilities, needs, demands and preferences (Kous & Polančič, 2019). A summary of the most known recommendations is presented in Table 2, while the description is presented below:

- **Web Content Accessibility Guidelines (WCAG)** are classified as guidelines and present the most known guidelines published by W3C (Kous et al., 2019), which describes how to make web content more accessible to all web users, especially to people with disabilities (W3C WAI, 2018a). The first version of WCAG was published in May 1999, while the next version (WCAG 2.0) was available in December 2008. Six years later, the last version (WCAG 2.1) was released (W3C WAI, 2019b).
- **ISO/IEC 40500:2012 Information technology — W3C Web Content Accessibility Guidelines (WCAG) 2.0** (International Organization for Standardization, 2012) is presented as a standard and is based on the recommendations defined in WCAG 2.0. In 2012, WCAG 2.0 was approved as ISO standard ISO/IEC 40500 (International Organization for Standardization, 2012). The recommendations, described in the standard, are the equivalent of WCAG 2.0. It is composed of four main principles, 12 guidelines and 61 success criteria, which are classified into three conformance levels (A, AA, AAA).
- **Standard EN 301 549 - Accessibility requirements for ICT (information and communications technology) products and services** (European Telecommunications Standards Institute, 2018) is classified as the first European Standard for accessible ICT, produced by the European Committee for Standardization (CEN), European Committee for Electrotechnical Standardization (CENELEC) and the European Telecommunications Standards Institute (ETSI). The primary purpose of the Standard EN 301 549 is to improve the accessibility of websites, software, and digital devices for public authorities and other public sector bodies during the public procurement process. The accessibility requirements are separated into nine sections: Generic requirements, ICT with two-way voice communication, ICT with video capabilities, Hardware, Web, Non-web documents, Software, Documentation and support services, and ICT providing relay or emergency

Table 2. Overview of the most known recommendations

Name	Type	Area	Last version - Year	Based on WCAG
Web Content Accessibility Guidelines	guideline	Around the world	2.1 (2018)	-
ISO/IEC 40500:2012	standard	Around the world	2012	WCAG 2.0
Standard EN 301 549	standard	EU	V2.1.2 (2018)	WCAG 2.1 (A, AA)
Stanca Act	law	Italy	2013	WCAG 2.0 (A, AA)
Section 508	standard	United States	2018	WCAG 2.0 (A, AA)
Design for all	concept	Around the world	N/A	no

service access. The web accessibility requirements presented in section nine are equivalent to the WCAG 2.1 at level A and AA.

- **Stanca Act** is an Italian law, dealing with the accessibility of information technology. It became operational in 2005 and the first version was compliant with WCAG 1.0. It was reformed in 2013 to shift to WCAG 2.0 level AA compliance (Barricelli, Sciarelli, Valtolina, & Rizzi, 2018). Currently, the Stanca Act is composed of 22 technical requirements[1].
- **Section 508** of the Rehabilitation Act is classified as a standard, requiring federal agencies to develop, procure, maintain and use ICT that is accessible for people with disabilities - regardless of whether or not they work for the federal government (United States Environmental Protection Agency, 2019). In January 2018, the technical requirements of Section 508 have changed to incorporate WCAG 2.0 A and AA success criteria (WebAIM, 2018).
- **Design for all** is defined as concepts. It is the intervention into environments, products, and services, which aims to ensure that anyone, including future generations, regardless of age, gender, capacities, or cultural background, can participate in social, economic, cultural and leisure activities with equal opportunities (Design for all foundation, 2019). It includes the eight criteria related to respectful, safe, healthy, functional, comprehensible, sustainable, affordable and appealing (Design for all foundation, 2019).

In summary, most of the above-mentioned recommendations are based on content presented in WCAG. Additionally, 25% of laws presented in Table 1 (The list of laws in countries around Europe) are based on WCAG content (such as Ireland - the Equal Status Acts 2000 to 2004, Italy - Stanca Law, United Kingdom - Equality Act 2010), 17% is presented as a derivation of WCAG (such as France - Order of 29 April 2015 on the general accessibility framework for public administrations, Germany - Federal Ordinance on Barrier-Free Information Technology), while others (58%) do not have any relation with WCAG 2.0. Therefore, the WCAG represents the *de facto* standard of recommendations for developing accessible websites. In the following subsection, the last version of WCAG 2.1 is presented.

Web Content Accessibility Guidelines (WCAG) 2.1

As mentioned in the previous subsection, the last version of WCAG was published in 2018. The topical version included all success criteria from WCAG 2.0 and added 17 additional success criteria that address mobile accessibility, people with low vision and people with cognitive and learning disabilities (W3C WAI, 2018e).

WCAG 2.1 is constructed around four main principles and each principle is supported by guidelines (W3C WAI, 2018c):

- principle Perceivable: information and user interface components must be presentable to users in ways they can perceive:
    - Guideline 1.1 Text alternatives: provide text alternatives for any non-text content so that it can be changed into other forms people need, such as large print, braille, speech, symbols or simpler language.
    - Guideline 1.2 Time-based media: provide alternatives for time-based media.
    - Guideline 1.3 Adaptable: create content that can be presented in different ways (for example, a simpler layout) without losing information or structure.
    - Guideline 1.4 Distinguishable: make it easier for users to see and hear content, including separating the foreground from the background.
- principle Operable: user interface components and navigation must be operable.
    - Guideline 2.1 Keyboard accessible: make all functionality available from a keyboard.
    - Guideline 2.2 Enough time: provide users enough time to read and use the content.
    - Guideline 2.3 Seizures and physical reactions: do not design content in a way that is known to cause seizures or physical reactions.
    - Guideline 2.4 Navigable: provide ways to help users navigate, find content, and determine where they are.
    - Guideline 2.5 Input modalities: make it easier for users to operate functionality through various inputs beyond the keyboard.
- principle Understandable: information and the operation of user interface must be understandable.
    - Guideline 3.1 Readable: make text content readable and understandable.
    - Guideline 3.2 Predictable: make web pages appear and operate in predictable ways.
    - Guideline 3.3 Input assistance: help users avoid and correct mistakes.
- principle Robust: content must be robust enough so that it can be interpreted by a wide variety of user agents, including assistive technologies.
    - Guideline 4.1 Compatible: maximize compatibility with current and future user agents, including assistive technologies.

Each guideline includes a different amount of success criteria (Appendix 1:
Table 8. List of Success criteria (WCAG 2.1)) that describes what must be realized to conform to these recommendations (W3C WAI, 2018e). These success criteria are divided into the following conformance levels:

- level A (minimum level) includes 34 success criteria,
- level AA (medium level) includes 16 success criteria,
- level AAA (highest level) includes 28 success criteria.

The overview of the WCAG 2.1 structure is presented in Figure 1.

Figure 1. WCAG 2.1

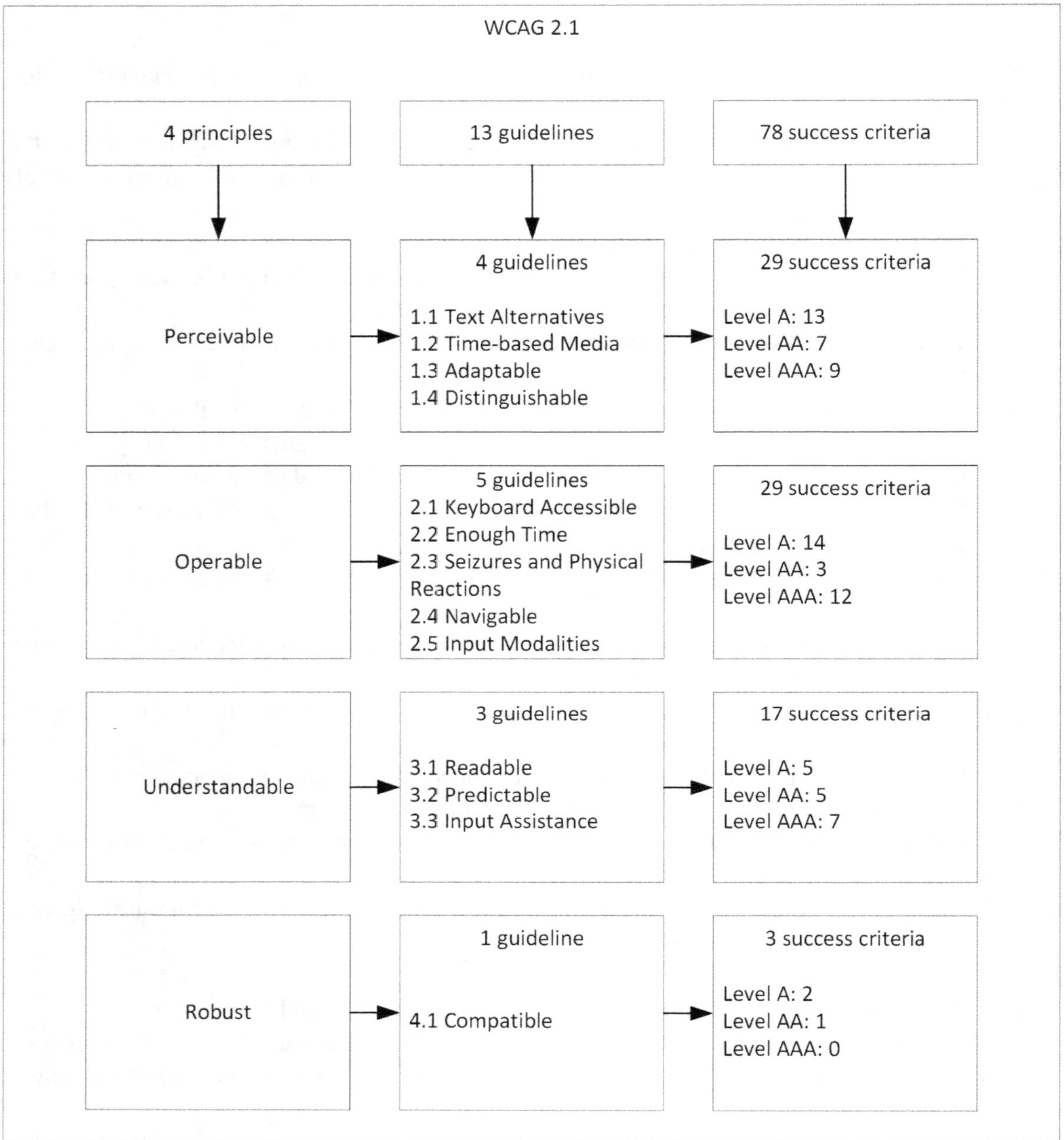

Assessment Tools

According to existing literature, several tools for automatically web accessibility evaluation exist (Schiavone & Paterno, 2015). The list with the most relevant tools is presented by W3C (W3C WAI, 2019a). From the existing list published on W3C WAI (2019a), we selected the tools which automatically check the groups of web pages or web sites and evaluate their compliance with WCAG 2.1. Table 3 represents the overview of the evaluation tools with classification based on different criteria (e.g., License, Type of tool, Supported formats).

As presented in Table 3, several vendors published their products for evaluating compliance with WCAG 2.1. Most of the tools (44%) have commercial and/or enterprise licenses, including a free and/or demo fixed-term license, which presents the version with limited functionalities. 33% of tools have only commercial and enterprise licenses, while only two tools (11%) are offered as free software and two tools (11%) are marked as open-source tools with a free license. The next category is the type of tool where 39% of tools are classified as an online service, 2% of them as browser plugins, while 33% are marked as both online service and browser plugins. For three of them, this information was unavailable. Different tools include support for different formats (e.g. WAI-ARIA, CSS, HTML, XHTML, SVG, PDF documents, Images, SMIL, Microsoft Office documents).

Based on the tool review and evaluation, we used the tool Multi-guideline Accessibility Usability Validation Environment (MAUVE v2.0) in our research. The tool is presented in Schiavone and Paterno (2015) and is used in similar research at other domains (Alshamari, 2016; Paterno & Schiavone, 2015).

RELATED WORK

Existing research reveals the popularity of accessibility topics in the science domain. Many researchers have investigated accessibility within different domains, such as in education, government, medicine, as well as tourism. The literature review of accessibility in the tourism domain shows that accessibility investigation is on the rise in scientific research, and much research on accessible tourism has been performed during recent years. Nevertheless, the literature review reveals that the automatic evaluation of web accessibility within tourism websites lack investigation.

The literature review was performed with the research string ("web accessibility" AND "tourism") within the following knowledge bases: IEEE Xplore, Web of Science, ACM Digital Library and Science Direct. Detailed information, including the number of obtained records separated per the included knowledge database, is presented in Table 4. An inquiry with the research string across the knowledge databases returned 91 results. Based on a review of the abstracts and main segments of the empirical part of the study, the results that did not investigate the automatic evaluation of tourism websites, were excluded. In the end, only six studies were similar to our research and they included relevant information for further review. The details of this research are presented below.

In 2003, Oertel, Haße, Scheermesser, Thio and Feil (2004) focused on the investigation of 16 official national tourism websites of the European Union via the automated tool Bobby. The research is based on studying the compliance of websites with 14 guidelines of WCAG 1.0 defined by the Web Accessibility Initiative. The results indicate that none of the investigated websites complied with WCAG 1.0. The websites of Denmark (78%), followed by Finland (68%) and the United Kingdom (68%), Ireland (66%), Belgium (65%) and Portugal (65%) were shown to be particularly compliant with WCAG 1.0. The lowest level of accessibility was seen with the website of the Netherlands (34%).

In 2005, Maswera, Dawson and Edwards (2005) investigated the accessibility and usability of tourist organizations from South Africa, Kenya, Zimbabwe and Uganda and compared results with tourist organizations from Europe. The research was based on the automatic evaluation of 318 websites from various sectors of the tourism industry with the tools Bobby and LIFT. The results of the comparison with WCAG 1.0 showed that no website was completely consistent with the mentioned guidelines.

Table 3. Overview of the evaluation tools with classification based on different criteria

No.	Name (Producer)	License	Type of tool	Supported formats
1	A11ygato v1.0.1 (Orange)	Free Software, Open Source	**Online Service:** Server installation	WAI-ARIA, CSS, HTML, XHTML, SVG, Images
2	Access Alchemy v2.0 (Level Access)	Commercial, Enterprise	**Browser plugins:** Chrome, Firefox	CSS, HTML, XHTML
3	Access Analytics v3.0 (Level Access)	Commercial, Enterprise	**Online Service:** Hosted service	CSS, HTML
4	Access Continuum v2.1 (Level Access)	Commercial, Enterprise	N/A	CSS, HTML, XHTML
5	Access Continuum Community Edition v2.1 (Level Access)	Free Software	N/A	CSS, HTML
6	A.Q.A. (UsableNet)	Commercial, Enterprise	**Browser plugin:** Chrome, **Online Service:** Online checker, Hosted service, Server installation	WAI-ARIA, CSS, HTML, XHTML
7	Crownpeak Digital Quality Management (D.Q.M.) (Crownpeak)	Commercial, Enterprise	**Browser plugins:** Chrome, Firefox, Edge / Explorer **Online Service:** Hosted service	CSS, HTML, PDF documents
8	CSS HTML Validator for Windows v2019 (AI Internet Solutions L.L.C.)	Commercial	N/A	WAI-ARIA, CSS, HTML, XHTML
9	Domain Accessibility Audit v0.1.2 (Michigan State University)	Free Software, Open Source	**Online Service:** Online checker, Server installation	CSS, HTML, XHTML
10	Dyno Mapper v3.0 (Indigo Design Company L.L.C.)	Trial or Demo, Commercial, Enterprise	**Online Service:** Online checker	CSS, HTML, XHTML
11	EqualWeb (EqualWeb)	Free Software, Trial or Demo, Commercial, Enterprise	**Browser plugins:** Any (bookmarklet), Chrome, Firefox, Edge, Opera **Online Service:** Online checker, Hosted service, Server installation	WAI-ARIA, CSS, HTML, XHTML, SVG, PDF documents, Images, SMIL, Microsoft Office documents
12	Little Forest index (LFi) v2.1 (Little Forest)	Trial or Demo, Commercial, Enterprise	**Online Service:** Online checker, Hosted service	WAI-ARIA, CSS, HTML, XHTML, SVG, PDF documents, Images, SMIL, Microsoft Office documents, Open Document Format (ODF)
13	Make-Sense v2.1 (Make-Sense)	Trial or Demo, Commercial, Enterprise	**Online Service:** Hosted service	WAI-ARIA, CSS, HTML, XHTML, SVG, Images, SMIL, Microsoft Office documents
14	MAUVE v2.0 (Human Interfaces in Information Systems Laboratory)	Free Software	**Browser plugins:** Chrome, Firefox **Online Service:** Online checker	WAI-ARIA, CSS, HTML, XHTML, Images
15	Rocket Validator (Lighthouses S.C.P.)	Trial or Demo, Commercial	**Online Service:** Online checker	WAI-ARIA, CSS, HTML
16	uTester v1.0018 (User1st)	Trial or Demo, Commercial, Enterprise	**Browser plugin:** Chrome **Online Service:** Online checker, Hosted service, Server installation	CSS, HTML, XHTML, WAI-ARIA, PDF documents, Images
17	WAVE (WebAIM)	Free Software, Trial or Demo, Commercial	**Browser plugins:** Chrome, Firefox **Online Service:** Online checker, Hosted service, Server installation	CSS, HTML, XHTML, Images
18	WP ADA Compliance Check Plugin for Wordpress v2.0 (AlumniOnline Web Services)	Free Software, Commercial	**Plugin**	CSS, HTML, XHTML, Images

Source: W3C WAI (2019a).

Table 4. Included knowledge bases

Digital library	Acronym	Web page	Search field	Number of records
Science Direct	SD	www.sciencedirect.com	"all fields"	50
IEEExplorer	IEEE	ieeexplore.ieee.org	"all metadata"	4
ACM Digital Library	ACM	dl.acm.org/search/advanced	"anywhere"	24
Web of Science	WoS	apps.webofknowledge.com	"all fields"	13

In 2019, Silva and Borges (2019) researched the automatic evaluation of 19 Portuguese institutional websites in the tourism area with an automatic validator named AccessMonitor. The investigation was based on a comparison with WCAG 2.0. The results indicated that none of the included websites were consistent with WCAG 2.0.

In 2019, Gonçalves, Camarinha, Abreu, Teixeira and Ferreira (2019) evaluated the accessibility of the most used websites in Portugal's tourism sector with the automated tools AChecker and Wave. The research investigated compliance with WCAG 2.0 for conformance level A, AA, and AAA. Although the results indicated that four websites were entirely consistent with WCAG 2.0., the highest frequency of errors were found to correspond to the principle Perceivable (89.71%), followed by the principle Understandable (7.03%), the principle Operable (2.66%) and the principle Robust (0.59%).

Another relevant research was performed by Vila, González and Darcy, and was based on a web accessibility investigation of the official national tourism organization websites around the world. In the mentioned research, a website accessibility evaluation tool named TAW was used in the analysis, according to AA and AAA levels of conformance to WCAG 2.0. The results of this investigation are presented in two separate papers (Domínguez Vila et al., 2018; Vila et al., 2019). In Domínguez Vila et al. (2018), the results present the focal issues that are related to the following guidelines: navigable, compatible, adaptability, text alternatives and also referred to other assistive technologies. In Vila et al. (2019), the results indicate that all reviewed websites included at least one case of non-compliance with the WCAG 2.0 guidelines.

To summarize, although the last version WCAG 2.1 was published in June 2018, all related research (in the tourism domain) included an investigation of web accessibility compliance with WCAG 2.0 or with the older version, WCAG 1.0. Additionally, in similar areas of research, the accessibility evaluation is usually conducted on the entry webpage and does not include an evaluation of other sub-pages on the website. Based on these deficiencies, we were motivated to achieve an insight into web accessibility at the tourism domain and to verify whether the official tourism websites of EU members are compliant with WCAG 2.1 on all three levels of conformance (level A, level AA and level AAA). The entry page, together with sub-pages on the website, were included in the investigation.

METHODOLOGY ADOPTED IN AN INVESTIGATION OF THE WEB ACCESSIBILITY OF TOURISM WEBSITES

As discussed in previous sections, the main research objective was focused on an investigation of website accessibility. Therefore, we investigated the compliance of tourism websites with WCAG 2.1 and identified the most commonly violated and overlooked WCAG 2.1 success criteria.

Investigated Websites

The official tourism website addresses were taken from the presentation of individual EU member states published on the official EU website. The research did not include two EU members because the tool was unable to get data from their websites. To ensure the validity of the research and to provide a common starting point, English versions of the websites were analyzed. Web accessibility was verified across multiple pages of the website, and so was not limited to the homepage only, as is common in similar research. The investigation included nine web pages within one website, more specifically an entry webpage and eight randomly selected web pages for the maximum of five levels of depth. As a result, 234 webpages were investigated.

Automatic Evaluation Tool

The accessibility evaluation was performed with the Multiguideline Accessibility Usability Validation Environment (MAUVE) (Pulina, Broccia, Paternò, Manca, & Schiavone, 2020). MAUVE is a system to evaluate the accessibility of websites by validating HTML and CSS code through guidelines, which are to be specified with an XML-compliant specification language called LWGD (Language for Web Guideline Definition) that maintains the guidelines separated from the underlying logic (Pulina et al., 2020). It can validate the accessibility of a single web page or website with both versions of WCAG, 2.0 and 2.1. At the end of the validated process, the system generates a report and sends it to the user's email. The report includes different evaluation analyses that are grouped by principles (such as perceivable, understandable, operable, robust), categories (such as Aria, color, content, Form, HTML, Image, Link, Media, Meta, Responsive, Table, Other) and by comparison between HTML and style. The MAUVE's web page for the validation of a website is presented in Figure 2.

Figure 2. The entry website for multipage evaluation in MAU

Operation

The investigation procedure was divided into the following four phases:

Phase 1: In the beginning, the extraction of the official tourism website addresses published on the official EU website was performed.

Phase 2: After that, the validation phase was begun. In this part, each web address was copied into the input field on MAUVE's website while specifying the remaining criteria: the maximum member of investigated webpages was set to 9, maximum depth to 5 and the comparison with WCAG 2.1 with level AAA was selected. Afterward, automatic validation began.

Phase 3: The next phase was reporting, where the report with an analysis was sent via email.

Phase 4: At the end of the investigation procedure, a detailed analysis of the received reports was performed. The most relevant results are presented in the next subsection.

RESULTS AND DISCUSSION

Accessibility Overview

Figure 3 presents the percentage of an overview of accessibility per investigated country. The results indicate how much each website is accessible in terms of the number of checkpoints successfully evaluated in relation to the total number of evaluated checkpoints.

Figure 3. Ranking of accessibility evaluation per countries

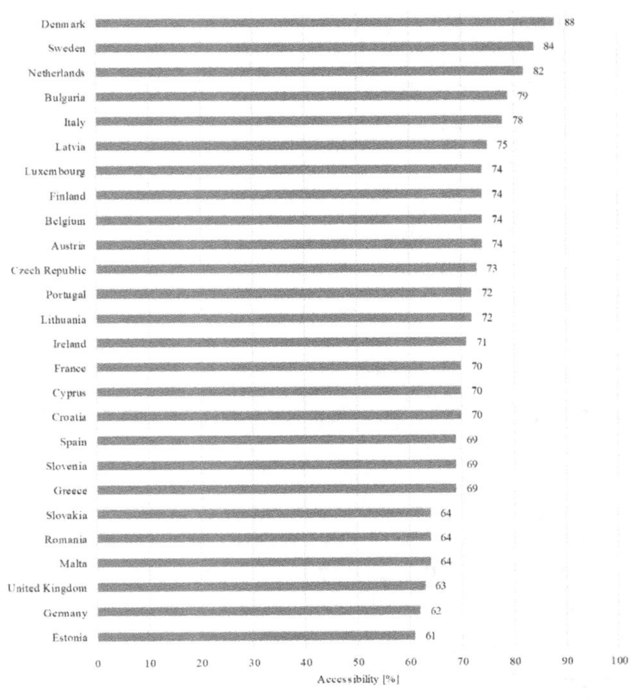

The results presented in Figure 3 show that none of the evaluated websites was consistent with WCAG 2.1. The highest level of accessibility was achieved by the website of Denmark with 88% and an average of 42 errors per page, followed by Sweden with 84% and an average of 24 errors per page, the Netherlands with 82% and an average of 80 errors per page and Bulgaria with 79% and an average of 36 errors per page. At the bottom of the rankings: the websites of the United Kingdom with 63% and an average of 172 errors per page, followed by Germany with 62% and an average of 314 errors per page and Estonia with the lowest accessibility rate (61%, with an average of 142 errors per page).

To summarize, the results indicate that none of the investigated websites were fully compliant with WCAG 2.1. The results are comparable with research performed by Oertel et al. in 2003 (Oertel et al., 2004), where the researchers evaluated the national official tourism websites within the European Union with the automated tool Bobby and checked their compliance with WCAG 1.0. In this research, the following counties were not included in the investigation: Bulgaria, Latvia, the Czech Republic, Lithuania, Croatia, Cyprus, Slovenia, Malta, Romania, Slovakia, and Estonia. The comparison between the results is presented in Figure 4.

Figure 4. Comparison with existing research from 2003
Legend: 1 – Denmark; 2 – Sweden; 3 – Netherlands; 4 – Italy; 5 – Austria; 6 – Belgium; 7 – Finland, 8 – Luxembourg, 9 – Portugal, 10 – Ireland;
11 – France; 12 – Greece; 13 – Spain; 14 – United Kingdom; 15 – Germany.

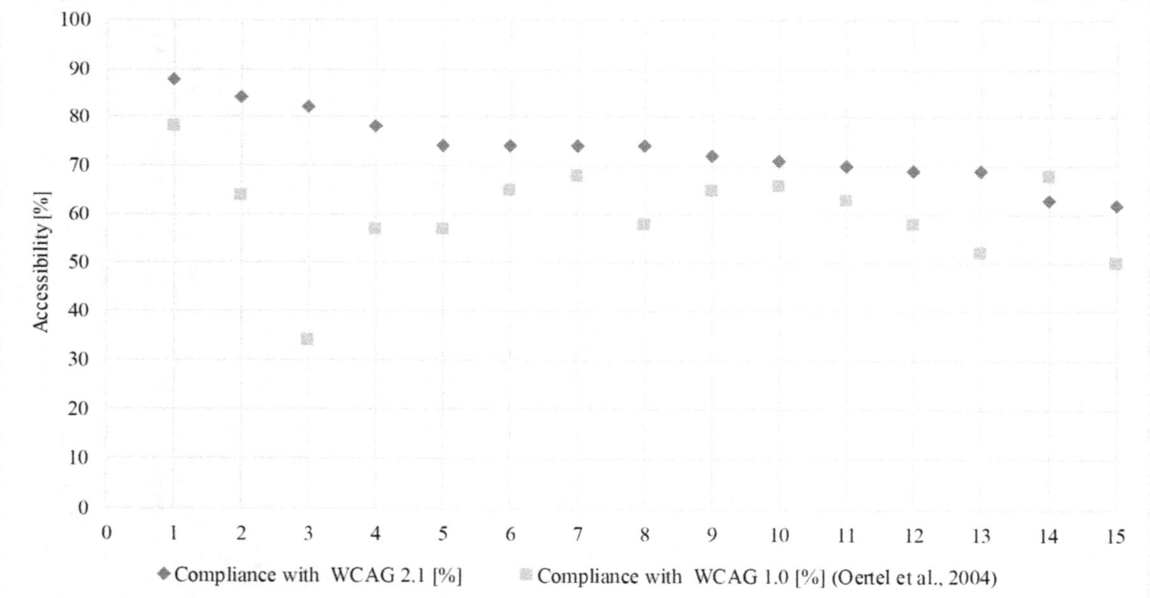

The results in Figure 4 show that the rate of accessibility per country has been improved for all (on average for 15%) countries except for the United Kingdom, where the percentage of accessibility decreased by 5%. The most improvements were detected on the website of the Netherlands, where accessibility increased from 34% (2003 - WCAG 1.0) to 82% (2020 - WCAG 2.1), followed by Italy (a difference of 21%), Sweden (a difference of 20%), Austria (a difference of 17%) and Spain (a difference of 17%).

Accessibility Divided by Principles

In total (considering all webpages) the automatic evaluation identified 305,411 errors, and Figure 5 shows the percentages of identified errors by WCAG principles.

Figure 5. Identified errors by principle

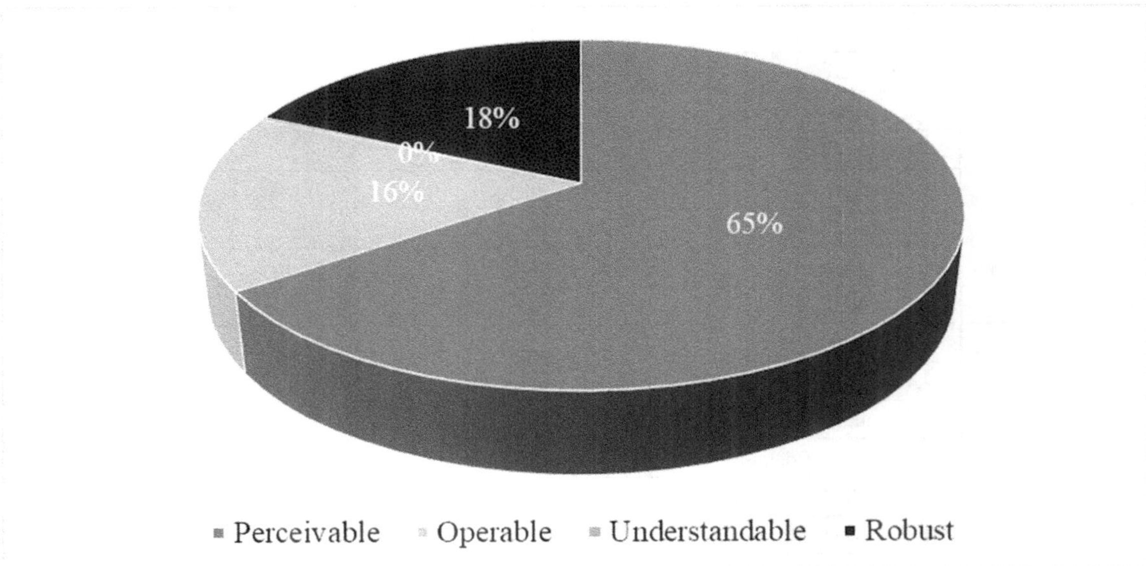

As can be seen in Figure 5, the highest frequency of identified errors is associated with the principle Perceivable (65%), which aligns with existing research (Gonçalves et al., 2019). For the principle Perceivable, the following success criteria were violated: *Reflow* (29% of all identified errors in Perceivable), *Identify Purpose* (27%), *Resize text* (16%), *Visual Presentation* (12%), *Non-text Content* (5%), *Info and Relationships* (4%), *Contrast (Enhanced)* (3%), *Contrast (Minimum)* (2%) and *Non-text Contrast* (1%). The second most frequently identified errors relate to principle Robust (18%), where the violation of success criteria *Name, Role, Value* (94% of all identified errors in Robust) and *Parsing* (6%) was found. The third most frequently identified errors belong to the principle Operable (16%), where the violation of success criteria *Link Purpose* (86% of all identified errors in Operable) and *Bypass Blocks* (14%) was identified. The total number of identified errors related to the principle Understandable is 796 (<1%) and represents a negligible value compared to the identified errors in the remaining principles.

Most Frequently Identified Errors Related to Success Criteria and Conformance Levels

Table 5 presents the detailed results of identified errors per individual success criteria divided into conformance levels. Regardless of the conformance levels, the most frequently identified error was related to *Reflow* (19% of all identified errors), followed by *Name, Role, Value* (17% of all identified errors),

Identify Purpose (17% of all identified errors), *Link Purpose* (14% of all identified errors), *Resize text* (11% of all identified errors) and *Visual Presentation* (8% of all identified errors).

Table 5. The most frequently identified errors divided into conformance levels

Conformance level	Success criteria	SUM of errors	% of errors	% of countries
A	4.1.2 Name, Role, Value	51996	17%	100%
	2.4.4 Link Purpose (In Context)	42746	14%	96%
	1.1.1 Non-text Content	10374	3%	92%
	1.3.1 Info and Relationships	8149	3%	54%
	2.4.1 Bypass Blocks	7084	2%	69%
	4.1.1 Parsing	3207	< 1%	75%
	3.2.2 Labels or Instructions	523	< 1%	69%
	3.1.1 Language of Page	187	< 1%	46%
	3.2.2 On Input	127	< 1%	27%
AA	1.4.10 Reflow	58531	19%	96%
	1.4.4 Resize text	32128	11%	88%
	1.4.3 Contrast (Minimum)	4982	2%	65%
	1.4.11 Non-text Contrast	1544	< 1%	77%
	1.3.5 Identify Input Purpose	513	< 1%	100%
	1.4.13 Content on Hover or Focus	300	< 1%	58%
	1.3.4 Orientation	194	< 1%	35%
	2.4.7 Focus Visible	122	< 1%	4%
AAA	1.3.6 Identify Purpose	52909	17%	100%
	1.4.8 Visual Presentation	24270	8%	81%
	1.4.6 Contrast (Enhanced)	5515	2%	85%
	2.5.6 Concurrent Input Mechanisms	10	< 1%	12%

As seen in Table 5, the most frequent error at conformance level A was related to the success criterion *Name, Role, Value,* which was violated on the websites of all countries. This success criterion is primarily intended to web developers and it states that the name and role of each user interface element on the web page should be programmatically determined so that assistive technologies can gather information about and interact with them (W3C WAI, 2018c). The second most frequently identified error was related to the success criterion *Link Purpose*, which was found on the websites of 96% of countries. This success criterion states that the purpose of a link should be determined from the link text or from the link text together with its programmatically determined link context (W3C WAI, 2018c). The third most frequently identified error was related to success criteria *Non-text Content* and *Info and Relationships*. Although both success criteria constitute only 3% of all identified errors, the violation of *Non-text Content* was found on the websites of 92% of countries and the violation of *Info and Relationships* was identified on the websites of 54% of countries. The success criterion *Non-text Content* includes the recommendation that all non-text objects should be present with a textual alternative with an equivalent purpose, while the intent of the success criterion *Info and Relationships* is to ensure the preservation of the information and relationships that are implied by visual or auditory formatting when the presentation format is changed (W3C WAI, 2018c). The violation of other success criteria in conformance level A constitute 2% (*Bypass Blocks*) or even less than 1% (*Parsing, Labels or Instructions, Language of Page, On Input*) of all identified errors.

From the conformance level AA, the most frequently identified error was the success criterion *Reflow* (19% of all identified errors), which was found on the websites of 96% of countries. This success criterion states that the content should be presented without a loss of information or functionality, and without requiring scrolling in two dimensions (W3C WAI, 2018c). The second most frequently identified problem was *Resize text* (11% of all identified errors). It was found that the text could not be resized up to 200 percent without assistive technology (W3C WAI, 2018c) on the websites of 86% countries. The violation of other success criteria in conformance level AA constitute 2% (*Contrast (Minimum)*) or even less than 1% (*Non-text Contrast, Identify Input Purpose, Content on Hover or Focus, Orientation, Focus Visible*) of all identified errors.

From the conformance level AAA, the most frequently identified error was related to the success criterion *Identify Purpose* (17% of all identified errors) as it was found on the websites of all investigated countries. This success criterion states that the purpose of many elements on a page should be programmatically determined so that user agents (e.g., assistive technologies) can extract and present that purpose to users using different modalities (W3C WAI, 2018c). The second most frequently identified problem was the success criterion *Visual Presentation* (8% of all identified errors) which was identified on the websites of 81% of the investigated countries. This success criterion states that for the visual presentation of blocks of text, different options should be available to ensure that the content is presented in such a manner that it can be perceived by different users without its layout interfering with its readability (W3C WAI, 2018c). For example, the foreground and background colors can be selected by the user, the text is not justified, the line spacing is at least a space-and-a-half between paragraphs and paragraph spacing is at least 1.5 times larger than the line spacing, etc (W3C WAI, 2018c). The violation of other success criteria to the conformance level AAA constitute 2% (*Contrast (Enhanced)*) or even less than 1% (*Concurrent Input Mechanisms*) of all identified errors.

The above findings are not directly comparable to the results of existing research, because none of them investigated compliance with WCAG 2.1, but it is evident that the order of the most frequently identified errors has changed. Similar research into WCAG 2.0 compliance usually reported that the most frequent error at conformance level A was related to the success criterion *Non-text alternative* (Kous, Kuhar, Rajšp, & Šumak, 2020), while in this research, the same success criterion was only 3% of all identified errors. The existing research reported that the most frequently identified error at conformance level AA was related to success criteria *Resize text* or *Contrast (Minimum)*, while in this research, the success criterion *Reflow* was identified as the most violated success criterion. The conclusion is that the comparison of websites with WCAG 2.1 shows the violation of other success criteria compared to WCAG 2.0 - in this case, the violation of the new success criteria (*Reflow, Identify Purpose*), which were absent in WCAG 2.0.

RECOMMENDATIONS FOR THE IMPROVEMENTs OF WEB ACCESSIBILITY

Some recommendations for web accessibility improvements related to the most frequently violated success criteria are presented below:

1. Reflow

The success criterion *Reflow* states that the content (except images, maps, diagrams, video, games, presentations, data tables) should be presented without any loss of information or functionality, and without requiring scrolling in two dimensions (W3C WAI, 2018c). Therefore, when the content in the browser is enlarged to 400%, the content should be presented in one column so that scrolling in more than one direction is not necessary (W3C WAI, 2018c). This way of presenting content enables the user to avoiding scrolling in the direction of reading, which makes it easier to read the content. The mentioned success criterion helps people with low vision who often need to enlarge content to easily read the information (W3C WAI, 2018c). The implementation of this success criterion can be provided through a responsive design that uses CSS media queries to reflow the page for different viewport sizes (Deque Systems, 2020b).

2. Name, Role, Value

The success criterion *Name, Role, Value* is technically-oriented and is primarily intended for web developers who develop or script their user interface (when the standard HTML is used according to specifications, this success criterion has already been met). Therefore, it states that for all user interface components (W3C WAI, 2018c):

- the name and role can be programmatically determined;
- states, properties, and values that can be set by the user can be programmatically set;
- notification of changes to these items is available to user agents, including assistive technologies.

The objective of this technically oriented recommendation is to allow assistive technology to understand the content so that it can convey equivalent information to the user through an alternate user interface and allow them to operate controls through assistive technology (W3C WAI, 2018b). This recommendation involves using standard, documented and supported features to expose these properties to assistive technology (W3C WAI, 2018b). It relies on the fact that these standard controls in standard browsers meet the requirements (W3C WAI, 2018b). Table 6 shows how to provide the correctly defined role, name values, and states for HTML elements, where the requirements are well-defined.

The implementation of this success criteria helps the wider population, especially people who use assistive technology, such as screen readers, screen magnifiers, and speech recognition software (W3C WAI, 2018c).

3. Identify Purpose

The success criterion *Identify Purpose* states that the purpose of user interface components, icons, and regions can be programmatically determined. The page's code should contain metadata that assistive technologies could interpret to give users additional information or modify the page for the user (Deque Systems, 2020a). Therefore, developers should identify the type and purpose of items on a web page. The implementation of this success criterion could help people who have different cognitive disabilities (related to memory, focus and attention, language, etc.) (W3C WAI, 2018c).

4. Link Purpose (In Context)

The success criterion Link Purpose (In Context) states that the purpose of each link should be determined from the link text alone or the link text together with its programmatically determined link context (W3C WAI, 2018c). This means that the purpose of a link should be clear and easy to understand for all users and the link text should describe its purpose without needing additional context. This success criterion is intended to help users understand the purpose of each link so they can decide whether they want to follow the link (W3C WAI, 2018c). Table 7 presents the recommendations with examples of how to provide the success criterion Link Purpose.

Table 6. HTML elements with accessibility API roles, names, values, and states

HTML element	Role	Name	Value	State
<a>	link	'title' attribute, text within <a> element or 'alt' attribute if image link. Concatenated if both text and image 'alt' attribute are provided	'href' attribute	
<button>	push button	text inside <button> element or 'title' attribute		
<fieldset>	grouping	text inside <legend> element within fieldset element		
<input type = "button", "submit", or "reset">	push button	'value' attribute		
<input type = "image">	push button	'alt' attribute or 'title' attribute		
<input type = "text">	editable text	<label> element associated with it or 'title' attribute	'value' attribute	
<input type = "password">	editable text	<label> element associated with it or 'title' attribute	value is purposefully hidden	
<input type="file">	editable text	<label> element associated with it or 'title' attribute	'value' attribute	
<input type="checkbox">	checkbox	<label> element associated with it or 'title' attribute		'checked' attribute
<input type="radio">	radio button	<label> element associated with it or 'title' attribute		'checked' attribute
<select>	list box	<label> element associated with it or 'title' attribute	<option> element with 'selected' attribute set to "selected"	
<textarea>	editable text	<label> element associated with it or 'title' attribute	text within <textarea> element	

Source: W3C WAI (2018b).

The proper implementation of this success criterion could provide the opportunity to people with motion impairment to skip links that are uninteresting for them. Therefore, they can avoid the keystrokes needed to visit the referenced content (W3C WAI, 2018c). It may also help people with cognitive limitations to not become disoriented by using the content they are not interested in as well as allow people with visual disabilities to determine the purpose of a link by exploring the link's context (W3C WAI, 2018c).

LIMITATIONS AND FUTURE WORK

Our findings should be interpreted by considering the following limitations. The first limitation is related to specific characteristics of tourism websites; therefore, the results cannot be generalized to other websites. The second limitation is related to the selected tool, i.e. If the websites were to be analyzed with another tool, the results might be different. Additionally, the results may also differ in the case of evaluating entry websites only. Our future work will be focused on detailed web accessibility evaluation with different automatic tools and comparing the obtained results. In this way, the accuracy of the tools may be investigated.

CONCLUSION

This chapter presents an insight into web accessibility within the tourism domain. The research investigated the accessibility of official tourism websites among 26 EU member states and verified if the websites were compliant with WCAG 2.1. The results, based on an automatic evaluation using the automated tool Multi-guideline Accessibility Usability Validation Environment (MAUVE), indicate that none of the websites achieved all of the recommendations of WCAG 2.1. Nevertheless, the results indicate that countries in the EU are aware of the importance of accessibility, which in general increased when compared to research performed in 2003. The most frequently identified errors were related to the success criteria (1) *Reflow*, (2) *Name, Role, Value*, (3) *Link Purpose (In Context)* and (4) *Identify Purpose*. The conclusion is that a comparison of websites with WCAG 2.1 shows the violation of other success criteria compared to WCAG 2.0 - in this case, the violation of the new success criteria (*Reflow*, *Identify Purpose*), which were absent in WCAG 2.0. According to the violations of WCAG 2.1 success criteria on these websites, recommendations on how to meet specific accessibility requirements were provided.

ACKNOWLEDGMENT

The authors acknowledge financial support from the Slovenian Research Agency (research core funding No. P2-0057).

REFERENCES

W3C WAI. (2018a). *Introduction to Web Accessibility*. Retrieved from https://www.w3.org/WAI/fundamentals/accessibility-intro/

Table 7. Recommendations for success criterion Link Purpose

Recommendations	Examples
A link contains text that gives a description of the information at that URI	A page contains the sentence, "There was much bloodshed during the Medieval period of history." Where "Medieval period of history" is a link.
A link is preceded by a text description of the information at that URI	A page contains the sentence "Learn more about the Government of Ireland's Commission on Electronic Voting at Go Vote!" where "Go Vote!" is a link.
Both an icon and text are included in the same link.	An icon of a voting machine and the text "Government of Ireland's Commission of Electronic Voting" are combined to make a single link. The alt text for the icon is null, since the purpose of the link is already described by the text of the link next to the icon.
A list of book titles	A list of books is available in three formats: HTML, PDF, and mp3 (a recording of a person reading the book). To avoid hearing the title of each book three times (once for each format), the first link for each book is the title of the book, the second link says "PDF" and the third says, "mp3."
News article summaries	A Web page contains a collection of news articles. The main page lists the first few sentences of each article, followed by a "Read more" link. A screen reader command to read the current paragraph provides the context to interpret the purpose of the link.

Source: W3C WAI (2018c).

W3C WAI. (2018b). *Techniques for WCAG 2.1*. Retrieved from https://www.w3.org/WAI/WCAG21/Techniques/#techniques

W3C WAI. (2018c). *Understanding WCAG 2.1*. Retrieved from https://www.w3.org/WAI/WCAG21/Understanding/

W3C WAI. (2018d). *Web Accessibility Laws & Policies*. Retrieved from https://www.w3.org/WAI/policies/

W3C WAI. (2018e). *Web Content Accessibility Guidelines (WCAG) 2.1*. Retrieved from https://www.w3.org/TR/WCAG21/

W3C WAI. (2019a). *Web Accessibility Evaluation Tools List*. Retrieved from https://www.w3.org/WAI/ER/tools/?q=wcag-21-w3c-web-content-accessibility-guidelines-21&q=groups-of-web-pages-or-web-sites

W3C WAI. (2019b). *What's New in WCAG 2.1*. Retrieved from https://www.w3.org/WAI/standards-guidelines/wcag/new-in-21/

Acosta, T., Acosta-Vargas, P., & Lujan-Mora, S. (2018). Accessibility of eGovernment Services in Latin America. In *2018 International Conference on eDemocracy eGovernment (ICEDEG)* (pp. 67–74). 10.1109/ICEDEG.2018.8372332

Ahmi, A., & Mohamad, R. (2015). Web Accessibility of the Malaysian Public University Websites. *Proceedings of the International Conference on E-Commerce*. Retrieved from www.icoec.my

Al-Khalifa, H. S., Baazeem, I., & Alamer, R. (2017). Revisiting the accessibility of Saudi Arabia government websites. *Universal Access in the Information Society*, *16*(4), 1027–1039. doi:10.100710209-016-0495-7

Alshamari, M. (2016). Accessibility Evaluation of Arabic E-Commerce Web Sites Using Automated Tools. *Journal of Software Engineering and Applications*, *09*(09), 439–451. doi:10.4236/jsea.2016.99029

Barricelli, B. R., Sciarelli, P., Valtolina, S., & Rizzi, A. (2018). Web accessibility legislation in Italy: A survey 10 years after the Stanca Act. *Universal Access in the Information Society*, *17*(1), 211–222. doi:10.100710209-017-0526-z

Darcy, S., & Dickson, T. J. (2009). A Whole-of-Life Approach to Tourism: The Case for Accessible Tourism Experiences. *Journal of Hospitality and Tourism Management*, *16*(1), 32–44. doi:10.1375/jhtm.16.1.32

Deque Systems. (2020a). *1.3.6 Identify Purpose (AAA).* Retrieved from https://dequeuniversity.com/resources/wcag2.1/1.3.6-identify-purpose

Deque Systems. (2020b). *1.4.10 Reflow (AA).* Retrieved from https://dequeuniversity.com/resources/wcag2.1/1.4.10-reflow

Design for all foundation. (2019). *Design for all.* Retrieved from http://www.designforall.org/

Domínguez Vila, T., Alén González, E., & Darcy, S. (2018). Website accessibility in the tourism industry: An analysis of official national tourism organization websites around the world. *Disability and Rehabilitation*, *40*(24), 2895–2906. doi:10.1080/09638288.2017.1362709 PMID:28793789

ENAT - European Network for Accessible Tourism. (2018). *What is "Accessible Tourism"?* Retrieved from https://www.accessibletourism.org/?i=enat.en.faq.744

Espadinha, C., Pereira, L. M., Da Silva, F. M., & Lopes, J. B. (2011). Accessibility of Portuguese Public Universities' sites. *Disability and Rehabilitation*, *33*(6), 475–485. doi:10.3109/09638288.2010.498554 PMID:20594034

EUR-Lex. (2018). *Accessibility of public sector websites and mobile apps.* Retrieved from https://eur-lex.europa.eu/eli/dir/2016/2102/oj

European Commission. (2018). *Access city award.* Retrieved from https://ec.europa.eu/social/main.jsp?catId=1141

European Network for Accessible Tourism. (2008). Rights of Tourists with Disabilities in the European Union Framework. *European Network for Accessible Tourism.*

European Telecommunications Standards Institute. (2018). *EN 301 549 V2.1.2 (2018-08) - Accessibility requirements for ICT products and services.* Author.

Eurostat. (2018). *1 in 4 people in the EU have a long-term disability.* Retrieved from https://ec.europa.eu/eurostat/web/products-eurostat-news/-/EDN-20181203-1?inheritRedirect=true

Gonçalves, M. J. A., Camarinha, A. P., Abreu, A. J., Teixeira, S., & Ferreira, A. (2019). Web Accessibility in the Tourism Sector: an analsis of the most used Websites in Portugal. *Advances in Tourism, Technology and Smart Systems, Proceedings of ICOTTS 2019.*

International Organization for Standardization. (2012). ISO/IEC 40500:2012 Information technology - W3C Web Content Accessibility Guidelines (WCAG) 2.0. International Organization for Standarization, 2012.

Ismail, A., & Kuppusamy, K. S. (2016). Accessibility of Indian universities' homepages: An exploratory study. *Journal of King Saud University - Computer and Information Sciences*. doi:10.1016/j.jksuci.2016.06.006

ISO 26800:2011. (2011). *ISO 26800:2011 - Ergonomics - General approach, principles and concepts*.

ISO 9241-112:2017. (2017). *ISO 9241-112:2017 Ergonomics of human-system interaction — Part 112: Principles for the presentation of information*.

ISO 9241-171. (2008). *ISO 9241-171:2008 Ergonomics of human-system interaction - Part 171 : Guidance on software accessibility. International Organization*. International Organization for Standardization, 2008.

ISO/IEC 30071-1:2019. (2019). *ISO/IEC 30071-1:2019 Information technology — Development of user interface accessibility — Part 1: Code of practice for creating accessible ICT products and services*.

ISO/TR 22411:2008. (2008). *ISO/TR 22411:2008 Ergonomics data and guidelines for the application of ISO/IEC Guide 71 to products and services to address the needs of older persons and persons with disabilities*.

Kous, K., Kuhar, S., Pavlinek, M., Heričko, M., & Pušnik, M. (2020). Web accessibility investigation of Slovenian municipalities' websites before and after the adoption of European Standard EN 301 549. *Universal Access in the Information Society*, (0123456789). Advance online publication. doi:10.100710209-020-00732-9

Kous, K., Kuhar, S., Pušnik, M., & Šumak, B. (2019). Comparative analysis of faculties ' websites accessibility based on an automatic evaluation. In *42nd International Convention on Information and Communication Technology, Electronics and Microelectronics (MIPRO)* (pp. 1498–1502). Croatian Society MIPRO. 10.23919/MIPRO.2019.8757202

Kous, K., Kuhar, S., Rajšp, A., & Šumak, B. (2020). Investigation of the accessibility of non-text content published on websites. In *MIPRO 2020 : 43nd International Convention* (pp. 1953–1958). Retrieved from http://docs.mipro-proceedings.com/proceedings/mipro_2020_proceedings.pdf

Kous, K., & Pavlinek, M. (2017). Longitudinalna raziskava dostopnosti spletnih mest slovenskih občin. *Uporabna Informatika*, *25*(3), 135–152.

Kous, K., & Polančič, G. (2019). Empirical insights of individual website adjustments for people with dyslexia. *Sensors (Basel)*, *19*(10), 1–25. doi:10.339019102235 PMID:31091827

Maswera, T., Dawson, R., & Edwards, J. (2005). Analysis of Usability and Accessibility Errors of E-Commerce Websites of Tourist Organisations in Four African Countries Tonderai Maswera. *World Wide Web Internet And Web Information Systems*. Retrieved from http://www.ifitt.org/admin/public/uploads/Maswera.pdf

Nunez, A., Moquillaza, A., & Paz, F. (2019). Web Accessibility Evaluation Methods: A Systematic Review. In A. Marcus & W. Wang (Eds.), *Design, User Experience, and Usability, Practice and Case Studies, 21st HCI International Conference, HCII 2019*. Orlando, FL: Academic Press.

Oertel, B., Haße, C., Scheermesser, M., Thio, S. L., & Feil, T. (2004). Accessibility of Tourism Web sites within the European Union. In *Information and communication technologies in tourism 2004* (pp. 358–368). Springer-Verlag New York Inc. doi:10.1007/978-3-7091-0594-8_34

Paterno, F., & Schiavone, A. G. (2015). The role of tool support in public policies and accessibility. *Interactions (New York, N.Y.)*, *22*(3), 60–63. doi:10.1145/2745395

Pulina, F., Broccia, G., Paternò, F., Manca, M., & Schiavone, A. G. (2020). *Multiguideline Accessibility and Usability Validation Environment.* Retrieved from https://mauve.isti.cnr.it/index.jsp

Rajšp, A., Kous, K., Kuhar, S., Šumak, B., & Šorgo, A. (2019). Preliminary review of jobs, skills and competencies for implementation of Digital Accessibility. In *Proceedings of the Central European Conference on Information and Intelligent Systems* (pp. 93–99). Retrieved from http://archive.ceciis.foi.hr/app/public/conferences/2019/Proceedings/ETICT/ETICT3.pdf

Schiavone, A. G., & Paterno', F. (2015). An extensible environment for guideline-based accessibility evaluation of dynamic Web applications. *Universal Access in the Information Society*, *14*(1), 111–132. doi:10.100710209-014-0399-3

Silva, F. M., & Borges, I. (2019). Digital Accessibility on Institutional Websites of Portuguese Tourism. In V. Ratten (Ed.), *Technological Progress, Inequality and Entrepreneurship* (pp. 67–85). Springer.

Teixeira, L., Eusébio, C., & Silveiro, A. (2019). Website Accessibility of Portuguese Travel Agents. In *14th Iberian Conference on Information Systems and Technologies (CISTI)* (pp. 1–6). 10.23919/CISTI.2019.8760949

United States Environmental Protection Agency. (2019). *What is Section 508?* Retrieved from https://www.epa.gov/accessibility/what-section-508

Vila, T. D., González, E. A., & Darcy, S. (2019). Accessibility of tourism websites: the level of countries ' commitment. *Universal Access in the Information Society*. doi:10.100710209-019-00643-4

WebAIM. (2018). *Section 508 Checklist.* Retrieved from https://webaim.org/standards/508/checklist

World Health Organization. (2018). *Disability and health.* Retrieved from https://www.who.int/newsroom/fact-sheets/detail/disability-and-healthTa

APPENDIX 1

Table 8. List of Success criteria (WCAG 2.1)

Principles	Guidelines	Success criteria	Level
Perceivable	1.1 Text Alternatives	1.1.1 Non-text Content	A
	1.2 Time-based Media	1.2.1 Audio-only and Video-only (Prerecorded)	A
		1.2.2 Captions (Prerecorded)	A
		1.2.3 Audio Description or Media Alternative (Prerecorded)	A
		1.2.4 Captions (Live)	AA
		1.2.5 Audio Description (Prerecorded)	AA
		1.2.6 Sign Language (Prerecorded)	AAA
		1.2.7 Extended Audio Description (Prerecorded)	AAA
		1.2.8 Media Alternative (Prerecorded)	AAA
		1.2.9 Audio-only (Live)	AAA
	1.3 Adaptable	1.3.1 Info and Relationships	A
		1.3.2 Meaningful Sequence	A
		1.3.3 Sensory Characteristics	A
		1.3.4 Orientation	AA
		1.3.5 Identify Input Purpose	AA
		1.3.6 Identify Purpose	AAA
	1.4 Distinguishable	1.4.1 Use of Color	A
		1.4.2 Audio Control	A
		1.4.3 Contrast (Minimum)	AA
		1.4.4 Resize text	AA
		1.4.5 Images of Text	AA
		1.4.6 Contrast (Enhanced)	AAA
		1.4.7 Low or No Background Audio	AAA
		1.4.8 Visual Presentation	AAA
		1.4.9 Images of Text (No Exception)	AAA
		1.4.10 Reflow	A
		1.4.11 Non-text Contrast	A
		1.4.12 Text Spacing	A
		1.4.13 Content on Hover or Focus	A

continued on following page

Table 8. Continued

Principles	Guidelines	Success criteria	Level
Operable	2.1 Keyboard Accessible	2.1.1 Keyboard	A
		2.1.2 No Keyboard Trap	A
		2.1.3 Keyboard (No Exception)	AAA
		2.1.4 Character Key Shortcuts	A
	2.2 Enough Time	2.2.1 Timing Adjustable	A
		2.2.2 Pause, Stop, Hide	A
		2.2.3 No Timing	AAA
		2.2.4 Interruptions	AAA
		2.2.5 Re-authenticating	AAA
		2.2.6 Timeouts	AAA
	2.3 Seizures and Physical Reactions	2.3.1 Three Flashes or Below Threshold	A
		2.3.2 Three Flashes	AAA
		2.3.3 Animation from Interactions	AAA
	2.4 Navigable	2.4.1 Bypass Blocks	A
		2.4.10 Section Headings	A
		2.4.2 Page Titled	A
		2.4.3 Focus Order	A
		2.4.4 Link Purpose (In Context)	A
		2.4.5 Multiple Ways	AA
		2.4.6 Headings and Labels	AA
		2.4.7 Focus Visible	AA
		2.4.8 Location	AAA
		2.4.9 Link Purpose (Link Only)	AAA
		2.4.10 Section Headings	AAA
	2.5 Input Modalities	2.5.1 Pointer Gestures	A
		2.5.2 Pointer Cancellation	A
		2.5.3 Label in Name	A
		2.5.4 Motion Actuation	A
		2.5.5 Target Size	AAA
		2.5.6 Concurrent Input Mechanisms	AAA

continued on following page

Table 8. Continued

Principles	Guidelines	Success criteria	Level
Understandable	3.1 Readable	3.1.1 Language of Page	A
		3.1.2 Language of Parts	AA
		3.1.3 Unusual Words	AAA
		3.1.4 Abbreviations	AAA
		3.1.5 Reading Level	AAA
		3.1.6 Pronunciation	AAA
	3.2 Predictable	3.2.1 On Focus	A
		3.2.2 On Input	A
		3.2.3 Consistent Navigation	AA
		3.2.4 Consistent Identification	AA
		3.2.5 Change on Request	AAA
	3.3 Input Assistance	3.3.1 Error Identification	A
		3.3.2 Labels or Instructions	A
		3.3.3 Error Suggestion	AA
		3.3.4 Error Prevention (Legal, Financial, Data)	AA
		3.3.5 Help	AAA
		3.3.6 Error Prevention (All)	AAA
Robust	4.1 Compatible	4.1.1 Parsing	A
		4.1.2 Name, Role, Value	A
		4.1.3 Status Messages	AA

[1]The technical requirements are related to: Conform to specification, No frames, Text equivalents, Color specific, Flicker, Visual contrast, Avoid server-side image maps, Text links for server-side image maps, Data table header elements, Data table header elements multiple levels, Use style sheets, Use relative rather than absolute units of measures, Layout table linearize and markup, Form labels, Script/object alternatives, Script/object accessible, Objects are accessible, Synchronized text equivalents for multimedia, Link text, Time limits, Links selectable and dimensions and Alternate accessible page (Barricelli et al., 2018).

Chapter 9
Competitiveness Factors of Accessible Tourism E-Travel Agencies

Lucía Rubio-Escuderos
University of Alicante, Spain

Hugo García-Andreu
University of Alicante, Spain

ABSTRACT

This chapter aims to detect the differentiating factors of the accessible e-travel agencies that enable them to be competitive in the market and, therefore, to promote the development of accessible products in the tourism sector. With this objective, an analysis of 14 accessible travel agencies' websites has been carried out. The analysis was guided by theoretical frameworks of competitiveness and accessible tourism. Taking into account the specific needs of people with disabilities, accessible e-travel agencies offer a wide range of specialized and tailor-made products and services with great quality while finding efficiency formulas to contain prices: being receptive in a specific destination; having a network of contacts in different parts of the world that can verify the specific requirements of the client; or being specialized in a specific destination/s, offering tourist packages that have previously been tested by other people with the disability for which it is offered.

1. INTRODUCTION

People with disabilities account for approximately 15% of the world's population (WHO, 2020) and accessible tourism has many benefits for these individuals (Agovino *et al.*, 2017; Daniels, Rodgers & Wiggins, 2005; Eichhorn, Miller & Tribe, 2013; Kastenholz, Eusébio & Figueiredo, 2015; Moura, Kastenholz & Pereira 2018; Pagán, 2015) and also for the tourism economy (Buhalis & Michopoulou, 2011; Domínguez, Fraiz & Alén, 2013). Nevertheless, when travelling, the requirements of people with disabilities are not always met correctly by tourism companies and, particularly, by travel agen-

DOI: 10.4018/978-1-7998-6428-8.ch009

cies (Agovino *et al.*, 2017; Altinay *et al.*, 2016; Bi, Card & Cole, 2007; Blichfeldt & Nicolaisen, 2011; McKercher *et al.*, 2003; UNWTO, 2016). Likewise, a lack of knowledge of the tourism sector about the needs of people with disabilities is detected (Bizjak, Knežević & Cvetrežnik, 2011; Buhalis *et al.*, 2005; Eichhorn & Buhalis, 2011; Loi & Kong, 2017; McKercher & Darcy, 2018). Due to the lack of accessible tourism facilities and services, people with disabilities are cautious when planning their trip and principally gather information from the internet and through "word of mouth" (Ray & Ryder, 2003). Furthermore, accessible information found *online* is mostly generic and non-specific and does not always support assistive technologies (Buhalis & Michopoulou, 2011; Michopoulou & Buhalis, 2013). This is one of the reasons why many people with disabilities decide to participate in tourist trips organised by associations (Bi *et al.*, 2007; Blichfeld & Nicholaisen, 2011).

Traditionally, people with disabilities have not trusted the information provided by conventional travel agencies (McKercher *et al.*, 2003). However, the emergence of the Internet has altered the tourism market and represented a challenge for travel agencies which have been obliged to reinvent themselves in order to survive. Some authors (Abrate *et al.*, 2019; Aguiar-Quintana, Moreno-Gil & Picazo-Peral, 2016; Buhalis & Law, 2008; Lacalle, 2013; Tsang, Lai & Law, 2010; Zhang *et al.*, 2006) highlight the importance and need to research the different survival strategies implemented by traditional travel agencies. Along these lines, in recent years, travel agencies specialised in accessible tourism have been described as 'the golden key in the future' (Özogul & Baran, 2016). This means that targeting individuals with accessibility needs is one of the keys for the survival of the travel agencies in the future, "providing appropriate product/service by targeting the individuals with accessibility needs together with a correct approach and strategy, the specialized travel agencies will be able to have a competitive advantage" (Özogul & Baran, 2016). In fact, accessible tourism travel agencies are indeed emerging. They seem to offer a large number of specialised products adapted to the needs of people with disabilities, thereby demonstrating that there are numerous opportunities for value creation in the travel agency sector. It should be noted that there is a profound lack of research on e-travel agencies in the field of accessible tourism and, specifically, on their competitiveness factors in the age of Information and Communication Technologies (ICTs). Agencies specialised in accessible tourism play a potentially fundamental role in creating value for travellers with special needs, as a "good travel agent" can greatly facilitate travel (McKercher *et al.*, 2003).

In order to obtain in-depth information about the characteristics of accessible e-travel agencies, an analysis of the information on the offer published on the accessible travel agencies' websites has been carried out. Websites constitute a valuable marketing tool, useful to both suppliers and consumers in disseminating information, communication and online purchasing (Buhalis & Law, 2008). This study seeks to detect the differentiating factors of the accessible e-travel agencies that enable them to be competitive in the market (Abrate *et al.*, 2019; Aguiar-Quintana, Moreno-Gil & Picazo-Peral, 2016; Tsang, Lai & Law, 2010; Zhang *et al.*, 2006) and, therefore, to promote the development of accessible products in the tourism sector.

This chapter begins by providing information about the online requirements of people with disabilities in the context of ICTs (section 2.1). It then describes the competitive factors of online/traditional travel agencies, how the emergence of ICTs has forced them to become more competitive and how are they addressing the market segment of people with disabilities (section 2.2). Section 3 outlines the data collection and analysis process and defines the categories and sub-categories used, and section 4 presents the results and discusses them in relation to the literature. The conclusions of the study are drawn in section 5.

2. LITERATURE REVIEW

2.1. Accessible Tourism and ICTs

Accessible tourism was originally defined as the process of removing the barriers that impede impaired/disabled people from fully enjoying a tourist experience. Currently, this concept has evolved to include the process of reshaping the whole tourist environment (in terms of physical, informative, online, attitudinal aspects, etc.) so it may be used by all population groups, regardless of their age, condition, capacity or impairment, in accordance with Universal Design (Darcy & Buhalis, 2011; Buhalis & Michopoulou, 2011; UNWTO, 2016). Universal Design is the design and composition of an environment so that it can be accessed, understood and used to the greatest extent possible by all people regardless of their age, size, ability or disability. Any building, product, or service in that environment should be designed to meet the needs of all people who wish to use it (The Centre for Excellence in Universal Design, 2020). In this sense, ICTs are no exception. Technological accessibility in tourism is a broad concept interrelated with information and online accessibility (Buhalis & Michopoulou, 2011). This is the practice of ensuring that information available on the Internet relating to tourism infrastructures and services is highly detailed and presented in formats that can be understood by people with visual, hearing and intellectual disabilities (Buhalis & Michopoulou, 2011; Eichhorn *et al.*, 2008; Darcy, 2010; Eichhorn & Buhalis, 2011; Michopoulou & Buhalis, 2013; Emrouzeh *et al.*, 2017; De Lemos, Vieira & Pinho, 2020 ; Mayordomo-Martínez *et al.*, 2019). As not all people with disabilities have the same requirements, the right kind of information about infrastructure and services should be adapted to their specific needs (Darcy, 2010; Eichhorn *et al.*, 2008; Michopoulou & Buhalis, 2013). Specifically, they prefer information formats that combine text, floorplans and photographs that include sizes and a description of the spaces and adaptations (Darcy, 2010; Michopoulou & Buhalis, 2013). Social media tools are also important when searching for information, since people with disabilities use them to expand their knowledge of accessible places and venues (Altinay et al., 2016). Social media enables the sharing of tourism knowledge and experiences and is also useful for raising awareness about the need for accessibility (Altinay *et al.*, 2016).

In addition to expanding the limits of accessibility, ICTs have also forced travel agencies to evolve and make considerable efforts to survive in a strongly competitive market (Aguiar-Quintana, Moreno-Gil & Picazo-Peral, 2016; Buhalis & Law, 2008; Tsang, Lai & Law, 2010). One specialisation strategy for travel agencies is to operate in the accessible tourism market (Özogul & Baran, 2016). In recent years, travel agencies specialised in accessible tourism have been emerging, specialised in assisting groups with accessibility needs such as people with disabilities.

2.2. COMPETITIVENESS OF TRAVEL AGENCIES IN THE CONTEXT OF ICTs

ICTs have completely changed the way tourism functions and, therefore, how business is conducted within this sector (Abrate *et al.*, 2019; Aguiar-Quintana, Moreno-Gil & Picazo-Peral, 2016; Buhalis & Law, 2008; Cortés & Vargas, 2018; Huang, Yung & Yang, 2011; Tsang, Lai & Law, 2010; Zhang *et al.*, 2006). With the emergence of ICTs, traditional travel agencies have been relegated by online travel agencies and the former have been forced to become increasingly competitive over the last twenty years (Buhalis & Law, 2008; Tsang, Lai & Law, 2010). According to the specialised literature, there are three main strategies for increasing the competitiveness of both online and traditional travel agencies (Aguiar-Quintana, Moreno-Gil & Picazo-Peral, 2016): (1) improving the quality of the services provided by the travel agency; (2) improving the marketing strategies from the demand side and (3) increasing the efficiency of the travel agency. When improving the quality of the services provided by the travel agency, the dimensions related to reliability and the response capacity of the travel agents are particularly relevant (Millan & Esteban, 2004), key to gain tourist's loyalty to the agency (Lai, 2014). With respect to the second strategy, technological innovations, increasingly demanding consumers, fierce competition, and changes in the tourist distribution system mean that online and traditional travel agencies need to pay attention to customers' details, investigate the existing and potential service quality expectations of customers, understand their needs, and design an online service system that meets those needs and expectations. Ultimately, it is vital that travel agencies focus on the customer if they want to survive and gain a competitive advantage (Abrate *et al.*, 2019; Bigné *et al.*, 2008; Tsang, Lai & Law, 2010; Zhang *et al.*, 2006). The third and last competitiveness strategy, is related to an increase in efficiency and the existing studies suggest that activities related to capital, labour, sales and marketing are those that principally determine this efficiency of the travel agencies (Aguiar-Quintana, Moreno-Gil & Picazo-Peral, 2016).

There are also studies that analyse competitiveness, differentiating between online and traditional travel agencies. The strengths of e-travel agencies compared to traditional agencies are considered to be: accessibility, search capability, user interfaces, online community (Zhang *et al.*, 2006) and the possibility of introducing consumer-generated content such as customer travel reviews or question and answer forums among users (Gretzel & Yoo, 2008). Tsang, Lai & Law (2010) identify six dimensions of perceived e-service quality for online travel agencies: a.Website Functionality, b.Information Quality and Content (accurate, up-to-date and concise nature, reliability and completeness of information), c.Fulfilment and Responsiveness (accuracy of service promises and the willingness to provide a prompt service), d.Safety and Security (trustworthiness), e.Appearance and Presentation and f.Customer Relationship (available customisation features and interactive communications). Most tourists book their accommodation through online travel agencies mainly because browsing is easier and better prices are offered. Online reviews, promotions and photographs are also important factors (Pinto & Castro, 2019).

Travel guaranty and assurance issues are still seen as strengths of traditional travel agencies (Zhang *et al.*, 2006). In this case, the customer seeks higher quality and a better quality/price relationship, as well as a wider variety of products and services, more assistance in certain aspects, trust, more information, convenience, greater security, and speed. The desired service is characterised by a high level of customer service, sensitivity, adaptability, trust, security and flexibility, with the human aspect being highly valued. However, the higher price is also a differentiating element, as well as an initial barrier, mainly for young and old customers (Aguiar-Quintana, Moreno-Gil & Picazo-Peral, 2016).

The study by McKercher *et al.* (2003) is one of the few that examine accessibility in travel agencies. One survival strategy implemented is for travel agencies to charge an extra fee for the additional services required by people with disabilities, although this also means that many people with lower incomes cannot afford these trips, so the authors do not recommend this solution. This suggestion is due to the results of the study that reveal that most people with disabilities consider their experience with travel agencies to be negative and do not trust the information they provide (McKercher *et al.*, 2003). The main reasons are: (1) the popularisation of standardised tourist packages that do not adapt to the needs of the individual customer, cheap and easy to sell, which is why agencies reduce costs and increase profits; and (2) their lack of knowledge about the reality and needs of people with disabilities. Many of them think that tourists with different types and/or degrees of disability have the same accessibility requirements, erroneously recommending tourist products to their customers.

For these reasons, people with disabilities usually travel with members of an association rather than with other groups (Bi *et al.*, 2007; Blichfeldt & Nicolaisen, 2011). These associations have been reported to provide relevant information and act as forums in which tourists with disabilities are able to learn from each other's experiences (Blichfeldt & Nicolaisen, 2011). They act as specialised travel agencies organising holiday packages, enabling people with disabilities to acquire a ''tourist'' role (Blichfeldt & Nicolaisen, 2011). However, some studies reveal a certain rejection among people with disabilities towards group trips, as they wish to see themselves as everybody else without placing the central focus of their identity on their disability (Eichhorn, Miller & Tribe, 2013).

In recent years, travel agencies have emerged which are specialised in offering services to groups with specific accessibility needs. However, it has been corroborated that there is a huge research gap with respect to e-travel agencies in the context of accessible tourism. Identifying and disseminating the competitiveness factors of these specialised agencies can lead the way for other travel agencies and therefore broaden the specialised accessible tourism offer.

3. METHOD

This study aims to explore the characteristics and competitive factors of accessible tourism travel agencies. With this objective, an analysis of the websites of fourteen travel agencies has been carried out.

3.1. Sample Selection

The criteria for selecting these agencies are that they must be travel agencies, with an online presence (with a website) and an online channel of communication and that they describe themselves as being specialised in accessible tourism. Furthermore, the agencies were also selected in accordance with two principal qualitative criteria: (1) they form part of a recognised network of accessible tourism, such as ENAT (European Network for Accessible Tourism) and/or RedEstable (Spanish network of accessible tourism), and/or (2) an assessment of the richness of the information provided on their user experience blogs and social media (Facebook, Instagram, etc). The fourteen companies analysed are: Travel-Xperience.com, Accesitravel.com, Trianaviajes.com, Travel for all, Accessible Madrid, Travelfast, Islandia 66, Ocio y turismo accesibles, Accessible Spain Travel, Barcelona Special Traveler, VAVAVA, Solacces Service, BCN ACCESTOUR and EDUCNATUR.

3.2. Analysis of Websites

An analysis of the websites has been conducted in order to seek and identify the competitiveness factors of the travel agencies specialised in accessible tourism. Therefore, this website analysis has been guided by two overlapping theoretical frameworks, that of competitiveness and that of accessible tourism. On the one hand, we considered whether the travel agencies analysed implemented one or all of the major competitiveness strategies, namely, improving the quality of the services provided by the supplier, improving the marketing strategies from the demand side and increasing efficiency. On the other hand, due to the broad and generic nature of the competitiveness strategies, the study requires a second theoretical framework derived from accessible tourism. This framework enables us to determine what the competitiveness strategies focus on or should focus on in the field of accessible tourism. For example, according to one of the competitiveness strategies of travel agencies, marketing should be customer-oriented (Abrate *et al.*, 2019; Aguiar-Quintana, Moreno-Gil & Picazo-Peral, 2016; Tsang, Lai & Law, 2010; Zhang *et al.*, 2006). In the case of accessible tourism, this is complemented with accessible companies having to meet a series of requisites in terms of training, information and the product offered (Darcy, 2010; Eichhorn *et al.*, 2008; Michopoulou & Buhalis, 2013). Therefore, the assessment of the customer-orientation of the accessible tourism e-travel agencies is based on the type of specific needs they cater to ("type of disability" of the check-list), whether the information provided on the website is accessible (richness of information) and which specialised services adapted to people with disabilities are offered (accessibility services). In order to evaluate the strategy for improving the quality of the services provided, the checklist includes "type of trip", "destination" and "activity". To assess the efficiency strategy, during the analysis we paid particular attention to the information about the way in which the agency adapted to the specific needs of its customers and whether this could influence the price. In this way we were able to observe how certain agencies claimed to offer a specialised and individualised product at competitive prices. This is an important aspect for travel agencies specialised in accessible tourism given that, according to the literature, planning a trip for people with disabilities is complex and travel agencies are able to offer competitive prices by providing standardised products for the general public (McKercher, *et al.,* 2003). Therefore, this survival strategy has considered a unique value of these agencies, referring to a "special value", observing the repetition of three patterns: a network of reliable contacts, accessible products that are standardised for certain already tested needs and being a receptive agency. With these reference frameworks we have created a checklist that contains specific aspects to observe in the websites. The study of the websites has used two types of analysis, the counting method (Law, Qi & Buhalis, 2010) and content analysis (Holsti, 1969). On the one hand, the counting method is used to evaluate the performance of the website or to determine the content richness. The information extracted using the counting method has been analysed with the help of comparative tables which evaluate whether the agency possesses or lacks a certain characteristic defined in the checklist categories. After elaborating the table in which the attributes of each of the travel agencies specialised in accessible tourism are determined, frequencies were constructed to find the attributes that are more or less repeated among the cases analysed. On the other hand, the content analysis comprises the series of interpretative procedures of communicative products (Raigada, 2002), in this case, of the content of the websites. The content analysis examined the specific terms with which the travel agencies express themselves with respect to each dimension analysed. The results are also presented in the form of quotes.

3.3. Description of Categories Analysed

Table 1 synthesizes the checklist corresponding to the categories and sub-categories of the website analysis.

Table 1. Categories and sub-categories of analysis

Category	Sub-category
Type of disability	Physical
	Sensory
	Others
Type of trip	Individual
	Groups
Type of activity	Water
	Adventure
	City
	Cruises
	Cultural
	Rural/Nature
	Theme parks
	Sun & Beach
	Wellness
	Others
Destination scope	National
	International
Special value	Receptive
	Contacts
	Tested by other clients, evaluators, etc.
Information richness	Activities
	Accommodation
	Routes
	Services
Specific characteristics of accessibility services	Tailor-made trips
	Reception
	Rent Technical Aids
	Repair TA
	Personal assistance
	Special tour-guides
	Consultancy
	Others

Source: authors.

Providing a description of the items shown in Table 1, "type of disability" refers to those disabilities to which their services are directed, that is, if they cover all disabilities or if they are specialised in one (or several) in particular. These disabilities are categorised in the analysis according to the type of impairment (Darcy & Buhalis, 2011) as:

- **Physical**: people with difficulties to carry out movements with varying degrees of restriction. Examples of requirements in tourism: continuous pathways, specialist equipment, circulation space…
- **Sensory**: limited capacity to see or hear at varying degrees or the inability to see or hear. Examples of requirements in tourism: sign language interpreters, telephone typewriters, information and communication in alternative formats, areas for guide dogs…
- **Others**: these can be communicative -limited capacity to use expressive or receptive language-, intellectual/mental -with lifelong limitations that result in a behavioural disorder or limited cognition -and/or hidden- comprise a wide range of illnesses that are not always so obvious or not seen but require special attention-. Examples of requirements in tourism: plain language text, dietary considerations, attendant support…

"Type of trip" indicates if they are aimed at organising group trips of people with disabilities or are oriented to the individual client and his/her immediate surroundings. In the case of "destination scope", if the trip has a national or international scope and whether it focuses on a single destination.

In reference to "type of activity", the most common activities encountered were categorised as:

- **"Water"**: adventure activities related to water, such as diving, canoeing…
- **"Adventure"**: tourism that normally takes place in destinations with specific geographical characteristics and landscapes and tends to be associated with physical activity and a closeness to nature. This experience may involve some type of real or perceived risk and may require significant physical and/or mental effort. It generally includes outdoor activities such as mountaineering, climbing, hiking… (UNWTO, 2019), in this case, we excluded the water activities in order to group them in a different category.
- **"City"**: visits to cities/urban spaces.
- Adapted **"Cruises"** packages.
- **"Cultural"**: activity in which the visitor's essential motivation is to experience the tangible and intangible cultural attractions/products in a tourist destination. These are related to a set of distinctive material, intellectual, spiritual and emotional features of a society that encompasses arts and architecture, historical and cultural heritage, culinary heritage, literature, music, creative industries and living cultures with their lifestyles, value systems, beliefs and traditions (UNWTO, 2019).
- **"Rural/Nature"**: activity in which the visitor's experience is related to a wide range of products generally linked to nature-based activities (UNWTO, 2019).
- Visits to **"theme parks"**.
- **"Sun & Beach"**. The agencies offer packages to coastal areas designed for people who want to enjoy beaches and sun.
- **"Wellness"**: activity that aims to improve and balance all of the main domains of human life including physical, mental, emotional, occupational, intellectual and spiritual. The primary motivation for the wellness tourist is to engage in preventive, proactive, lifestyle-enhancing activities such as fitness, healthy eating, relaxation, pampering and healing treatments (UNWTO, 2019).
- **Others**: other kind of activities not categorised for the analysis.

In addition, it was also taken into account whether the activities are the main attraction of the agency's product.

It was considered for the present analysis that "special Value/Efficiency" correspond to the value as an accessible agency, that is, the ability to adapt its offer to the specific needs of its clients, trying to ensure that this does not incur a significant increase in prices. We have observed that there are three ways in which agencies claim that they test the products to meet their costumer's needs: they are receptive agencies at the destination, they have a wide network of contacts or they offer standardised products already tested by other people (customers, evaluators…) with special needs. The analysis also assesses the "richness of the information" presented on the website, that is, description of its services and its products (routes, accommodation, activities ...), formats in which it is presented (text, photographs and plans) and whether it offers the opportunity for customers to exchange information between one another on the website.

Last, "Specific characteristics of accessibility services" are defined as follows:

- **"Tailor-made trips"**, according to the specific needs/requirements of the agency's clients with disabilities.
- **"Reception"** at the destination.
- **"Rent Technical Aids"**. If the agency offers a renting service of technical aids for people with disabilities at the destination.
- **"Repair TA"**. If the agency offers a repair service of technical aids for people with disabilities at the destination.
- **"Personal assistance"**. If the agency offers personal assistance for people with disabilities at the destination.
- **"Special tour-guides"**. If the agency offers special tour-guides for people with disabilities (especially for people with sensory and/or mental/intellectual impairments) at the destination.
- **"Consultancy"**. If the agency offers a consultancy service of accessibility for tourist companies.
- **"Others"**.

4. RESULTS

During the results analysis process, the information collected from the agencies' websites was organised according to the categories and sub-categories defined. For every main category (disability, type of trip, destinations, activities, value, information and services), a table is presented summarising the results. In the tables, each column represents a case or agency and the rows the categories with their corresponding sub-categories. The "X"s mean that the agency offers that particular product or service or has a certain characteristic (sub-categories).

The results from the content analysis of the travel agencies' websites are divided into four main themes: market segment addressed, destinations and activities they offer, value and quality of the information and services they provide.

4.1. Market Segment Addressed (Disability and Type of Trip).

The selected travel agencies either address the needs generated by a wide number of disabilities (physical, sensory, mental/intellectual or even hidden...), (7), or declare themselves to be more specialised in one type of disability (6). However, they all (14) claim to meet the needs of people with physical impairments.

Table 2. Summary of results according to type of disability

Agency	1	2	3	4	5	6	7	8	9	10	11	12	13	14
DISABILITY														
Physical	X	X	X	X	X	X	X	X	X	X	X	X	X	X
Sensory		X	X	X						X	X		X	X
Others		X	X	X		X				X	X		X	X

Source: authors.

The following quote from the information published on a agency's webpage shows that one of its main commitments is to adapt to the different adaptability needs that different types of disabilities may have, trying to offer great quality to these people who they consider as "excluded" or "forgotten" through innovation and great knowledge of the people with disabilities' market to face their "business challenge":

Our objective is to create and offer unique and fully satisfactory experiences to all people with all kinds of disabilities (sensory, physical, intellectual, organic...) and people with reduced mobility problems (elderly people, temporarily disabled people such as those with injuries or convalescing...). A project born form a passion to offer top quality products and services to groups that are usually forgotten or excluded. A project that uses creativity and innovation as the best tools with which to address business challenges. The agency has arisen from an in-depth knowledge of the needs and desires of the disabled tourist and with the vocation to offer experiences that, until now, were unimaginable for many people with disabilities in Barcelona and Catalonia.

-Information provided on Barcelona Special Traveler's website-

Some of them are specialised in groups of people with a certain type of disability (12), but most are oriented to the individual client and his/her relatives (11). Although some of them attend groups, they also prepare customised trips according to customer preferences (9), (except one that only has pre-designed accessible packages for certain disabilities).

Table 3. Summary of results according to type of trip

Agency	1	2	3	4	5	6	7	8	9	10	11	12	13	14
TYPE TRIP														
Individual	X	X		X	X	X	X	X	X	X	X		X	
Groups	X	X	X	X	X	X		X	X	X	X	X		X

Source: authors.

They especially highlight their commitment to providing versatility and adapt to individual needs, organising family trips, romantic or "personal experiences", for individuals or groups, national or international, offering great quality for the same prices as a standard trip:

We prepare tailor-made trips, adapted and accessible for everyone, not for that reason, more expensive, because we start from the basis that all people have the right to enjoy leisure and tourism under the same conditions and at the same prices, always with quality warranty. We organize national and international trips, vacations, getaways ... tailored to measure, for people with functional disabilities, both individually and in groups, all our vacation packages created to suit our clients

-Information provided on Accessitravel.com's webpage-

According to the literature, people with disabilities do not always have the same needs, so the services should be universal and/or adapt to personal characteristics (Darcy & Buhalis, 2011; Buhalis & Michopoulou, 2011; UNWTO, 2016), and provide very detailed and reliable information presented in formats that can be understood by people with visual, hearing and intellectual disabilities (Buhalis & Michopoulou, 2011; Eichhorn *et al.*, 2008; Darcy, 2010; Eichhorn & Buhalis, 2011; Michopoulou & Buhalis, 2013; Emrouzeh *et al.*, 2017; De Lemos, Vieira & Pinho, 2020 ; Mayordomo-Martínez *et al.*, 2019).

It is worth remembering that McKercher *et al.* (2003) indicate that, due to the standardised tourist packages that do not adapt to the needs of the individual customer, which are easy to sell and cheap, enabling agencies to reduce costs and increase profits and the low level of awareness among travel agents of the needs of people with disabilities, travel agencies were not an option for tourists with disabilities. However, accessible travel agencies have specific characteristics that differ from those of ordinary online/traditional travel agencies. This is because competitive prices are not reached exclusively through greater customer self-management (Pinto & Castro, 2019; Zhang *et al.*, 2006). The difference between a traditional travel agency and an online one is the price-service ratio (Aguiar-Quintana, Moreno-Gil & Picazo-Peral, 2016; Pinto & Castro, 2019; Tsang, Lai & Law, 2010; Zhang *et al.*, 2006). The accessible e-travel agencies try to differentiate themselves by offering a comprehensive service adapted to the needs of people with disabilities that is either previously designed for a specific type of disability or for groups or totally customised without this greatly affecting the price:

During these years of experience as a continuously evolving company, we created our groups-accessible department. The negotiation of special conditions for groups (company groups, unions, public bodies, federations and associations belonging to the third sector, etc.) is the mission of this department. It is responsible for providing the members of the groups, through express agreement, with exclusive products at unbeatable prices, creating an accessible product, exclusively adapted to third sector groups.

-Information provided on Triana viajes's website-

Therefore, the efforts made by travel agencies to be more competitive and survive in a market with an increasingly growing supply due to ICTs (Aguiar-Quintana, Moreno-Gil & Picazo-Peral, 2016; Buhalis & Law, 2008; Tsang, Lai & Law, 2010), should include an emerging and developing strategy based on the specialisation of meeting the demands of the market segment of people with disabilities.

4.2. Destinations and Activities

Although for many people with disabilities specialised group trips with disability associations can be the best way of travelling due to the facilities they provide in the planning and development of the trip (Blichfeld & Nicholaisen, 2011), in other cases, these individuals see tourism as a means to forge their personal identity in a way that shifts the focus away from their physical/medical condition (Eichhorn, Miller & Tribe, 2013). Accessible travel agencies take this motivation into account when offering individuality and a wide range of destinations and differentiating activities on demand. Although one of the agencies analysed specialises in groups, it offers experiences in different parts of the world and the possibility of participating in 'any' desired activity in its adapted version (water activities, adventure, visits to cities or rural areas, cultural tourism, trips to beaches, visits to theme parks, wellness, wine tourism…). The activities organised by the accessible tourism e-travel agency therefore, can be diverse: from cultural visits in cities or cruises to extreme sports. Many of these agencies advertise adapted adventure activities and sports in nature.

Table 4. Summary of results according to type of activity

Agency	1	2	3	4	5	6	7	8	9	10	11	12	13	14
ACTIVITIES														
Water	X		X	X		X		X	X	X				
Adventure	X	X	X	X		X		X		X			X	
Cities	X	X	X	X	X	X		X	X	X	X	X	X	X
Cruises		X	X			X							X	
Cultural	X	X	X	X	X	X		X	X	X			X	X
Rural/Nature	X	X	X	X		X	X	X	X	X			X	X
Theme parks	X	X	X	X	X	X								
Sun & Beach	X	X	X	X	X	X		X	X	X	X	X	X	X
Wellness		X	X	X		X		X		X	X	X	X	
Others	X	X	X	X		X		X	X	X			X	X

Source: authors.

In comparison with a regular trip organised by a disability association, the main advantage of the accessible tourism travel agencies is the wide range of adapted activities that the user can request at his/her own convenience. The activities are offered either within tour packages or as the main attraction of the trip (if they advertise a cruise trip or adventure activities in a certain destination). Also, some agencies are specialised in a specific type of tourism (rural, family or adventure) or offer different types of activities depending on the season:

In the three important holiday periods of the year, we offer trips designed to make the most of your HOLIDAYS. At Christmas we seek to offer a family atmosphere in spaces of maximum comfort where we all live again the excitement of these important dates. For Easter, we focus on offering tourist trips in which to discover new places and live exciting experiences. In summer, like everyone else, we look for a vacation that makes us forget the daily routine. Adapting to all tastes; beach, pool, mountain ... No matter how many times you repeat with us, there will always be a new destination waiting for you!

-Information provided on Educnatur's webpage-

These agencies have the possibility of taking travellers to other countries. As Table 5 shows, 7 of the analysed agencies that organise trips to international destinations.

Table 5. Summary of results according to destination scope

Agency	1	2	3	4	5	6	7	8	9	10	11	12	13	14
DESTINATION														
National	X	X	X	X	X	X		X	X	X	X	X	X	X
International	X	X	X	X		X	X	X						

Source: authors.

The main requirement to organise a trip abroad adapted to the individual needs of people with disabilities is to verify the information of the suppliers (accommodation, tourist attractions, transportation, etc.) and search for accessible routes (Darcy, 2010; Michopoulou & Buhalis, 2013). For this reason, there are different cases: the agency has reliable contacts in the destinations, such as professionals with disabilities who have tested their products; they specialise in a specific destination with tested products. In any case, they all assure that the client's requirements will be met and that they will be available 24 hours during the trip to resolve any unforeseen events.

100% guarantee after all trips and experiences have been tested by wheelchair professionals

-Information provided on Travel-Xperience's website-

One of the travel agencies analysed is specialised in a country as a tourist destination and offers a differentiated service from the others. It offers packages of pre-designed accessible rural routes of adapted hotels throughout the country for people with reduced mobility. The customer can take these routes using an adapted rental car:

Iceland is easy to get to know with our CocheYnoche programs. These include the rental of the adapted VW Caddy vehicle (up to 5 people) and accommodation with breakfast in a combination of small hotels and rural tourism/farms with accessible rooms. Every day you know where you have to go to sleep and you can travel relaxed and at your own pace. An easy way to discover Iceland: glaciers, hot springs, geysers, active volcanoes, vast lava deserts and black sands.

-Information provided on Islandia66's website-

4.3. Special Value of Accessible e-Travel Agencies

The literature on travel agencies indicates, on the one hand, that a differentiating feature of online travel agencies is the ease with which the user can access and search the content, enabling a greater self-management of the trip and, therefore, more economical prices (Pinto & Castro, 2019; Zhang *et al.*, 2006). On the other hand, the traditional travel agencies are valued for offering greater security, quality, assistance, more information, adaptability and a more personalised service among other aspects, with the inconvenience of a higher price (Aguiar-Quintana, Moreno-Gil & Picazo-Peral, 2016; Zhang et al., 2006). However, agencies operating in the accessible tourism market are faced with the added difficulty of looking for accessible and trustworthy information as well as accommodation, transport, activities and products adapted to the individual needs of people with disabilities (Buhalis & Michopoulou, 2011; Eichhorn *et al.*, 2008; Darcy, 2010; Eichhorn & Buhalis, 2011; Michopoulou & Buhalis, 2013). Nevertheless, as we have seen in the previous sections, there is a wide variety of activities and destinations aimed at a broad market segment with special needs.

Most of the accessible tourism travel agencies analysed are receptive (9), that is, they provide their services in the destination where they are physically established. They stand out for being experts of the destination and always being available for the client. In the case of people with disabilities, it is a very convenient service. These agencies can check any aspect of the trip at first hand and organise any adapted activity in the destination, without the need for external contacts and a higher cost.

Table 6. Summary of results according to special value

Agency	1	2	3	4	5	6	7	8	9	10	11	12	13	14
VALUE														
Receptive		X			X	X			X	X	X	X	X	X
Contacts	X	X	X	X		X		X						
Tested by Other clients, evaluators...	X					X	X	X						

Source: authors.

Some of the receptive agencies also organise trips to other destinations (2). They have a network of contacts in different parts of the destination country or abroad who can verify the specific requirements of the clients. Another strategy is being specialised in a specific destination/s and offering tour packages that have previously been tested by others with the disability for which it is offered (4). Most products (accommodation, routes, attractions ...) tested by themselves or by other customers are added to listings and search engines on the company's website with information on specific accessibility features.

We organise custom-made trips to any destination in the world. We work directly with receptive agents in the destination with whom we establish highly competitive prices. Our experts will be at your service, contributing all of their experience in the country which constitutes added value for the final customer.

-Information provided on Travelfast's website-

4.4. Information Richness and Services Provided

The richness of the information on the website is an important requirement for people with disabilities. However, this richness is not obtained in all aspects. Many of the agencies promise on their website that this information will be provided after the client get in touch with the agency. According to the specialised literature (Darcy, 2010; Eichhorn *et al.*, 2008; Michopoulou & Buhalis, 2013), the main requirement is detailed information on all aspects of the trip in different formats (text, photographs and plans) to allow the customer to assess the product in accordance with their own perception of their capabilities. The information presented in more detail refers to services and activities. Some agencies publish information in text format and photographs along with opinions from other users about the accommodation.

Table 7. Summary of results according to information richness

Agency	1	2	3	4	5	6	7	8	9	10	11	12	13	14
INFORMATION														
Activities	X	X		X	X		X	X	X		X			
Accommodation	X	X					X	X	X		X			
Routes	X				X	X								
Services	X		X	X	X	X			X	X	X	X	X	X

Source: authors.

Most either have a blog and/or strong presence on social media, where users with disabilities share their travel experiences with the agency, valuing the services offered by the company and the activities, accommodation and the transportation hired. The possibility of introducing consumer-generated online content through the website is another characteristic of online travel agencies (Gretzel & Yoo, 2008) and this option is a very convenient resource for people with disabilities as they use social media (Altinay *et al.*, 2016) and "word of mouth" (Ray & Ryder, 2003) to obtain information about accessible places. The following quote narrates the experience of a client posted on the blog linked to the agency's website:

For the first time we decided to take a trip that included several countries, something that we were very concerned about due to using different modes of transport, different hotels, renting cranes and shower chairs, etc. But once again, (the agency) showed us that they are there to ease all our worries and show us that you can travel wherever you want. When we arrived at the stations, the assistants were already waiting for us, when we arrived at the hotels, the crane and shower chair were already in the room.

- Customer testimonial posted on the blog of Travel-Xperience-

It is also possible to find search engines on the website of accessible tourism resources, a characteristic element of online travel agencies.

We already know that it is not easy to find hotels with accessible swimming pools. Click on the button and you will find a selection of them in different parts of Spain.

-Information provided on Ocio y turismo accesibles's website-

The agencies analysed provide a wide range of specialised services for people with reduced mobility:

At Accesible Spain Travel we provide everything that a traveller with reduced mobility might need during a trip to Spain: accessible hotel accommodation, accessible transportation, accessible city tours, adapted vehicle hire, accessible train tickets, accessible restaurants, accessible shopping, accessible daily itineraries, travel insurance, and more. We have researched and audited personally all hotels, tours itineraries, restaurants and facilities that you can see on our portfolio to ensure that they are barrier free and meet sufficient access needs.

-Information provided on Accessible Spain's website-

Accessible agencies offer an assistance service at airports and train stations (13), the rental of orthopaedic aids -10- (transfer cranes, electric wheelchair, etc.) and a repair service of the tourist's own technical aids in the destination (3). Six of the agencies offer personal assistance during the trip by monitors or caregivers specialised in bedding, food and personal hygiene. They also provide adapted private vehicles with tour guides (9).

Table 8. Summary of results according to Specific characteristics of accessibility services

Agency	1	2	3	4	5	6	7	8	9	10	11	12	13	14
SERVICES														
Tailor-made trips	X	X	X	X	X	X		X	X	X	X	X	X	X
Reception	X	X	X	X	X	X		X	X	X	X	X	X	X
Rent tech. aids	X			X	X	X	X		X	X	X	X	X	
Repair TA				X	X						X			
Personal assistance				X		X					X	X	X	X
Special tour-guides			X	X	X	X	X		X	X	X			X
Consultancy			X	X	X	X		X	X	X				
Others	X		X	X	X		X				X			X

Source: authors.

The specific services offered to people with sensory disabilities (visual and auditory) focus on tourist guides at the destination trained in facilitating perception and communication with the environment. These guides accompany the tourists and describe the surroundings during the trip or translate through sign language. Many of them also offer accessibility consultancy to other tourist companies as they consider themselves as "experts of accessibility".

We carry out audits and prepare reports on Universal Accessibility. We have a technical consulting team that carries out projects, reports and routes on accessibility. Our clients are companies and public bodies that want to adapt and improve both their environment and their services. We have designed web/ mobile apps with accessible and geolocable urban cultural routes in Madrid, Toledo, Alcalá de Henares and Vejer de la Frontera. We implement creative, innovative and sustainable accessibility solutions that help improve the image of our clients, their reputation, the quality of their services and the profitability of their businesses. Our premise is that the needs of all people in the physical environment are covered, including those with reduced mobility.

-Information provided on AccessibleMadrid.com's website-

5. CONCLUSIONS

The tourism industry is not yet fully prepared to meet the specific needs of people with disabilities. Within this margin for improvement there are many business opportunities in which travel agencies can play a central role. Although most people with disabilities travel with associations, we can observe the emergence of travel agencies specialised in accessible tourism. In a time when traditional travel agencies are being relegated by the self-management of travel and when competition in tourism industry is fiercer, the search for specialisation is becoming crucial. This study analyses the opportunities for specialisation and value creation of accessible e-travel agencies. The results reveal that in the accessible tourism specialisation these travel agencies have found a formula with which to improve their competitiveness. These travel agencies identified the accessible tourism segment as a market with potential in which to develop specific improvement actions in line with the three major strategies for increasing competitiveness identified in the literature. Taking into account the specific needs and expectations of people with disabilities has enabled travel agencies to improve the quality of their products and services while finding formulas to contain prices. They can offer products and services ranging from accessible travel packages aimed at people with certain disabilities to completely tailor-made trips where the user chooses what he/ she wants to do, and the agency ensures that this activity will meet absolutely all the needs and expectations of its costumer. Travel agencies specialised in accessible tourism allow the client to choose who they want to be accompanied with (in a group, with their family/friends, caretaker or even alone), where to go (some are specialised in a single destination and others promise to take the tourist to any part of the world) and what activities to do during the trip (from a visit to a museum to adventure activities).

It should be noted that people with disabilities need services adapted to the specific requirements of each person and require reliable information provided by knowledgeable professionals. In this sense, the information published on the website of the agencies does not always meet all of the standards, leading the client to contact the company staff so that they can find out the details of the trip. In any case, many agencies have a user experiences blog on their website and spaces for dialogue between company-user and/or user-user, in addition to links to the company's social media accounts. We can also find search engines on the websites for adapted travel products. The specific services for people with disabilities that are usually offered are: reception at the destination, personal assistance during the trip, special tour guides (for sensory or mental/intellectual disabilities) rental of orthopaedic aids and/or the repair service of the tourist's own technical aids in the destination. For many of their clients it would be impossible to travel if it were not for these specific services.

Planning a trip is not an easy task for people with disabilities, since the information offered on the internet is not always reliable. The arduous planning required and the considerable service offered can be translated into high costs at the customer's expense. However, as we have seen in this study, accessible e-travel agencies offer competitive prices based on three efficiency-related strategies: they are receptive agencies in a specific destination; they have a network of contacts in different parts of the world that can verify the specific requirements of the client; or they are specialised in a specific destination/s, offering tourist packages that have previously been tested by other people with the disability for which it is offered.

In conclusion, accessible e-travel agencies have particular characteristics that give them a competitive advantage over others: they combine the advantages of being an online agency (providing travel information on the internet, costumer-generated content or ease of booking) and those of being a traditional agency, with greater security, personalised service and customised trips. Considering the concept of universality, the fact that they are specialised in knowing and meeting the requirements of people with disabilities does not mean that they cannot serve groups without disabilities, but rather that they can broaden their spectrum of clients and increase their business possibilities.

This study contributes to furthering the knowledge about the emerging niche market of accessible tourism e-travel agencies and how this specialisation can constitute a survival option in the immensely competitive market of tourism intermediaries. The analysis therefore, could be considered as a starting point for researching in greater depth those agencies specialised in meeting the requirements of people with disabilities. This would include the study of the supply offered in different countries or more specific analyses based on the experience of the travel agencies themselves. In terms of methodology, the study of the accessible supply through the analysis of websites can be applied to other different kinds of tourist companies such as hotels that have chosen to specialise in the market of people with disabilities, from which we can learn about inclusive practices in the sector and business specialisation options.

REFERENCES

Abrate, G., Bruno, C., Erbetta, F., & Fraquelli, G. (2019). Which Future for Traditional Travel Agencies? A Dynamic Capabilities Approach. *Journal of Travel Research*, *59*(5), 777–791. doi:10.1177/0047287519870250

Agovino, M., Casaccia, M., Garofalo, A., & Marchesano, K. (2017). Tourism and disability in Italy. Limits and opportunities. *Tourism Management Perspectives*, *23*, 58–67. doi:10.1016/j.tmp.2017.05.001

Aguiar-Quintana, T., Moreno-Gil, S., & Picazo-Peral, P. (2016). How could traditional travel agencies improve their competitiveness and survive? A qualitative study in Spain. *Tourism Management Perspectives*, *20*, 98–108. doi:10.1016/j.tmp.2016.07.011

Altinay, Z., Saner, T., Bahçelerli, N. M., & Altinay, F. (2016). The role of social media tools: Accessible tourism for disabled citizens. *Journal of Educational Technology & Society*, *19*(1), 89–99.

Bi, Y., Card, J. A., & Cole, S. T. (2007). Accessibility and attitudinal barriers encountered by Chinese travellers with physical disabilities. *International Journal of Tourism Research*, *9*(3), 205–216. doi:10.1002/jtr.603

Bigné, J. E., Küster, I., Andreu, L., & Blesa, A. (2008). Orientación al mercado, resultados e indicadores básicos de competitividad. Interrelación en las agencias de viajes. *Revista Española de Investigación de Marketing. ESIC*, *12*(1), 97–122.

Bizjak, B., Knežević, M., & Cvetrežnik, S. (2011). Attitude change towards guests with disabilities: Reflections from tourism students. *Annals of Tourism Research*, *38*(3), 842–857. doi:10.1016/j.annals.2010.11.017

Blichfeldt, B. S., & Nicolaisen, J. (2011). Disabled travel: Not easy, but doable. *Current Issues in Tourism*, *14*(1), 79–102. doi:10.1080/13683500903370159

Buhalis, D., & Law, R. (2008). Progress in information technology and tourism management: 20 years on and 10 years after the Internet. The state of eTourism research. *Tourism Management*, *29*(4), 609–623. doi:10.1016/j.tourman.2008.01.005

Buhalis, D., & Michopoulou, E. (2011). Information-enabled tourism destination marketing: Addressing the accessibility market. *Current Issues in Tourism*, *14*(2), 145–168. doi:10.1080/13683501003653361

Cortés Bello, R., & Vargas Martínez, E. E. (2018). Prospectiva en Agencias de Viajes: Una Revisión de la Literatura (Prospective in Travel Agencies: A Revision of Literature). *Turismo e Sociedade*, *22*, 45–64. doi:10.18601/01207555.n22.03

Daniels, M. J., Rodgers, E. B. D., & Wiggins, B. P. (2005). "Travel Tales": An interpretive analysis of constraints and negotiations to pleasure travel as experienced by persons with physical disabilities''. *Tourism Management*, *26*(6), 919–930. doi:10.1016/j.tourman.2004.06.010

Darcy, S. (2010). Inherent complexity: Disability, accessible tourism and accommodation information preferences. *Tourism Management*, *31*(6), 816–826. doi:10.1016/j.tourman.2009.08.010

Darcy, S., & Buhalis, D. (2011). Conceptualising disability. In D. Buhalis & S. Darcy (Eds.), *Accessible tourism: Concepts and issues* (pp. 21–45). Channel View Publications.

De Lemos, F. F., Vieira, D., & Pinho, T. (2020). Digital Marketing Strategies for Accessible Tourism. In Digital Marketing Strategies for Tourism, Hospitality, and Airline Industries (pp. 71-92). IGI Global. doi:10.4018/978-1-5225-9783-4.ch004

Domínguez, T., Fraiz, J. A., & Alén, E. (2013). Economic profitability of accessible tourism for the tourism sector in Spain. *Tourism Economics*, *19*(6), 1385–1399. doi:10.5367/te.2013.0246

Eichhorn, V., & Buhalis, D. (2011). Accessibility: A key objective for the tourism industry. Accessible tourism: Concepts and issues. In Accessible tourism: Concepts and issues. Bristol: Channel View Publications.

Eichhorn, V., Miller, G., Michopoulou, E., & Buhalis, D. (2008). Enabling access to tourism through information schemes? *Annals of Tourism Research*, *35*(1), 189–210. doi:10.1016/j.annals.2007.07.005

Eichhorn, V., Miller, G., & Tribe, J. (2013). Tourism: A site of resistance strategies of individuals with a disability. *Annals of Tourism Research, 43*, 578–600. doi:10.1016/j.annals.2013.03.006

Emrouzeh, M. P., Dewar, K., Fleet, G., & Bourgeois, Y. (2017). Implementing ICT for Tourists with Disabilities. In *Proceedings of the 2017 International Conference on E-Education, E-Business and E-Technology* (pp. 50-53). ACM. 10.1145/3141151.3141152

Gretzel, U., & Yoo, K. H. (2008). Use and Impact of Online Travel Reviews. In P. O'Connor, W. Höpken, & U. Gretzel (Eds.), *Information and Communication Technologies in Tourism*. Springer. (Original work published 2008)

Holsti, O. R. (1969). *Content Analysis for the Social Sciences and Humanities. Reading. MA: Addison-Wesley (content analysis)*. Publishing Company.

Huang, L., Yung, C. Y., & Yang, E. (2011). How do travel agencies obtain a competitive advantage?: Through a travel blog marketing channel. *Journal of Vacation Marketing, 17*(2), 139–149. doi:10.1177/1356766710392737

Kastenholz, E., Eusébio, C., & Figueiredo, E. (2015). Contributions of tourism to social inclusion of persons with disability. *Disability & Society, 30*(8), 1259–1281. doi:10.1080/09687599.2015.1075868

Lacalle, L. (2013). Agencias de viajes en España. Una industria convulsa. *Papers de turismo*, (54), 122–138.

Lai, K. W. (2014). The role of service quality, perceived value, and relationship quality in enhancing customer loyalty in the travel agency sector. *Journal of Travel & Tourism Marketing, 31*(3), 417–442. doi:10.1080/10548408.2014.883346

Law, R., Qi, S., & Buhalis, D. (2010). Progress in tourism management: A review of website evaluation in tourism research. *Tourism Management, 31*(3), 297–313. doi:10.1016/j.tourman.2009.11.007

Loi, K. I., & Kong, W. H. (2017). Tourism for all: Challenges and issues faced by people with vision impairment. *Tourism Planning & Development, 14*(2), 181–197. doi:10.1080/21568316.2016.1204357

Mayordomo-Martínez, D., Sánchez-Aarnoutse, J. C., Carrillo-de-Gea, J. M., García-Berná, J. A., Fernández-Alemán, J. L., & García-Mateos, G. (2019). Design and Development of a Mobile App for Accessible Beach Tourism Information for People with Disabilities. *International Journal of Environmental Research and Public Health, 16*(12), 2131. doi:10.3390/ijerph16122131 PMID:31208146

McKercher, B., & Darcy, S. (2018). Re-conceptualizing barriers to travel by people with disabilities. *Tourism Management Perspectives, 26*, 59–66. doi:10.1016/j.tmp.2018.01.003

McKercher, B., Packer, T., Yau, M. K., & Lam, P. (2003). Travel agents as facilitators or inhibitors of travel: Perceptions of people with disabilities. *Tourism Management, 24*(4), 465–474. doi:10.1016/S0261-5177(02)00107-3

Michopoulou, E., & Buhalis, D. (2013). Information provision for challenging markets: The case of the accessibility requiring market in the context of tourism. *Information & Management, 50*(5), 229–239. doi:10.1016/j.im.2013.04.001

Millan, A., & Esteban, A. (2004). Development of a multiple-item scale for measuring customer satisfaction in travel agencies services. *Tourism Management*, *25*(5), 533–546. doi:10.1016/j.tourman.2003.07.002

Moura, A. F. A., Kastenholz, E., & Pereira, A. M. S. (2018). Accessible tourism and its benefits for coping with stress. *Journal of Policy Research in Tourism, Leisure & Events*, *10*(3), 241–264. doi:10.1080/19407963.2017.1409750

Özogul, G., & Baran, G. G. (2016). Accessible tourism: The golden key in the future for the specialized travel agencies. *Journal of Tourism Futures*, *2*(1), 79–87. doi:10.1108/JTF-03-2015-0005

Pagán, R. (2015). The contribution of holiday trips to life satisfaction: The case of people with disabilities. *Current Issues in Tourism*, *18*(6), 524–538. doi:10.1080/13683500.2013.860086

Raigada, J. L. P. (2002). Epistemología, metodología y técnicas del análisis de contenido. *Sociolinguistic Studies*, *3*(1), 1–42.

Ray, N. M., & Ryder, M. E. (2003). "Ebilities" tourism: An exploratory discussion of the travel needs and motivations of the mobility-disabled''. *Tourism Management*, *24*(1), 57–72. doi:10.1016/S0261-5177(02)00037-7

Shaw, G., & Coles, T. (2004). Disability, holiday making and the tourism industry in the UK: A preliminary survey. *Tourism Management*, *25*(3), 397–403. doi:10.1016/S0261-5177(03)00139-0

The Centre for Excellence in Universal Design. (2019). *What is Universal Design?* Retrieved from: http://universaldesign.ie/What-is-Universal-Design/

Tsang, K., Lai, M. T., & Law, R. (2010). Measuring e-service quality for online travel agencies. *Journal of Travel & Tourism Marketing*, *27*(3), 306–323. doi:10.1080/10548401003744743

UNWTO. (2016). *Manual on Accessible Tourism for All: Principles, Tools and Best Practices*. UNWTO Publications.

UNWTO. (2019). *UNWTO Tourism Definitions*. UNWTO Publications.

WHO - World Health Organization. (2020). *10 facts on disability*. Retrieved from: https://www.who.int/features/factfiles/disability/en/

Zhang, Y. J., Driscoll, D. A., Lazarony, P. J., Lin, B., & Zhang, R. (2006). What affects travellers' preference of e-travel or traditional travel agencies: An empirical study. *International Journal of Services and Standards*, *2*(4), 385–399. doi:10.1504/IJSS.2006.010471

KEY TERMS AND DEFINITIONS

Accessible Information: Very detailed information about accessibility of spaces and services offered, accompanied by pictures, descriptions, sizes and floorplans, published in accessible formats (easy-language, compatible for assistive technology, etc.).

Competitiveness: Is the ability of a company to develop competitive advantage/s that differentiate/s it from its competitors, obtaining an advantageous position.

Efficiency: Is the ability to achieve an objective with the least possible resources or in the shortest time possible.

Online Travel Agency: Is a tourist company that differs from a traditional agency in the possibility of its customers to book travel related services directly on its website.

Receptive: Located in the destination where it offers its services.

Tailor-Made Trip: Trip that is organized according to the needs and tastes of the individual client.

Traditional Travel Agency: Is a tourist company dedicated to organization, intermediation, commercialization, and fulfillment of tourist's projects, plans and itineraries between their customers and travel providers. The term "traditional" refers to characteristics related to personal treatment between the company and the client.

Chapter 10
Museum Website Accessibility Examined Using Automatic Diagnostic Tools:
A Study in Portugal

Celeste Eusébio
https://orcid.org/0000-0002-2220-5483
GOVCOPP, DEGEIT, University of Aveiro, Portugal

Maria João Carneiro
GOVCOPP, DEGEIT, University of Aveiro, Portugal

Leonor Teixeira
https://orcid.org/0000-0002-7791-1932
IEETA, DEGEIT, University of Aveiro, Portugal

Diana Lemos
DEGEIT, University of Aveiro, Portugal

ABSTRACT

This chapter examines the website accessibility of museums located in Portugal (N = 575), based on the web content accessibility guidelines (WCAG 2.0), and on the three type of conformance levels (A, AA, and AAA), using two web diagnostic tools: AccessMonitor and TAW (Test de Accessibilidad Web). The results show that the average accessibility of museum websites is 5.80 on a scale from 1 to 10. However, while according to AccessMonitor, there are 14 websites (2.43% of the total) with a level of accessibility of 10 (excellent web accessibility practices); there are also some websites with a global index lower than 3 (classified as bad practice). The results obtained through TAW show that an average of 58.43 problems were identified in each website analyzed. These results highlight that several improvements should be introduced in the websites of Portuguese museums in order to make information available to all people.

DOI: 10.4018/978-1-7998-6428-8.ch010

INTRODUCTION

Despite all citizens having the right to participate in leisure and tourism activities (Kastenholz, Eusébio & Figueiredo, 2015; Yau, McKercher & Packer, 2004), there are some groups in our society that face several constraints to participation in these activities, mainly people with disabilities (PwD) (Smith, 1997). The literature on leisure constraints highlights three types (intrapersonal, interpersonal and structural) (Nyaupane & Andereck, 2008) which impede access to leisure activities. Some of these travel constraints prevent the creation of preferences (specifically intrapersonal constraints) while others, specifically the interpersonal and structural, make the transformation of preferences into participation difficult (Devile & Kastenholz, 2018). Some studies have used leisure constraints theory to examine travel constraints of PwD (e.g. Daniels, Rodgers & Wiggins, 2005; Devile & Kastenholz, 2018) revealing that PwD face considerable difficulties when participating in recreation and tourism activities. Of these three types of constraints, the absence of information (a structural type) prevents PwD from participating in recreation and tourism activities. However, there is a limited number of studies which have examined this kind of travel constraints.

Although several studies and data show that PwD are an important growth market, the majority of tourism supply agents have neglected them (Kastenholz et al., 2015). Museums are an important cultural attraction both for PwD and people without disabilities, contributing to increasing their cultural knowledge (Mesquita & Carneiro, 2016). Due to the high levels of risk that PwD face when visiting cultural attractions such as museums, these consumers need more information about the accessibility of these spaces (Handa, Dairoku, & Toriyama, 2010; Mangani & Bassi, 2019; Poria, Reichel & Brandt 2009). Thus, the lack of reliable information concerning accessibility may eventually prevent some PwD from visiting museums. With the proliferation of the Internet, the way in which people obtain information to plan their tourism trips changed (Dinis, Eusébio & Breda, 2020; Kim, Xiang, & Fesenmaier, 2015; Scholl-Grissemann & Schnurr, 2016). The Internet is nowadays one of the most important information sources used by PwD and people without disabilities to plan their trips (Almeida-Santana & Moreno-Gil, 2007; Luo et al., 2004; Zajadacz, 2014). Therefore, the websites of tourism supply agents, including the websites of museums, are important sources of information and should be accessible to the largest possible range of people, regardless of their needs, preferences, skills and personal access conditions (Handa et al., 2010; Langa et al., 2012; Mills, Han, & Clay, 2008; Williams & Rattray, 2005).

To increase website accessibility levels, the Web Content Accessibility Guidelines (WCAG) were developed by the World Wide Web Consortium (W3C, 2018) to help designers and developers to create websites which are more accessible to PwD. These guidelines are taken as a reference at a worldwide level.

Studies that examine the website accessibility level of the museums are of utmost importance to increase the participation of PwD in these cultural activities. However, only a limited number of studies analyzing the accessibility level of websites' museums (e.g. Langa et al., 2012; Leoporin & Norscia, 2008; Petrie et al., 2005). Moreover, the majority of studies published only examines the website accessibility of a very small number of museums. Despites, in Portugal there is already regulation that defines the requirements of websites' accessibility of public organizations (Decree-Law 83/2018), no study is known that examines the accessibility level of museums' websites.

Therefore, in order to increase knowledge in this field, it is the intention of this chapter to examine the website accessibility of the museums located in Portugal based on the Web Content Accessibility Guidelines (WCAG 2.0) and using two web diagnostic tools: AccessMonitor and TAW (Test de Accesibilidad Web).

To accomplish the aforementioned aims, this chapter is structured into four sections. After this introduction, a literature review on travel constraints of PwD when visiting museums is presented, followed by a literature review concerning web accessibility. In the third section, the methodology used to collect and to analyze the data is described. Section four presents the results obtained through the use of two automatic evaluation tools (AccessMonitor and TAW) in terms of web accessibility of museums located in Portugal. Finally, the chapter ends with the most important conclusions of this study and some contributions which may increase the accessibility level of museums' websites. Limitations and suggestions for further research are also presented in this section.

LITERATURE REVIEW

People With Disability and Museums

It is already widely recognized that museums should be inclusive and provide conditions to ensure the participation and engagement of all people in their exhibitions (Seale, Carrizosa, Rix, Sheehy & Hayhoe, 2020). People are also aware that some important efforts have already been made by museum managers in order to increase the accessibility of these cultural attractions (Mesquita & Carneiro, 2016; Poria et al., 2009; Zakaria, 2020). Nevertheless, the literature reveals that PwD experience several constraints while visiting museums. Although several researchers recognize the importance of developing research to help increase the accessibility of museums, the research that examines the constraints to visiting museums as perceived by PwD is still very limited.

The small set of research on accessibility of museums focused on supply or demand provides some relevant insights into difficulties that PwD experience when they want to visit a museum. Some constraints to visiting these tourism attractions seem to be related to physical access to the museum space. These include architectural barriers and other physical features such as narrow spaces, lack of facilities (e.g. toilets) with adequate accessibility for PwD (Zakaria, 2020), lack of handrails (Mesquita & Carneiro, 2016; Zakaria, 2020) or of tactile paths (Kusayama, 2005; Mesquita & Carneiro, 2016), as well as inexistence or inappropriateness of access ramps (Poria et al., 2009; Zakaria, 2020). Other problems, especially for people with wheelchairs, are that some counters are excessively high, and some objects are displayed very high up. While the first issue makes contact with staff behind the counter difficult, the second prevents a good appreciation of the exhibition (Poria et al., 2009).

The limited access to exhibitions, specifically, the inexistence of opportunity to touch objects or replicas, and even the lack of multisensory tours are also recognized as a constraint, especially for people with visual impairments. Their participation can become broadly limited, without many opportunities for engagement if non-visual stimuli are not provided (Handa et al., 2010; Mesquita & Carneiro, 2016; Zakaria, 2020).

Staff attitudes can also be a constraint at times. Some staff do not speak in a normal way with PwD (for example, they talk very loud or use very simple words) inferring that these people will have difficulties in understanding them, they do not talk to them directly or even avoid communicating with them (Poria et al., 2009). This may partially result from the lack of training of staff to deal with PwD, which has been observed in several museums (Handa et al., 2010; Kusayama, 2005).

Another major constraint to visiting museums which PwD experience is the lack of appropriate and reliable information. In some studies (e.g. Poria et al., 2009), this issue is considered an even greater barrier than physical access. The literature reveals that providing reliable information about the accessibility of the museum, i.e. on their exhibitions and facilities (Handa et al., 2010; Poria et al., 2009) and conveying information in an appropriate way for PwD (Handa et al., 2010; Mesquita & Carneiro, 2016; Poria et al., 2009), is crucial for a good visit. Nevertheless, sometimes during the visit, the information provided by the staff concerning the accessibility of the museum is not accurate (e.g. Poria et al., 2009). Moreover, some museums do not always convey information properly in such a way that it is accessible for all. Some relevant issues that have been identified in this scope are the lack of accessible audio-guides (Handa et al., 2010; Mesquita & Carneiro, 2016; Zakaria, 2020), of information in large characters (Mesquita & Carneiro, 2016; Zakaria, 2020), of Braille information (Handa et al., 2010; Mesquita & Carneiro, 2016), or of tactile information (Zakaria, 2020), as well as the inexistence of sign language services and assistive devices for people with hearing impairments (Zakaria, 2020).

Some research remarks, however, that PwD not only demand an appropriate provision of information while at the museum, but also prior to arriving there, while planning the trip (Handa et al., 2010; Langa et al., 2012; Poria et al., 2009). This is in line with research on PwD undertaken in the field of tourism, which points out that these people, due to the constraints they face, need more information for planning the trip and feel highly constrained when this information does not exist or when they have various difficulties in accessing this information (Daniels et al., 2005; Devile & Kastenholz, 2018; Yau et al., 2004). Some studies on museums reveal that PwD find it very important to have useful and complete information on the website which may be consulted before the visit, with some highlighting the importance of having details about physical obstacles that may be encountered (Handa et al., 2010; Poria et al., 2009), probably to be aware of the main difficulties that may be faced during the visit and to adopt strategies to overcome them. However, it is also emphasized that museum web pages should not only convey information on accessibility but should also be accessible for PwD (Handa et al., 2010). Therefore, not having information on the accessibility of the museum provided in an accessible way through the museum's website, may potentially prevent some PwD from visiting museums. Despite the importance of the website as a major information source for PwD (Devile & Kastenholz, 2018), particularly when these people want to visit museums (Handa et al., 2010; Langa et al., 2012; Mangani & Bassi, 2019), only a limited number of studies were found related to museums, accessibility and websites which examine a set of museum websites (e.g. Langa, et al., 2012; Leporini & Norscia, 2008; Mangani & Bassi, 20019; Petrie et al., 2005). However, some of them only analyze whether the museums have online information for PwD (e.g. Mangani & Bassi, 2019) and others examine a low number of websites (e.g. Leporini & Norscia, 2008). The results of published studies show that, on average, only a small proportion of the museums analyzed convey information for PwD, revealing a low level of concern for this group, and several failures concerning web accessibility are detected, with many websites violating the WCAG guidelines.

Web Accessibility

The literature defines web accessibility as the degree to which a website allows access to its content to the largest possible range of users, including those users who have visual, auditory, motor or cognitive disabilities or people who experience some kind of environmental constraints (Akgül & Vatansever, 2016; Mankoff, Fait, & Tran, 2005). To correctly access information on the web, some people need to use assistive technologies (AT), such as alternative pointing devices, voice recognition, screen readers, refreshable Braille displays, and alternate keyboards (Buhalis & Michopoulou, 2011). In order to ensure web accessibility, on the one hand allowing correct visualization and understanding of all web content and, on the other hand the connection with AT, it is important to ensure that these web solutions are well designed and implemented. To promote this accessibility, the World Wide Web Consortium (W3C) (W3C, 2018) in a Web Accessibility Initiative (WAI) created the Web Content Accessibility Guidelines (WCAG), an international standard which defines the accessibility of web content to support developers (designers and programmers) in making their websites more accessible for PWD (Akgül & Vatansever, 2016).

The WCAG are presented in several versions, WCAG 1.0 being the first one, published in May 1999. They correspond to a set of guidelines that have checkpoints distributed in 3 priorities and success criteria that determine the conformance level. WCAG 2.0 was published in December 2008 and WCAG 2.1 in June 2018 (W3C, 2018). In February 2020 the first public working draft of WCAG 2.2 was announced (W3C, 2020). The last three versions are based on a set of guidelines distributed in four different principles: (i) perceivable; (ii) operable; (iii) understandable and (iv) robust.

As regards specifics, WCAG 2.0 (W3C, 2016), which was approved as an ISO/IEC 40500 international accessibility standard in October 2012, is composed of 12 guidelines grouped into the four principles—perceivable, operable, understandable and robust, representing the version most used to evaluate web accessibility according to WCAG guidelines (Akgul & Vatansever, 2016; Domínguez Vila, González & Darcy, 2018, 2019; Mankoff et al., 2005; W3C, 2018).

Perceivable principle is responsible for ensuring that the information displayed on the user interface is in a format which is perceivable to users, independently of their condition. Operable principle ensures that the navigation process is operable. Understandable principle guarantees that the web content and user interface are comprehensible. Finally, Robust principle is responsible for the robustness of the solution in terms of facilitating the integration of a wide variety of user agents, including AT. To evaluate web content according to these four principles, the WCAG has available a set of guidelines for each principle that are evaluated according to one of three levels of conformity: (i) A—the minimum level with basic accessibility; (ii) AA—intermediate accessibility; and (iii) AAA—high accessibility (W3C, 2018).

WCAG 2.0 guidelines have been used to examine web accessibility condition in different tourism domains (e.g. Akgül & Vatansever, 2016; Domínguez Vila et al., 2018, 2019; Kuzma et al., 2009; Mills et al., 2008; Shi, 2006), but there are few studies available in the literature which focus on evaluations of museum websites and no study is found regarding Portuguese museums. However, considering that tourism in Portugal is a relevant and growing economic activity and due the high number of museums in this country, it is very important to ensure that all people, regardless of their conditions, may access the information available in the museums' websites.

METHODOLOGY

Data Collection Methods

To analyze the accessibility level of the museum websites, a multistage procedure was used. First, a database with the websites of Portuguese museums was created through an online search on the Portugal Museums website (https://museusportugal.org/) and on the Directorate General for Cultural Heritage (http://www.patrimoniocultural.gov.pt/pt/museus-e-monumentos/rede-portuguesa/). Additionally, an intensive search on Google was carried out to collect links to the websites of some museums. Finally, this database was complemented with a search on the websites of the city councils. The database contains information about the name of the museum, the website link and the localization of the museum. A total of 681 museums were identified. However, only 575 have a website (83.92% of the total). Therefore, this study focuses only on the museums that have a website.

Data Analysis Methods

The accessibility level of Portuguese museum websites was examined based on the Web Content Accessibility Guidelines (WCAG) 2.0, considering the three conformance levels (A, AA and AAA). In order to select the most suitable tools to carry out this analysis, a search on the World Wide Web Consortium (W3C) (W3C, 2019) and on studies published on website accessibility (e.g. Akgül & Vatansever, 2016; Adepoju & Shehu, 2016; Domínguez Vila et al., 2019; Kuzma, Yen & Oestreicher, 2009; Mills et al., 2008; Shi, 2006; Silveiro, Eusébio & Teixeira, 2019;) was undertaken. From this search two automatic evaluation tools were selected to use in this research: (i) AccessMonitor and (ii) TAW ('Test de Accesibilidad Web'). These automatic evaluation tools examine the homepage of the websites based on the WCAG 2.0.

The AccessMonitor tool was developed by the Portuguese Foundation for Science and Technology. This automatic evaluation tool examines web accessibility according to WCAG 2.0. The website link is inserted in the AccessMonitor page (https://accessmonitor.acessibilidade.gov.pt/amp/) and an accessibility report and a global accessibility index is automatically generated. The errors found on the websites examined are presented by the three degrees of compliance (A = basic accessibility; AA = intermediate accessibility and AAA = high accessibility). The global accessibility index generated varies between 1 (very poor web accessibility practice) and 10 (excellent web accessibility practice).

In order to complement the information obtained with AccessMonitor, TAW, also an automatic online tool used in some studies published on website accessibility (e.g. Akgül & Vatansever, 2016; Adepoju & Shehu, 2016; Shi, 2006, Silveiro et al.,2019), was adopted. This automatic evaluation tool (https://www.tawdis.net/) also checks the level of accessibility of websites taking WCAG 2.0 guidelines into account, but the results are categorized into three groups: (i) "problems" (corrections are necessary); (ii) "warnings" (points to the need for a review by a technician); and "not reviewed" (a complete manual review is required). However, the "not reviewed" data were not examined in this study.

The accessibility level of the Portuguese museums' websites was examined using the two automatic evaluation tools mentioned (AccessMonitor and TAW) during the month of July and August 2019. The results obtained were analyzed using the data analysis Software Package for the Social Science (SPSS). Descriptive statistics were used to evaluate the accessibility level of the websites analyzed.

RESULTS AND DISCUSSION

Sample Profile

In this chapter, a sample of 575 websites of museums located in Portugal were examined. As can be observed in Table 1, more than half of the museums (52%) are located in the Central Region of Portugal and in the Lisbon Metropolitan Area. On the other hand, the lowest number of museums analyzed is of those located in the Autonomous Region of Madeira (3.7%).

Table 1. Location of the analyzed museums

NUTS II	Number	%
North	67	19.80
Central Region	172	29.90%
Lisbon Metropolitan Area	133	23.10%
Alentejo Region	67	11.70%
Algarve	33	5.70%
Autonomous Region of the Azores	35	6.10%
Autonomous Region of Madeira	21	3.70%
Total	575	100.00%

Concerning ownership of the website, the data presented in Table 2 reveals that only 32.2% of the museums analyzed have their own website. In the majority of the museums (67.8%) analyzed, the website is owned or developed by another entity, mainly by city councils.

Table 2. Type of website according to the owner

Ownership of the website	Number	%
Museum with own website	185	32.20%
Museum with website of another entity	390	67.80%
Total	575	100%

Accessibility Level Using AccessMonitor

Table 3 shows the global index obtained with AccessMonitor, and the different types of errors according to the conformance level (A, AA, and AAA). Results reveal that the museum websites analyzed do not achieve all the necessary requirements to be accessed by PwD. The global index varies between 2.6 and 10 (excellent practice of web accessibility), with an average of 5.8 and a standard deviation (SD) of 1.806. On average, 7.51 errors were detected on each website examined, with compliance level A showing the highest average of errors per website (mean = 4.98, SD = 3.457), revealing that a large number of the websites analyzed do not meet the minimum requirements to be accessed by PwD. Moreover, some errors were also detected concerning AA and AAA conformance levels (Table 3).

Table 3. Results of AccessMonitor - Global Index and errors (A, AA, AAA)

Indicators	N	Minimum	Maximum	Mean	Standard deviation
AccessMonitor Index	575	2.6	10	5.800	1.806
Errors—Level A	575	0	16	4.980	3.457
Errors—Level AA	575	0	7	0.970	0.927
Errors—Level AAA	575	0	3	1.560	0.781
Number total of errors	575	0	22	7.510	4.190

A detailed analysis of the global index obtained with AccessMonitor (Figure 1) shows that the majority of the museum websites analyzed (47%) have regular accessibility practice, 26% of the websites have good practice, 5% very good practice and 9% have excellent practice. Moreover, of the 575 websites analyzed, 14 of them (2.43%) have the highest global accessibility index (10), which is considered very positive and demonstrates the concern that exists in this type of cultural attractions for a certain democratization of access to culture. However, 13% of the websites analyzed (75) have bad practice. Therefore, considerable improvements must be introduced in these websites to increase their accessibility level, such that everyone (people with disabilities and people without disabilities) has access to the information that is disseminated through the museums' websites.

Figure 1. AccessMonitor global index of museums' websites

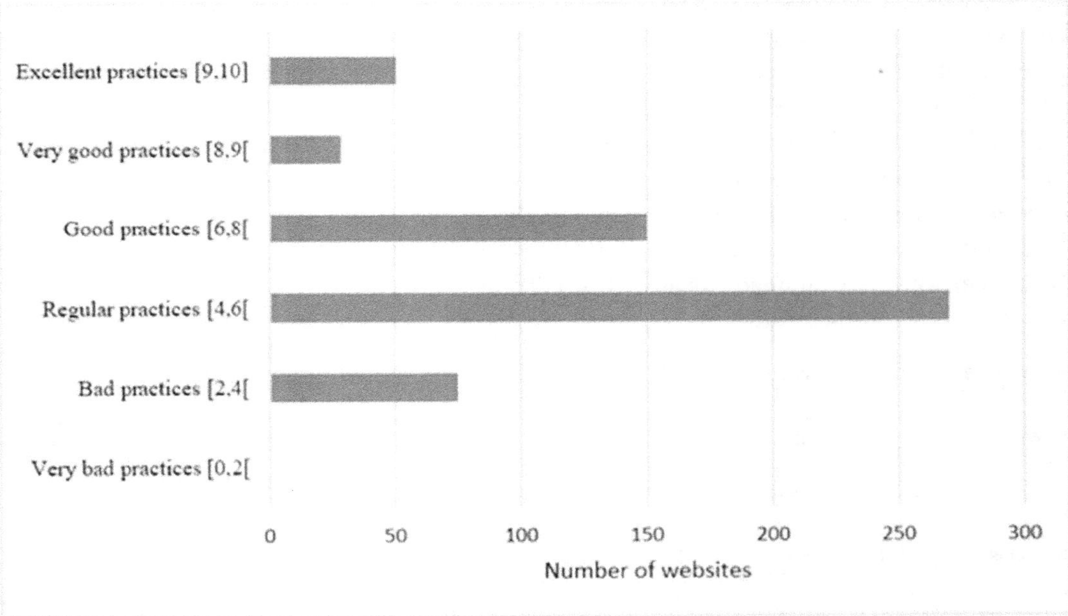

Accessibility Level Using TAW

To complement the results obtained with AccessMonitor to examine the web accessibility of Portuguese museums, in the Table 4 the problems and warnings identified with TAW, according to the four principles of the WCAG 2.0 (perceivable, operable, understandable and robust) and the three conformance levels (A, AA and AAA). Globally, although some websites have no problems or warnings (the minimum value is zero), a great number of websites have various problems, with a mean for each website of 58.36, and warnings, with a mean for each website analyzed of 242.28 (more than four times the number of problems). These results highlight that a great number of interventions should be made on the museums' websites to overcome the problems identified and that a human review should also be carried out to solve the warnings identified. These findings corroborate the results of other empirical studies that reveal a low level of accessibility or various failures in accessibility on museum websites (Langa et al., 2012) and of other suppliers of the tourism sector (Silveiro et al., 2019; Williams & Rattray, 2005).

In terms of the problems identified, the robust principles reveal the highest mean of problems for each website examined (24), showing that the content of a great number of museum websites are not robust enough to be interpreted in a trustworthy manner by a wide variety of user agents, including AT. This kind of technologies are of utmost relevance, for example, for people with visual disabilities to have access to the website content or tetraplegic people who need additional devices to interact with the web content either through input or output mechanisms. Considering the studies of Paciello (2000) and Buhalis & Michopoulou (2011) some PwD need to use AT such as alternative pointing devices, voice recognition, screen readers, refreshable braille displays, and/or alternate keyboards for the correct visualization of web content. Thus, it is important to ensure compatibility through the success criteria of the robustness principle. In view of the results presented in this study, the designers and developers of the museum websites must modify the websites to meet the robust principle, assuring that user agents, including AT, accurately interpret and parse the content in order to make it interpretable by the user.

The perceivable principle occupies the second position in terms of mean of problems for each website (mean = 17.16), closely followed by the operable principle (mean = 15.18). Consequently, various museums' websites need intervention to be perceivable and the user interface components and navigation must be operable by all people, regardless of their disabilities. However, the results obtained also show that on a great number of museum websites analyzed, the information and the operation of the user interface is understandable, given that, of all the principles it is the one that presents the lowest number of problems per website examined (mean = 1.98) (Table 4).

Table 4. Results of TAW – Types of problems and warnings in the museums' websites

TAW	Principles	N	Minimum	Maximum	Mean	Stand. Deviation
Problems	Perceivable	574	**0**	311	17.160	30.006
	Operable	574	0	184	15.180	30.845
	Understandable	575	0	68	1.980	3.932
	Robust	575	0	304	24.000	42.578
	Total	574	0	370	58.360	66.458
Warmings	Perceivable	575	0	1209	113.890	137.030
	Operable	575	0	697	47.560	67.672
	Understandable	575	0	224	6.810	10.621
	Robust	575	0	1724	74.020	164.676
	Total	575	0	1876	242.280	263.960

Concerning the warnings identified (Table 4), the perceivable principle emerges as that which has the highest average of incidents for each website examined (mean = 113.89 and a SD = 137.030), followed by the robust principle (mean = 74.02, SD = 164.676) and the operable principle (mean = 47.56, SD = 67.672). Again, the understandable principle is that which has the lowest average of incidents (mean = 6.81, SD = 10.621). These results confirm the need for programmers and museum website designers to introduce changes on websites to make their content perceivable for all people. Furthermore, the contents should also be robust enough to be interpreted by various users, including AT, and the user interface components and navigation should be operable by all, independently of their functional limitations.

A detailed analysis of the problems identified in the different guidelines that belong to each principle (Table 5) provides relevant insights to identify types of interventions that must be made in the Portuguese museum websites to increase their accessibility level. Of all the guidelines analyzed, it is possible to observe in Table 5 that those with the greatest number of problems per website are, in decreasing order: "compatible" (in the robust principle), "navigable" (operable principle), "text alternative" and "adaptable" (both in the perceivable principle). No problems were detected in: "keyboard accessible navigable", "enough time" and "seizures" guidelines. Therefore, changes must be introduced in the websites analyzed to increase their compatibility with user agents, especially including AT, and to help the users to navigate and to find the content that they need. Moreover, improvements in the adaptable guidelines are needed to ensure that all the information is available in a form that can be perceived by all users, independently of their functional limitations. Finally, all websites must provide text alternatives to all non-text content to ensure it is all available for all people (such as people with visual disabilities).

Table 5. Type of problems identified with TAW in the museum websites

Principles	Guidelines	Minimum	Maximum	Mean	Stand. Deviation
Perceivable	1.1. Text alternative	0	128	9.560	10.351
	1.2. Time-based media	0	4	0.010	0.168
	1.3. Adaptable	0	308	7.280	22.502
	1.4. Distinguishable	0	153	0.350	6.749
Operable	2.1. Keyboard accessible navigable	0	0	0.000	0.000
	2.2. Enough time	0	0	0.000	0.000
	2.3. Seizures	0	0	0.000	0.000
	2.4. Navigable	0	184	15.150	30.824
Understandable	3.1. Readable	0	1	0.280	0.449
	3.2. Predictable	0	3	0.230	0.495
	3.3. Input assistance	0	1	1.470	3.813
Robust	4.1 Compatible	0	304	24.050	42.603

The results presented in Table 6 reinforce the results described in Table 5, concerning the guidelines of WCAG 2.0 which are not accomplished by a considerable number of Portuguese museums' websites. Again, "compatible" and "navigable" emerge as the guidelines with most incidents (warnings) for each website analyzed. However, in the case of perceivable principles the "adaptable", "distinguishable" and "text alternatives" guidelines are those with most warnings by each website analyzed, in decreasing order. Thus, to overcome incidents related to "distinguishable" guidelines, the programmers and web designers must introduce changes on the websites to make it easier for users to see and hear content. The use of contrast, resized text and audio control are some of success criteria related to this guideline.

Table 6. Type of warnings identified with TAW in the museums' websites

Principles	Guidelines	Minimum	Maximum	Mean	Stand. deviation
Perceivable	1.1. Text alternative	0	234	24.580	36.125
	1.2. Time-based media	0	58	0.180	2.498
	1.3. Adaptable	0	1060	49.930	76.948
	1.4. Distinguishable	0	474	39.460	56.897
Operable	2.1. Keyboard accessible navigable	0	198	2.490	14.399
	2.2. Enough time	0	0	0.000	0.000
	2.3. Seizures	0	3	0.005	0.125
	2.4. Navigable	0	697	45.020	66.155
Understandable	3.1. Readable	0	1	0.000	0.059
	3.2. Predictable	0	6	0.010	0.250
	3.3. Input assistance	0	218	6.800	10.420
Robust	4.1 Compatible	0	1724	74.280	164.905

CONCLUSIONS AND IMPLICATIONS

So that PwD have access to tourism experiences, it is important to ensure that all activities in the tourism industry are accessible and have accessible information. Museums are cultural attractions that play a very important role in cultural tourism and, in many situations, visiting museums is the main reason for a tourist trip (Boniface & Cooper, 2001). However, literature has identified some barriers in museums that hinder or prevent PwD from enjoying these spaces, whether physical, attitudinal or communication barriers. Communication barriers in particular can have an impact on other barriers, since they are related to the way the information, including information on accessibility, is communicated and reaches everyone, regardless of their condition. Thus, it is crucial to ensure that the means of communication used are accessible, with the web being the most widely used channel of communication today. However, given the complex nature of stakeholders involved, which include different suppliers and end-users with distinct needs, this task involves great challenges.

This work aimed to evaluate the accessibility level of the information and content communicated through the websites of a set of museums. To this end, a sample of 575 websites of museums located in Portugal were analyzed, based on WCAG 2.0 guidelines, using two automatic evaluation tools—TAW and AccessMonitor. In general, although websites achieve a median score that ranks them as "regular practice", both tools used in the present analysis detected a high number of problems and warnings at level A (lowest level of compliance) which reveals that many of the museums' websites do not meet the basic requirements of web accessibility, increasing the difficulties and barriers faced by PwD in accessing the websites' contents. Some problems are also detected at level AAA (the highest-level conformance).

In more detail, the results show that the average accessibility of museum websites is 5.80, on a scale from 1 to 10. However, according to AccessMonitor, there are 14 websites in our sample with a level of accessibility of level 10 (excellent web accessibility practices), corresponding to 2.43% of the total websites analyzed. Nevertheless, there are also some websites with a global index lower than 3 (classified as bad practice). The results obtained through TAW show that an average of 58.43 problems were identified in each website analyzed. Moreover, a deeper analysis of the problems and warnings identified reveals that the biggest problems of the websites analyzed are related to "perceivable (e.g. non-text content of pictures, no audio description or media alternative, no sign language) and robust (i.e. the websites should be able to be interpreted by a great variety of users including AT). These results highlight that several improvements should be introduced in the websites of the Portuguese museums to make information available to all people. So, there is an urgent need to introduce changes in websites to make the content of the websites perceivable for all people and interpreted by several types of AT such as alternative pointing devices, voice recognition, screen readers, refreshable Braille displays, and alternate keyboards.

The present results highlight that, despite the web development advances and the emergences of specific regulation regarding web accessibility, the museums still present a high number of incidents. Therefore, the results found in this study are in line with the findings obtained in a study conducted by Domínguez Vila et al. (2019), who analyzed official tourism websites located in Northern European countries, reporting that perceptible and robust principles were the most critical. Moreover, these results reinforce the findings obtained in Silveira et al.'s (2019) study, which was conducted to evaluate the web accessibility of a set of tourism agencies located in the Central Region of Portugal, as well as the results of the studies carried out by Shi (2006) and Mills et al. (2008), who concluded that the websites analyzed failed in at least one WCAG criterion.

In order to improve the accessibility level of Portuguese museum websites, it is important to comply with the requirements endorsed by the WCAG and make the web content more accessible (W3C, 2018). In this sense, it is crucial to consider some recommendations, such as: (i) providing text alternatives in the case of non-textual content in order to facilitate the switch to other formats according to the specific needs of the user (such as large print, Braille, speech, symbols or simpler language); (ii) providing a mechanism to request an update of content instead of automatically updating, as automatic updates will make the context change and, consequently, may cause disorientation of the disabled user; and (iii) providing descriptive titles and labels on some HTML components such as tables, images, videos and text blocks, among others. Related to robustness, it is important to introduce improvements in HTML tags, in order to maximize the compatibility with devices usually used by PwD, including AT.

Due to the lack of studies on web accessibility based on WCAG in the area of museums, this study provides important theoretical contributions to increase knowledge on web accessibility of museums. On the other hand, from a practical perspective, these results can help web designers and programmers, since it contributes a diagnosis on the level of website accessibility and consequently a set of recommendations aforementioned to improve the accessibility level of museums' web contents.

Despite the contributions of this study, particularly in the diagnosis and global evaluation of the accessibility level of museums' websites, some limitations have been found. The first is related to the type of methods used to analyze the websites, which were exclusively automatic. So, since this analysis establishes the presence of valid attributes in HTML tags, and considers only the homepage, it does not take into account the perspective of the people that experience those situations. Thus, it is important to use other evaluation mechanisms that involve users, such as usability tests and task analysis in order to understand the people that experience these conditions (i.e. domain experts), as well as to perceive the main difficulties that can occur in the interactions between the users and the web applications in a real context.

ACKNOWLEDGMENT

This work was also financially supported by the project POCI-01-0145-FEDER-030376, funded by FEDER, through COMPETE2020 – Programa Operacional Competitividade e Internacionalização (POCI), and by national funds (OE), through FCT/MCTES.

References

W3C. (2016). *How to Meet WCAG 2 (Quick Reference).* Available: https://www.w3.org/WAI/WCAG21/quickref

W3C. (2018). *Web Content Accessibility Guidelines (WCAG) 2.1.* Available: https://www.w3.org/TR/2018/REC-WCAG21-20180605

W3C. (2019). *Web Accessibility Evaluation Tools List.* Available: https://www.w3.org/WAI/ER/tools

W3C. (2020). *Web Content Accessibility Guidelines (WCAG) 2.2 in Development.* Available: https://www.w3.org/blog/2020/02/wcag-22-in-development/

Akgül, Y., & Vatansever, K. (2016). Web accessibility evaluation of government websites for people with disabilities in Turkey. *Journal of Advanced Management Science*, *4*(3), 201–210. doi:10.12720/joams.4.3.201-210

Almeida-Santana, A., & Moreno-Gil, S. (2017). New trends in information search and their influence on destination loyalty: Digital destinations and relationship marketing. *Journal of Destination Marketing & Management*, *6*(2), 150–161. doi:10.1016/j.jdmm.2017.02.003

Boniface, B., & Cooper, C. (2001). Worldwide destinations casebook: The geography of travel and tourism: Second edition. In Worldwide Destinations Casebook: The Geography of Travel and Tourism: Second Edition. doi:10.4324/9780080949758

Buhalis, D., & Michopoulou, E. (2011). Information-enabled tourism destination marketing: Addressing the accessibility market. *Current Issues in Tourism*, *14*(2), 145–168. doi:10.1080/13683501003653361

Daniels, M. J., Rodgers, E. R., & Wiggins, B. P. (2005). Travel tales': An interpretive analysis of constraints and negotiations to pleasure travel as experienced by persons with physical disabilities. *Tourism Management*, *26*(6), 919–993. doi:10.1016/j.tourman.2004.06.010

Devile, E., & Kastenholz, E. (2018). Accessible tourism experiences: The voice of people with visual disabilities. *Journal of Policy Research in Tourism, Leisure & Events*, *10*(3), 265–285. doi:10.1080/19407963.2018.1470183

Dinis, M. G., Eusébio, C., & Breda, Z. (2020). Assessing social media accessibility: The case of the Rock in Rio Lisboa music festival. *International Journal of Event and Festival Management*, *11*(1), 26–46. doi:10.1108/IJEFM-02-2019-0012

Domínguez Vila, T., Alén González, E., & Darcy, S. (2018). Website accessibility in the tourism industry: An analysis of official national tourism organization websites around the world. *Disability and Rehabilitation*, *40*(24), 2895–2906. doi:10.1080/09638288.2017.1362709 PMID:28793789

Domínguez Vila, T., González, A., & Darcy, S. (2019). Accessible tourism online resources: A Northern European perspective. *Scandinavian Journal of Hospitality and Tourism*, *19*(2), 140–156. doi:10.1080/15022250.2018.1478325

Handa, K., Dairoku, H., & Toriyama, Y. (2010). Investigation of priority needs in terms of museum service accessibility for visually impaired visitors. *British Journal of Visual Impairment*, *28*(3), 221–234. doi:10.1177/0264619610374680

Kastenholz, E., Eusébio, C., & Figueiredo, E. (2015). Contributions of tourism to social inclusion of persons with disability. *Disability & Society*, *30*(8), 1259–1281. doi:10.1080/09687599.2015.1075868

Kim, H., Xiang, Z., & Fesenmaier, D. R. (2015). Use of the internet for trip planning: A generational analysis. *Journal of Travel & Tourism Marketing*, *32*(3), 276–289. doi:10.1080/10548408.2014.896765

Kusayama, K. (2005). Access to museums for visually challenged people in Japan. *International Congress Series*, *1282*, 877–880. doi:10.1016/j.ics.2005.05.068

Kuzma, J., Yen, D., & Oestreicher, K. (2009). Global e-government Web Accessibility : An Empirical Examination of EU, Asian and African Sites. In *International Conference on Information and Communication Technologies and Accessibility*, (pp. 1–6). Hammamet, Tunisia: Academic Press.

Langa, L., Herman, A., Hillsgrove, D., Jones, B., Li, C. M., Matthai, Z., Schlaffer, J., Slack, J., Sizemore, B., Weir, R. Jr, Fagan, J., Lazar, J., Henderson, H., Knauff, S., Adesina, T., & Baumgart, D. M. (2012). Museum web accessibility: A Pilot assessment of Mid-Atlantic museums. *International Journal of the Inclusive Museum*, *4*(1), 15–28. doi:10.18848/1835-2014/CGP/v04i01/44360

Leporini, B., & Norscia, I. (2008). "Fine-Tuning" image accessibility for museum websites. *Journal of Universal Computer Science*, *14*(19), 3250–3264.

Luo, M., Feng, R., & Cai, L. A. (2004). Information search behavior and tourist characteristics. *Journal of Travel & Tourism Marketing*, *17*(2), 269–193.

Mangani, A., & Bassi, L. (2019). Web information, accessibility and museum ownership. *International Journal of Tourism Policy*, *9*(4), 265–281. doi:10.1504/IJTP.2019.105486

Mankoff, J., Fait, H., & Tran, T. (2005). Is your web page accessible?: A comparative study of methods for assessing web page accessibility for the blind. In *Proc. of the SIGCHI Conf. on Human factors in Computing Systems*, (pp. 41–50). 10.1145/1054972.1054979

Mesquita, S., & Carneiro, M. J. (2016). Accessibility of European museums to visitors with visual impairments. *Disability & Society*, *31*(3), 373–388. doi:10.1080/09687599.2016.1167671

Mills, J. E., Han, J.-H., & Clay, M. J. (2008). Accessibility of hospitality and tourism websites: A challenge for visually impaired persons. *Cornell Hospitality Quarterly*, *49*(1), 28–41. doi:10.1177/1938965507311499

Mills, J. E., Han, J. H., & Clay, M. J. (2008). Accessibility of hospitality and tourism websites: A challenge for visually impaired persons. *Cornell Hospitality Quarterly*, *49*(1), 28–41. doi:10.1177/1938965507311499

Nyaupane, G. P., & Andereck, K. L. (2008). Understanding travel constraints: Application and extension of a leisure constraints model. *Journal of Travel Research*, *46*(4), 433–439. doi:10.1177/0047287507308325

Petrie, H., King, N., & Weisen, M. (2005). The accessibility of museum web sites: Results from an English investigation and international comparisons. In J. Trant & D. Bearman (Eds.), *Museums and the Web 2005: Proceedings. Archives & Museum Informatics, 2005*. Available https://www.archimuse.com/mw2005/papers/petrie/petrie.html

Poria, Y., Reichel, A., & Brandt, Y. (2009). People with disabilities visit art museums: An exploratory study of obstacles and difficulties. *Journal of Heritage Tourism*, *4*(2), 117–129. doi:10.1080/17438730802366508

Scholl-Grissemann, U., & Schnurr, B. (2016). Room with a view: How hedonic and utilitarian choice options of online travel agencies affect consumers' booking intentions. *International Journal of Culture, Tourism and Hospitality Research*, *10*(4), 361–376. doi:10.1108/IJCTHR-06-2016-0062

Seale, J., Garcia Carrizosa, H., Rix, J., Sheehy, K., & Hayhoe, S. (2020). A participatory approach to the evaluation of participatory museum research projects. *International Journal of Research & Method in Education*, 1–21. Advance online publication. doi:10.1080/1743727X.2019.1706468

Shi, Y. (2006). The accessibility of Queensland visitor information centres' websites. *Tourism Management*, *27*(5), 829–841. doi:10.1016/j.tourman.2005.05.012

Silveiro, A., Eusébio, C., & Teixeira, L. (2019). Heterogeneity in accessibility of travel agency websites: A study in the central Portugal region. *RISTI - Revista Ibérica de Sistemas e Tecnologias de Informação, 35*, 18-34.

Smith, R. W. (1987). Leisure of disabled tourists: Barriers to participation. *Annals of Tourism Research*, *14*, 376–389. doi:10.1016/0160-7383(87)90109-5

Williams, R., & Rattray, R. (2005). UK hotel web page accessibility for disabled and challenged users. *Tourism and Hospitality Research*, *5*(3), 255–268. doi:10.1057/palgrave.thr.6040024

Yau, M. K.-S., McKercher, B., & Packer, T. L. (2004). Traveling with a disability - More than an access issue. *Annals of Tourism Research*, *31*(4), 946–960. doi:10.1016/j.annals.2004.03.007

Zajadacz, A. (2014). Sources of tourist information used by deaf people. Case study: The Polish deaf community. *Current Issues in Tourism*, *17*(5), 434–454. doi:10.1080/13683500.2012.725713

Zakaria, N. N. (2020). Barriers to social inclusion with the Egyptian museums; New approach towards disability. *Curator*, *63*(1), 115–130. doi:10.1111/cura.12353

KEY TERMS AND DEFINITIONS

Accessible Tourism: Tourism for all that enables people with various access requirements to have access to tourism products, services, and environments, without constraints.

AccessMonitor: An automatic validator available on-line, developed by the Portuguese Foundation for Science and Technology, that aims to analyze website accessibility, considering the version 2.0 of the WCAG.

Museums: Organizations devoted to the acquisition, conservation, study, exhibition, and educational interpretation of tangible and intangible heritage having scientific, historical, or artistic value.

People With Disabilities: People suffering from any restriction or lack of ability (resulting from an impairment) to perform an activity in the manner or within the range considered normal for a human being.

TAW: An automatic tool available on-line for analyzing website accessibility, considering the version 2.0 of the Web Accessibility Guidelines (WCAG 2.0).

WCAG: Web Content Accessibility Guidelines published by the Web Accessibility Initiative of the World Wide Web Consortium, reporting a set of recommendations for making Web content more accessible, mainly for people with disabilities, and available in different versions.

Chapter 11
The Accessibility of Museum Websites:
The Case of Barcelona

Ariadna Gassiot Melian
University of Girona, Spain

Raquel Camprubí
University of Girona, Spain

ABSTRACT

Tourism accessibility has emerged as a topic to be discussed and studied by both academics and professionals. In this line, museums have started to adapt their offer to their diverse audience, including people with disabilities and people with special needs. In this regard, museum websites, as a powerful information channel to promote visitation, must also be adapted to make its information accessible. Therefore, this chapter aims at exploring accessibility and accessible information of museums' websites using a holistic approach. As a case study, 45 museums in the city of Barcelona are considered and accessibility of tourism websites is assessed by means of content analysis, taking as reference four categories and 36 items that have emerged from previous literature review. Findings reveal that museums websites are still far from being considered accessible, and improvements in several areas are required.

INTRODUCTION

The analysis of accessibility of tourism activities was first focused on the experiences of people with disabilities while travelling and the functional elements linked to these experiences. Nowadays, due to the recent research on the area, this definition goes far beyond this first idea and accessible tourism is considered 'a form of tourism that enables people with access requirements to function independently and with equity and dignity through the delivery of universally designed tourism products, services, and environments' (Darcy & Buhalis, 2011).

DOI: 10.4018/978-1-7998-6428-8.ch011

People's behavior when using tourism information has shifted dramatically following the ubiquitous access to the Internet and there is diversity of behavioral tendencies that determine the level of usage and interaction with information sources (Chang & Caneday, 2011). For example, previous studies (Gronflaten, 2009) state that socio-demographic characteristics condition information search behavior. Due to the diversity of patterns when using the Internet and the information sources, it is important to evaluate and compare how people search for information, select information sources, among other informational behaviors, to fully meet their needs.

People with disabilities have the same needs and desires for tourism as others (Blichfeldt & Nicolaisen, 2011). Because of accessibility needs, they may behave differently when searching and selecting information and information sources. For instance, access requirements, differences in skills and other characteristics may lead to longer searches, different selected sources, etc. In order to analyze these behavioral patterns, and when ensuring a good provision of information for all, two concepts must be addressed. First, the information accessibility, based on its format or design. Second, the information on accessible products and services in terms of accuracy or credibility.

Cultural tourism offer is important to develop destinations all over the world, and particularly in Europe. There is an increasing number of people travelling for cultural motivations and they are particularly interested in cultural heritage (Velasco, 2009). Museums can be defined as one of these cultural attractions for the consumption and delight of cultural tourists. Museums should be prepared to meet the needs of cultural tourists while they ensure its social function (Vacas, 2005). In this context, this type of offer must be accessible to all groups of the population.

During the pre-trip planning process, in order to ensure museums accessibility to people with special access needs, the information process must be addressed. As it is stated above, Internet has become one of the most important information channels and includes different information sources, such as websites (Llodrà-Riera, Martínez-Ruiz, Jiménez-Zarco & Izquierdo-Yusta, 2015). Thus, the main aim of the book chapter is to explore accessibility and accessible information of museums' websites.

This chapter is structured as follows: first, literature review is given, where the main concepts related to the objective of the chapter are defined –i.e. accessible tourism and museums, and information search–; second, methodological design and steps are described; third, main results are given; fourth, main conclusions are provided.

LITERATURE REVIEW

Accessible Tourism and Museums

Accessible tourism research has recently become an area of interest in the tourism field. In the last decades, studies addressing different types and degrees of disability and the tourism activity have helped in broadening the diversity of the field, by including 10-20 per cent of the population who had previously been excluded from our understanding (Darcy, Mckercher, & Schweinsberg, 2020). We have moved from to separate concepts, i.e. tourism and accessibility, to an integrative concept, accessible tourism, which facilitates the emergence of the collaborative processes when defining and managing tourism destinations around the world (Darcy et al., 2020)

It is well-known that this market segment does not only include people with disabilities, but any person with special access needs. Many tourism subsectors (e.g. accommodation services, restaurants, service providers, etc.) have been working on improving accessibility standards in order to ensure the provision of equal tourism experiences to everybody, but what is the state of the art of museums and exhibition centers?

According to the American Heritage Dictionary, museums can be understood as 'an institution for the acquisition, preservation, study and exhibition of works of artistic, historical or scientific value (cited in Dillenburg, 2011). These entities are sometimes considered the main attraction to tourists into many cities (Moreno Gil & Ritchie, 2009), so access to them may be ensured to all groups of the population, and this in turn will keep a destination attractive and competitive for all.

Online accessibility of museum resources, either through exhibitions or collections, is the main incentive of digital heritage programs (Müller, 2002), while accessibility of physical museum resources is the incentive of on-site heritage programs. Yet, the value of museums and, consequently, its attractiveness and competitiveness depend not only on the cultural significance of the objects they exhibit, but also on other elements such as their accessibility and interpretation (Sheng & Lo, 2010). Accessibility and interpretation, as part of the holistic understanding of museums function, can be linked to the provision and use of information sources. While communication has no strategic value for some museums and it is not consciously planned or managed, neither is it considered crucial to engage in dialogue with their publics (Capriotti, 2013), other museums focus their communication and information strategies on long-term and planned efforts. The latter ones, from a strategic point of view, would be more prepared to offer this integrated vision of accessibility and interpretation.

It has been more than 20 years since the Internet and digitalization changed informational and societal rules and behaviors. Specifically, from the offer point of view, some years ago, ICT changed the profile of museums, as they have moved from being mere information interpreters to information providers (Müller, 2002). At the beginning of Internet, ICT use and implementation could be considered a choice for museums. Some then well-established museums were moving cautiously to the Net, while new museums already started virtual initiatives with, for example, a website (Müller, 2002). Nowadays, ICT are not understood as a choice anymore but a matter of survival in the market, as they play a key role in their cultural communication framework through the provicion of new methods and tools (Sylaiou & Papaioannou, 2019). Traditional museums and exhibition centers have been challenged to meet the needs of a demand that has significantly changed. Thus, digital objects, online visitors and virtual communication are defining both the online and the on-site museums (Müller, 2002) and, more recently, ICT have resulted in online and on-site applications of the museum (Sylaiou & Papaioannou, 2019).

Information Search

How ICT have dramatically changed the behavior of consumers on information search has been previously discussed. For example, the most frequently used information sources to retrieve information about a tourism destination are search engines, recommendations from friends and relatives, and the official website (Coromina & Camprubí, 2016).

For people with disabilities, information quality and precision become basic to plan a successful visit to a museum. The market with disabilities has the following three types of requirements: physical accessibility, information about accessibility and information accessible online (Buhalis & Michopoulou, 2011). For the purpose of this study, this integrative view of tourists with special access needs behavior is considered. In other words, these tourists' needs and preferences will be analyzed through physical requirements to access the websites, but also accessibility to information and quality and quantity of information about accessible products and services offered.

While ensuring physical accessibility is a really important element for on-site museum visitors, information about accessibility and accessible products and services and accessibility to this information can be critical for both physical and online museum visitors. In other words, even though facilities are accessible, unless museums provide relevant and reliable information about accessibility and unless this information is accessible, people with special access needs will not succeed in planning or engaging to a specific trip or visit (Michopoulou & Buhalis, 2013).

Museum Virtual Spaces and Websites

The idea of museum visitors is often linked to physical guests. However, online museum visitors may exceed physical visitors (Hawkey, 2004). Online museums have been developed, modernized and thought as complementary to physical museums and are a place to provide educational opportunities for all (Stylianou-Lambert & Stylianou, 2010). We may think that, for people with disabilities, this virtual access is especially important to provide them with equal opportunities, as the immediate access to artworks and information from a personal environment may minimize accessibility issues they encounter in different unfamiliar physical environments.

Story-based environments, such as museums and exhibition centers, can be generated for all user types (Dyson & Moran, 2000). These stories may be, at a first moment, segregated, static, and with isolated objects, but the aim of any cultural entity is to make them dynamic, contextualized and in the form of engaging stories (Dyson & Moran, 2000). As a consequence, to guarantee equality of opportunities, museums should communicate these scenarios as something accessible and interpretable by all groups of the population, regardless their abilities, knowledge, skills, etc.

For example, in this virtual context, the World Wide Web can transport audiences into unexpected realms, including being connected to the arts and culture in any museum and gallery experience (Lisney, Bowen, Hearn, & Zedda, 2013). Through a combination of video, audio and virtual world technologies, participants can be moved into another space, such as a museum, and have real experiences like meeting and communicating with avatars or objects in the physical space (Lisney et al., 2013). Despite the romanticized and utopian understanding of the Web as a 'decentralized, open, transnational and translocal space, and a space in which all of the connected could represent themselves and their interest on an equal basis' (Kluitenberg, 2008, p.290), it could be reconsidered a highly privatized and occupied space with very specific forces (Stylianou-Lambert & Stylianou, 2010). People with special access needs have the same rights and desires to be involved and actively participate in these virtual experiences. However, as a consequence of this unbalance of roles, responsibilities and power, and mainly due to accessibility constraints, it may become a challenging process.

Even though the presence of all these circumstances, it is clear that the provision of virtual and web-based resources has broadened and increased the museums audience, so a greater number and more diverse profiles must be addressed. There has been some research about how different are the professional and the general public online or on-site visitors. For example, Walsh et al. (2020) characterize Liverpool online museum users and define that most of them are part of the 'general public' and 'non-professional groups'. Furthermore, the 'general public' tends to visit the website to pass some time, and it is something similar to the situation of physical museums. However, both the 'general public' and 'non-professional' user groups mainly visit the official websites to prepare the visit and for content information. As a consequence, online visitors' needs should be addressed in a completely different way. Müller (2002) suggests seven features necessary for the development of online exhibitions: space, time, links, storytelling, interactivity, production values and accessibility.

In addition, when assessing the relationship between virtual and on-site visitors behavioral patterns, Pallud & Straub (2014) also suggest that visitors who have a good experience with a museum website and who intend to return to this website also develop a stronger interest in the museum's collection. In contrast, Pallud & Straub (2014) find that visitors who enjoyed the website but who do not plan to return to the website are less likely to make the effort to visit the physical museum. Therefore, accessible and well-designed websites make people return to these virtual spaces and, in turn, they are more likely to visit the museum in person.

Instead of giving a comparison of online and on-site visitor, other studies have focused on analyzing online visitors profile. For instance, Thomas & Carey (2005) suggest that online visitors mainly look for general information on a museum's website, despite the website's available interactive possibilities such as web-based activities, forums, etc.

In a nutshell, people with disabilities not being an exception in the group of virtual museum visitors must be understood as a diverse group of the population. For example, differences in abilities may lead to differences in virtual spaces interpretation and accessibility requirements. Furthermore, aspects such as the degree and the type of disability must be considered to explore these differences in this market segment, always letting every individual assess for himself or herself whether a facility, service or virtual space is accessible to him or her (Figueiredo, Eusébio, & Kastenholz, 2012; Michopoulou & Buhalis, 2013).

Among other information sources, museum websites are considered to be reliable sources of information. Mainly due to museums' ownership of the Intellectual Property Rights to a large quantity of quality content, they have gained reputation for quality, authority, and objectivity (Stylianou-Lambert & Stylianou, 2010).

For any museum, websites have become a useful tool to set effective product communication and promotion goals, as well as they are a good source of information for further physical visits to the museums or to make people repeat. Pallud & Straub (2014) highlight that museum websites serve as an information source or communication channel for potential visitors to find information about physical infrastructures and resources in the museums. Marty (2007) also indicates that the use of museum websites is complementary to their visits to physical museums.

From a strategic point of view, it is clear that developing attractive and resourceful websites by using ICT has become critical for museums. These websites, for instance, like other commercial websites, have a key role to attract physical visitors to the museum, as previously discussed in this chapter. In other words, a website visitor should not be considered a mere online visitor, but a potential on-site visitor with information requirements about the product, the activities, a convenient order process, among other issues (Chiou, Tseng, Hwang & Heller, 2010).

Indeed, museum websites represent an alternative for people who may be interested in museological content but who are unable or not willing to visit the physical setting (Pallud & Straub, 2014). It is possible that the same people who visit museums also visit museum websites, and this raises questions concerning the degree of accessibility of online museums (Stylianou-Lambert & Stylianou, 2010) and museum websites.

Museum Websites Accessibility

Even though the ICT paradigm shift and development from the situation on the 1990s and the 2000s to nowadays, websites are still considered relevant as a tourism communication and dissemination tool. As stated above, it is important to make this information source accessible for all. For this purpose, website accessibility is addressed in this section.

Accessibility is often seen to be a technical issue. However, new conceptualizations of accessibility and universal design go beyond this idea. Consequently, solutions about accessibility, in order to surmount all types of constraints, need to go far beyond technical adaptations and include a holistic point of view and discussion with the users (Lisney et al., 2013). From this point of view, digital accessibility needs to as inclusive as possible with the aim of reaching a wide audience (Lisney et al., 2013).

To start with, from a technical point of view, there are some initiatives, such as the Web Accessibility Initiative (WAI) that develop guidelines to minimize and surmount constraints that people with disabilities may find when accessing the net. For example, these guidelines include recommendations about information format (e.g. braille interfaces, audio and video transcripts, etc.), simple and consistent navigation mechanisms, among other issues (Müller, 2002).

Due to the diversity of access needs, skills or knowledge, these virtual experiences when visiting a website can result in completely different realities. Consequently, the analysis of website accessibility must consider these differences across population group. For example, Dincer et al. (2019) analyze Istanbul museum websites for people with visual impairments and highlight their differences in terms of perceptions, needs and challenges. Darcy (1998) already anticipated the importance of objective measurements through the use of reliable assessments and reporting mechanisms. This objectivity should be based on individuals' judgements instead of others assessments. It is important, however, to provide people with information about accessibility to let them subsequently assess by themselves their experience in both physical or virtual environment. So, considering the naturally diverse group of people with disabilities, some aspects to be addressed to assess website accessibility are described. The process of moving objects from a physical to a virtual environment redefines meanings and values and websites and ICT have blurred the distinction between the authentic and the virtual experiences (Müller, 2002).

According to the Disability Rights Commission (2004), partially sighted users, physically impaired users, hearing impaired users and dyslexic users share the following problems when accessing and using websites: unclear and confusing layout of pages, confusing and disorienting navigation systems, inappropriate use of colors and poor contrast between content and background, and graphics and text size that are too small. Consequently, page layout simplicity, navigation mechanisms, content-background color contrast, and font size and options must be checked when assessing website accessibility.

Furthermore, according to the Disability Rights Commission (2004), both partially sighted and hearing-impaired users face constraints linked to complex language and terminology. As a consequence, language complexity and alternatives must be addressed. For the latter ones, lack of audio-based information (subtitles, sign language, etc.) may reduce their information consumption.

For both partially sighted and blind users, incompatibility between accessibility software and webpages may be a challenge. What is more, blind users often do not find links, form elements and frames labelling, and face cluttered and complex page structures (Disability Rights Commission, 2004). Again, simple website design must be considered when checking website accessibility.

Michopoulou & Buhalis (2013) define the following major challenges in terms of accessible information: interoperability, content integration, and personalization. Consequently, Michopoulou & Buhalis (2013) set the following requirements linked to accessibility to physical environment, accessibility information provision and accessible information online. Accessibility information provision includes: information on accessible facilities; richness of information (the higher the degree of disability, the more information required); information reliability; objective measurements; and information accuracy. Regarding accessible information, the following aspects are considered: content integration to minimize the effort and time in travel information gathering process; accessible design to be able to access online content; and customization to enable them to decide the tradeoffs of the travel attributes.

According to Weisen et al. (2005), the five most frequent problems about websites are: poor page design (in terms of layout), or cluttered and complex pages with dense blocks of text, no clear identification of headings, etc.; ambiguously named links that lead to unexpected contents and navigation problems; color scheme and contrast –mainly challenging for dyslexic and partially sighted–; no skip navigation link; and external navigational links with information about other facilities when the websites are integrated into the host institutions websites. These issues may lead to confusing and segregating virtual experiences when accessing a website.

Regarding the content format and presentation on the websites, people expect them to be given in a simple, familiar and understandable way (Marty & Twidale, 2004). Museum personnel, due to their knowledge and experience may offer authentic content on museum websites. However, there is the risk to impose requirement to the users of the museum websites (Marty & Twidale, 2004).

Dyson and Moran (2000) evaluate seven museum websites using a formal and an informal method. The websites formal evaluation includes aspects linked to provision of: contextual information; aids on searching and its searching means; user information identifying sites which target specific audiences; catalogues or collections describing the nature of searches of collections; structural context with the access to collection; collection search mechanisms with methods available; and search return or records on search outcomes nature. Informal evaluation comprises factors such as: learnability consistency, predictability, efficiency accessibility, and orientation.

In addition, and beyond certain all these technological and technical issues of access, there is the problem of language. For example, the largest virtual museums online are based in the UK, USA, Canada, and Australia. These museums spend a lot of economic and staff resources on online environments development, promotion and evaluation, leaving non-English speakers with little alternative (Stylianou-Lambert & Stylianou, 2010). In parallel, in other countries where English is not the official language, website contents are obviously developed in the official language and, depending on resources available, these contents are translated to other languages.

In conclusion, accessibility must be analyzed from a holistic and integrative view. There is not an agreement about items to be considered to evaluate websites and website accessibility (Law, Qi & Buhalis, 2010; Park, Gretzel & Sirakaya-Turk, 2007), but it is clear that these must go beyond technical aspects and include aspects such as information formats, information about accessible products and services, among other issues.

METHODOLOGY

Study Site

This study is based on the analysis of 45 museums and exhibition centers located in Barcelona (Table 1). The sample is extracted from 'Turisme de Barcelona' official website (Table 1). This entity could be considered a body set up to promote tourism in the city. Consequently, both private and public museums and exhibition centers are promoted through this website. The initial sample was formed by a total of 54 museum and exhibition spaces. However, 9 were later eliminated since they correspond to decentralized spaces that belong to another museum or exhibition center included in the sample, such as Refugi 307 or El Call which belong to the History Museum of Barcelona, among others.

Table 1. Description of the sample

Id.	Museum	Number of visitors	Ownership
1	La Virreina. Centre de la Imatge	158.255	PU
2	Arts Santa Mònica	173.680	PU
3	Centre de Cultura Contemporània de Barcelona	383.866	PU
4	Museu Marítim de Barcelona	334.987	PU
5	Museu de Cera	189.320	PR
6	Museu d'Història de Catalunya	218.681	PU
7	Museu Picasso	1.108.120	PR
8	Museu Frederic Marès	68.138	PR
9	Museu de la Xocolata	138.335	PR
10	Museu Europeu d'art Modern - MEAM	175.760	PR
11	El Born Centre de Cultura i Memòria	1.643.943	PU
12	Museu de les Cultures del Món	160.676	PU
13	Hash Marihuana & Hemp Museum	17.693	PR
14	Museu de l'Eròtica	70.658	PR
15	La Pedrera	990.112	PR
16	Museu Egipci de Barcelona	253.931	PR
17	Fundació Antoni Tàpies	83.543	PU
18	Museu de la Música	131.174	PU
19	Museu del Modernisme de Barcelona	21.805	PR
20	Fundació Alorda Derksen	n.a.	PR
21	Museu de l'Eròtica	3.440	PR
22	Fundació Suñol	17.529	PR
23	CaixaForum	1.379.636	PR
24	Museu d'Arqueologia de Catalunya	35.351	PU
25	Museu Nacional d'Art de Catalunya	847.281	PU
26	Fundació Joan Miró	872.623	PR
27	Museu de ciències naturals	327.103	PU
28	Fundació Fran Daurel	215.115	PR
29	Museu Olímpic i de l'Esport Joan A. Samaranch	114.176	PU

Continued on following page

Table 1. Continued

Id.	Museum	Number of visitors	Ownership
30	Col·lecció de Carrosses Fúnebres	n.a.	PR
31	Museu Etnològic	160.676	PU
32	Camp Nou Experience	1.785.903	PR
33	Reial Monestir de Santa Maria de Pedralbes	n.a.	PU
34	CosmoCaixa	1.930.920	PR
35	Museu Palmero	n.a.	PR
36	Museu Blau	327.103	PU
37	Museu del Disseny de Barcelona	267.153	PU
38	Museu Can Framis. Fundació Vila Casas	n.a.	PR
39	Museu Agbar de les Aigües	n.a.	PR
40	MUHBA	974.284	PU
41	MACBA	713.627	PU
42	Palau Robert	715.000	PR
43	Centre d'Art Contemporani Fabra i Cots	10.003	PR
44	La Capella	58.376	PR
45	Casa Museu Gaudí	322.603	PU

n.a. = Not available; PU = Public; PR = Private

Study Method

A content analysis was conducted in order to assess accessibility standards and accessible products and services information (Domínguez Vila, Darcy, & Alén González, 2015) on official museum websites from the city of Barcelona. Content analysis can be defined as 'an observational research method that is used to systematicall y evaluate the symbolic content of all forms of recorded communications' (Kolbe & Burnett, 1991).

A total of 36 items (Table 2), classified into 4 categories or study areas, were considered in the analysis. In order to structure the analysis and to generate the list of items to be assessed, previous literature on website evaluation (Morrison, Taylor, Morrison & Morrison, 1999; Doolin, Burgess & Cooper, 2002; Park et al., 2007; Law et al., 2010), museums website evaluation (Dyson & Moran, 2000; Cunliffe, Kritou & Tudhope, 2001; Wilson, 2011; Marty & Twidale, 2004); and website accessibility evaluation (Domínguez Vila et al., 2015; Shi, 2006) was considered. After these considerations, the following categories aroused (Table 2):

1. Website Design. This category includes elements linked to website usability, such as website design homogeneity, the presence of a search engine or the presence of different font size.
2. Products General Information. This category includes information on museum products and services characteristics, such as museum contact details, opening hours, exhibition typology (permanent or temporal), among other details.
3. Information on Products and Services Accessibility. This section is focused on determining whether the product that has been offered and promoted by the museum through the website is accessible for all groups with special access needs, including those with different types and degrees of dis-

Table 2. Analyzed items

Categories	Items	Authors
Web design	Speed in information processing	Morrison *et al.* (1999); Doolin *et al.*, (2002); Park *et al.* (2007); Law *et al.* (2010)
	Homogeneous page layout	
	Simple backgrounds	
	Search engine	
	Site map	
	Different font size	
	Return button	
	Contrast of colors	
	Number of languages	
	Languages	
Product general information	Address	Dyson & Moran, 2000; Cunliffe *et al.*, 2001; Wilson, 2011; Marty & Twidale, 2004
	Contact	
	Location map	
	Public transport connection	
	GPS coordinates	
	Information of the product	
	Museum history	
	Schedules fit tourism activity	
	Opening days fit tourism activity	
	Guided tours	
	Temporary exhibitions	
	Permanent exhibitions	
	Educational offer	
	Different sale points	
	Social networks	
Product accessibility information	Accessibility section on the web	Domínguez Vila, Darcy, & Alén González, 2015; Shi, 2006
	Accessible products information	
	Accessible tourism products	
	Accessible tourism network	
	Accessibility certification	
	Types of disability	
	Number of disabilities considered	
	Distinction of disability type	
Information format	Types of formats (Text, pictures, multimedia and audio)	Domínguez Vila, Darcy, & Alén González, 2015; Shi, 2006
	Information format combination	
	Accessible formats	

ability. For each of these groups, the following items have been considered: is there information about product accessibility? If there is a specific accessibility section on the website, are there specific products for each type of disability (mental, sensory, visual, mobility, etc.)? Do they refer to specific aids and formats of information for each type of disability?

4. Information Format Accessibility. In this section, items about the format of the information provided are assessed: Is the format interpretable by people with special access needs? Are there link descriptions? Is there the possibility of automatic voice reading without the use of own software or hardware? Are there alternative formats of information, such as videos with sign language or subtitles?

Content analysis was carried out considering guidelines provided by Camprubi & Coromina (2016), Kolbe & Burnett (1991) and Holsti (1968), in order to guarantee objectivity, validity and replicability of the obtained results. Content analysis was conducted by two coders following the rules and procedures established previously. Each website was analyzed twice, in order to validate the collected information. When inconsistences came up, the matter was discussed until an agreement was reached.

Results

Websites in museums are a communication tool that should be useful for museum's visitors to organize their visit. Ideally, museums' websites must be capable to provide enough and appropriate information about the museum to potential visitors, regardless their skills or abilities. Findings in this section illustrate the current situation in Barcelona's museums in terms of usability and accessibility; and strengths and weaknesses in each of the four evaluated categories are discussed.

Website Design

Kim and Fessenmaier (2008) state that websites should be user-friendly so that users can easily navigate with no difficulties. Results (Table 3) show that general museum's websites are characterized by a suitable speed in information processing (95.6%), and to present a homogenous page layout (95,6%). Simplicity of background (91.5%) is also a common feature in museums' websites. In a smaller number

Table 3. 'Website design' category description

Website design	Yes (%)	No (%)
Speed in information processing	95.6	4.4
Homogeneous page layout	95.6	4.4
Simple backgrounds	91.1	8.9
Search engine	66.7	33.3
Site map	31.1	68.9
Different font size	6.7	93.3
Return button	2.2	97.8
Contrast of colours	0	100

of museums' websites, the existence of search engine (66.7%) or a site map (31.1%) are detected; and only in a few websites, website design includes the possibility to change font size (6.7%) or a return button (2.2%); and the color contrast options are not found in any of the analyzed museums.

According to Doolin et al. (2002), to provide information in some languages proves the interest to attract international markets. In this case, findings show websites can be consulted in an average of 3.32 languages (standard deviation = 1.588). The most common languages are Catalan, Spanish and English (73.2%); however, there is a group of museums that also add French to the 3 languages previously mentioned (73.3%). Only 3 museums (7.3%) provide information in a number of languages above the average: Camp Nou Experience Museum (10 languages), Catalonia National Art Museum (9 languages) and Hash Marihuana & Hemp Museum (6 languages).

Results also show that a reduced number of museums (7.3%) provide information in the two co-official languages (Catalan and Spanish); or in Spanish and English (2.4%). Only a museum provides the information in a single language (Catalan), targeting its offer to this market segment. Globally, it can be considered that in terms of usability, the analyzed websites fulfil the necessary requirements to attract visitors. According to Pallud and Straub (2014, p.367), 'a well-designed website does indeed induce visitors to return to the website and arouse their interest to visit the museum'. However, the general website design is not the single variable that will influence visitors' decision-making, but there are other issues related to content and information provided that need to be considered too. All these issues are analyzed in the following sections.

General Information of the Product

The general information of the product in a museum's website is essential (Table 4). In this way, the visitor can obtain information related to the characteristics of the product, and make the preparation of the visit easier. In this sense, all the museums provide the contact details. The address of the museum is usually mentioned, as well as another information to contact them, for example: a telephone number, an e-mail address or an information form. Quite generally, the following elements are also provided: the location map (93.3%) or public transport connection options (89%). Only in a few cases GPS coordinates are provided.

Regarding the details about the offered product, there are the following information elements provided: a description of the product (97.8%), and, less frequently, the references about the museums' history (91.1%). Concerning the schedule and open days, it has been found that, in some cases, there are not well planned to suit tourism visitors (11%). The main reasons are: these museums close during summer or public holidays or the schedule is very restrictive. Focusing on the exhibitions, it has been found that some of the analyzed museums only offer permanent exhibitions, while others do not have a permanent collection and only offer temporary exhibitions. In most of the cases, museums provide information about the guided tours (84.4%), or the educational offer for schools and other educational centers (80%). It must be highlighted that only 57.8% of the museums offer the possibility to buy the tickets in different sales points and, therefore, the rest of the museums (42.2%) only sell tickets in the physical ticket counters of the museum.

Focusing on communication in social networks, it has been found that most of the museums include a direct link to a range of social networks on their website (95.6%), using an average of 4.9 social networks (standard deviation = 2.4). Museums usually use these means to provide information about their products, activities, etc. Among other social networks identified, the following are used by the museums:

firstly, Facebook and Twitter (76%, in both cases) as the most used; and secondly, Youtube (54%) and Instagram (52%) used by more than half of the museums analyzed. Some museums also use other social networks such as Pinterest (24%), Google+ (24%), Flickr (22%) and Vimeo (20%) and, in a few cases, they use less well-known social networks, such as Issuu or Diigo.

Table 4. 'Product General Information' category description

Items	Yes (%)	No (%)
Contact details		
Address	100.0	0.0
Contact	100.0	0.0
Location map	93.3	6.7
Public transport connection	89.0	11.0
GPS coordinates	4.4	95.6
Product details		
Information of the product	97.8	2.2
Museum history	91.1	8.9
Schedules fit tourism activity	91.1	8.9
Opening days fit tourism activity	88.9	11.1
Guided tours	84.4	15.6
Temporary exhibitions	82.2	17.8
Permanent exhibitions	80.0	20.0
Educational offer	80.0	20.0
Different sale points	57.8	42.2
Communication		
Social networks	95.6	4.4

Information About the Accessibility of the Product

Focusing on the aspects directly related with the aim of this study, it can be observed that the accessibility in terms of product is still little present in the analyzed museums (Table 5). It has been found that only 42.2% of museums include a section about accessibility on the website, being lower the number of museum centers that include information on accessible products (35.6%). It has been found that only 42.2% of the museums include a section about accessibility on their website, and being lower the number of museums that include information about their accessible products (35.6%). In fact, very few museums are part of an accessible tourism network or association (22.2%), and it is only La Pedrera that holds and communicates an accessibility certification.

It is relevant to highlight that the 44.4% of the museums analyzed do not mention, or make any reference to disability types or degrees (Table 6). Conversely, among the 63.6% of the museums that mention some type of disability, it has been found that a 26.7% consider the 4 different types of disabilities (mobility, visual, hearing and cognitive impairments); although the number of museums that consider a single type of disability (17.8%) is also substantial. In general, as it can be observed in Table 5, it has been found that the mentions to mobility impairments are the most frequent (53.3%), being in second place the visual (40%) and hearing (37.8%) impairments; and, lastly, the cognitive disabilities (33.3%).

Nevertheless, the fact of making some reference or consideration to the different audiences with some kind of disability does not mean that there are specific products for these groups. Therefore, results show that only 22% of the museums offer accessible products. It is noteworthy to mention that only three museums (7.2%) offer 2 or more accessible products, and the remaining cases (14.6%) offer a single adapted product. The most common example of accessible products is guided tours.

Table 5. 'Information on products and services accessibility' category description

Information on products and services accessibility	Yes (%)	No (%)
General accessibility issues		
Accessibility section on the web	42.2	57.8
Accessible product information	35.6	64.4
Networking		
Accessible tourism network	22.2	77.8
Accessibility certification	2.2	97.8
Types of disability		
Distinction of disability type	44.4	55.6
Mobility impairment	53.3	46.7
Visual impairment	40.0	60.0
Hearing impairment	37.8	62.2
Cognitive impairment	33.3	66.7

Table 6. Disabilities and accessible products considered on websites

Disabilities and accessible products	Frequency	%
Number of disabilities considered (mean = 1.66; s.d.. = 1.712)		
None	20	44.4
1 type of disability	8	17.8
2 type of disability	3	6.7
3 type of disability	2	4.4
4 type of disability	12	26.7
Total	41	100.0
Number of accessible products (mean = 0.41; s.d. = 1.095)		
None	32	78.0
1 product	6	14.6
2 products or more	3	7.2
Total	41	100.0

Accessibility of Information Format

The tourist experience starts long before the consumption of the tourism product. So, the existence of products adapted to the demand with disabilities is not enough for them. Information is also a relevant issue that will influence these tourist experiences. Therefore, in order to offer information adapted to the wide range of special needs of visitors with disabilities, to facilitate the process of preparing the visit to the museum center is a relevant factor.

Results about information format (Table 7) show that the text is the main communication element or format in all the websites, and it is usually complemented by pictures or images (97.8%); and near half of the cases also include other multimedia material and formats (48.9%). In a reduced number of cases, there are also audios (6.7%). Nevertheless, the use of more than one format does not represent an adaptation to the special needs of people that have some kind of impairment. Due to this reason, the level of format accessibility is included in this analysis. In particular, results suggest that there are few cases which display information with a high level of accessibility. Results in Table 7 show that 26.7% of the museums offer an alternative image description; or a link description (24.4%). Results also reveal that, among the museums that also have videos on their websites (48.9%), only as small part of them are accessible through video subtitles (11.1%) or sign language (4.4%), being mutually exclusive.

Table 7. 'Accessibility of the information format' category description

Accessibility of Information Format	Yes (%) n=41	No (%) n=41
Types of formats		
Text	100.0	0.0
Pictures	97.8	2.2
Multimedia	48.9	51.1
Audio	6.7	93.3
Information format combination	48.9	51.1
Accessible formats		
Alternative image description	26.7	73.3
Link description	24.4	75.6
Video subtitles	11.1	88.9
Automatic reading	6.7	93.3
Sign language on videos	4.4	95.6

Patterns on Accessibility Management

Until now, findings have pointed out a relative low level of accessibility in both information issues and products offered to visitors. Thus, findings support previous studies that reveal that this information is scarce and inadequate (Mangani & Bassi, 2019). In order to better understand the profile and patterns of museums regarding accessibility issues, a bivariate analysis has been carried out. Two variables have been used to characterize behavioral patterns of museums and exhibition centers: ownership (public or private) and number of visitors.

In the case of the ownership, the 51.9% of the museums and exhibition centers are public and the 48.1% are private. Exploration of relationship between ownership and other variables related to accessibility in this study has shown a positive relationship between the type of ownership and their belonging to an accessibility network (Table 8). In particular, results show that public museums and exhibition centers are more engaged to accessibility networks than private ones. The museum community can play a crucial role in creating public-private partnerships and fight against social exclusion (Zakaria, 2020), so action to promote private entities participation in this type of alliances should be promoted. Other variables about information provision included in the analysis do not show statistically significant relationship with the type of ownership of the museum. However, previous studies (Mangani & Bassi, 2019) state that public museums provide more information regarding people with special access needs than private entities.

Table 8. Table of contingencies of accessibility network by ownership

	Accessibility network		
Ownership	**No**	**Yes**	**Total**
Public	40.00%	90.00%	51.10%
Private	60.00%	10.00%	48.90%
Total	100.00%	100.00%	100.00%

V-Cramer = 0.416; P-value = 0.005

Figure 1. Number of visitors

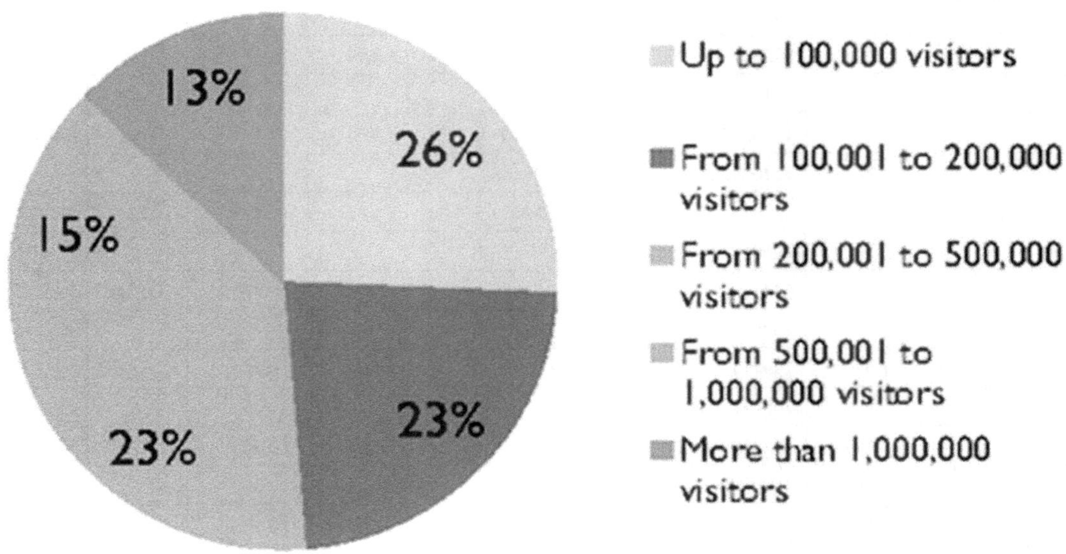

Regarding the number of visitors (Figure 1), results show that the 26% of the cases receive less than 100,000 visitors per year; in contrast with the 13% that receive more than a million of visitors per year.

The relationship analysis between the number of visitors and accessibility variables included in this study (Table 9) have suggested significant relationships between the level of accessibility management and adaptation to disabled visitors and the number of visitors as it is reported. In general terms, the higher the number of visitors the museum has, the better the adaptation of the website to visitors with special access needs is, especially regarding mobility, hearing and cognitive impairments. Mean comparisons of number of visitors between the group of museums that include an accessibility section on their website and those who do not show statistically significant difference (Table 9). Furthermore, those museums with information about accessible products on their webpages receive, on average, more visitors than those that do not include this type of information on their websites. Regarding format, museums that include sign language in videos, link description and video subtitles have, on average, more visitors than those that do not consider these features. In addition, there are significant differences of number of visitors between the museums that consider the wide range of impairments and the museums that do not make differences between the types of disabilities when providing information about accessible products.

Table 9. Descriptive statistics, mean comparison and t-tests of accessibility items linked to the number of visitors

Items		Mean	N	Std. Dev.	eta	Levene (sig)	T	t-test (p-value)
Accessibility section on the web	Yes	730006.3	18	624220.1	0.507	0.000	-3.375	0.003*
	No	202879.3	21	240145.9				
Accessible products information	Yes	660837.5	14	606275.2	0.310	0.105	-1.984	0.055
	No	325954.2	25	441859.0				
Sign language in videos	Yes	1388007.5	2	562709.2	0.423	0.861	-2.836	0.007*
	No	395258.49	37	479697.1				
Link description	Yes	891885.9	11	634966.5	0.539	0.014	-3.061	0.009*
	No	271065.5	28	354597.2				
Video subtitles	Yes	1256282.2	5	421409.9	0.600	0.799	-4.557	0.000*
	No	327034.35	34	426268.8				
Distinction of type of disabilities	Yes	737958.9	18	615459.5	0.521	0.000	-3.508	0.002*
	No	196062.8	21	243413.1				
Mobility impairment	Yes	634323.5	22	600887.1	0.413	0.000	-3.018	0.005*
	No	202674.2	17	262375.3				
Hearing impairment	Yes	799717.6	16	625904.8	0,569	0.000	-3.658	0.002*
	No	200221.6	23	233843.9				
Cognitive impairment	Yes	762470.9	14	599255.5	0.457	0.018	-2.772	0.012*
	No	269039.5	25	388221.1				
Visual impairment	Yes	698140.9	16	585780.4	0.406	0.023	-2.528	0.018*
	No	270883.6	23	404576.4				

*p-value < 0,05.

Table 10. Correlations between number of visitors and website accessibility items

Variables		Number of visitors	Number of disabilities	Number of accessible products
Number of visitors	Pearson correlation	1	,519**	0,258
	Sig. (bilateral)	-	0,001	0,113
	N	39	39	39

Thus, there is a positive relationship and significant relationship between the number of visitors and the number of types of disabilities included on the website (Table 10). In other words, the more visitors a museums has, the more diverse the range of disabilities addressed and considered on their website is. In particular, Picasso Museum, according to the items analyzed, could be considered the most accessible museum, and it receives more than 1.1 million visitors per year; by contrast, La Capella, which is the least accessible exhibition center receives around 58,000 visitors per year. This aspect could be linked to the available human and economic resources in these museums.

CONCLUSIONS

According to Law et al. (2010), the existing tourism literature simply does not have any commonly agreed-upon standards or techniques for website evaluation (Law et al., 2010). This chapter aims at exploring different approaches on website evaluation, especially focusing on museum websites and their accessibility. To do so, a holistic approach is considered, and factors about information format, information about accessibility and accessibility information are included in the analyses.

Results show there is a trend to include information on accessibility on the websites. However, this information is not integrated in the mainstream offer, but in a separate section focusing on people with special access needs. Principles of universal design recommend to integrate information on accessible products and services on the general information, as separate information may lead to segregated experiences.

In addition, there are few museums that include a section about accessibility on the website, and even less include information about accessible products on them. Surprisingly, once museums include information about accessibility, a good proportion addresses the 4 types of disabilities (mobility, hearing, visual and cognitive). However, it is not as surprising that mobility impairments references are the most common ones and hidden or mental disabilities are the least considered types of disabilities or impairments on museums websites. Although Dyson and Moran (2000) point out the need to develop web interfaces to meet the needs of different audiences in museums, and as highlighted in the results of this chapter, these interfaces are currently far from meeting this wide diversity of access needs.

Considering the new perspectives of universal design and the new needs and uses of information sources due to the access of the Internet, the provision of adequate and accessible information may be crucial for any tourism attraction, such as museums. Consequently, this study may help museum owners and managers when communicating their offer to everybody. From a theoretical point of view, this investigation may contribute to the understanding of the diversity of needs and behaviors related to the information sources, such as websites. According to Pallud & Straub (2014), it is crucial, for any organization, to align its strategy and its website design. This will, in turn, positively influence the physical visits at the museum. It is well-known that museums, especially those that are public, should develop activities and offer closely linked to their naturally social objectives. Among these social objectives, we can find the need to provide all groups of the population, regardless their skills or abilities, with accessible information and experiences. Thus, a website with accessible and appropriate information according to all types and degrees of abilities and a website with information about accessible products could be considered a well-designed website from this point of view. As a consequence, this good website design will be determinant in potential physical visits to the museum, as visitors who have a positive experience with a museum website and who are willing to return to this website also develop a higher interest to the museums and are more willing to visit the museum physically (Pallud & Straub, 2014).

Furthermore, all these requirements analyzed should be considered essential for people with special access needs when planning a physical visit to the museum. If their information needs are met and they are accessible and easy to interpret, people will more likely visit the museum physically. In contrast, the failure to incorporate them into the website design will result in dissatisfaction and rejection (Michopoulou & Buhalis, 2013).

Museums and information professionals recognize the need to conduct usability analyses of museums websites (Marty & Twidale, 2004). From a managerial point of view, these analyses can be relevant to understand museum positioning in the people with special access needs market and to understand the lacks and flaws that need to be addressed to offer accessible tourism products and services, as well as accessible information. Despite this being a good starting point for museums to improve accessibility standards by implementing technical, technological or informational measures, more efforts to ensure equality of experiences need to be made regarding awareness and willingness, considered to be the main potential barriers in the process (Lisney et al., 2013).

Finally, the objective of this chapter is to analyze Barcelona's museum websites. Therefore, findings are limited to this type of institutions, leaving aside other types of tourist attractions also relevant for the destination discovery and enjoyment by tourists, such as monuments. In addition, only museums from the city of Barcelona where included in the sample, providing a limited picture about the reality of accessible information. In this sense, the reality encountered in this city may be completely different from what is found in other destinations around the world. Therefore, it would be interesting to analyze these scenarios in other tourism destinations. In addition, after giving an insight of website accessibility, it would be of interest to go far beyond this analysis and analyze the quality of both physical and virtual museum experiences from the point of view of the user and let them personally evaluate these encounters, as it has been done in previous literature (e.g. Pallud & Straub, 2014).

REFERENCES

Ajuntament de Barcelona. (2019). *Anuari Estadístic de la Ciutat de Barcelona 2019*. Retrieved from http://www.bcn.cat/estadistica/catala/dades/anuari/cap06/index.htm

Blichfeldt, B. S., & Nicolaisen, J. (2011). Disabled travel: Not easy, but doable. *Current Issues in Tourism, 14*(1), 79–102. doi:10.1080/13683500903370159

Buhalis, D., & Michopoulou, E. (2011). Information-enabled tourism destination marketing: Addressing the accessibility market. *Current Issues in Tourism, 14*(2), 145–168. doi:10.1080/13683501003653361

Camprubí, R., & Coromina, L. (2016). Content analysis in tourism research. *Tourism Management Perspectives*, *18*, 134–140. doi:10.1016/j.tmp.2016.03.002

Capriotti, P. (2013). Managing Strategic Communication in Museums. The case of Catalan museums. *Communicatio Socialis*, *26*(3), 98–116.

Chang, G., & Caneday, L. (2011). Web-based GIS in tourism information search: Perceptions, tasks, and trip attributes. *Tourism Management*, *32*(6), 1435–1437. doi:10.1016/j.tourman.2011.01.006

Chiou, C. K., Tseng, J. C., Hwang, G. J., & Heller, S. (2010). An adaptive navigation support system for conducting context-aware ubiquitous learning in museums. *Computers & Education*, *55*(2), 834–845. doi:10.1016/j.compedu.2010.03.015

Coromina, L., & Camprubí, R. (2016). Analysis of tourism information sources using a Mokken Scale perspective. *Tourism Management*, *56*, 75–84. doi:10.1016/j.tourman.2016.03.025

Cunliffe, D., Kritou, E., & Tudhope, D. (2001). Usability evaluation for museum web sites. *Museum Management and Curatorship*, *19*(3), 229–252. doi:10.1080/09647770100201903

Darcy, S. (1998). *Anxiety to access: Tourism patterns and experiences of New South Wales people with a physical disability*. Tourism New South Wales.

Darcy, S., & Buhalis, D. (2011). Introduction: from disabled tourists to accessible tourism. *Accessible tourism: Concepts and issues*, 1-20.

Darcy, S., Mckercher, B., & Schweinsberg, S. (2020). From tourism and disability to accessible tourism: A perspective article. *Tourism Review*, *75*(1), 140–144. doi:10.1108/TR-07-2019-0323

Dillenburg, E. (2011). What, if anything, is a museum. *The Exhibitionist*, *30*(1), 8–13.

Disability Rights Commission. (2004). *The Web: Access and Inclusion for Disabled People*. London: TSO. Available at: http://www.drc-gb.org/publicationsandreports/report.asp 11.

Domínguez Vila, T., Darcy, S., & Alén González, E. (2015). Competing for the disability tourism market – A comparative exploration of the factors of accessible tourism competitiveness in Spain and Australia. *Tourism Management*, *47*, 261–272. doi:10.1016/j.tourman.2014.10.008

Doolin, B., Burgess, L., & Cooper, J. (2002). Evaluating the use of the Web for tourism marketing: A case study from New Zealand. *Tourism Management*, *23*(5), 557–561. doi:10.1016/S0261-5177(02)00014-6

Dyson, M. C., & Moran, K. (2000). Informing the design of web interfaces to museum collections. *Museum Management and Curatorship*, *18*(4), 391–406. doi:10.1080/09647770000501804

Figueiredo, E., Eusébio, C., & Kastenholz, E. (2012). How Diverse are Tourists with Disabilities? A Pilot Study on Accessible Leisure Tourism Experiences in Portugal. *International Journal of Tourism Research*, *14*(6), 531–550. doi:10.1002/jtr.1913

Gronflaten, O. (2009). Predicting Travelers' Choice of Information Sources and Information Channels. *Journal of Travel Research*, *48*(2), 230–244. doi:10.1177/0047287509332333

Hawkey, R. (2004). *Learning with digital technologies in museums, science centres and galleries.* Institute of Museum and Library Services 2006. Status of Technology and Digitization in the Nation's Museums and Libraries. Accessed 29 April 2020 at https://telearn.archives-ouvertes.fr/hal-00190496/

Holsti, O. R. (1968). Content analysis. The handbook of social psychology, 2, 596-692.

Istanbullu Dincer, F., Ozcit, H., Cifci, I., Sezer, B., Kahraman, O. C., & Sahinoglu, S. (2019). Accessible Museums for Visually Impaired: A Case Study from Istanbul. *Journal of Tourismology*, *5*(2), 113–126. doi:10.26650/jot.2019.5.2.0032

Kim, H., & Fesenmaier, D. R. (2008). Persuasive design of destination web sites: An analysis of first impression. *Journal of Travel Research*, *47*(1), 3–13. doi:10.1177/0047287507312405

Kluitenberg, E. (2008). *Delusive Spaces: Essays on Culture, Media and Technology. NAi Publishers.* Institute of Network Cultures.

Kolbe, R. H., & Burnett, M. S. (1991). Content-analysis research: An examination of applications with directives for improving research reliability and objectivity. *The Journal of Consumer Research*, *18*(2), 243–250. doi:10.1086/209256

Law, R., Qi, S., & Buhalis, D. (2010). Progress in tourism management: A review of website evaluation in tourism research. *Tourism Management*, *31*(3), 297–313. doi:10.1016/j.tourman.2009.11.007

Lisney, E., Bowen, J. P., Hearn, K., & Zedda, M. (2013). Museums and Technology: Being Inclusive Helps Accessibility for All. *Curator (New York, N.Y.)*, *56*(3), 353–361. doi:10.1111/cura.12034

Llodrà-Riera, I., Martínez-Ruiz, M. P., Jiménez-Zarco, A. I., & Izquierdo-Yusta, A. (2015). A multidimensional analysis of the information sources construct and its relevance for destination image formation. *Tourism Management*, *48*, 319–328. doi:10.1016/j.tourman.2014.11.012

Mangani, A., & Bassi, L. (2019). Web information, accessibility and museum ownership. *International Journal of Tourism Policy*, *9*(4), 265–281. doi:10.1504/IJTP.2019.105486

Marty, P. F. (2007). Museum Websites and Museum Visitors: Before and After the Museum Visit. *Museum Management and Curatorship*, *22*(4), 337–360. doi:10.1080/09647770701757708

Marty, P. F., & Twidale, M. B. (2004). Lost in gallery space: A conceptual framework ofr analysing the usability flaws of museum Web sites. *First Monday*, *9*(9), 1–13. doi:10.5210/fm.v9i9.1171

Michopoulou, E., & Buhalis, D. (2013). Information provision for challenging markets: The case of the accessibility requiring market in the context of tourism. *Information & Management*, *50*(5), 229–239. doi:10.1016/j.im.2013.04.001

Moreno Gil, S., & Ritchie, J. R. B. (2009). Understanding the museum image formation process: A comparison of residents and tourists. *Journal of Travel Research*, *47*(4), 480–493. doi:10.1177/0047287508326510

Morrison, A. M., Taylor, S., Morrison, A. J., & Morrison, A. D. (1999). Marketing small hotels on the World Wide Web. *Information Technology & Tourism*, *2*(2), 97–113.

Müller, K. (2002). Museums and Virtuality. *Curator (New York, N.Y.)*, *45*(1), 21–33. doi:10.1111/j.2151-6952.2002.tb00047.x

Pallud, J., & Straub, D. W. (2014). Effective website design for experience-influenced environments: The case of high culture museums. *Information & Management*, *51*(3), 359–373. doi:10.1016/j.im.2014.02.010

Park, Y. A., Gretzel, U., & Sirakaya-Turk, E. (2007). Measuring web site quality for online travel agencies. *Journal of Travel & Tourism Marketing*, *23*(1), 15–30. doi:10.1300/J073v23n01_02

Sheng, J., & Lo, A. (2010). Evaluating the tourism potential of public museums in Hangzhou: A supply-side perspective. *Journal of Travel & Tourism Marketing*, *27*(3), 287–305. doi:10.1080/10548401003744719

Shi, Y. (2006). The accessibility of Queensland visitor information centres' websites. *Tourism Management*, *27*(5), 829–841. doi:10.1016/j.tourman.2005.05.012

Stylianou-Lambert, T., & Stylianou, E. (2010). A third space: reconsidering issues of neutrality and accessibility in the virtual art museum. *EVA London 2010: Electronic visualisation & the arts. Proceedings of a conference held in London*, 66-71. 10.14236/ewic/EVA2010.12

Sylaiou, S., & Papaioannou, G. (2019). ICT in the Promotion of Arts and Cultural Heritage Education in Museums. In *Strategic Innovative Marketing and Tourism* (pp. 363–370). Springer. doi:10.1007/978-3-030-12453-3_41

Thomas, W., & Carey, S. (2005). Actual/Virtual Visits: What Are The Links? In J. Trant & D. Bearman (Eds.), *Museums and the Web 2005: Proceedings*. Archives & Museum Informatics.

Vacas, T. (2005). Los museos del "Paseo del Arte: Espacio turístico cultural de ámbito internacional. *Estudios Turísticos*, *165*, 97–129.

Velasco, M. (2009). Gestión Turística del Patrimonio Cultural: Enfoques para un desarrollo sostenible del Turismo cultural. *Cuadernos de Turismo*, *23*, 237–253.

Walsh, D., Hall, M. M., Clough, P., & Foster, J. (2020). Characterising online museum users: A study of the National Museums Liverpool museum website. *International Journal on Digital Libraries*, *21*(1), 75–87. doi:10.100700799-018-0248-8

Weisen, M., Petrie, H., King, N., & Hamilton, F. (2005). Web Accessibility Revealed: The Museums, Libraries and Archives Council Audit. *Ariadne*, (44), 1–10.

Wilson, R. J. (2011). Behind the scenes of the museum website. *Museum Management and Curatorship*, *26*(4), 373–389. doi:10.1080/09647775.2011.603934

Chapter 12
Assistive Technologies in Museums for People With Visual Impairments

Susana Vasconcelos Mesquita
https://orcid.org/0000-0003-2415-3960
DEGEIT, University of Aveiro, Portugal

Maria João Carneiro
GOVCOPP, DEGEIT, University of Aveiro, Portugal

ABSTRACT

Museums are important cultural places where technology, and specifically assistive technology, offers a prime opportunity for people with visual impairments to access the exhibition and the space. People with visual impairments represent a large group of the population but still experience several constraints during their museum visits. Nevertheless, there are few studies on the importance of assistive technologies in museums. This chapter aims to discuss the relevance of technology in museums and to identify guidelines to implement assistive technologies in order to improve the experience of people with visual impairments during their visits to museums. A reflection on the potential of assistive technologies for people with visual impairments and some guidelines are provided in order to promote more accessible and inclusive museums.

INTRODUCTION

It is recognised that technology has become central to many tourism stakeholders, ranging from transportation companies, accommodation and restaurants to visitor attractions, including cultural heritage and, more specifically, museums (Green et al., 2014). Cultural heritage is known to be one of the main contributors to tourism development and technology and internet tools allow heritage to be made more accessible and appealing to people, especially to people with disabilities (PwD) (ICOM, 2016; Vaz et al., 2018).

DOI: 10.4018/978-1-7998-6428-8.ch012

People with visual impairments (PwVI) are one of the most marginalised groups and still experience several limitations to their leisure activities (Richards et al., 2010). In the museum context, some changes are being implemented by museums to ensure the participation of PwD, including PwVI, in social activities (Andrade et al., 2015; Vaz, 2020). Some researchers emphasise that assistive technology (AT), which refers to products, devices or equipment that are used to maintain, increase or improve the functional capabilities of PwD, are of major importance in achieving this aim. However, few studies approach the topic of assistive technologies in museums and PwD.

This chapter aims to discuss the relevance of technology in museums, as well as to identify guidelines to provide AT in these attractions in order to enhance the experiences of PwVI in museums. The chapter begins by exploring the concepts of disabilities and technologies to PwVI. Then, it proceeds with a discussion of different technologies that museum managers may adopt to help minimise or eliminate constraints during their visit. Some examples of museums that already adopt AT will be mentioned to show how these technologies are being implemented in museums in order to make visits to these attractions more accessible and appealing to PwVI. Finally, the chapter ends with a reflection on the potential of AT for PwVI and with the provision of some guidelines on how to adopt this kind of technology.

LITERATURE REVIEW

People With Visual Impairments

Although there is no consensus on the definition of disability and impairment, disability is "any restriction or lack (resulting from an impairment) of ability to perform an activity in the manner or within the range considered normal for a human being" (Mandal, 2019, p. 1). Impairment is any loss or abnormality of psychological, physiological or anatomical structure or function (Mandal, 2019; Poria et al., 2009).

Disability is considered a minor issue by many people (Ochoggia, 2003; Poria et al., 2009; Shakespeare 2018, WHO,2016); however, PwD already represent 15% of the world population, which means that there are more than 1 billion PwD in the world (Shakespeare, 2018; The World Bank, 2016; WHO, 2018). Between 110 million and 190 million adults suffer from significant difficulties in functioning. Almost everyone will be affected by a disability at some point in life, temporarily or permanently. Moreover, almost everyone has a relative or friend with a disability.

There are several types of disabilities and several ways to classify disabilities (Buhalis & Darcy, 2011; Yau et al., 2004). Disabled World (2016) classified disabilities into the following subcategories: (i) mobility and physical impairments; (ii) spinal cord disability; (iii) head injuries – brain disability; (iv) vision disability; (v) hearing disability; (vi) cognitive or learning disabilities; (vii) psychological disorders; and (viii) invisible disabilities.

Visual impairment is defined as the limitation of actions and functions of the visual system (Mandal, 2019). "Visual impairment refers to a significant functional loss of vision that cannot be corrected by medication, surgical operation, or ordinary optical lenses such as spectacles" (WHO, 2016). People may have low vision if they cannot see well enough to do things like reading, driving, recognising people's faces, telling colours apart or watching television or screens clearly (National Eye Institute, 2019).

PwVI already represent a big group of the population. Two thousand people need glasses or other vision devices (WHO, 2016). According to the World Health Organization (2016), 285 million people are visually impaired while 39 million are blind and 246 million suffer from low vision. Data also show that about 90% of the world's visually impaired live in developing countries and that 82% of them are aged 50 and above, numbers that are expected to increase in the next decades. When talking about PwVI, one must consider those that are blind and those that suffer from low vision.

According to the International Classification of Diseases 11, 2018 version (WHO 2018), vision impairment is classified into two groups, distance and near presenting vision impairment. The first is distance vision impairment and it can be divided into: (i) mild – presenting visual acuity worse than 6/12; (ii) moderate – presenting visual acuity worse than 6/18; (iii) severe – presenting visual acuity worse than 6/60; and blindness – presenting visual acuity worse than 3/60. The second is near vision impairment, which means presenting near visual acuity worse than N6 or M.08 with existing correction. When people suffer from moderate and severe visual impairment they are grouped under the term "low vision". Total blindness is rare, and it is known that blind people do not usually lose all perception of light, which is often enough to detect the location, motion and brightness of light (Hicks et al., 2013). The ICF systemises distance vision impairment into groups highlighting peoples' vision acuity in the better eye (WHO, 2019).

The leading causes of visual impairment are diverse (WHO, 2019): uncorrected refractive errors, cataracts (general clouding of the vision and, as the whole eye lens is affected, the blurring of vision may be diffuse until totally lost); age-related macular degeneration (a central area of woolly or cottony opacity obscures the central part of the vision); glaucoma (due to the rise of normal fluid pressure inside the eyes with the type of vision usually like a tunnel); diabetic retinopathy (diabetes leads to damage of the smaller arteries and blood vessels at the back of the eye over the retina; it is the most common diabetic eye disease and a leading cause of blindness in adults); myopia or near-sightedness or short-sightedness means that a person can see nearby objects clearly but distant objects appear blurred. High myopia may leads to vision impairment, corneal opacity and trachoma. The causes of vision impairment present some variations across countries. Concerning developing countries, the main causes of visual impairments are cataracts and uncorrected refractive errors, while in developed countries the most prevalent causes are diabetic retinopathy and glaucoma (Cupples et al., 2012).

PwVI suffer from a lot of daily difficulties, including the ability to work and to develop personal relationships (Fernandes et al., 2019) and experience several constraints to visiting museums (Mesquita & Carneiro, 2016; Vaz 2020). However, PwVI are a heterogeneous group, also due to their moment of sight loss (Burton, 2003). Those who were born blind or totally lost vision before the age of six experience more constraints than those who are late blind, who benefit from years of normal prior visual experiences when performing tasks (Vaz, 2020).

Assistive Technologies

Several models that have been proposed across the time have revealed different perspectives on disability. According to two of these models – the social model and the biopsychosocial model – the participation of PwD in several activities and in society in general is frequently inhibited, at least in part, by the complex social environments and attitudes imposed by society (Buhalis & Darcy, 2011) or by the interaction between individuals and the previously referred environments (Balakrishnam et al., 2019).

These models may play a role in creating more accessible environments for PwD that facilitate and enhance the interaction these people have with the various environments they use and where they perform various activities. AT is crucial in this context. Hence, it is argued that, in this area, designers and engineers have two areas of concern: (i) design for all; and (ii) design of AT (Hersh & Johnson, 2008).

AT, also known as assistive devices and adaptive technology, derives from many disciplines and is specially driven by technology (Bhowmick & Hazarika, 2017). Lately, a large number of technological advances have taken place. These changes in information and communications technology (ICT) allowed the design and the development of a large number of interfaces that assist humans with disabilities. ICT covers any product that will store, retrieve, manipulate, transmit or receive information electronically in a digital form. Since 1980, technology has become a key aspect for PwD in overcoming the constraints of society. ICT run from websites, interactive ordering systems (e-Table technology) and interactive mobile platforms to social media channels (Avci, 2019). Although the concept of AT is largely used there is no accepted definition (Hersh & Johnson, 2008), AT is any product or service designed to enable independence of disabled and older people (NHS, 2020).

The concept of AT was introduced in 1988 in the USA (Assistive Technology Act, 1998) and in this document the US government recognised that technology is essential for the inclusion of PwD. AT is defined as "technology designed to be utilised in an AT device or service" while an assistive device can be regarded as "any item, piece of equipment, or product system, whether acquired commercially, modified, or customised, that is used to increase, maintain, or improve the functional capabilities of an individual with disabilities" (ATA, 2004, p. 4). In the same line of thought of previously mentioned researchers and organisations, Deloitte & Touche (2003, p. 11) state that "Assistive Technology refers to products, devices or equipment that are used to maintain, increase or improve the functional capabilities of PwD. Assistive Technology can help to compensate functional limitations and enable PwD to participate in the activities of daily life, including employment and training". Although there are several definitions of AT, they all have something in common, since all remark that this kind of technologies aim to facilitate and improve the daily life of PwD. Moreover, another common aspect to all these definitions is that there is always an item or piece of equipment that allows PwD to have full inclusion and integration in society (Fernandes et al., 2019).

In the present work the authors will adopt the following definition, which highlights the broad types of AT that exist and the types of barriers that it may help to overcome:

Assistive Technology is a generic or umbrella term that covers technologies, equipment, devices, apparatus, services, systems, processes and environmental modifications used by disabled and/or elderly people to overcome the social, infrastructural and other barriers to independence, full participation in society and carrying out activities safely and easily. (Hersh & Johnson, 2008, p. 196)

The quality of a person's life is greatly affected by visual impairments (Tadic et al., 2015). AT (e.g. electronic wheelchairs, environmental control devices, augmentative communication systems, magnifying devices, screen reading, and speech software) helps PwD to become more independent in daily life, improving the quality of life (Penton, 2015). Example of this are the "Kinesic Mouse, software that enables users with mobility disabilities to control PCs with facial expressions and head movements, virtual reality wearables, and augmentative and alternative communication (AAC) apps", among others (Freeman et al. 2016, p. 26). AT is based on standard ICT and its progress pushes forward disability concerns and inclusion of PwD (Miesenberger et al., 2006). The more ICT and AT enter the lives of PwD, the greater the importance of the subject of accessibility becomes in the tourism industry (Pühretmair & Nussbaum, 2011).

AT has long been a constant in the lives of PwD including PwVI. According to the ICF (2001), "assistive technologies are considered to be an environmental factor", specially designed for improving the functioning of a disabled person.

International Standard ISO 9999 provides a classification of existing AT. AT includes 11 different classes of assistive products for PwD. Priority assistive products are those products that are essential to maintain or improve an individual's functioning.

Table 1. Classes of assistive products for PwD

Class of products	Class identifier
Aids for personal medical treatment	Class 04
Aids for training in skills	Class 05
Orthoses and prostheses	Class 06
Aids for personal care and protection	Class 09
Aids for personal mobility	Class 12
Aids for housekeeping	Class 15
Furnishings and adaptations to homes and other premises	Class 18
Aids for communication, information and signalling	Class 21
Aids for handling products and goods	Class 24
Aids and equipment for environmental improvement, tools and machines	Class 27
Aids for recreation	Class 30

Source: ISO 9999:2016

Today, according to the World Health Organization (2018) more than 1 billion people in the world need AT. Among them, 970 million need glasses and low vision aids, 75 million people need wheelchairs, 150 million need mobility aids, 35 million people need prostheses or orthoses, 94 million people need hearing aids and 150 million need cognitive aids. Despite this situation, only 1 in 10 people have access to the product they need. In the next years, growth of AT for PwD is expected, changing their lives along with the lives of the elderly (Bhowmick & Hazarika, 2017). In 2030, according to the same study, more than 2 billion people will need AT (WHO, 2018).

Assistive Technologies and PWVI

Wiener et al. (2010) mentioned five needs that people with blindness have: (i) detection of obstacles in the travel path; (ii) travel surface information, including textures and discontinuities; (iii) distant object and cardinal direction information for projection of a straight line; (iv) landmark location and identification information and (v) information enabling self-familiarisation and mental mapping of an environment.

Regarding AT, several components are needed for non-visual access to information for PwVI (D'Andrea & Siu, 2015): (i) the device (hardware); (ii) programs (software); and (iii) accessible media that is compatible with the previous components mentioned. For PwVI, AT is related to "technologies, equipment, devices, apparatus, services, systems, processes and environmental modifications" (Hersh & Johnson, 2008, p. 196).

There are areas of function in which PwVI need assistance, namely orientation and mobility, as well as communication (Lancioni et al., 2019). Regarding orientation, the literature reveals that some issues such as the lack of appropriate light and signage (Handa et al., 2010), as well as of tactile paths (Mesquita & Carneiro, 2016) are major constraints to visiting a museum for PwVI. Moreover, concerning communication, the inexistence of information in macro characters or in relief (Mesquita & Carneiro, 2016) and in Braille (Poria et al., 2009; Handa et al., 2010) also inhibits some PwVI to visit museums. The lack of devices to enlarge the font size of the information provided at the museum also seem to be a major constraint in this scope (Mesquita & Carneiro, 2016). Furthermore, the lack of multisensory experiences also makes museum visits less attractive to this market since the absence of these experiences greatly restrict the information these people can get from the museum environment (Poria et al., 2009; Handa et al., 2010).

Several AT have been developed to help PwVI overcome many orientation and communication constraints and these technologies have undergone great changes over time. Hence, an effective way to minimise these constraints could be the use of orientation technology such as orientation systems relying on auditory direction cues (Lancioni et al., 1998; Uslan et al., 1988). Thus, navigation assistance is recommended (Dincer et al., 2019). New solutions are becoming more discrete and more sophisticated. In recent years a technological revolution has taken place. This revolution has been pushed by the innovation of ICT and by the role technology has achieved in people's daily life (Pühretmair & Nussbaum, 2011). AT has appeared to facilitate input or enhance output of PwD to interact, for example, with computers (WHO, 2020).

According to WHO (2018), by 2030, 2 billion people will need at least one AT. Improving accessibility for PwVI by providing them, whether at work or in their leisure time, with the opportunity to navigate independently in the environment, means enhancing their quality of life. Concerning PwVI, some priority assistive products are referred to by WHO (2016):

- Alarm signallers with sound;
- Audio players with DAISY capability;
- Braille displays (note takers);
- Braille writing equipment/braillers;
- Communication software;
- Filters;
- Global positioning system (GPS) locators;
- Digital, hand-held magnifiers;
- Optical magnifiers;
- Recorders;
- Screen readers;
- Spectacles for low vision and short distance;
- Talking and touching watches;
- White canes.

Among these assistive products we find important AT that can contribute to a more inclusive life for PwD (Table 2).

Table 2. Assistive technology for people with visual impairments

Accessibility to:	Access method		
	Visual	**Tactile**	**Auditory**
Print media	Optical devices • Hand-held/stand magnifiers • Telescopes and monoculars Non-optical devices • Large print • Reading stands Video magnifiers • Video magnifiers • Scanning and OCR • Interactive whiteboards	Braille • Braille embosser • Refreshable Braille display Tactile graphics	Electronic book readers • DAISY • ePub • Text PDF's Digital talking books Text to speech engines • Encoding and formatting for navigation Scanning systems • Stand-alone electronic reading devices • Computer-based reading devices and software
Computers	Screen magnification • High resolution monitors • Software to adjust contrast, magnification Scanning/masking/highlighting text features	Refreshable Braille displays • External or peripheral, connected via Bluetooth or USB • Built into a portable notetaker	Screen reading software with keyboard navigation • Native to device • Specialised
Mobile and multimedia	Screen magnification Font, colour, contrast adjustment Hi-resolution camera Virtual reality Augmented reality	3D printed materials Haptic displays Tactile graphics Refreshable Braille displays connected via Bluetooth	Enhance visitors' physical access concerning orientation, mobility and wayfinding Quantity of information must be balanced Provide audio navigation indications Touch-activated screen reader Image description (alt tags) Video (audio) description Talking tools Digital voice recorders
Written communications	Manual tools Electronic tools • General or specialised word processing programs with magnification or contrast adjustment • Word prediction software • Computer/PDA/notetaker With QWERTY or Braille keyboard	Manual devices Electronic tools • Electronic Braille writers • Computer, laptop • Portable notetaker with QWERTY or Braille keyboard • Braille translation software • Braille	Talking portable word processing device Word processor + screen reader software Voice dictation software
Producing materials in alternate formats	Scanner (to convert from print to digital) Copier (to enlarge print on paper) Printer	Braille • Scanning + OCR • Word processing software • Braille translation software • Embosser Models • 3D printer • Computer assisted Drawing (CAD) software Tactile graphics production • Tactile graphics machine + capsule paper • Thermoform machine + molds • Tactile drawing board + stylus	Digital/analogue audio recording devices Scanning + OCR system Screen readers Word processing with accessible formatting Image description (alt tags/text) Video (audio) description

Source: Adapted from Presley and D'Andrea (2009) and D'Andrea and Sui (2015).

As Table 2 shows, there is a wide variety of AT to help PwD, including PwVI. Hence, today the market presents a large number of AT. Despite the huge variety of devices and strategies that exist, it is difficult to identify the ideal solution for a particular visual impairment (Gamble et al., 2006). For example, Braille displays and/or audio output supported by on-screen keyboards allow information to be accessible for PwVI, making information that was not readable accessible to this group (Pühretmair &

Nussbaum, 2011). Nowadays, screen-reader software is also an important website-assistive technology (Andronico et al., 2006; Sahib et al., 2012; Sorrell et al., 2017) that translates the message into Braille or into audible language (Gooda et al., 2015; Harper & DeWaters, 2008). This AT is important to help PwVI to form a mental picture of the message that is being transmitted (Sorrell et al., 2017).

It is often important to use ICT and AT that perform different and complementary roles, for instance input media, like different types of keyboards, voice input through microphone and output media like loudspeakers, audio communication, spoken text or state information (Michopoulou & Buhalis, 2011).

The Importance of Assistive Technologies in Museums

The definition of a museum is dynamic and has evolved, alongside society, being updated according to the changes of the community that it serves (ICOM, 2016). The latest definition states that "a museum is a nonprofit, permanent institution in the service of society and its development, open to the public, which acquires, conserves, researches, communicates and exhibits the tangible and intangible heritage of humanity and its environment for the purposes of education, study and enjoyment" (ICOM, 2016).

Museums are important cultural places, visited each year by millions of people (Ambrose & Paine, 2018). There is an international consensus that museums should be "inclusive to all" as inclusion in museums is a matter of social justice. It is considered that every person has the right to visit museums, despite his/her condition. Museums are dynamic places which maintain their inner functions while exploring new ways of meeting the needs of people, including those with visual disabilities (Black, 2005; ICOM, 2016). One of the current concerns of museums is to be equally accessible to all. PwD require a large amount of information but the reality shows that the higher the degree of disability, the less information is provided (Pühretmair & Nussbaum, 2011).

In recent decades, technology has been put forward as one of the most important ways of minimising constraints experienced by PwVI during their visits to museums (Falk & Dierking, 2016). Museums have been using technologies to improve visitors' experiences in order to make visits more encompassing, more engaging, and more collaborative (Metatla et al., 2018). The choice of technology used in museums depends on what museums want their exhibit to do and who the museum's audience target is (Ambrose & Paine, 2018). Regardless of this finding, there are still limited empirical studies concerning PwVI in museums. Most of the studies of PwD in museums highlight physical access (Argyropoulos & Kanari, 2015; Dincer et al., 2019) rather than sensory access, which is of prime importance for PwVI, as some areas in which PwVI often need assistance are orientation and mobility, as well as communication (Poria et al., 2009).

One of the most important technological changes present in museums is the interactive exhibit, increasingly present not just in science centres, but in museums of all types (Ambrose & Paine, 2018). A large number of devices are now considered "interactive", since everything that involves the visitor doing something is "interactive". According to Freeman et al. (2016, p. 26), AT have the "potential to increase access and enable opportunities for disabled persons" since they can help museums serve different publics and reduce accessibility barriers (Vaz, 2020).

PwVI use AT during their interactive experiences in museums to operate computers, to better communicate or just to visit the museum independently.

Educational environments (e.g. museums) have been using technologies to improve visitors' experiences in order to make visits more accessible (Metatla et al., 2018). In order to interact with ICTs, PwVI usually need to use additional technology that include a wide number of inputs and outputs, such as different types of keyboards, voice input through microphone, loudspeakers, audio communication, spoken text or state information (Michopoulou & Buhalis, 2011).

PwVI's access to web information is supported by different assistive information technology, like screen readers, special web browsers or standards of accessible web design, among others (Bártek & Kopeček, 2006; Metatla et al., 2018; Pühretmair & Nussbaum, 2011). When talking about visual impairments it is important to consider that there are people with total blindness and people with low vision (Crow, 2008), and that these two groups may have different needs. For example, as they cannot see at all, people who are totally blind usually rely on devices known as screen readers that generate text contents which are audible to the users. People with low vision often use screen magnifiers to enlarge areas of the screen in order to have more readable information (Crow, 2008). Nowadays, there is a range of AT for PwVI that may be used in museums as will be discussed in the next section.

Visits to Museum for PWD Based on Assistive Technologies

Technology provides access to communication and navigation tools that remove barriers, allowing independence, as well as inclusion in the daily life of PwD. Nowadays much of our time is spent using technology. Therefore, museums professionals have to use this new actuality in order to attract visitors' attention and to increase their motivations to visit these spaces (Chelini, 2012; Vaz, 2018). The growth of technologies provides new ways for museums to present exhibitions in different formats such as new ways to present interactive installations and digital document preservation, media and web presentation and graphical animations, among others (Giannoulis et al., 2001). ICT has become an essential tool in the 21st century, changing the way people live, work, learn or experience leisure. Internet and web-based technologies have been employed to provide information, offering better services to museum visitors. The use of higher-end technology (augmented reality, virtual reality, wearable technology, robotics and artificial intelligence) results in a creative, fascinating and sophisticated use of technology. According to Vaz et al. (2018), augmented reality and virtual reality are not the same thing; the first is usually defined as a combination of virtual and real objects to increase the significance of the environment (Graham et al., 2013). It allows digital information to be added to images and real-life contexts, while virtual reality refers to computer-generated environments that replicate places, presence of people and objects, or fictional worlds, allowing realistic sensory experiences by a full immersion in a digital environment (Adams, 2016; Freeman et al., 2016; Guerra et al., 2015; Jung et al., 2016; Vaz et al., 2018). This concept of virtual reality is a variation of augmented reality and appears with wearable computers and with developments in ICT (Avci, 2019). According to Jhajharia et al. (2014, p. 5700), the terms "wearable technology" and "wearable device" refer to "electronic technologies or computers that are incorporated into items of clothing and accessories which can comfortably be worn on the body". This equipment performs calculation or data collection on a device. Wearable technology refers to computer-based devices that take the form of a portable and lightweight accessory (smartwatches, Google Glass to see digital content, smartglasses embodied with sensors to measure heart rate, breathing and brain waves interacting with the exhibit) that can be worn by users, and can include tools to track social media interactions, movement and vital signs or perform productive tasks like sending emails (Charara, 2016; Johnson et al., 2015; Vaz et al., 2018).

When one talks about the new digital tools presented in museum visits, it is important to refer to robotics. According to Tung and Law (2017), robotics has its origins in industrial automation and the International Federation of Robotics (IFR, 2020) defines a service robot as "a robot that performs useful tasks for humans or equipment excluding industrial automation application". Robotic devices are slowly becoming a reality in 21st century museums competing with other digital interactive forms (Brynjolfsson et al., 2014). According to Claudio et al. (2017), with the inclusion of this tool in museums, we can provide new types of visits such as digital storytelling.

In the field of tourism, which is the context of analysis in this chapter, many researchers (Calver & Page, 2013; Kempiak et al., 2017; López Sintas et al., 2014) are likely to identify three main stages in the consumption experience – the pre-visit, the on-site visit and the post-visit (Kempiak et al., 2017; Waligo, 2013). When thinking about PwD, one must consider these three phases in order to understand the way museums can encourage this public to visit museums and enhance the visit to a museum, making it participative, engaging and appealing. In this section, the intention is to discuss how AT can be adopted in museums to make the visits of PwVI more engaging and appealing, illustrating the way that these technologies are already being adopted by some museums with some examples. Information on the way AT are being implemented in specific museums was obtained through the websites of these museums.

Concerning museums and PwD, the pre-visit phase is of great importance as the visit experience begins before entering the museum. During this stage visitors plan their visit and access to museums' information and to the exhibition contents (Vaz, 2020). For PwD it is highly important to access relevant contents as well as reliable, accurate and up-to-date information (Pühretmair & Nussbaum, 2011). The lack of accurate and updated information when searching about museum accessibility may be a cause of anxiety and dissatisfaction for many visitors who have visual impairments (Poria et al., 2010). On its website, the Smithsonian American Art Museum, in Washington, provides information on the acceptance of service dogs and on verbal description tours (Smithsonian American Art Museum, 2020).

Apart from having information available on accessibility of museums, it is crucial that all the information is accessible. The sources on the internet can sometimes cause several constraints among PwD. Inaccessible web pages are the result of the inaccessible design of these pages. To overcome some of these barriers, PwD use AT that enhances their interrelation with computers (Pühretmair & Nussbaum, 2011).

Nowadays, there are several rules for creating accessible web pages. One of the most comprehensive is the Web Content Accessible Guidelines (WCAG) of the WAI (Web Accessibility Initiative). The WAI is part of the World Wide Web Consortium (W3C) and the goal of WAI is to provide straightforward standards for web content accessibility that meet the needs of PwD (D'Andrea & Siu, 2015). Accessibility, according to WAI, can present three levels ranging from A to AAA level (W3C, 2020). Level AA is strongly recommended to remove significant barriers. Museums' websites must follow WCGA recommendations in order to provide understandable information, text alternatives for non-text content, audio-described multimedia content, and facilities for easy navigation and content finding (Vaz, 2020).

To control computers, people with physical disabilities use devices like alternative keyboards and/or alternative pointing devices such as joysticks, trackballs and head or eye trackers combined with on-screen keyboards (Pühretmair & Nussbaum, 2011).

PwVI can use Braille displays and/or audio outputs supported by screen reader software (Kim et al., 2016). The screen will verbalise the content. The website of the Tiflológico museum, a museum created by the Spanish National Organization for the Blind (ONCE), in Madrid, can be seen with high or normal contrast (ONCE, 2020).

The growth of social media websites like Twitter, Facebook, Pinterest and Instagram can have an important positive effect on PwVI. These promote people's relationships with others who have the same conditions as them, leading to an increased awareness of the accessible aspects of multiple exhibitions, as well as accessible services and assistance provided by museums (Johnson et al., 2015; Vaz, 2020).

Through the tools present in their websites, various museums (e.g. Louvre Museum, Victoria & Albert Museum, Museum of London) make it possible to download maps and take an interactive, virtual tour through the museum, with 360º controllable images and 3D exploration (https://www.louvre.fr/en/,https://www.vam.ac.uk,https://www.museumoflondon.org.uk/museum-london).

All these different possibilities can help PwD prepare their visit and access important information about exhibitions.

According to some authors, digital media in museums could be more accessible for PwVI by avoiding the frequent use of graphics on their websites due to the difficulty for screen readers to decode the content (Pühretmair & Nussbaum, 2011; Reich et al., 2011). Descriptive alt-text (alternative text) labelled graphics, more detailed and descriptive directions (walking directions from bus stop and drop-off areas, descriptions of the museum building and nearby landmarks) are also recommended (Meliones & Sampson, 2018; Pühretmair & Nussbaum, 2011).

The on-site visit phase begins with people having access to the exhibition. Blind and low-vision visitors experience wayfinding difficulties (Coughlan et al., 2006). Therefore, it is important to design the exhibitions in a way that these people can feel more independent when visiting a museum (Poria, 2009). Traditional travel and navigation aids for people with blindness have been long (white) canes and guide dogs for overcoming mobility challenges (Meliones & Sampson, 2018). Some limitations led to the development of electronic assistive travel aids. Nowadays technology is becoming one of the ways to support positive museum experiences. The introduction of GPS systems was a step towards integrating navigation-assistance systems for PwVI. In 2004, Loadstone GPS represented an early effort for satellite navigation for these groups. Since then, several other products have been introduced in the market, but a substantial number of problems related to navigation instructions still remain (Meliones & Sampson, 2018). Dozens of devices were designed in recent decades with the aim of detecting obstacles. These systems make use of laser, infrared, ultrasonic sensors or cameras, among others, and transmit the information to the users through audio, tactile or haptics interfaces (Chanana et al., 2017). In the Tiflológico museum there is also a system of detection and location of nearby objects that produces sounds, using Beepcons – intelligent guide beacons – which may be used by having a mobile phone (ONCE, 2020).

Some pedestrian navigation systems for PwVI have been developed in order to enhance the ability of this group to navigate without the help of guides or staff. The BlindHelper system is "a smartphone application which interacts with a small embedded system responsible for reading simple user controls, high-accuracy global positioning system (GPS) tracking of pedestrian mobility in real time" (Meliones & Sampson, 2018, p. 1). With this app, PwVI can identify near-field barriers and traffic light status along the route. As a result, the information is communicated to the smartphone app, which issues commands and instructions or undertakes other actions (Meliones & Sampson, 2018).

Some museums provide more technology in specific areas of the museums so that PwVI may have more access to information on the museum and its exhibitions. The Victoria and Albert Museum (2020), in London, provides some AT, namely Windows computers and settings that can be adapted to fit the needs of visitors, in some parts of the museum – in the National Art Library and in the Prints & Drawings Study Room.

Some examples of AT are highly descriptive audio guides, among a variety of delivery devices. Audio guides are the most common solution, with some including both navigation and descriptions of selected collection artworks in the exhibit which can be particularly helpful in helping to "see" the exhibits (Vaz, 2020; Reich et al., 2011). Some museums offer touch tours and detailed audio guide descriptions for these tours in order to make the visit to the museum more participative and engaging. This is the case of the Louvre Museum, which provides a touch tour in its Touch Gallery (Louvre, 2020). In the Victoria and Albert Museum (2020), visitors also have access to audio descriptions of historic locations, of some parts of the museum (namely the Furniture Gallery) and of objects that can be touched in touch tours. Radio frequency identification tagging is recommended as audio guides with numbers are difficult to use for some visitors (Reich et al., 2011).

Some museums like the San Francisco Museum of Modern Art (SFMOMA) developed an app designed to keep phones in pockets and eyes on the art (Chun, 2016).

Many visitors use their own devices when visiting museums which allow museums to produce successful applications (apps) which engage visitors with the exhibits (Freeman et al., 2016; McMulla, 2015). In some cases, based on a virtual map, the app uses location-aware technology on visitor's phone's to triangulate its position with the contents of the exhibits (Chun, 2016). In the Tiflológico museum, the works exhibited in the temporary exhibition hall have QR codes so that PwVI can have all the information accessible through mobile phones (ONCE, 2020).

Other museums use an app that lets visitors explore some masterpieces with a level of detail invisible to the naked eye (giga-pixel resolution). Some museums use augmented reality which allows, for example, skins to be added to dinosaur skeletons to give life to animals or to restore destroyed objects, among other possibilities (Czikk, 2013).

Portable devices like smartphones, tablets and smartwatches have become essential to improving the visits, as they include built-in cameras and sensors which allow them to work as magnifying glasses or as flashlights which might help PwD in dimly lit situations (Freeman et al., 2016; Johnson et al., 2011; Reich et al., 2011). Some apps were developed in order to answer visitors' questions during the onsite visit, for example, helping people to get answers about the works of art in the museum (Vaz et al., 2018).

With the growth of technologies such as screen readers and audio voice recognition devices, the use of Braille has seen a decline (Knochel et al., 2018). However, the principle – "a translation of letterforms into embossed dot patterns that can be felt through the fingertips" (Knochel et al., 2018, p. 8) – led to the appearance of tactile graphics using embossing printers and microcapsule paper technologies. Most of the time, museums depend on visual and verbal descriptions so, 3D scanning and printing may help PwVI to have an experience with the objects of the exhibition (Andrade et al., 2015; Knochel et al., 2018). With time, Braille embossers became cheaper and a wide range of portable devices were designed to allow six-key (Braille) or QWERTY entry, and an internal processer that could be printed.

The post-visit phase is also very important, since it is when visitors remember the visit, share some part of their experiences and, possibly, recommend the museum (Dinis et al., 2019; McMillen & Alter, 2017). For these purposes visitors can, for example, use social media platforms to post their comments. According to Dinis et al. (2019) the growing interest in social media highlights the need to have accessible information on social media. Social media play a great role providing inclusion for PwD, allowing them to participate, learn and connect to the world (Altinay et al., 2016). McMillen and Alter (2017) state that social media have the power to drive museum visitation and to influence positively a museum's credibility. When someone with a disability shares information about an exhibition with a friend on social media, this is more effective than other types of advertising strategies.

This reinforces the importance of museums exploring the potential of this kind of platforms. If PwD are provided with good access to museum websites, they can take advantage of this to enrich their knowledge and further study specific contents that were less explored or even clarify doubts with the institution (McMillen & Alter, 2017; Vaz, 2020).

The analysis of the museums' websites suggests that museums still only use a small number of assistive technologies that are restricted to audioguides, computers, screen reading programmes such as Jaws and/or refreshable Braille displays, recognition devices, applications and optical magnifiers. Therefore, museums should invest in the adoption of more assistive technologies to complement those already existing in museums or, if they already have these technologies, they must inform visitors to their websites that they have them.

FUTURE RESEARCH DIRECTIONS

In the present chapter, the analysis of assistive technologies adopted in museums was limited to the identification of assistive technologies in museums' websites. However, in order to have more in-depth knowledge on the adoption of these technologies by museums, further research should be carried out based on primary data collection, either using qualitative approaches – interviews with museum managers and observation of museums – or quantitative ones – survey questionnaire directed to museum managers – in several regions and countries. Moreover, it would also be important to carry out studies with PwVI, asking them to provide their perceptions on assistive technologies adopted by museums and to report their satisfaction when using them.

CONCLUSIONS

The present chapter shows that museums face several challenges to provide accessibility for those with special needs to experience a better visit. Technologies, and internet in particular, nowadays represent important tools for museums to increase opportunities for the general public and for PwD, including PwVI. AT play a major role in overcoming the limitations that PwVI experience during their leisure activities. More than 1 billion people in the world need AT and this number will continue to increase in the next decade.

AT for PwVI may encompass technologies, equipment, devices, apparatus, services, systems, processes and environmental modifications. Orientation and mobility, together with communication, are some areas in which PwVI need assistance and where AT are crucial. This chapter reveals the existence of a variety of technologies that may be used in museums and which are already used by some of them, during the three main stages of the consumption experience – pre-visit, on-site visit and post-visit – to overcome some limitations in these fields.

As far as orientation and mobility are concerned, the use of orientation technology such as orientation systems relying on auditory direction cues is of great relevance. This chapter also reveals that new solutions are becoming more discrete and more sophisticated. Making use of GPS systems, laser, infrared, ultrasonic sensors and cameras, among other technologies, has proved to be very useful to create more efficient technological systems in this scope. All these features and the development of apps have opened new opportunities for PwVI to move around the museum more independently, and also find specific places and interesting objects.

Regarding communication, this chapter emphasises the relevance of developing efforts to offer appropriate services to PwVI from the planning stages of the trip, to persuade potential visitors to actually visit the museum. In this scope, it is of major importance not only to ensure that websites and other technological platforms used to communicate with potential visitors offer information on how the museums are accessible, but also to ensure that these platforms are accessible to these people. Some of these concerns are also of great importance to ensure a pleasant post-visit experience, where visitors may keep searching for more information on the museums and their exhibitions, and in which they may also help to promote the museums through positive word of mouth.

Furthermore, to communicate with PwVI when they are visiting the museum, it is necessary to use additional technology that may be widely varied, such as different types of keyboards, voice input through microphone, loudspeakers, audio communication, spoken text or state information. It is essential to find technologies and devices that may complement each other in an appropriate way since often more than one technology or device is needed. Highly descriptive audio guides and apps which engage visitors with the exhibits can be of major importance in offering emotional and appealing visits to this market. The use of higher-end technology results in a creative, fascinating and sophisticate use of technology. Despite the huge number of different devices and strategies that exist, it can be difficult to identify the ideal solutions for someone who has visual impairments. A lot of strategies are referred to that may be used to induce a better experience in the museums, ranging from the use of augmented reality, virtual reality and wearable technology, to robotics and artificial intelligence.

The chapter provides significant implications for museum managers regarding the wide variety of assistive technologies that can be adopted by museums to make these tourist attractions more accessible and appealing to PwDs, and shows the importance of motivating these managers to explore assistive technologies that do not seem to be much explored yet. The chapter also suggests the need to encourage museum managers to inform potential visitors that museums have these technologies, through the museum's website.

In this chapter it is widely recognised that PwVI use AT during their interactive experiences in museums to operate computers, to better communicate or just to visit the museum independently. AT are important tools to help enhance the experience that PwD and, in particular PwVI, have during their visits to museums. A lot of work still needs to be done as people with some kind of disability will increase in the next decades and technology evolves very fast, offering new opportunities to facilitate and reinvent the experience of visiting a museum for PwVI.

ACKNOWLEDGMENT

This work was also financially supported by the project POCI-01-0145-FEDER-030376, funded by FEDER, through COMPETE2020 – Programa Operacional Competitividade e Internacionalização (POCI), and by national funds (OE), through FCT/MCTES.

References

Adams, E. (2016). *Most Virtual Reality Is Not Virtual Reality. Here's Why*. Retrieved March 25, 2020, from http://gearpatrol.com/2016/07/18/virtual-reality-vs-360-video/

Altinay, Z., Saner, T., Bahçelerli, N. M., & Altinay, F. (2016). The role of social media tools: Accessible tourism for disabled citizens. *Journal of Educational Technology & Society*, *19*(1), 89–99.

Ambrose, T., & Paine, C. (2018). *Museum basics: The international handbook*. Routledge. doi:10.4324/9781315232898

Anderson, J. D., Lee, D. J., & Archibald, J. K. (2007). Embedded stereo vision system providing visual guidance to the visually impaired. In *2007 IEEE/NIH Life Science Systems and Applications Workshop* (pp. 229-232). IEEE. 10.1109/LSSA.2007.4400926

Andrade, E. C., Delgado, R., Hauptner, B., Kerschbaum, F., List, R., Pobežalová, P., . . . Wögerbauer, S. (2015). *Inclusive technologies in museums: for a better access to culture for blind and visually impaired people*. Vienna, Austria: Project AMBAVis. Accessed on April 10, 2020, from http://www.ambavis.eu/wp-content/ uploads/2016/11/AMBAVis Broschuere_BF_V01_EN_bf.pdf

Andronico, P., Buzzi, M., Castillo, C., & Leporini, B. (2006). Improving search engine interfaces for blind users: A case study. *Universal Access Information*, *5*(1), 23–40. doi:10.100710209-006-0022-3

Argyropoulos, V. S., & Kanari, C. (2015). Re-imagining the museum through "touch": Reflections of individuals with visual disability on their experience of museum-visiting in Greece. *Alter*, *9*(2), 130–143. doi:10.1016/j.alter.2014.12.005

Assistive Technology Act. (2004). Accessed on April 10, 2020, from https://www.congress.gov/108/plaws/publ364/PLAW-108publ364.pdf

Avci, E. (2019). *Enhancing the Cultural Tourism Experience Through Augmented Reality*. Co-Editors, 215.

Balakrishnan, A., Kulkarni, K., Moirangthem, S., Kumar, C. N., Math, S. B., & Murthy, P. (2019). The Rights of Persons with Disabilities Act 2016: Mental Health Implications. *Indian Journal of Psychological Medicine*, *41*(2), 119–125. doi:10.4103/IJPSYM.IJPSYM_364_18 PMID:30983658

Barnsley Hospital NHS Foundation Trust. (2017). Accessed March 8, 2020 from https://www.barnsley-hospital.nhs.uk/assistive-technology/about-assistive-technology-at/

Bártek, L., & Kopeček, I. (2006, July). Web pages for blind people—generating web-based presentations by means of dialogue. In *International Conference on Computers for Handicapped Persons* (pp. 114-119). Berlin: Springer. 10.1007/11788713_18

Bhowmick, A., & Hazarika, S. M. (2017). An insight into assistive technology for the visually impaired and blind people: State-of-the-art and future trends. *Journal on Multimodal User Interfaces*, *11*(2), 149–172. doi:10.100712193-016-0235-6

Black, G. (2005). *The engaging museum: Developing museums for visitor involvement*. Psychology Press.

Brynjolfsson, E., & McAfee, A. (2014). *The second machine age: Work, progress, and prosperity in a time of brilliant technologies*. WW Norton & Company.

Buhalis, D., & Darcy, S. (2011). *Accessible tourism: Concepts and issues*. Channel View Publications.

Burton, H. (2003). Visual cortex activity in early and late blind people. *The Journal of Neuroscience: The Official Journal of the Society for Neuroscience*, *23*(10), 4005–4011. doi:10.1523/JNEUROSCI.23-10-04005.2003 PMID:12764085

Calver, S. J., & Page, S. J. (2013). Enlightened hedonism: Exploring the relationship of service value, visitor knowledge and interest, to visitor enjoyment at heritage attractions. *Tourism Management*, *39*, 23–36. doi:10.1016/j.tourman.2013.03.008

Chanana, P., Paul, R., Balakrishnan, M., & Rao, P. V. M. (2017). Assistive technology solutions for aiding travel of pedestrians with visual impairment. *Journal of Rehabilitation and Assistive Technologies Engineering*, *4*, 2055668317725993. doi:10.1177/2055668317725993 PMID:31186934

Charara, S. (2016). *Hussein Chalayan has turned emotion sensing into a thing of beauty: Biometric wearables at the Design Museum's new Fear and Love exhibition*. Accessed March 7, 2020, from https://www.wareable.com/wearable-tech/hussein-chalayan-emotions-design-museum-3603

Chelini, M. (2012). Novas tecnologias para... novas (?) expografias. *Revista Museologia & Interdisciplinaridade*, *1*(2), 59–71. doi:10.26512/museologia.v1i2.12655

Chun, R. (2016). *The SFMOMA's New App Will Forever Change How You Enjoy Museums*. Retrieved October 29, 2016, from https://www.wired.com/2016/05/sfmoma-audio-tour-app/

Claudio, G., Luca, G., & Luce, L. M. (2017). Interaction design for cultural heritage. A robotic cultural game for visiting the museum's inaccessible areas. *The Design Journal, 20*(sup1), S3925-S3934.

Coughlan, J., Manduchi, R., & Shen, H. (2006). Computer vision-based terrain sensors for blind wheelchair users. In *International Conference on Computers for Handicapped Persons* (pp. 1294-1297). Springer. 10.1007/11788713_186

Crow, K. L. (2008). Four types of disabilities: Their impact on online learning. *TechTrends*, *52*(1), 51–55. doi:10.100711528-008-0112-6

Cupples, M. E., Hart, P. M., Johnston, A., & Jackson, A. J. (2012). Improving healthcare access for people with visual impairment and blindness. *BMJ (Clinical Research Ed.)*, *344*(1), 1–5. doi:10.1136/bmj.e542 PMID:22290101

Czikk, J. (2013). *New ScopifyROM App Gives Royal Ontario Museum Visitors an Augmented Experience*. Retrieved March 15, 2020, from https://betakit.com/new-scopifyrom-app-gives-royal-ontario-museum-visitors-an-augmented-experience/

D'Andrea, F. M., & Siu, Y. T. (2015). Students with visual impairments: Considerations and effective practices for technology use. In D. Edyburn (Ed.), Efficacy of Assistive Technology Interventions (Advances in Special Education Technology, Volume 1) (pp. 111-138). Emerald Group Publishing Limited.

Deloitte and Touche. (2003). *Access to Assistive Technology in the European Union*. European Commission.

Dincer, F. I., Ozcit, H., & Cifc, İ, I., Sezer, B., Kahraman, O. C., & Sahinoglu, S. (2019). Accessible Museums for Visually Impaired: A Case Study from Istanbul. *Journal of Tourismology*, *5*(2), 113–126.

Dinis, M. G., Eusébio, C., & Breda, Z. (2020). Assessing social media accessibility: The case of the Rock in Rio Lisboa music festival. *International Journal of Event and Festival Management*, *11*(1), 26–46. doi:10.1108/IJEFM-02-2019-0012

Disabled World. (2019). *Disabilities: Definition, Types and Models of Disability*. Accessed on April 18, 2020, from https://www.disabled-world.com/disability/types/

Falk, J. H., & Dierking, L. D. (2016). *The museum experience revisited*. Routledge.

Fernandes, H., Costa, P., Filipe, V., Paredes, H., & Barroso, J. (2019). A review of assistive spatial orientation and navigation technologies for the visually impaired. *Universal Access in the Information Society*, *18*(1), 155–168. doi:10.100710209-017-0570-8

Freeman, A., Becker, S. A., Cummins, M., McKelroy, E., Giesinger, C., & Yuhnke, B. (2016). *Nmc horizon report: 2016 museum edition*. The New Media Consortium.

Gamble, M. J., Dowler, D. L., & Orslene, L. E. (2006). Assistive technology: Choosing the right tool for the right job. *Journal of Vocational Rehabilitation*, *24*(2), 73–80.

Giannoulis, G., Coliou, M., Kamarinos, G. S., Roussou, M., Trahanias, P., Argyros, A., . . . Haehnel, D. (2001). Enhancing museum visitor access through robotic avatars connected to the web. Proc of museums and the web.

Graham, M., Zook, M., & Boulton, A. (2013). Augmented reality in urban places: Contested content and the duplicity of code. *Transactions of the Institute of British Geographers*, *38*(3), 464–479. doi:10.1111/j.1475-5661.2012.00539.x

Green, J., Pridmore, T., & Benford, S. (2014). Exploring attractions and exhibits with interactive flashlights. *Personal and Ubiquitous Computing*, *18*(1), 239–251. doi:10.100700779-013-0661-3

Guerra, J., Pinto, M., & Beato, C. (2015). Virtual Reality-Shows A new vision for tourism and heritage. *European Scientific Journal*, *11*(9), 49–54.

Handa, K., Dairoku, H., & Toriyama, Y. (2010). Investigation of priority needs in terms of museum service accessibility for visually impaired visitors. *British Journal of Visual Impairment*, *28*(3), 221–234. doi:10.1177/0264619610374680

Harper, K. A., & DeWaters, J. (2008). A quest for website accessibility in higher education institutions. *The Internet and Higher Education*, *11*(3-4), 160–164. doi:10.1016/j.iheduc.2008.06.007

Hersh, M. A., & Johnson, M. A. (2008). On modelling assistive technology systems–Part I: Modelling framework. *Technology and Disability*, *20*(3), 193–215. doi:10.3233/TAD-2008-20303

Hicks, S. L., Wilson, I., Muhammed, L., Worsfold, J., Downes, S. M., & Kennard, C. (2013). A depth-based head-mounted visual display to aid navigation in partially sighted individuals. *PLoS One*, *8*(7), 667–695. doi:10.1371/journal.pone.0067695 PMID:23844067

ICOM. (2016). *Museum Definition*. International Council of Museum http://archives.icom.museum/definition.html

ICOM. (2018). https://icom.museum/wp-content/uploads/2018/07/ICOM-code-En-web.pdf

International Federation of Robitics (IFR). (2020). *Service Robots*. https://ifr.org/service-robots

International Standard Organization. (2016). *ISO 9999:2016 Assistive products for persons with disability — Classification and terminology.* https://www.iso.org/standard/60547.html

Jhajharia, S., Pal, S. K., & Verma, S. (2014). Wearable computing and its application. *International Journal of Computer Science and Information Technologies*, *5*(4), 5700–5704.

Johnson, L., Adams, S., & Witchey, H. (2011). *The NMC Horizon Report: 2011 Museum Edition.* http://www.nmc.org/pdf/2011-horizon-report-museum.pdf

Johnson, L., Becker, S., Estrada, V., & Freeman, A. (2015). *NMC Horizon Report: 2015 Higher Education Edition*. The New Me.

Jung, T., tom Dieck, M. C., Lee, H., & Chung, N. (2016). Effects of virtual reality and augmented reality on visitor experiences in museum. In *Information and communication technologies in tourism 2016* (pp. 621–635). Springer. doi:10.1007/978-3-319-28231-2_45

Kempiak, J., Hollywood, L., Bolan, P., & McMahon-Beattie, U. (2017). The heritage tourist: An understanding of the visitor experience at heritage attractions. *International Journal of Heritage Studies*, *23*(4), 375–392. doi:10.1080/13527258.2016.1277776

Kennard C (2013) A depth-based head-mounted visual display to aid navigation in partially sighted individuals. *PLoS One, 8*(7).

Kim, H. K., Han, S. H., Park, J., & Park, J. (2016). The interaction experiences of visually impaired people with assistive technology: A case study of smartphones. *International Journal of Industrial Ergonomics*, *55*, 22–33. doi:10.1016/j.ergon.2016.07.002

Knochel, A. D., Hsiao, W. H., & Pittenger, A. (2018). Touching to see: Tactile learning, assistive technologies, and 3-D printing. *Art Education*, *71*(3), 7–13. doi:10.1080/00043125.2018.1436320

Lancioni, G. E., O'Reilly, M. F., Oliva, D., & Bracalent, S. (1998). Guiding a person with blindness and intellectual dis- ability in indoor travel with fewer auditory cues. *Journal of Visual Impairment & Blindness*, *92*(9), 609–614. doi:10.1177/0145482X9809200904

Lancioni, G. E., Singh, N. N., O'Reilly, M. F., Alberti, G., Chiariello, V., Campanella, C., Grillo, G., & Tagliente, V. (2019). A Program Based on Common Technology to Support Communication Exchanges and Leisure in People With Intellectual and Other Disabilities. *Behavior Modification*, *43*(6), 879–897. doi:10.1177/0145445519850747 PMID:31104482

López Sintas, J., García Álvarez, E., & Pérez Rubiales, E. (2014). Art museum visitors: Interaction strategies for sharing experiences. *Museum Management and Curatorship*, *29*(3), 241–259. doi:10.1080/09647775.2014.919175

Louvre. (2020). https://www.louvre.fr/en/touch-gallerya-new-tour-sculpting-body

Mandal. (2019). https://www.news-medical.net/health/Types-of-visual-impairment.aspx

McMillen, R., & Alter, F. (2017). Social media, social inclusion, and museum disability access. *Museums & Social Issues*, *12*(2), 115–125. doi:10.1080/15596893.2017.1361689

McMulla. (2015). *Shake up the gallery: how iPads are changing the way we visit museums*. Accessed on March 27, 2020, from https://www.alphr.com/apple/1001182/shake-up-the-gallery-how-ipads-are-changing-the-way-we-visit-museums

Meliones, A., & Sampson, D. (2018). Blind MuseumTourer: A system for self-guided tours in museums and blind indoor navigation. *Technologies*, *6*(1), 1–31. doi:10.3390/technologies6010004

Mesquita, S., & Carneiro, M. J. (2016). Accessibility of European museums to visitors with visual impairments. *Disability & Society*, *31*(3), 373–388. doi:10.1080/09687599.2016.1167671

Metatla, O., Serrano, M., Jouffrais, C., Thieme, A., Kane, S., Branham, S., . . . Bennett, C. L. (2018). Inclusive education technologies: Emerging opportunities for people with visual impairments. *Extended Abstracts of the 2018 CHI Conference on Human Factors in Computing Systems*, 1-8.

Michopoulou, E., & Buhalis, D. (2011). Technology platforms and challenges. Accessible Tourism Concepts and Issues, 287-299.

Miesenberger, K., Klaus, J., Karshmer, A., & Zagler, W. (Eds.). (2006). *Computers Helping People with Special Needs: Proceedings of the 10th International Conference* (Vol. 4061). Springer Science & Business Media. 10.1007/11788713

Ochoggia, R. (2003). Persons with Disabilities Bill 2002: Implications concerning visual disabilities for academic library and Information Services in Kenya. Kenya. *New Library World*, *104*(7/8), 307–312. doi:10.1108/03074800310488086

ONCE. (2020). http://museo.once.es/home.cfm?id=5&nivel=1

Penton, V. M. (2015). Assistive Technology Provision for People with Disabilities in Newfoundland and Labrador, Canada. In D. Edyburn (Ed.), Efficacy of Assistive Technology Interventions (Advances in Special Education Technology, Vol. 1), (pp. 139-162). Emerald Group Publishing Limited.

Poria, Y., Reichel, A., & Brandt, Y. (2009). People with disabilities visit art museums: An exploratory study of obstacles and difficulties. *Journal of Heritage Tourism*, *4*(2), 117–129. doi:10.1080/17438730802366508

Poria, Y., Reichel, A., & Brandt, Y. (2010). The Flight Experiences of People with disabilities: An Exploratory Study. *Journal of Travel Research*, *49*(2), 216–227. doi:10.1177/0047287509336477

Presley, I., & D'Andrea, F. M. (2009). *Assistive technology for students who are blind or visually impaired: A guide to assessment*. American Foundation for the Blind.

Puhretmair, F., & Nussbaum, G. (2011). *Web design, assistive technologies and accessible tourism. Accessible Tourism Concepts and Issues*. Channel View Publications.

Reich, C., Lindgren-Streicher, A., Beyer, M., Levent, N., Pursley, J., & Mesiti, L. A. (2011). *Speaking out on art and museums: A study on the needs and preferences of adults who are blind or have low vision*. Boston: Museum of Science, Boston and Art Beyond Sight.

Richards, V., Pritchard, A., & Morgan, N. (2010). (Re) Envisioning tourism and visual impairment. *Annals of Tourism Research*, *37*(4), 1097–1116. doi:10.1016/j.annals.2010.04.011

Sahib, N., Tombros, A., & Stockman, T. (2012). A Comparative Analysis of the Information-Seeking Behavior of Visually Impaired and Sighted Searchers. *Journal of the American Society for Information Science and Technology*, *63*(2), 377–391. doi:10.1002/asi.21696

Sahib, N., Tombros, A., & Stockman, T. (2015). Evaluating a Search Interface for Visually Impaired Searchers. *Journal of the Association for Information Science and Technology*, *63*(11), 2235–2248. doi:10.1002/asi.23325

SFMOMA. (2016). *The New SFMOMA Announces Transformed Digital Strategy: Supported by Bloomberg Philanthropies, SFMOMA Puts Mobile at the Forefront*. San Francisco Museum of Modern Art. Retrieved from https://s3-us-west-2.amazonaws.com/sfmomamedia/media/pressreleases/downloads/SFMOMA_Digital_Engagement_Press_Release_5_2_2016.pdf

Shakespeare, T. (2018). *Disability: the basics*. Routledge.

Smithsonian American Art Museum. (2020). https://americanart.si.edu/visit/accessibility

Sorrell, M., Norton, D., McAdams, J., Winterling, R., & Dipple, K. (2017). Creating an online scientific art exhibit formatted for people with a visual impairment. *Journal of Web Librarianship*, *11*(2), 105–123. doi:10.1080/19322909.2017.1300788

Tadić, V., Hundt, G. L., Keeley, S., & Rahi, J. S. (2015). Seeing it my way: Living with childhood onset visual disability. *Child: Care, Health and Development*, *41*(2), 239–248. doi:10.1111/cch.12158 PMID:24913734

Taheri, B. (2011). *Unpacking visitor engagement: Examining drivers of engagement in museums* (Doctoral dissertation). University of Strathclyde.

The World Bank. (2016). *Disability-overview*. Retrieved from The World Bank: http:// www.worldbank.org/en/topic/disability/overview#1

Tung, V., & Law, R. (2017). The potential for tourism and hospitality experience research in human-robot interactions. *International Journal of Contemporary Hospitality Management*, *29*(10), 2498–2513. doi:10.1108/IJCHM-09-2016-0520

Uslan, M., Peck, A., & Waddell, W. (1988). Audible Traffic Signals: How Useful Are They. *ITE Journal*, 37–43.

Vaz, R. (2020). Co-Creating an Integrative Framework to Enhance the Museum Experience of Blind and Visually Impaired Visitors. In J. M. F. Rodrigues, C. M. Q. Ramos, P. J. S. Cardoso, & C. Henriques (Eds.), *Handbook of Research on Social Media Applications for the Tourism and Hospitality Sector* (pp. 164–191). IGI Global. doi:10.4018/978-1-7998-1947-9.ch011

Vaz, R., Fernandes, P., & Veiga, A. (2018). Interactive technologies in museums: How digital installations and media are enhancing the visitors' experience. In *Handbook of Research on Technological Developments for Cultural Heritage and eTourism Applications* (pp. 30–53). IGI Global. doi:10.4018/978-1-5225-2927-9.ch002

Victoria and Albert Museum. (2020). https://www.vam.ac.uk/info/disability-access

Waligo, V. M., Clarke, J., & Hawkins, R. (2013). Implementing sustainable tourism: A multistakeholders involvement management framework. *Tourism Management*, *36*, 342–353. doi:10.1016/j.tourman.2012.10.008

Wiener, W., Welsh, R., & Blasch, B. (2010). Foundations of orientation and mobility (3rd ed. Vol. 1). New York: AFB (American Foundation for the Blind) Press.

World Health Organization. (2010). *New estimates of visual impairment and blindness.* Retrieved from: http://www.who.int/blindness/publications/globaldata/ en/

World Health Organization. (2016). *World Report on Disability.* Retrieved from https://apps.who.int/iris/bitstream/handle/10665/70670/WHO_NMH_VIP_11.01_eng.pdf?sequence=1WHO

World Health Organization. (2016). *Disability and health.* Accessed on December 5, 2016, from https://www.who.int/mediacentre/factsheets/fs352/en/

World Health Organization. (2017). *Vision impairment and blindness.* Accessed on March 24, 2020, from http://who.int/mediacentre/factsheets/fs282/en/

World Health Organization. (2018). *International classification of diseases: the global standard for diagnostic health information.* Accessed on March 24, 2020 from https://icd.who.int/en/

World Health Organization. (2019). *Blindness and vision impairment.* Accessed on March 24, 2020. from https://www.who.int/news-room/fact-sheets/detail/blindness-and-visual-impairment

World Wide Consortium. (2020). *Web Content Accessibility Guidelines 1.0.* Accessed on April 1, 2020 from https://www.w3.org/TR/WCAG21/#orientation

Yau, M. K.-S., McKercher, B., & Packer, T. L. (2004). Traveling with a disability. *Annals of Tourism Research*, *31*(4), 946–960. doi:10.1016/j.annals.2004.03.007

KEY TERMS AND DEFINITIONS

Accessible: Easy to approach, use, enter, reach.

Assistive Technology: An adaptive technological tool, designed for and used by someone living with a disability to engage in daily life.

Disability: The lack or restriction (resulting from an impairment) of ability to perform an activity in the manner or within the range considered normal for a human being.

Museums: A place where objects of permanent value are kept and displayed.

People With Visual Impairments: People with limitation of actions and functions of the visual system.

Technology: The branch of knowledge that deals with the creation and use of technical means.

Wearable Technology: An advanced electronic device that is worn or carried on the body.

Chapter 13
Virtual Reality (VR) and Augmented Reality (AR) Technologies for Accessibility and Marketing in the Tourism Industry

Meltem Altinay Ozdemir
https://orcid.org/0000-0002-3002-6127
Mugla Sitki Kocman University, Turkey

ABSTRACT

The study investigated the roles of virtual reality (VR) and augmented reality (AR) for accessibility and marketing in tourism. Literature is reviewed in this exploratory study and examples of VR and AR for accessibility and marketing in the tourism industry are presented. The review was conducted with two perspectives: accessibility role and marketing role of VR and AR technologies in the tourism industry. As a result of reviewing, these technologies have both strengths and weaknesses for accessibility in tourism. VR has both strengths such as a secure environment, alternative access, and weaknesses such as user neglect, high cost. Similarly, AR has both strengths such as knowledge enrichment, enhancement of experience, and weaknesses such as high cost, lack of security. The high cost of both technologies restricts accessibility and marketing. Also, VR and AR have benefits as reducing strategy costs, attracting tourists with gamification strategies, ease of brand promotion, building brand loyalty, collecting data, and customized product development in tourism marketing.

INTRODUCTION

Virtual reality (VR) and Augmented Reality (AR) technologies, pioneers of modern times, are now used in all fields and sectors, from communication to automotive, military to tourism, though previously entertainment-oriented, despite reaching out to previous years. Among emerging technologies such as

DOI: 10.4018/978-1-7998-6428-8.ch013

VR and AR can be used for various purposes, such as knowledge enrichment, experience enhancement, learning, and training facilitation (Jung et al., 2016; Kounavis et al., 2012). However, it should be noted that VR and AR can be used as an impressive, open, and marketing tool when evaluated for service providers. Such technologies already have significant implications for many sectors, from services such as tourism, health (Nichol, 2017) to the military (Livingston et al., 2011). Also, these technologies are more important for people who need accessible information. Whether what are their needs, they may access such information from where they stay. With this in mind, it can be said that immersive technologies provide accessibility before marketing or other functions. As one of the possibilities for virtual reality technology, visitors with limited mobility can "experience the inaccessible" with VR (Tom Dieck et al., 2019). These tools, which create great potential for accessible tourism (Bec et al., 2019; Guttentag, 2010; Tecau et al., 2019; Tussyadiah, Jung, et al., 2018) play an important role in reducing barriers to access to tourism (Guttentag, 2010; Plimmer et al., 2006). In this case, studies (Bogicevic et al., 2019; Didehbani et al., 2016; Standen & Brown, 2005; Tecau et al., 2019) have shown that people with disabilities (PwD) use and benefit from virtual reality (VR) technology to detect and avoid physical obstacles in tourism activities. Furthermore, after VR experience, behaviors, preferences, or visiting intentions of PwD have changed positively (Gibson & O'Rawe, 2018; Tussyadiah, Wang, et al., 2018).

Tourism research shows that VR and AR applications often address: "*enhance the experience*" (Rancati et al., 2016; Tom Dieck & Jung, 2018); "*facilitating accessibility*" (Jung & Han, 2014; Mesároš et al., 2016; N.-J. Shih et al., 2019; Wang & Fesenmaier, 2013; Waruwu et al., 2015); and "*marketing*" (Dadwal & Hassan, 2016; Shabani et al., 2019; Tussyadiah et al., 2017). Tourists' experience in virtual environments of destinations and services (hotel room, tour, museum, restaurants, etc.) through smart hardware such as VR headsets, AR glasses can make it easier for the deciding process before purchasing. Tourists, who previously access information about a destination on the internet, now acquire this information in virtual environments in the modern era -even having pre-experience opportunities. With reality technology, PwD can have a secure and controlled environment where they can encounter real-life scenarios and implement skills and talents that can help them better integrate themselves into the real world (Didehbani et al., 2016; Lorenzo et al., 2016; Parsons & Cobb, 2011; Standen & Brown, 2005). They have a chance to decide before purchasing whether the holiday or destination in question is suitable for them. Therefore recently, tourists are interested in VR and AR tools to provide pre-experience with hotels, museums, destinations before making travel decisions (Loureiro et al., 2020; Nayyar et al., 2018). For destination organizations and tourism businesses, VR and AR help to increase the attractiveness of destinations and tourist products (accommodation businesses, food and beverage businesses, museums, ancient cities, festivals, activities, etc.), enable marketing and promotional costs to decrease.

These innovations, in particular, ensure accessible tourism and enable both tourists with and without disabilities to experience tourist products in the same way. Since they ensure accessible information, VR and AR provide promotion and marketing of nearly all products (Celtek, 2015; Mascho & Singh, 2014; Nayyar et al., 2018). Functionally, these tools give users the chance to enhance their experiences by providing accessible information. Besides, considering the consumption society of the modern age and the future with the widespread use of digital technologies, it can be said that the marketing approaches of young generations are guided by technology. Therefore, it may be an advantage in marketing activities that tourism organizations and service manufacturers appeal to young tourists as the target audience. The study focused on the roles of VR and AR technologies for accessibility and marketing in tourism. In an exploratory review article, literature is reviewed and examples of VR and AR for accessibility and marketing in the tourism industry are presented.

THEORETICAL BACKGROUND

The concepts and scope of VR and AR technologies, also their accessibility and marketing roles are explained in this section of the study. Virtual and augmented reality technologies are seen as the most changing technologies in the world in the 21st century. These technological developments stimulate the senses with images created in digital environments and make it possible to have temporary experiences in a virtual environment that is a reflection of reality (Hale & Stanney, 2019; Schmalstieg & Hollerer, 2016). These innovations, which enable virtual and real knowledge to be combined, often contribute to the enrichment of real.

Virtual reality is a simulation model that gives its users a feeling of realism, enabling mutual communication with a dynamic environment created by computers. According to Lavalle (2019), users sense especially visually, and perceive a feeling that they are "in the real world". Virtual reality is an immersive 3D simulation environment that allows users to have a feeling of being in the real world (telepresence). The most important feature of it is that it gives users a feeling of reality. Users should be able to go wherever they want in this environment created by computers, that is, *they feel that control is in their hands*. This can only be achieved through mutual interaction. Virtual reality applications should include minimal VR headsets (Pimentel & Teixeira, 1994). It is a technology that absorbs users with sensory stimuli (vision/sounds), separating them from reality to gain a superficial and improved experience.

Expressing a fully synergistic world that may or may not mimic the real world, VR (Griffin et al., 2017) also forms the infrastructure of AR technologies. However, AR technology has different features than VR technology. This is because, unlike VR, AR is the realization of real-life (Shabani et al., 2019). It's a mix of real-world and VR. AR aims to combine what is happening in the real and virtual environments. Azuma et al. (2001) report that AR technologies have three basic functions. These are combining real and virtual objects in a real environment, interactive and real-time action, and integration of real and virtual objects. By combining the information produced in computer systems with the environment we are in real-time and physically (Azuma et al., 2001), AR establishes a seamless bridge between reality and virtuality. Augmented reality means "enlarging" the real world by combining video, audio, graphics, GPS data produced by devices such as camera speakers, computers, the smartphone with real data (Tussyadiah, Jung, et al., 2018). This technology provides a great benefit to users, especially in the completion of information deficiencies and in the transmission of information (Shih et al., 2019; William, Yao & Nurse, 2017).

Virtual reality and AR technologies, which are among the determinants of the digital age, have been applied and continue to be applied in various fields of industry (Kotler et al., 2019; Lorenz et al., 2018; Paulo Lima et al., 2017; Rancati et al., 2016) and education (Martin et al., 2018). According to a study, it is predicted that the number of people using mobile AR technologies worldwide would be 2.4 billion by 2023. It is estimated that this number would be 1.96 billion in 2021 and 85 million of this will be made up of US users only[1]. It is possible to predict that if the number continues to increase in this way, AR using would become widespread in the future and may even be used even in areas not used today. Immersive technology, including Mix Reality (MR) technologies, is expected to have a consumer market value of $ 6.3 billion by the end of 2020. Virtual reality and AR technologies constitute the biggest revenue of the market. In addition, in the estimations made effects of Covid-19 outbreak, which was declared as a pandemic in 2020, on market was taken into consideration. With these estimates, these technologies will turn into a $7.9 billion market by 2023. Major brands such as Samsung, Sony, Google and Facebook are investing heavily in these technologies to gain a share from this market[2].

Virtual and Augmented Reality Technologies to Accessibility

Individuals with restricted access capabilities can have access to several products with VR and AR technology accessibility features. Accessible features generally refer to features that make resources accessible to users who do not have access or find it difficult to access for reasons such as language barriers. For example, accessibility features may include content enlargement, audio narration, changing one or more colors of the content (for example, for color-blind users), content translation from one language to another, and other accessibility features (Verma et al., 2019). Virtual world accessibility can increase motivation among people with a mobility disability (Guo et al., 2014). De Oliveira et al. (2017), who researched the ease of access of AR technologies for physically disabled people, discussed mobile AR and tag technology to facilitate the navigation of physically disabled people in indoor environments. AR-reliability tags were put at points of interest for experiment testing. Users can find the best route to a particular location by way of voice or touch commands on a mobile app. Similarly, wheelchair users can get support from AR technologies to see the obstacles they may encounter and avoid collisions (Zolotas et al., 2018). Immersive technologies such as AR offer visual motivation by aiding PwD with navigation and mental modeling. Augmented reality and virtual reality are seen as increasingly popular innovations, providing safe and robust mobility testing environments (Devigne et al., 2017; Ktena et al., 2015). Therefore, it can be said that more virtual applications will be created in the coming years, new technologies will be produced and the world will continue to become more advanced digitally.

Virtual and Augmented Reality Technologies in Marketing

The growing VR and AR market offers an opportunity and an alternative channel for digital marketing[3] . The benefits of using VR in a marketing strategy are handled by Brown (2016)[4] as a *unique experience, avoiding ad blockers* and *ready-made social platforms.* Creating a perception of reality by appealing to senses in the digital environment, VR experiences have immersive, impressive, and permanent features. Customer interaction is one of the main areas that brands focus on. The most important thing for brands and marketers is to curate brand/ad content and feed them to customers on social media[5]. Also, these technologies are seen as new platforms for e-commerce, especially to challenge traditional "virtual brick and mortar[6]" retailers. Currently, VR and AR tools used in marketing include[78]:

- Virtual staging (environment);
- 3D architectural visualization;
- 3D virtual 360 tours;
- Flythrough videos;
- Professional photography;
- Augmented reality apps;
- 3D modeling (high poly and low poly);
- 360° video;
- Computer simulation;
- Games (Video Game and Digital Set Design for Video);
- 3D Animation / Motion Graphics;
- AR Geolocalization apps.

Considering the characteristics and benefits of AR technology, it will be easier to introduce and sell new products to customers through this technology in the field of marketing (Bule & Peer, 2013). Mirror applications (Carmigniani et al., 2011) in which customers can control a physical object with their hand from a distance, remote control in work carried out in this area. In recent years, AR has been used as a tool to increase e-marketing in many ways due to its unique features and capabilities. Augmented reality is seen as a new trend in marketing and sales strategies that enable brands to offer their customers unique experiences with ease of touch on their mobile devices. Brands integrate this technology into digital marketing strategies to *"feel what they can expect from a product or service they buy"*[9].

In literature, studies (Bulearca & Tamarjan, 2010; Shabani et al., 2019) that draw attention to VR and AR technologies, are addressed are especially experiential marketing, and reveal that these technologies provide customers *with a time-saving, convenience, practicability* and *fun experience.* The technological experience also strengthens people's selves-confidence who needs access, to experience in a virtual environment, before purchasing any service or product. Stores built for marketing by VR Show Rooms (e.g. IKEA Shoe Room) allow consumers to feel and see products in an interactive environment without buying them[10]. The ability of AR and VR technologies is ideal for sales and marketing departments to build relationships and meaningful connections between the company and the consumers[11].

THE ROLE OF VR AND AR FOR ACCESSIBILITY AND MARKETING IN THE TOURISM INDUSTRY

This study aims to examine the role of VR and AR technologies for accessibility and marketing in tourism and their e through reviewing. The study is explorative and conceptual. It reviews literature and examines VR and AR tools using for tourism accessibility and marketing. The paper reviewed studies reached with the keywords "accessibility and augmented reality in tourism", "accessibility and virtual reality in tourism", "tourism marketing and augmented reality", "tourism marketing and virtual reality" in Google Scholar and Scopus database. The review was conducted between February 15 and March 22 in 2020, with two perspectives: accessibility role (1) and marketing role (2) of VR and AR technologies in tourism.

The Role of VR and AR in Accessible Tourism

The meaning of the "accessibility" concept expands even more and reaches an unlimited dimension with the development of immersive technologies such as VR and AR. All disabled and non-disabled tourists can access destinations and hotels they want to go to, and even activity experiences that will stimulate their senses before making holiday decision. More importantly, in the face of these technologies, all individuals use the same ways to access tourism experience while having equal rights and equal skills. Thanks to immersive technologies, PwD can ensure "access" to tourism by deciding on the suitability of the environment by personally experiencing without a vehicle. They can access museums, artworks, and historical remains in a virtual environment. Tourists with physical disabilities can easily visit multi-story historical buildings without elevators. Thus, VR and AR are among the most important technical elements that provide "accessible" tourism and information. This information highlights the "accessibility" dimension as well as enrichment of experience, full immersion, engagement, social interaction, personalization, and image formation, as reported in the study of Tussyadiah et al. (2018).

Unfortunately, disabled people that want to travel face many difficulties such as architecture or landscape, transportation, and negative attitudes (Goodall et al., 2004). Physical access barriers can sometimes be easily eliminated with tourism providers' cooperation, but in some places, it may be impossible to change physical features to accommodate visitors with disabilities, such as historic heritage sites, due to protection requirements, or huge cost-preventing costs (Goodall et al., 2004). At this point, the importance of "*accessible information*" revealed by Buhalis and Darcy (2010) becomes evident. They stated that *accessible information* is used to determine opportunities for PwD, and this information helps them make environmentally informed decisions about the suitability of the facilities (Hano, 2011, p. 38). Accessibility requirements vary depending on access dimensions, level of support requirements, and auxiliary equipment used. For instance, Ray and Ryder's (2003) study investigated expectations and problems in the holiday options, vacation planning, and travel stages of physically disabled people. They stated that even the facilities reported as "*accessible*" are not truly accessible and do not trust agencies in this regard.

However, in such cases, AR and VR technologies that enable experiential marketing can provide alternative access forms to disabled visitors (Goodall et al., 2004; Guttentag, 2010). For instance, disabled visitors can reach the second floor of the house Shakespeare's Birthplace, located in Stratford-upon-Avon in the UK with XR. They can visit the Fushimi Inari temple at the top of a mountain in Kyoto, Japan using a VR 3D experience. It is even possible to experience adventure activities in a virtual environment.

Virtual reality can provide a safe and controlled environment where PwD can experiment with real-world situations and implement skills and abilities that can assist in enhancing real-world integration (Didehbani et al., 2016; Lorenzo et al., 2016; Parsons & Cobb, 2011; Standen & Brown, 2005). In this way, studies (Bogicevic et al., 2019; Didehbani et al., 2016; Standen & Brown, 2005; Tecau et al., 2019) have shown that PwD uses and benefit from VR technologies in tourism activities to recognize and avoid physical barriers. Thus, PwD can positively change their behavior, preferences and visit intentions after VR experience (Gibson & O'Rawe, 2018; Tussyadiah, Wang, et al., 2018). Tourists with limited mobility have the chance to "experience the inaccessible" with VR as one of the opportunities of reality technology (Tom Dieck et al., 2019). Reality-based technology, thereby offering important opportunities for accessible tourism (Guttentag, 2010; Tecau et al., 2019; Tussyadiah, Jung, et al., 2018), allow access to places where physical or perceptual barriers to access cannot be easily removed (Guttentag, 2010; Plimmer, Pottinger & Goodall, 2006). For example, by enabling product experiences on tours, destinations, museums, etc. before purchasing, VR can be used to provide an emotional experience for visitors (Barnes, 2016). In this case, we can give the main **strengths of the accessible role of VR technologies** as follows:

- Providing a safe and controlled environment for PwD to integrate into the real world;
- Guiding PwD to recognize physical barriers in tourism activities;
- Enabling places that are not open to access/visit to be viewed and experienced in virtual environment-alternative access;
- Enabling the emotional experience, enhancing the drive to visit.

Marasco and Balbi (2019), focusing on tourists with limited mobility, emphasized that the challenges associated with immersive environments that provide ways to enhance the experiences of tourists are neglected. Concerning these challenges, Mascho and Singh (2014) emphasized in their study that there is still no general awareness of VR and AR and that, when they do not receive support, users are struggling with a lack of technological skills. Besides the cost barrier, consensus, availability is seen as an obstacle for the mass market, according to the study. For barriers to accessibility, Fogli et al. (2018) suggested that the alternative would be "availability of various types of users", rather than content that "suits everyone". Plimmer, Pottinger, and Goodall (2006) draw attention to barriers faced by many PwD in accessing VR technology, reflecting on concerns such as the need for support, lack of knowledge, and the failure to see virtual reality as a viable substitute. However, **weaknesses of the accessible role of VR technologies** can be described as:

- Negligence-Overlooking the difficulties users may encounter in accessing VR technologies;
- Lack of users' technological capabilities;
- Economic insufficiency;
- Usability-unidimensionality;
- Lack of information support;
- Lack of interaction;
- Not a substitute for real experience.

AR provides visitors with an innovative way to provide enhanced information in museums (Jung et al., 2016). This technology provides an excellent environment for visualizing and interacting with cultural heritage sites, including both buildings and artifacts (Čejka et al., 2020). Mesároš et al. (2016), focusing on AR experiences provided through smartphones, have created the NosfeRAtu AR game in Slovakia's historic Orava Castle. They, therefore, wanted to ensure that tourists had in-depth knowledge about the museum. Another example of AR is La Lonja, Valencia AR, supported by 3D and marks in Puyuelo's (2013)study. According to the model, building elements are controlled by visitors interacting with marks in front of the camera. The effect is achieved by moving the pointers near or away from the webcam in 3D content viewing (Puyuelo et al., 2013). According to a study by Čejka et al. (2020), via AR, divers set up a prototype device with an augmented reality guide to display old lost buildings at an underwater archeological site in Baiae, Italy. The key **strengths of the accessible role of AR** in such applications can be clarified as follows:

- Knowledge enrichment;
- Enhancement of experience;
- Ease of use;
- Experience sharing and social interaction;
- Advanced technology;
- Learning and experimentation.

One of the most important problems in the design of technological innovations such as AR is the "user group" factor. According to Sheehy et al. (2019), this factor reflects an awareness that PwD or learning difficulties should have an important role in the implementation of new technological developments as a solution to access problems. For this reason, to eliminate "design difficulty for accessibility" stated in Laurìa's (2016) study, it is very important to obtain data from user groups in the development of AR technologies (Sheehy et al., 2019). The **weaknesses of the accessible role of AR**, while similar to VR, can be given as:

- High cost;
- Breach of confidentially;
- Lack of security;
- Addiction;
- Health issues;
- Lack of hardware;
- Lack of perception.

The Role of VR and AR in Tourism Marketing

Marketing in the tourism industry is a key tool to introduce tourists to various destinations and convince them to visit a place or stay in a particular hotel. Marketing strategies have improved significantly in recent years due to technological developments. Therefore, recently, the tourism industry has begun to understand opportunities for digital content to be placed in the real environment of tourists (Jung et al., 2015). There are fun ways to offer an unforgettable experience using VR and AR marketing for tourism, ranging from taking users to flights to let them visit hotel rooms.

Although the adoption of VR is still new, an increasing number of companies and organizations are evolving, as they integrate VR elements into their marketing strategies. Many travel organizers have already adopted VR technologies to provide a preview of the hotel features, cruise ships, and travel experiences (Vitaliev, 2016). Besides, immersive experiences offered through VR are likely to positively affect the destination image, with an intangible experience to potential visitors (Griffin et al., 2017). Virtual reality has the potential to positively change destination image and meet expectations of tourists by creating an emotional connection for tourists before experience. In this case, tourists' purchasing behavior will likely reveal (Jung & Han, 2014). Tourism destination marketing or advertising based on VR and AR tools can attract customers' attention. The environmental features and content of tourism destinations allow for preliminary information before travel by integrating the real environment into a virtual environment or introducing it in a fully virtual environment. This can be achieved with VR and AR technologies.

Tourists can be convinced by promoting destination culture or museums in virtual environments and providing virtual experience with VR to tourists before visiting. Augmented reality applications, on the other hand, can increase the emotional bonds of tourists, which can help repeat sales and visits. There are five characteristics of tourism products that Kotler et. al. (2006) mentioned as "augmented products" in their works. These are *accessibility* (geographic location and working hours), *atmosphere* (visual, auditory, sniffing, and tactile dimensions), and *customer interaction with service organization* (consolidation, consumption, and separation), *customer engagement,* and *customers' interaction with each other*. They pointed out that these five features are very important in the use of AR technologies in tourism. For example, a foreign tourist needs to show the location of the hotel and other businesses on a 3D map. Creating a virtual environment focused on quality and giving real feelings to the tourist may be important in sales of touristic products.

Another use of AR technologies in marketing is customer relationship management. According to Kotler et al. (2006), all communication process from tourist-service combination to tourist-service separation can be provided with AR. At the same time, tourists can interact with other tourists in the AR environment. It is possible to examine the benefits of AR to tourists in six ways as the *solution of language differences, navigation without disappearing, effective decision-making process, seeing/knowing product content*, *convenience in transportation* and *virtual tour in the real environment*[12]. Augmented reality and virtual reality tools in tourism marketing identified through the study of Shabani et al. (2019) can be listed as follows.

- Interactive travel guides;
- Social travel gamification;
- Virtual city tours;
- Real-time navigation and guidance;
- Augmented hotel environment and room reservation;
- Interactive hotel rooms;
- Augmented language translation;
- Augmented access to food and entertainment centers;
- Participatory goal management;
- Augmented transportation;
- Augmented services in restaurants;
- Relive historical sites and events;
- AR scanners specific to destinations;
- Museum interaction.

Augmented reality allows digital content such as online information, graphics or other images to overlap as seen through a real-world camera. Therefore, AR can be highly used in the tourism industry. For example, when tourists point a camera on their smartphone or tablet to an AR image of a product, they can easily access interactive 2D or 3D visual superposition or video shows of that product. These augmented images can be the cover or page of a magazine, billboards at a bus stop, a daily newspaper, or other things in a person's environment (Berryman, 2012). The use of AR in marketing and promotional strategies may give tourism destination marketers a range of advantages. AR campaign provides customers bases with comparatively long-lasting goals, at the same time it allows both advertisers and customers to become more engaged with products by improving their ability to make meaningful selling and purchasing choices, respectively (Shabani et al., 2019).

Companies like YouVisit and Littlstar make virtual vacations an accessible dream. For a carnival ship Alaska, YouVisit has gathered a huge virtual travel media library. You can think of this as the ultimate version of Marriott's VR Postcards. Littlstar, another company that produces VR media on a large scale, offers YouVisit good competition. Although Littlstar is a full-fledged company that produces content for different industries, it has a good travel media library to give YouVisit good competition. Such companies will emerge over time as VR increases[13].

VR and AR Applications in Tourism Marketing: Some Examples

Destination, hospitality and aviation applications in tourism marketing of VR and AR technologies are exampled in this section.

Destination marketing

While destinations around the world are considering using VR for promotional purposes, only a few have implemented a marketing strategy that includes real VR content. Destination British Columbia in Canada, seen as a pioneer of VR enforced DMOs (Destination BC). Another big DMO with a VR concept (on a national basis) is Tourism Australia and SAT (South Africa Tourism Board). Destination British Columbia (Destination BC) DMO is the first of its kind in North America using VR as a promotional tool with the "*Wild Experience in VR*" campaign[14]. DMO displays stunning 360° video footage from the Great Bear Rainforest, a diverse and protected area along the Pacific coast of BC. The images are captured in both first-person and third-person view and follow the general tourist experiences for visitors. Viewers travel along the coastline on a commercial whale-watching boat, and then have the option of visiting a sea lion colony or hiking in the mountains, adding an element of interaction. The destination then partnered with BC Arc'teryx in 2017 and created the "*Hut Magic*" VR experience. Hut Magic (e.g. Arc'teryx[15]) is a VR storytelling experience shot in the Coast Mountains near Smithers[16]. South African Tourism get a campaign, named South Africa: a 5-minute holiday, for VR Travel Experience in 2016. Campaign realized in cooperated with Somethin' Else, UM London and Tribe by integrated into Oculus DK2, and Samsung Gear VR devices. Using VR headsets, "*South Africa: a 5-minute holiday*" experience allowed viewers to go down Table Mountain, swim with sharks, kite surfing, paragliding, feeding elephants, spot penguins at Boulders Beach, as well as sample South African street food without leaving England[17]. In 2016, Tourism Australia produced several VR videos that allow online users to experience some of the country's most famous sites such as the Great Ocean Road, Vlasoff Cay (Great Barries Reef), Sydney Harbour. 360-degree videos of 17 locations across the country were developed as a part of the global "*Nothing similar to Australia*" campaign which focuses on aquatic and coastal experiences (OECD, 2018).

Destinations usually rely on 360° experiences in AR. Some have only 360° video or photos, while storyteller video guide. Although 3D video guides that tell about the attractions before coming to a destination can be more attractive for tourists, more and more auditory and visual reality AR videos are being supported, leading more tourists to want to go to adopt the technologies and visit such destinations. The use of AR technology in marketing has very significant potential due to this attractiveness. There are available applications for this such as World VR tour, Voice map, Visit a city, seven wonders among others. Augmented reality tools include GPS autoplay, storytellers, walking tour trails, drone videos, and 360 virtual videos. These apps are mostly used by tourists in destinations or on holidays, offering advice like tourism businesses, e.g., restaurants, bars, activity centers, among others. As such, in this context, these applications have a marketing role as well as their benefits for tourists. An interactive and highly dynamic experience is achieved as mobile AR applications support the addition of new layers to their realities. Besides, tourists can also benefit from other facilities or services and adjust their navigation with comments of the destinations or locations of their choice, as these applications are mostly on mobile devices such as smartphones or tablets with GPS functions (Carmigniani et al., 2011; Kounavis et al., 2012). In addition to basic VR and AR technologies, it is compatible with rapidly developing technologies such as *geotagging, geofencing, geotargeting,* and is becoming more and more stable. DMOs responsible for destination management needs to adopt these strategies in terms of visiting destinations. Although the interactive, tactile and auditory functions of destinations (other than DMOs such as BC, SAT and Tourism Australian) are limited, AR and VR applications are generally used for promotional purposes.

Figure 1. YouVisit destination virtual tour examples

Figure 1 shows examples of destinations (Minova and New York) contents produced by YouVisit[1819]. In YouVisit applications, storyteller (A) offers a virtual environment consisting of 3D map (B), 360-degree videos and photos. Also, it offers the chance for VR users to enter with the QR code (C) from the destination web address.

Hospitality marketing

Many hotels (e.g., Renaissance Hotels, Ludlow Hotel, Grand Velas Riviera Maya, and Isleta El Espina) have these VR and AR Technologies[20]. They usually offer these technologies within the scope of "*virtual tours*". YouVisit producing virtual tours of various hotels also use AR technologies effectively with interactive videos and offers 360° experience of hospitality businesses. For instance, there is all department of Hilton Waikiki Beach Hotel in a virtual environment (as seen in Figure 2). Tourists who research holidays have a chance to look real environment of the hotel by virtual and 360° videos. Photo A is a 360º illustration of a hotel room in Hilton Waikiki Beach. You can look at every side of the room in virtual and you can see room view (e.g. photo B). In this augmented tour, there is a hotel guide (e.g. photo C) speaking in four language options. Also, the hotel offers VR tours to tourist who has VR headsets, allowing them to watch in their mobile phone after reading a QR code (photo D)[21].

JW Marriot hotel launched "VRoom Service" in 2015 in cooperation with Samsung Electronics America, as the first of its kind in the travel industry. Guests can call a special VRoom Service extension or use Marriott's Mobile Request app to request the Samsung Gear VR headset and the accompanying headphones. The program was first tested for two weeks at the New York Marriott Marquis and London Marriott Park Lane facilities.

Marriott also launched its new virtual travel content platform "VR Postcards". Virtual reality postcards are intimate and immersive travel stories that users experience in 360 3D with a VR headset. Stories follow a real traveler; viewers immerse the destination and listen to personal stories of travelers about why

Figure 2. 360° virtual video of Hilton Waikiki Beach Hotel (USA)

Figure 3. 360° Beijing VR Postcard- Samsung Gear VR headset
Source: Framestore.com (2014)[22]

travel is important to them. The first three VR Postcards were shot in the Andes Mountains in Chile, an ice cream shop in Rwanda and the busy streets of Beijing (Figure 3)[23].

Las Vegas MGM Resorts offers VR gaming to its guests, as well as a mobile application that offers online check-in, online table reservation and a digital card. The "*Virtual Reality Powered by Zero Latency*" arena immerses up to eight players in a fully interactive digital universe and allows them to physically navigate over 2000 square meters of space while collaborating with other players on heartbeat challenges. Teams can choose from three different games for a 30-minute experience, depending on what type of adventure they're looking for. This app is integrated into MGM Grand's tech-savvy game hall LEVEL UP. Virtual reality games with a walking focus also take place in and around the hotel. It is a strange impression to see users around the hotel with gaming equipment[24]. Even if these VR games are not for marketing purposes, such technologies inside the hotel will convince tourists. Nowadays, called the digital era, such applications can be described as a necessity.

Maybe, after the COVID-19 pandemic in 2020, technologies such as AR and VR will become more of a necessity than an option for tourism businesses. –The author.

Post-COVID-19 recovery will likely include integrating new era technologies related to hotels, booking engines, mobile applications, digital concierge services, automatic check-in and check-out and keyless entry systems. For example, Crowne Plaza Greater Noida, a business hotel in Noida (India), starts a VR Experience Center, which is in the hotel lobby. Now, the hotel is taken to clients and business partners through portable Oculus glasses. Virtual Reality Center includes five steps: virtual and physical inspection, showcase multiple decor setups, global distribution through WAVE, virtual booking process and scalable. The hotel has adopted this application to increase sales and distribution[25]. This expands the base of the new customer while allowing the clients to experience the hotel, from any location. This technology also allows the product to be carried to potential customers and a technologically advanced way to discover a hotel like never before[26]. Virtual Reality technology has helped the hotel touch its new customer base with potential guests who can experience the hotel from their home. VR technology is an important tool for gaining new business for hotels.

Aviation marketing

Inflight VR, which creates virtual reality systems for the air transport industry, offers passengers a headset that includes immersive entertainment, including shopping, travel experiences, 3D cinemas and games to provide impressive VR content to passengers in the passenger lounge at Rome Fiumicino Airport. Lufthansa tried Avegant's Glyph video glasses at the business hall at Frankfurt Airport and gave passengers a cinematic experience. In 2017, glasses were awarded the Innovation Award at the CES technology fair[27].

Lufthansa airline combines real-time data, the position of the aircraft and 3D images, and makes the passengers fly into the clouds before they even get on board with the VR moving map experience created in the VR environment. This virtual environment runs on Samsung Gear VR using a smartphone application that can access flight data over a wireless local network with a built-in server created by 3Spin. Users can explore the cockpit of the Airbus A330 and check information from flight speed, height, external temperature and flight time. Lufthansa uses VR to persuade passengers to *buy premium (first/*

business) seats instead of economy class. Lufthansa passengers, who first experienced this in-flight entertainment, were flying by an LH630 Flight from Frankfurt to Dubai in 2018[28][29].

The Japanese airline All Nippon Airways (ANA)[30] has recently announced that they are upgrading their 777-300ER cabins and premiering their updated business class concept at a new product launch event on 24 July 2019 in Tokyo using virtual reality. ANA offers users a Business Class VR Tour (as seen in Figure 4) where they can experience firsthand the unique features and comfort of the cabin by entering one of the largest business-class living environments in the world. Users wearing Oculus Rift S VR headset can open and close the cabin doors, change lighting conditions, and watch an in-flight movie by intuitive hand control. This is a very innovative and practical marketing tool for ANA, as this means that new cabin elements do not need to travel by plane to experience flight. Allowing users to experience sitting features through VR is fast, effective, and most importantly, memorable[31].

Figure 4. ANA Business Class VR Tour
Source: mbryonic.com (2019)[31]

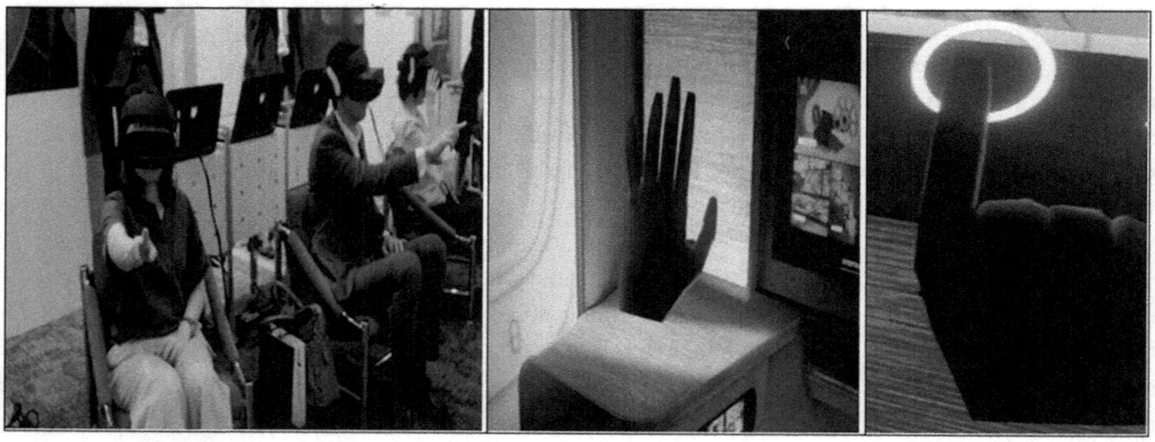

Inflight VR, also integrates VR technologies into Qatar Airways in the aerospace industry (airline, airport), while cooperating with other organizations (Iberia, Sun Express, Star Alliance, FlixBus). In the case of Iberia, Inflight VR provided the airline with approximately 21 devices per plane over the six-month trial period. The devices were presented online flights from Madrid centers to New York, Miami and Tel Aviv[32]. British Airways, which uses in-flight VR headsets for the first time in the UK, has offered VR options to its passengers flying from London to New York in 2019 in award-winning films, documentaries and travel programs in 2D, 3D and 360-degree formats[33].

Tourism Marketing Benefits of VR and AR Technologies

The sense of vision is very important especially in tourism, where experiences depend on visual stimulation (Guttentag, 2010). Therefore, VR seems to be the perfect choice for the tourism industry with visually stunning destinations. The great potential of VR in tourism promotion has been determined by other researchers (Çeltek, 2019; Griffin et al., 2017; Guttentag, 2010; He et al., 2018; Stangl & Weismayer, 2008; Tussyadiah et al., 2017; Wei, 2019; Yung & Khoo-Lattimore, 2019) ve AR (Affan et al.,

2018; Çeltek, 2016; Han et al., 2018; Hu & Tsai, 2017; Jung & Han, 2014; Lim et al., 2011; Nayyar et al., 2018; Tussyadiah, Jung, et al., 2018).

According to the **advantages of VR technologies'** in tourism marketing, visiting a museum's website will *increase one's interest* in visiting a real museum and will serve as indirect evidence that visiting destinations in VR can *encourage real visits.* Virtual reality can serve as a marketing tool for destinations. Virtual reality systems allow tourists to preview destinations and experience related attractions and facilities. Tourists can feel the atmosphere of the destination before deciding which destination (Guttentag, 2010).

Advantages of AR technologies: the ability of customers to point the cameras of their mobile phones to any object or image and visualize/interact with a unique visual profile such as brand logo, bookmarks, tourism catalogs, magazine cover, posters, signs, brochures, business cards and so on (Berryman, 2012; Carmigniani & Furht, 2011). Augmented reality is also useful for tourists, museums, heritage areas, cities, especially as it is possible to *organize and transmit layers of information.* Augmented reality provides *interactive annotations* integrated with map-based services and additional information, provides access to location-based information about immediate surroundings of tourists, and enables access to a variety of well-timed and updated content (Çeltek, 2019). Thus, AR combines with technologies such as geotagging, saving time for tourists and also marketing other tourism services that will make them feel special. Augmented information systems not only *provide valuable and important information* about a tourist attraction or destination but also *maximize their experience* and *provide entertainment opportunities* (Kounavis et al., 2012). Augmented reality also can provide *customized content and services to all users, including tourists*, to their specific needs. In other words, AR can function in preliminary requests and display content accordingly when tourists visit sites at a destination. Businesses can effectively use AR technologies information for their marketing purposes. However, it provides *less costly advertising* to tourism businesses and provides *easier access to potential customers.* Next, the **benefits of VR and AR technologies** are presented, adapted from Çeltek's (2019) study that determines the benefits of smart technologies in marketing, in tourism marketing:

- Reducing strategy cost;
- Attracting tourist with gamification strategies;
- Efficiency;
- Ease of brand promotion;
- Competitiveness;
- Big data analysis and innovation;
- Improving customer service;
- Real-time update;
- Informing about unknown products;
- Building brand loyalty;
- Collecting data about customers;
- Tracking visitor movements;
- Connection with social networks, smart posters and target-tagging;
- Increasing interest;
- Transformation of tourism distribution from customer-business to business-customer;
- Creating emotional brand-customer loyalty in a virtual and augmented environment;
- Enhance customer perception about businesses;

- Customized product development;
- Ensure that customers become confident in travel (e.g., lack of language and directions);
- Change customer's perception about touristic product positively (by augmented and virtual experience associated with the real environment);
- Faultless service flow by including the virtual and augmented environment in the digital platform;
- Increasing sales of expensive luxury products.

DISCUSSION and CONCLUSION

Consumption of VR and AR tools has increased with the development of smart devices. Nonetheless, it is suspected that tourism businesses are a little late in carrying reality technologies forward. *The world is no longer the same.* Although knowledge was previously on websites, today it is found in virtual environments. For tourism businesses to be effective in their strategies, they should first pursue technology and then the customer. They will thus be one step ahead of their customers. As stated in previous studies (Loureiro et al., 2020; Nayyar et al., 2018), reality technologies have become effective tools for tourists in tourist product preference. In the studies (Gibson & O'Rawe, 2018; Tussyadiah, Wang, et al., 2018) conducted, however, it was found that reality technologies enhance the motivation of PwD by allowing barriers to be recognized in the real world in the virtual environment and providing "accessible information" as mentioned by Guttentag (2010). It is thought that reality technologies will take place more in the tourism industry if problems such as high costs are eliminated. These tools, the main aim of which is to enhance the experience (Rancati et al., 2016; Tom Dieck & Jung, 2018), can be strengthened in time by equipping them with smart technologies. For example; receiving orders from a waiter in a virtual environment by activating VR glasses in restaurants and monitoring how the ordered food is cooked; similar to Trailscape VR experience[34] developed by Capra Merrell, enabling wheelchair tourists to choose a hotel by visiting a virtual hotel environment; presenting immersive reality technology-supported games that provide hotel and region orientation by establishing technology departments of hotels; developing free or low cost tools that provide access to alternate routes within business and region through mobile AR. Furthermore, in post Covid-19 pandemic, the use of reality technologies is likely to increase. Tourist businesses damaged by pandemic will enable mobile apps assisted by digital concierge services, digital check-in and check-out systems and keyless entry systems to reduce touch. How reality technologies will gain momentum for tourism in post pandemic is a curious issue of this study such as Kwok & Koh (2020).

According to review analysis, VR and AR have both strengths and weaknesses for accessibility in tourism. Although VR has strengths such as *a safe and controlled environment, alternative access, barrier recognition*, and *emotional motivation*; it also exhibits weaknesses such as *negligence, lack of knowledge, high cost, unidimensionality*, and *lack of interaction*. Similarly, while AR provides *knowledge enrichment, enhancement of experience, ease of use, social interaction, learning* and *experimentation*; it indicates weaknesses such as *high cost, breach of confidentially, lack of security, addiction, health problems* and *lack of hardware*. The high cost of both technologies restricts their use for accessibility or different purposes. However, one-dimensional designs are considered to be a significant weakness, provided that all users are the same or identical. Nevertheless, it is remarkable that individuals' motivations, preferences and visit intentions may change after VR and AR experience, thanks to opportunities they offer to tourists who need access. The most functional features for accessibility of tourism activities are that VR offers a virtual experience before purchasing and that AR offers navigation and location tagging. In short, VR can check whether the environment is reliable for PwD to overcome barriers without going to a destination, hotel, or tour in a virtual environment. On the other hand, they are aware that AR navigations will enable them to create a route based on obstacle conditions.

The main benefits of AR and VR marketing are *reducing strategy cost, attracting tourist with gamification strategies, ease of brand promotion, competitiveness, informing about unknown products, building brand loyalty, collecting data about customers, connection with social networks, smart posters and target-tagging, the transformation of tourism distribution from customer-business to business-customer, customized product development and Increasing the sales of expensive luxury products.* As a result, it was observed that these technologies were used in the tourism industry for accessibility and promotional purposes in the digital era as well as ensuring the accessibility of tourism and information.

Theoretical Contributions and Managerial Implications

This study contributes to the literature by explaining the accessible role of VR and AR technologies as well as their marketing roles. On the other hand, while the strengths and benefits of reality technologies are mostly discussed in previous research, the weaknesses stated in this study have brought a different perspective to the issue. Because of the high-cost problem, businesses cannot use rich content apps and typically prefer mobile apps. Because of rapid technological advances, it is necessary to develop reality technologies at a lower cost. The growing number of consumers interested in VR and AR applications worldwide does not mean that all consumers have reality technologies. Therefore, first of all, consumers need to have access to this hardware. For example, the establishment of VR and AR centers in countries with potential marketing destinations around the world would encourage this process.

Providing videos, pictures and photos of the destination or products marketed for tourism businesses as reservation, with VR functionality on the website of the business can provide the user with all accessible information they need in a single experience. This can save businesses time and provide powerful transformation. Besides, these experiences provide important advantages for tourists who need accessibility. Also, important, official destination organizations (for marketing purposes) use VR technologies to persuade tourists before visiting. It is observed that tourism businesses interact with tourists with integrated augmented technologies such as posters and data matrix, and try to increase demand and accessibility. Destinations using AR technologies use this with geotagging technologies that enable navigation-supported tourist tracking. Besides, the high investment required by such technologies makes businesses a little abstaining in this regard. Despite, many businesses cooperate with technology companies and use VR and AR technologies within the limitation of their investments. Investment and hardware are seen as proportional. Therefore, these products offered for many tourism businesses such as destinations, hotels, museums, and airlines are different. At this point, the "*investment in technology*" issue emerges. To solve this issue, it would be essential for businesses to invest in a technology center or department (e.g. Lasvegas MGM Grand's tech game arcade). Progressive technology will support this initiative. Thus, through working in collaboration with technology experts who will closely monitor their technological advances, businesses will become open to technological innovations. Besides content marketing through VR is an important way to serve potential customers directly from VR headsets or AR glasses. Virtual reality and AR applications must be shared on social media sites for maximum exposure. Tourism businesses should keep in mind that content should be as interesting as possible where online users can continue or experience the experience. Content marketers should always stay on top of evolving technology trends, taking advantage of these technologies for tourism businesses.

LIMITATIONS AND FUTURE RESEARCH

This study investigated the roles of VR and AR technologies inaccessibility to tourism and marketing in tourism. The study focused on accessible tourism and tourism marketing strategies rather than the general use of VR and AR technologies in the tourism industry. According to tourism businesses' use, these technologies' uses for accessible and marketing purposes are not known exactly. Thus, this issue is a limitation of the study. Another limitation stems from the fact that research was carried out with an exploratory and reviewing approach. Therefore, it is not possible to reach definitive judgments. The delimitations of research are to focus only on AR and VR technologies used for accessible tourism and in tourism marketing. Although the uses of AR and VR technologies are widely known, their accessible and marketing roles are not clear. For this reason, it should be much research on this issue. This study tried to explain the accessibility and marketing roles of AR and VR technologies in the tourism industry. Future researchers may benefit from this study for general information about this issue. At the same time, they can realize their study with quantitative than qualitative method. Many current innovations of these technologies can be examined in terms of behavior, expectations, and satisfaction of tourists. Especially, it is advised that the accessible role of these technologies should be investigated with perspectives of PwD.

REFERENCES

Affan, B. N., Suryanto, A., & Arfriandi, A. (2018). Implementation of Augmented Reality as Information and Promotion Media on Dieng Tourism Area. *INSIST*, *3*(1), 128–133. doi:10.23960/ins.v3i1.128

Azuma, R., Baillot, Y., Behringer, R., Feiner, S., Julier, S., & MacIntyre, B. (2001). Recent advances in augmented reality. *IEEE Computer Graphics and Applications*, *21*(6), 34–47. doi:10.1109/38.963459

Barnes, S. (2016, Nov. 3). Understanding virtual reality in marketing: Nature, implications, and potential. *Implications and Potential.*

Bec, A., Moyle, B., Timms, K., Schaffer, V., Skavronskaya, L., & Little, C. (2019). Management of immersive heritage tourism experiences: A conceptual model. *Tourism Management*, *72*, 117–120. doi:10.1016/j.tourman.2018.10.033

Berryman, D. R. (2012). Augmented Reality: A Review. *Medical Reference Services Quarterly*, *31*(2), 212–218. doi:10.1080/02763869.2012.670604 PMID:22559183

Bogicevic, V., Seo, S., Kandampully, J. A., Liu, S. Q., & Rudd, N. A. (2019). Virtual reality presence as a preamble of tourism experience: The role of mental imagery. *Tourism Management*, *74*, 55–64. doi:10.1016/j.tourman.2019.02.009

Bule, J., & Peer, P. (2013). *Interactive augmented reality marketing system in World Usability Day 2013.* Retrieved from http://eprints.fri.uni-lj.si/2505/

Bulearca, M., & Tamarjan, D. (2010). Augmented reality: A sustainable marketing tool. *Global Business and Management Research: An International Journal*, *2*(2), 237–252.

Carmigniani, J., & Furht, B. (2011). Augmented Reality: An Overview. In B. Furht (Ed.), *Handbook of Augmented Reality* (pp. 3–46). Springer New York. doi:10.1007/978-1-4614-0064-6_1

Carmigniani, J., Furht, B., Anisetti, M., Ceravolo, P., Damiani, E., & Ivkovic, M. (2011). Augmented reality technologies, systems and applications. *Multimedia Tools and Applications*, *51*(1), 341–377. doi:10.100711042-010-0660-6

Čejka, J., Zsíros, A., & Liarokapis, F. (2020). A hybrid augmented reality guide for underwater cultural heritage sites. *Personal and Ubiquitous Computing*, 1–14.

Celtek, E. (2015). Augmented Reality Advertisements in Tourism Marketing. In N. O. Taşkıran & R. Yılmaz (Eds.), *Handbook of Research on Effective Advertising Strategies in the Social Media Age* (pp. 125–146). IGI Global. doi:10.4018/978-1-4666-8125-5.ch007

Çeltek, E. (2016). Augmented Reality Advertisements in Tourism Marketing. In Mobile Computing and Wireless Networks: Concepts, Methodologies, Tools, and Applications (pp. 1082–1102). IGI Global.

Çeltek, E. (2019). Advantages of Augmented Reality, Virtual Reality, QR Code, Near Field Communication, Geo-Tagging, Geo-Fencing, and Geo-Targeting for Marketing Tourism. In D. Simões (Ed.), *Smart Marketing With the Internet of Things* (pp. 94–113). IGI Global. doi:10.4018/978-1-5225-5763-0.ch005

Dadwal, S. S., & Hassan, A. (2016). The Augmented Reality Marketing. In Mobile Computing and Wireless Networks: Concepts, Methodologies, Tools, and Applications (pp. 63–80). IGI Global. doi:10.4018/978-1-4666-8751-6.ch004

De Oliveira, L. C., Andrade, A. O., De Oliveira, E. C., Soares, A., Cardoso, A., & Lamounier, E. (2017). Indoor navigation with mobile augmented reality and beacon technology for wheelchair users. *2017 IEEE EMBS International Conference on Biomedical & Health Informatics (BHI)*, 37–40. 10.1109/BHI.2017.7897199

Devigne, L., Babel, M., Nouviale, F., Narayanan, V. K., Pasteau, F., & Gallien, P. (2017). Design of an immersive simulator for assisted power wheelchair driving. *2017 International Conference on Rehabilitation Robotics (ICORR)*, 995–1000. 10.1109/ICORR.2017.8009379

Didehbani, N., Allen, T., Kandalaft, M., Krawczyk, D., & Chapman, S. (2016). Virtual Reality Social Cognition Training for children with high functioning autism. *Computers in Human Behavior*, *62*, 703–711. doi:10.1016/j.chb.2016.04.033

Fogli, D., Sansoni, D., Trivella, E., Arenghi, A., & Passamani, I. (2018). Advanced Interaction Technologies for Accessible and Engaging Cultural Heritage. In *Lecture Notes of the Institute for Computer Sciences, Social-Informatics and Telecommunications Engineering* (pp. 364–373). LNICST. doi:10.1007/978-3-319-76111-4_36

Gibson, A., & O'Rawe, M. (2018). Virtual Reality as a Travel Promotional Tool: Insights from a Consumer Travel Fair. In Augmented Reality and Virtual Reality (pp. 93–107). Springer.

Goodall, B., Pottinger, G., Dixon, T., & Russell, H. (2004). Heritage property, tourism and the UK Disability Discrimination Act. *Property Management*, *22*(5), 345–357. doi:10.1108/02637470410570734

Griffin, T., Hwan, S., Lee, M., & Guttentag, D. (2017, June). *Virtual Reality and Implications for Destination Marketing* (Paper presentation). Travel and Tourism Research Association: Advancing Tourism Research Globally, Quebec, Canada.

Guo, R., Samaraweera, G., & Quarles, J. (2014). A unique way to increase presence of mobility impaired users—Increasing confidence in balance. *2014 IEEE Virtual Reality (VR)*, 77–78.

Guttentag, D. A. (2010). Virtual reality: Applications and implications for tourism. *Tourism Management, 31*(5), 637–651. doi:10.1016/j.tourman.2009.07.003

Hale, K. S., & Stanney, K. M. (2019). *Virtual Reality*. Cambridge University Press., http://vr.cs.uiuc.edu/vrbookbig.pdf

Han, D.-I., Tom Dieck, M. C., & Jung, T. (2018). User experience model for augmented reality applications in urban heritage tourism. *Journal of Heritage Tourism, 13*(1), 46–61. doi:10.1080/1743873X.2016.1251931

Hano, K. (2011). Accessible Tourism: Concepts and Issues. *Tourism Planning & Development, 8*(4), 467–468. doi:10.1080/21568316.2011.603886

He, Z., Wu, L., & Li, X. (2018). When art meets tech: The role of augmented reality in enhancing museum experiences and purchase intentions. *Tourism Management, 68*, 127–139. doi:10.1016/j.tourman.2018.03.003

Hu, P. Y., & Tsai, P. F. (2017). Mobile outdoor augmented reality project for historic sites in Tainan. *2016 International Conference on Advanced Materials for Science and Engineering (ICAMSE)*, 509-511. 10.1109/ICAMSE.2016.7840184

Jung, T., Chung, N., & Leue, M. C. (2015). The determinants of recommendations to use augmented reality technologies: The case of a Korean theme park. *Tourism Management, 49*, 75–86. doi:10.1016/j.tourman.2015.02.013

Jung, T., & Han, D. (2014). Augmented Reality (AR) in Urban Heritage Tourism. *E-Review of Tourism Research, 5*.

Jung, T., Tom Dieck, M. C., Lee, H., & Chung, N. (2016). Effects of Virtual Reality and Augmented Reality on Visitor Experiences in Museum. In A. Inversini & S. Roland (Eds.), *Information and Communication Technologies in Tourism 2016* (pp. 621–635). Springer New York. doi:10.1007/978-3-319-28231-2_45

Kotler, P., Bowen, J. T., Makens, J. C., & Baloglu, S. (2006). Bradford & Bingley audits age diversity in the workplace. *Human Resource Management International Digest, 14*(6), 27–28. doi:10.1108/09670730610690358

Kotler, P., Kartajaya, H., & Hooi, D. H. (2019). Marketing 4.0: Moving From Traditional to Digital. In P. Kotler, H. Kartajaya, & H. D. Huan (Eds.), *Asian Competitors:Marketing for Competitiveness in the Age of Digital Consumers* (pp. 99–123). World Scientific. doi:10.1142/9789813275478_0004

Kounavis, C. D., Kasimati, A. E., & Zamani, E. D. (2012). Enhancing the Tourism Experience through Mobile Augmented Reality: Challenges and Prospects. *International Journal of Engineering Business Management, 4*, 10. doi:10.5772/51644

Ktena, S. I., Abbott, W., & Faisal, A. A. (2015). A virtual reality platform for safe evaluation and training of natural gaze-based wheelchair driving. *2015 7th International IEEE/EMBS Conference on Neural Engineering (NER)*, 236–239.

Kwok, A. O. J., & Koh, S. G. M. (2020). COVID-19 and Extended Reality (XR). *Current Issues in Tourism*, 1–6.

Laurìa, A. (2016). Accessibility to cultural heritage as a design challenge. In Accessibility as a key enabling knowledge for enhancement of cultural heritage. Franco Angeli Edizioni.

Lavalle, S. M. (2019). Virtual reality. In *Virtual Reality*. Cambridge University Press. http://vr.cs.uiuc.edu/vrbookbig.pdf

Lim, Y., Park, Y., Heo, J., Yang, J., Kang, M., & Byun, Y.-C. (2011). A Smart Phone Application Based on AR for Jeju Tourism. *2011 First ACIS/JNU International Conference on Computers, Networks, Systems and Industrial Engineering*, 271–272. 10.1109/CNSI.2011.101

Livingston, M. A., Rosenblum, L. J., Brown, D. G., Schmidt, G. S., Julier, S. J., Baillot, Y., Swan, J. E., Ai, Z., & Maassel, P. (2011). Military Applications of Augmented Reality. In *Handbook of Augmented Reality* (pp. 671–706). Springer New York. doi:10.1007/978-1-4614-0064-6_31

Lorenz, M., Knopp, S., & Klimant, P. (2018, October). Industrial Augmented Reality: Requirements for an Augmented Reality Maintenance Worker Support System (Paper presentation). *2018 IEEE International Symposium on Mixed and Augmented Reality Adjunct (ISMAR-Adjunct)*, 151-153. 10.1109/ISMAR-Adjunct.2018.00055

Lorenzo, G., Lledó, A., Pomares, J., & Roig, R. (2016). Design and application of an immersive virtual reality system to enhance emotional skills for children with autism spectrum disorders. *Computers & Education*, *98*, 192–205. doi:10.1016/j.compedu.2016.03.018

Loureiro, S. M. C., Guerreiro, J., & Ali, F. (2020). 20 years of research on virtual reality and augmented reality in tourism context: A text-mining approach. *Tourism Management*, *77*, 104028. doi:10.1016/j.tourman.2019.104028

Marasco, A., & Balbi, B. (2019). Designing Accessible Experiences for Heritage Visitors Through Virtual Reality. *Ereview of Tourism Research*, *17*(3), 426–443.

Martin, J., Bohuslava, J., & Igor, H. (2018). Augmented Reality in Education 4.0. *2018 IEEE 13th International Scientific and Technical Conference on Computer Sciences and Information Technologies (CSIT)*, 231–236.

Mascho, E., & Singh, N. (2014). Virtual tourism: Use of "second life" for destination marketing. *Anatolia*, *25*(1), 140–143. doi:10.1080/13032917.2013.815644

Mesároš, P., Mandicák, T., Mesárošová, A., Hernandez, M. F., Kršák, B., Sidor, C., Štrba, L., Molokác, M., Hvizdák, L., Blišt'an, P., & Delina, R. (2016). Use of Augmented Reality and Gamification techniques in tourism. *Ereview of Tourism Research*, *13*(1/2), 366–381.

Nayyar, A., Mahapatra, B., Nhuong Le, D., & Suseendran, G. (2018). Virtual Reality (VR) & Augmented Reality (AR) technologies for tourism and hospitality industry. *International Journal of Engineering & Technology*, *7*(2.21), 156.

Nichol, P. B. (2017). Healthcare VR innovations are healing patients. *CIO (Framingham, Mass.)*.

OECD. (2018). *OECD Tourism Trends and Policies 2018*. OECD Publishing. https://www.oecd.org/cfe/tourism/2018-Tourism-Trends-Policies-Highlights-ENG.pdf

Parsons, S., & Cobb, S. (2011). State-of-the-art of virtual reality technologies for children on the autism spectrum. *European Journal of Special Needs Education*, *26*(3), 355–366. doi:10.1080/08856257.2011.593831

Paulo Lima, J., Roberto, R., Simões, F., Almeida, M., Figueiredo, L., Marcelo Teixeira, J., & Teichrieb, V. (2017). Markerless tracking system for augmented reality in the automotive industry. *Expert Systems with Applications*, *82*, 100–114. doi:10.1016/j.eswa.2017.03.060

Pimentel, K., & Teixeira, K. (1994). Virtual Reality: Through the New Looking Glass. *Bulletin of Science, Technology & Society*, *14*(4), 224–224.

Plimmer, F., Pottinger, G., & Goodall, B. (2006, October). *Accessibility Issues for Heritage Properties: a Frame of Mind?* Shaping the Change. *XXIII FIG Congress*, Munich, Germany.

Puyuelo, M., Higón, J. L., Merino, L., & Contero, M. (2013). Experiencing Augmented Reality as an Accessibility Resource in the UNESCO Heritage Site Called "La Lonja", Valencia. *Procedia Computer Science*, *25*, 171–178. doi:10.1016/j.procs.2013.11.021

Rancati, E., Scuotto, V., & Codignola, F. (2016). *Augmented Reality for Enhancing the Customer Experience in Museums : An Exploratory Analysis. Toulon-Verona Conference. Excellence in Services.*

Ray, N. M., & Ryder, M. E. (2003). "Ebilities" tourism: An exploratory discussion of the travel needs and motivations of the mobility-disabled. *Tourism Management*, *24*(1), 57–72. doi:10.1016/S0261-5177(02)00037-7

Schmalstieg, D., & Hollerer, T. (2016). *Augmented reality: principles and practice*. Addison-Wesley Professional.

Shabani, N., Munir, A., & Hassan, A. (2019). E-Marketing via Augmented Reality: A Case Study in the Tourism and Hospitality Industry. *IEEE Potentials*, *38*(1), 43–47. doi:10.1109/MPOT.2018.2850598

Sheehy, K., Garcia Carrizosa, H., Rix, J., Seale, J., & Hayhoe, S. (2019). Inclusive Museums and Augmented Reality: Affordances, Participation, Ethics, and Fun. *The International Journal of the Inclusive Museum*, *12*(4), 67–85. doi:10.18848/1835-2014/CGP/v12i04/67-85

Shih, N.-J., Diao, P.-H., & Chen, Y. (2019). ARTS, an AR Tourism System, for the Integration of 3D Scanning and Smartphone AR in Cultural Heritage Tourism and Pedagogy. Sensors, 19(17), 3725.

Standen, P. J., & Brown, D. J. (2005). Virtual reality in the rehabilitation of people with intellectual disabilities [review]. *Cyberpsychology & Behavior*, *8*(3), 272–282. doi:10.1089/cpb.2005.8.272 PMID:15971976

Stangl, B., & Weismayer, C. (2008). Websites and virtual realities: a useful marketing tool combination? An exploratory investigation. In Information and communication technologies in tourism 2008 (pp. 141–151). Springer. doi:10.1007/978-3-211-77280-5_13

Tecau, A. S., Bratucu, G., Tescaşiu, B., Chiţu, I. B., Constantin, C. P., & Foris, D. (2019). Responsible tourism-integrating families with disabled children in tourist destinations. *Sustainability (Switzerland)*, *11*(16), 4420. doi:10.3390u11164420

Tom Dieck, M. C., & Jung, T. (2018). A theoretical model of mobile augmented reality acceptance in urban heritage tourism. *Current Issues in Tourism*, *21*(2), 154–174. doi:10.1080/13683500.2015.1070801

Tom Dieck, M. C., Jung, T., & Michopoulou, E. (2019). Experiencing Virtual Reality in Heritage Attractions: Perceptions of Elderly Users. In M. C. Tom Dieck & T. Jun (Eds.), *Augmented Reality and Virtual Reality The Power of AR and VR for Business* (pp. 89–98). Springer. doi:10.1007/978-3-030-06246-0_7

Tussyadiah, I. P., Jung, T. H., & tom Dieck, M. C. (2018). Embodiment of Wearable Augmented Reality Technology in Tourism Experiences. *Journal of Travel Research*, *57*(5), 597–611. doi:10.1177/0047287517709090

Tussyadiah, I. P., Wang, D., & Jia, C. (2017). Virtual Reality and Attitudes Toward Tourism Destinations. In R. Schegg & B. Stangl (Eds.), *Information and Communication Technologies in Tourism 2017* (pp. 229–239). Springer International Publishing. doi:10.1007/978-3-319-51168-9_17

Tussyadiah, I. P., Wang, D., Jung, T. H., & tom Dieck, M. C. (2018). Virtual reality, presence, and attitude change: Empirical evidence from tourism. *Tourism Management*, *66*, 140–154. doi:10.1016/j.tourman.2017.12.003

Verma, S., Venugopala, N., Rao, P., & Balasubramanian, K. (2019). *Snapping, Virtual Inking, and Accesibility in Augmented Reality*. https://patentscope.wipo.int/search/zh/detail.jsf;jsessionid=111B01279B2DD3C7C81BC846BEA6C84A.wapp1nA?docId=US298181772&_cid=P10-KCM60G-37278-23

Vitaliev, V. (2016). VR around the... world. *Engineering & Technology*, *11*(3), 33. doi:10.1049/et.2016.0312

Wang, D., & Fesenmaier, D. R. (2013). Transforming the Travel Experience: The Use of Smartphones for Travel. In Information and Communication Technologies in Tourism 2013 (pp. 58–69). Springer Berlin Heidelberg.

Waruwu, A. F., Agung Bayupati, I. P., & Gede Darma Putra, I. K. (2015). Augmented Reality Mobile Application of Balinese Hindu Temples: DewataAR. *International Journal of Computer Network and Information Security*, *7*(2), 59–66. doi:10.5815/ijcnis.2015.02.07

Wei, W. (2019). Research progress on virtual reality (VR) and augmented reality (AR) in tourism and hospitality. *Journal of Hospitality and Tourism Technology*, *10*(4), 539–570. doi:10.1108/JHTT-04-2018-0030

Yung, R., & Khoo-Lattimore, C. (2019). New realities: A systematic literature review on virtual reality and augmented reality in tourism research. *Current Issues in Tourism*, *22*(17), 2056–2081. doi:10.1080/13683500.2017.1417359

Zolotas, M., Elsdon, J., & Demiris, Y. (2018). Head-Mounted Augmented Reality for Explainable Robotic Wheelchair Assistance. *2018 IEEE/RSJ International Conference on Intelligent Robots and Systems (IROS)*, 1823–1829. 10.1109/IROS.2018.8594002

KEY TERMS AND DEFINITIONS

Accessibility: The level to which as people as possible have access to a product, device, service, or environment according to their necessity.

Accessible Information: Information that is equally available and easy to access.

Accessible Tourism: A tourism approach that ensures the accessibility of tourism products, activities, and destinations for all; regardless of limitations and conditions.

Augmented Reality (AR): A semi-virtual model that provides its users a chance of enlarging the real world and enhancing their knowledge by combining real data with virtual data via smart devices.

Immersive Experience: Users' interaction with the real environment in a virtual environment (i) and integration of virtual content with the physical environment (ii).

People With Disabilities (PwD): People who have limited abilities and physical and mental impairments to carry out normal day-to-day activities.

Virtual Reality (VR): A simulation model that gives its users a feeling of realism, enabling mutual communication with a dynamic environment created by computers.

ENDNOTES

1 Statista.com. (2020a). *Global mobile augmented reality users 2015-2023.* https://www.statista.com/statistics/1098630/global-mobile-augmented-reality-ar-users/.

2 Statista.com. (2020b). *Global immersive technology consumer market revenue 2018-2023.* https://www.statista.com/statistics/936078/worldwide-consumer-immersive-technology-market-revenue/.

3 Shirer, M. and Torchia, M. (2017). *Worldwide Spending on Augmented and Virtual Reality Forecast to Reach $13.9 Billion in 2017, According to IDC..* https://web.archive.org/web/20180319084625/

4 Brown, E. (2016). *5 Ways to Use Virtual Reality in Your Marketing Strategy in 2018.* https://blog.hootsuite.com/vr-marketing/

5 Kathotia, A. (2019). *AR & VR – An Insightful Guide for the Modern Digital Marketer.* https://www.polepositionmarketing.com/emp/ar-vr-a-guide-for-the-modern-digital-marketer/.

Virtual brick and mortar means virtual stores that allow consumers to shop without going to the store and have a better in-store experience (Kathotia, 2019).

6 Dronova, T. (2019). *Should You Consider Visual Marketing Tools For Your Business?* Arvrjourney. https://arvrjourney.com/should-you-consider-visual-marketing-tools-for-your-business-9d5dc43b9233.

7 Galileo Creators. (2020). *Augmented Reality.* https://galileocreators.com/.

8 Mhakre, N. (2019). *How Brands Have-Integrated AR & VR In Their Digital Marketing Strategy?* Digital Agency Network. https://digitalagencynetwork.com/how-brands-have-integrated-ar-vr-in-their-digital-marketing-strategy/.

9 Steampowered.com. (2016). *IKEA VR Experience.* https://store.steampowered.com/app/447270/IKEA_VR_Experience/.

1[0] Samsung. (2020). *VR in sales and marketing: The next two years.* https://www.samsung.com/uk/business/vr-for-business-solutions/vr-in-sales-and-marketing-the-next-two-years.

1[1] Pehlivan, G. (2020). *Artırılmış Gerçeklik Turizm Sektörünü Nasıl Etkiliyor?* https://codemodeon.com/tr/artirilmis-gerceklik-turizm-sektorunu-nasil-etkiliyor/.

12 Anurag. (2017). *6 Pathbreaking Ways of Using Virtual Reality in Travel.* https://www.newgenapps.com/blog/virtual-reality-in-travel-impact/.

13 Anderson, S. (2015). *DBC Wild Within VR Experience.* https://www.youtube.com/watch?v=S3bYI2hx5s.

14 Arc'teryx. (2017). *Arc'teryx 360 - Hut Magic.* https://www.youtube.com/watch?v=10k8SfFCj8E

15 DBC. (2017). New "Hut Magic" virtual reality experience showcases BC ski. https://www.destinationbc.ca/content/uploads/2018/07/17-11_DBC_Directions.pdf.

16 Benjamin, K. (2016). *South African Tourism promotes destination with VR activity.* https://www.campaignlive.co.uk/article/south-african-tourism-promotes-destination-vr-activity/1409083.

17 YouVisit.com. (2020a). *New York City.* https://www.youvisit.com/tour/nyc.

18 YouVisit.com. (2020b). *The Village at Winova.* https://www.youvisit.com/tour/villageatwinona.

19 YouVisit.com. (2020c). *Hotels and Resorts Tours.* https://www.youvisit.com/hotels-and-resorts.

20 YouVisit.com. (2020d). *Hilton Waikiki Beach Tour.* https://www.youvisit.com/hotels-and-resorts.

21 Framestore.com. (2014). *Beijing VR Postcard.* VR Postcards-Marriott. https://www.framestore.com/work/vr-postcards?language=en.

22 Bethesda, M. D. (2015). *Marriott Hotels Introduces The First Ever In-Room Virtual Reality Travel Experience.* https://news.marriott.com/news/2015/09/09/marriott-hotels-introduces-the-first-ever-in-room-virtual-reality-travel-experience.

23 Mgmgrand.mgmresort.com. (2020). *Virtual Reality.* https://mgmgrand.mgmresorts.com/en/entertainment/virtual-reality.html.

24 BW Bureau. (2020). *Crowne Plaza Greater Noida launches World's First Hotel VR Experience Center.* http://bwhotelier.businessworld.in/article/Crowne-Plaza-Greater-Noida-launches-World-s-First-Hotel-VR-Experience-Center/09-03-2020-185809/.

25 Hotelierindia. (2020). *From social distancing to contactless check-in and dining - Interactive technology will be key in dealing with pandemic issues.* https://www.hotelierindia.com/business/10763-from-social-distancing-to-contactless-check-in-and-dining-interactive-technology-will-be-key-in-dealing-with.

26 Vertexvr.nl. (2018). *Airports tap into virtual reality for passenger entertainment.* https://www.vertexvr.nl/airports-tap-virtual-reality-passenger-entertainment/.

27 Carlton, B. (2018). *Lufthansa Airlines Uses VR To Give Flyers A Glass Bottom Airplane Experience.* Vrscout. https://vrscout.com/news/lufthansa-airlines-glass-bottom-vr/.

28 Maceda, C. (2018). *Dubai-bound flyers first to try Lufthansa's in-flight VR prototype.* Gulf News. https://gulfnews.com/business/aviation/dubai-bound-flyers-first-to-try-lufthansas-in-flight-vr-prototype-1.2261726.

29 Mwanalushi, K. (2020). *CABIN, CABIN NEWS All Nippon Airways introduces virtual reality training across business.* https://www.aviationbusinessnews.com/cabin/all-nippon-airways-virtual-reality-training-safety/.

30 mbryonic.com. (2019). *ANA Business Class VR Tour.* https://mbryonic.com/portfolio/ana/.

31 Loh, C. (2020). *Qatar Airways Looks To Virtual Reality For In Flight Entertainment.* https://simpleflying.com/qatar-airways-virtual-reality/.

32 Media Centre British Airways. (2019). *British Airways To Trial Virtual Reality Entertainment in The Skies.* https://mediacentre.britishairways.com/pressrelease/details/86/2019_319/11449.

33 Framestore. (2015). *Merrell 'Trailscape', Framestore VR Studio.* https://www.youtube.com/watch?v=efd6WhPmTyU.

Compilation of References

Abdul Rahman, A. A., Zaaba, Z. F., & Hussain, A. (2018). Usable tourism application: MyTouristPlanner (MTP). *2018 International Conference on Promising Electronic Technologies, ICPET 2018*, 31-36. 10.1109/ICPET.2018.00012

AbilityNet. (2020). *Digital Accessibility*. https://abilitynet.org.uk/accessibility-services/what-is-digital-accessibility

Abramo, L., Cecchini, S., & Ullmann, H. (2020). Addressing health inequalities in Latin America: The role of social protection. *Ciencia & Saude Coletiva, 25*(5), 1587–1598. doi:10.1590/1413-81232020255.32802019 PMID:32402029

Abrate, G., Bruno, C., Erbetta, F., & Fraquelli, G. (2019). Which Future for Traditional Travel Agencies? A Dynamic Capabilities Approach. *Journal of Travel Research, 59*(5), 777–791. doi:10.1177/0047287519870250

ABS. (2018). Disability, Ageing and Carers, Australia: Summary of Findings. Australian Bureau of Statistics.

Accessible Tourism Guide. (2019). Retrieved from https://www.buenosaires.gob.ar/sites/gcaba/files/c0_en_opt.pdf

Acosta, T., Acosta-Vargas, P., & Lujan-Mora, S. (2018). Accessibility of eGovernment Services in Latin America. In *2018 International Conference on eDemocracy eGovernment (ICEDEG)* (pp. 67–74). 10.1109/ICEDEG.2018.8372332

Adams, E. (2016). *Most Virtual Reality Is Not Virtual Reality. Here's Why*. Retrieved March 25, 2020, from http://gearpatrol.com/2016/07/18/virtual-reality-vs-360-video/

Adams, R., & Downey, C. (2016). Edited Platforms: Anticipating Future Consumption. Journal of Marketing Theory and Practice, 24(2), 224-235.

Affan, B. N., Suryanto, A., & Arfriandi, A. (2018). Implementation of Augmented Reality as Information and Promotion Media on Dieng Tourism Area. *INSIST, 3*(1), 128–133. doi:10.23960/ins.v3i1.128

Agapito, D., Valle, P., & Mendes, J. (2014). The sensory dimension of tourist experiences: Capturing meaningful sensory-informed themes in Southwest Portugal. *Tourism Management, 42*, 224–237. doi:10.1016/j.tourman.2013.11.011

Agarwal, Y. (2019). Accessibility information in New Delhi for "EasenAccess" Android-based app for persons with disability: An observational study. *Disability and Rehabilitation. Assistive Technology, 14*(7), 645–662. doi:10.1080/17483107.2018.1471743 PMID:29902941

Age, U. K. (2020). *One Digital*. https://www.ageuk.org.uk/our-impact/programmes/one-digital/

Agovino, M., Casaccia, M., Garofalo, A., & Marchesano, K. (2017). Tourism and disability in Italy. Limits and opportunities. *Tourism Management Perspectives, 23*, 58–67. doi:10.1016/j.tmp.2017.05.001

Aguiar-Quintana, T., Moreno-Gil, S., & Picazo-Peral, P. (2016). How could traditional travel agencies improve their competitiveness and survive? A qualitative study in Spain. *Tourism Management Perspectives, 20*, 98–108. doi:10.1016/j.tmp.2016.07.011

Ahmi, A., & Mohamad, R. (2015). Web Accessibility of the Malaysian Public University Websites. *Proceedings of the International Conference on E-Commerce*. Retrieved from www.icoec.my

Aiello, L. (2012). Role of meta-search engine in the relationship tourist/e-retailer. In *Proceedings - ICIDT 2012, 8th International Conference on Information Science and Digital Content Technology* (*vol. 3*, pp. 749–756). Academic Press.

Aitchison, C. (2003). From leisure and disability to disability leisure: Developing data, definitions and discourses. *Disability & Society*, *18*(7), 955–969. doi:10.1080/0968759032000127353

Ajuntament de Barcelona. (2019). *Anuari Estadístic de la Ciutat de Barcelona 2019*. Retrieved from http://www.bcn.cat/estadistica/catala/dades/anuari/cap06/index.htm

Akgül, Y., & Vatansever, K. (2016). Web Accessibility Evaluation of Government Websites for People with Disabilities in Turkey. *Journal of Advanced Management Science*, *4*(3), 201–210. doi:10.12720/joams.4.3.201-210

AlDairi, A., & Tawalbeh, L. (2017). Cyber security attacks on smart cities and associated mobile technologies. *Procedia Computer Science*, *109*, 1086–1091. doi:10.1016/j.procs.2017.05.391

Al-Khalifa, H. S., Baazeem, I., & Alamer, R. (2017). Revisiting the accessibility of Saudi Arabia government websites. *Universal Access in the Information Society*, *16*(4), 1027–1039. doi:10.100710209-016-0495-7

Allan, M. (2015). Accessible tourism in Jordan: Travel constraints and motivations. *European Journal of Tourism Research*, *10*, 109–119.

Allwinkle, S., & Cruickshank, P. (2011). Creating smarter cities: An overview. *Journal of Urban Technology*, *18*(2), 1–16. doi:10.1080/10630732.2011.601103

Almeida-Santana, A., & Moreno-Gil, S. (2017). New trends in information search and their influence on destination loyalty: Digital destinations and relationship marketing. *Journal of Destination Marketing & Management*, *6*(2), 150–161. doi:10.1016/j.jdmm.2017.02.003

Alper, M., Ellcessor, E., Ellis, K., & Goggin, G. (2015). Reimagining the Good Life with Disability: Communication, New Technology, and Humane Connections. In H. H. Wang (Ed.), Communication and the "Good Life" (pp. 197-212). New York: International Communication Association Theme Book Series, vol 2 - Peter Lang.

Alshamari, M. (2016). Accessibility Evaluation of Arabic E-Commerce Web Sites Using Automated Tools. *Journal of Software Engineering and Applications*, *09*(09), 439–451. doi:10.4236/jsea.2016.99029

Altinay, Z., Saner, T., Bahçelerli, N. M., & Altinay, F. (2016). The role of social media tools: Accessible tourism for disabled citizens. *Journal of Educational Technology & Society*, *19*(1), 89–99.

Ambrose, I., Darcy, S., & Buhalis, D. (2012). *Best Practice in Accessible Tourism: Inclusion, Disability, Ageing Population and Tourism*. Channel View Publications. doi:10.21832/9781845412548-029

Ambrose, T., & Paine, C. (2018). *Museum basics: The international handbook*. Routledge. doi:10.4324/9781315232898

Anderson, J. D., Lee, D. J., & Archibald, J. K. (2007). Embedded stereo vision system providing visual guidance to the visually impaired. In *2007 IEEE/NIH Life Science Systems and Applications Workshop* (pp. 229-232). IEEE. 10.1109/LSSA.2007.4400926

Andrade, E. C., Delgado, R., Hauptner, B., Kerschbaum, F., List, R., Pobežalová, P., . . . Wögerbauer, S. (2015). *Inclusive technologies in museums: for a better access to culture for blind and visually impaired people*. Vienna, Austria: Project AMBAVis. Accessed on April 10, 2020, from http://www.ambavis.eu/wp-content/ uploads/2016/11/AMBAVis Broschuere_BF_V01_EN_bf.pdf

Andronico, P., Buzzi, M., Castillo, C., & Leporini, B. (2006). Improving search engine interfaces for blind users: A case study. *Universal Access Information*, *5*(1), 23–40. doi:10.100710209-006-0022-3

Anonymous. (2002). Delta brings new display screens to San Francisco and Tampa airports. *Airline Industry Information,* 1. Retrieved from http://ezproxy.cardiffmet.ac.uk/login?url=https://search-proquest-com.ezproxy.cardiffmet.ac.uk/docview/210525066?accountid=15588

Anonymous. (2013). New interactive touch-screen directories work with mobile application to put MSP airport information at your fingertips. *Targeted News Service.* Retrieved from http://ezproxy.cardiffmet.ac.uk/login?url=https://search-proquest-com.ezproxy.cardiffmet.ac.uk/docview/1321651369?accountid=15588

Anonymous. (2016). Northern rail to install customer information screens across Cheshire, UK. *Progressive Digital Media Transportation (Incl Airports, Roadways, Railways, Shipping, Automotive & Logistics) News.* Retrieved from http://ezproxy.cardiffmet.ac.uk/login?url=https://search-proquest-com.ezproxy.cardiffmet.ac.uk/docview/1756502983?accountid=15588

Argyropoulos, V. S., & Kanari, C. (2015). Re-imagining the museum through "touch": Reflections of individuals with visual disability on their experience of museum-visiting in Greece. *Alter*, *9*(2), 130–143. doi:10.1016/j.alter.2014.12.005

Armitage, R., & Nellums, L. B. (2020). The COVID_19 response must be disability inclusive. *The Lancet. Public Health*, *5*(5), 257. doi:10.1016/S2468-2667(20)30076-1 PMID:32224295

Assistive Technology Act. (2004). Accessed on April 10, 2020, from https://www.congress.gov/108/plaws/publ364/PLAW-108publ364.pdf

Attitude is Everything. (2018). *State of Access Report 2018.* Available at: http://www.attitudeiseverything.org.uk/uploads/general/State_of_Access_Report_2018.pdf

Avci, E. (2019). *Enhancing the Cultural Tourism Experience Through Augmented Reality.* Co-Editors, 215.

Avis, A. H., Card, J. A., & Cole, S. T. (2005). Accessibility and attitudinal barriers encountered by travelers With Physical Disabilities. *Tourism Review International*, *8*(3), 239–248. doi:10.3727/154427205774791591

Aydemir, B., & Kılıç, S. N. (2017). Dünyada ve Türkiye'de Üçüncü Yaş Turizmi. *Yüzüncü Yıl Üniversitesi Sosyal Bilimler Enstitüsü Dergisi*, *1*(3), 1–11.

Ayeh, J. K. (2015). Travellers' acceptance of consumer-generated media: An integrated model of technology acceptance and source credibility theories. *Computers in Human Behavior*, *48*, 173–180. doi:10.1016/j.chb.2014.12.049

Azevedo Guedes, A. L., Carvalho Alvarenga, J., Dos Santos Sgarbi Goulart, M., Rodriguez, R., Vicente, M., & Pereira Soares, C. A. (2018). Smart cities: The main drivers for increasing the intelligence of cities. *Sustainability*, *10*(9), 3121. doi:10.3390u10093121

Azuma, R., Baillot, Y., Behringer, R., Feiner, S., Julier, S., & MacIntyre, B. (2001). Recent advances in augmented reality. *IEEE Computer Graphics and Applications*, *21*(6), 34–47. doi:10.1109/38.963459

Baggio, R., & Cooper, C. (2010). Knowledge transfer in a tourism destination: The effects of a network structure. *Service Industries Journal*, *30*(10), 1757–1771. doi:10.1080/02642060903580649

Balakrishnan, A., Kulkarni, K., Moirangthem, S., Kumar, C. N., Math, S. B., & Murthy, P. (2019). The Rights of Persons with Disabilities Act 2016: Mental Health Implications. *Indian Journal of Psychological Medicine*, *41*(2), 119–125. doi:10.4103/IJPSYM.IJPSYM_364_18 PMID:30983658

Bannister, D. (2017). Get your digital communications right. *Airport Business, 31*, 10-13. Retrieved from http://ezproxy.cardiffmet.ac.uk/login?url=https://search-proquest-com.ezproxy.cardiffmet.ac.uk/docview/1912912014?accountid=15588

Barbu, C., Kuhn, K., Spirescu, A., Lamorte, L., & Scanu, C. (2015). 3cixty: An innovative way to explore cities. *2015 IEEE 1st International Smart Cities Conference, ISC2 2015.* 10.1109/ISC2.2015.7366178

Barile, F., Calandra, D. M., Caso, A., Dauria, D., Di Mauro, D., Cutugno, F., & Rossi, S. (2014). ICT solutions for the OR.C.HE.S.T.R.A. project: From personalized selection to enhanced fruition of cultural heritage data. *10th International Conference on Signal-Image Technology and Internet-Based Systems, SITIS 2014*, 501-507. 10.1109/SITIS.2014.12

Barnes, S. (2016, Nov. 3). Understanding virtual reality in marketing: Nature, implications, and potential. *Implications and Potential.*

Barnes, C., & Mercer, G. (2005). Disability, work, and welfare. Challenging the social exclusion of disabled people. *Work, Employment and Society, 19*(3), 527–545. doi:10.1177/0950017005055669

Barnsley Hospital NHS Foundation Trust. (2017). Accessed March 8, 2020 from https://www.barnsleyhospital.nhs.uk/assistive-technology/about-assistive-technology-at/

Barricelli, B. R., Sciarelli, P., Valtolina, S., & Rizzi, A. (2018). Web accessibility legislation in Italy: A survey 10 years after the Stanca Act. *Universal Access in the Information Society, 17*(1), 211–222. doi:10.100710209-017-0526-z

Bártek, L., & Kopeček, I. (2006, July). Web pages for blind people—generating web-based presentations by means of dialogue. In *International Conference on Computers for Handicapped Persons* (pp. 114-119). Berlin: Springer. 10.1007/11788713_18

Baudoin, G., & Venard, O. (2010). Information, communication and localization environment for travelers with sensory disabilities in public transports. In *2010 5th International ICST Conference on Communications and Networking in China* (pp. 1-7). IEEE. 10.4108/chinacom.2010.116

Bec, A., Moyle, B., Timms, K., Schaffer, V., Skavronskaya, L., & Little, C. (2019). Management of immersive heritage tourism experiences: A conceptual model. *Tourism Management, 72*, 117–120. doi:10.1016/j.tourman.2018.10.033

Beck, J., Rainoldi, M., & Egger, R. (2019). Virtual reality in tourism: A state-of-the-art review. *Tourism Review, 74*(3), 586–612. doi:10.1108/TR-03-2017-0049

Beck, M. (2010). *Theory and Practice of Therapeutic Massage* (5th ed.). CENGAGE Learning.

Benfares, C., El Bouzekri El Idrissi, Y., & Amine, A. (2016). Smart city: Recommendation of personalized services in patrimony tourism. *Colloquium in Information Science and Technology, CIST*, 835-840. 10.1109/CIST.2016.7805003

Berryman, D. R. (2012). Augmented Reality: A Review. *Medical Reference Services Quarterly, 31*(2), 212–218. doi:10.1080/02763869.2012.670604 PMID:22559183

Beutel, M. C., Gökay, S., Kluth, W., Krempels, K., Ohler, F.G., Samsel, C., Terwelp, C., & Wiederhold, M. (2016). *Information Integration for Advanced Travel Information Systems*. Academic Press.

Bhattacharya, D., Painho, M., Mishra, S., & Gupta, A. (2017). Mobile traffic alert and tourist route guidance system design using geospatial data. *International Archives of the Photogrammetry, Remote Sensing and Spatial Information Sciences - ISPRS Archives, 42*(4W3), 11-18. 10.5194/isprs-archives-XLII-4-W3-11-2017

Bhowmick, A., & Hazarika, S. M. (2017). An insight into assistive technology for the visually impaired and blind people: State-of-the-art and future trends. *Journal on Multimodal User Interfaces, 11*(2), 149–172. doi:10.100712193-016-0235-6

Bieger, T., & Laesser, C. (2001). Segmenting Travel on the Sourcing of Information. In *Consumer Psychology of Tourism, Hospitality and Leisure* (pp. 153–167). Cabi Publishing. doi:10.1079/9780851995359.0153

Bigné, E., Sanz, S., & Ruiz, C. (2010). Why some internet users don't buy air tickets online. In U. Gretzel, R. Law, & M. Fuchs (Eds.), *Information and Communication Technologies in Tourism 2010* (pp. 209–221). Springer. doi:10.1007/978-3-211-99407-8_18

Bigné, J. E., Küster, I., Andreu, L., & Blesa, A. (2008). Orientación al mercado, resultados e indicadores básicos de competitividad. Interrelación en las agencias de viajes. *Revista Española de Investigación de Marketing. ESIC, 12*(1), 97–122.

Bi, Y., Card, J. A., & Cole, S. T. (2007). Accessibility and attitudinal barriers encountered by Chinese travellers with physical disabilities. *International Journal of Tourism Research, 9*(3), 205–216. doi:10.1002/jtr.603

Bizjak, B., Knežević, M., & Cvetrežnik, S. (2011). Attitude change towards guests with disabilities: Reflections from tourism students. *Annals of Tourism Research, 38*(3), 842–857. doi:10.1016/j.annals.2010.11.017

Black, G. (2005). *The engaging museum: Developing museums for visitor involvement.* Psychology Press.

Blichfeldt, B. S., & Nicolaisen, J. (2011). Disabled travel: Not easy, but doable. *Current Issues in Tourism, 14*(1), 79–102. doi:10.1080/13683500903370159

Blocksidge, D. (2003). *Dictionary of disability terminology.* http://www.dpa.org.sg/wp-content/uploads/2013/01/Dictionary.pdf

Boes, K., Gretzel, U., Zhong, L., Koo, C., Buhalis, D., & Inversini, A. (2016). Smart tourism destinations: Ecosystems for tourism destination competitiveness. *International Journal of Tourism Cities, 2*(2), 108–124. doi:10.1108/IJTC-12-2015-0032

Bogart, K. R., Rottenstein, A., Lund, E. M., & Bouchard, L. (2017). Who self-identifies as disabled? An examination of impairment and contextual predictors. *Rehabilitation Psychology, 62*(4), 553–562. doi:10.1037/rep0000132 PMID:28581320

Bogicevic, V., Seo, S., Kandampully, J. A., Liu, S. Q., & Rudd, N. A. (2019). Virtual reality presence as a preamble of tourism experience: The role of mental imagery. *Tourism Management, 74*, 55–64. doi:10.1016/j.tourman.2019.02.009

Boniface, B., & Cooper, C. (2001). Worldwide destinations casebook: The geography of travel and tourism: Second edition. In Worldwide Destinations Casebook: The Geography of Travel and Tourism: Second Edition. doi:10.4324/9780080949758

Borges Tiago, M., Couto, J., Tiago, F., & Dias Faria, S. (2016). Baby Boomers turning grey: European profiles. *Tourism Management, 54*, 13–22. doi:10.1016/j.tourman.2015.10.017

Borges, I., Costa, E., Sousa Pinto, A., & Abreu, A. (2020). Infoaccessibility on the websites of inbound markets of Portugal destination. In Á. Rocha, A. Abreu, J. de Carvalho, D. Liberato, E. González, & P. Liberato (Eds.), *Advances in Tourism, Technology and Smart Systems. Smart Innovation, Systems and Technologies* (Vol. 171, pp. 105–117). Springer. doi:10.1007/978-981-15-2024-2_10

Bouskela, M., Casseb, M., Bassi, S., De Luca, C., & Facchina, M. (2016). *The Road toward Smart Cities: Migrating from Traditional City Management to the Smart City.* Retrieved from: https://publications.iadb.org/publications/english/document/The-Road-toward-Smart-Cities-Migrating-from-Traditional-City-Management-to-the-Smart-City.pdf

Bowtell, J. (2015). Assessing the value and market attractiveness of the accessible tourism industry in Europe: A focus on major travel and leisure companies. *Journal of Tourism Futures, 1*(3), 203–222. doi:10.1108/JTF-03-2015-0012

Bryman, A. (2016). *Social Research Methods* (5th ed.). Oxford University Press.

Brynjolfsson, E., & McAfee, A. (2014). *The second machine age: Work, progress, and prosperity in a time of brilliant technologies*. WW Norton & Company.

Budd, L., & Ison, S. (2020). Supporting the needs of special assistance (including PRM) passengers: An international survey of disabled air passenger rights legislation. *Journal of Air Transport Management*, *87*, 101851. Advance online publication. doi:10.1016/j.jairtraman.2020.101851

Buenos Aires City. (2019). *Mapa Interactivo de lugares accesibles*. Retrieved from: www.buenosaires.gob.ar/copidis/accesibilidad/mapa-interactivo-de-lugares-accesibles

Buenos Aires City. (2020). *Communes Map*. Retrieved from: https://mapa.buenosaires.gob.ar/mapas/?lat=-34.620818&lng=-58.366615&zl=11&modo=transporte&map=comunas

Buhalis, D., Eichhorn, V., Michopoulou, E., & Miller, G. (2006). *Accessibility market and stakeholder analysis*. OSSATE, University of Surrey. Retrieved from: https://www.accessibletourism.org/resources/ossate_market_analysis_public_final.pdf

Buhalis, D. (1998). Strategic use of information technologies in the tourism industry. *Tourism Management*, *19*(5), 409–421. doi:10.1016/S0261-5177(98)00038-7

Buhalis, D. (2019). Technology in tourism-from information communication technologies to eTourism and smart tourism towards ambient intelligence tourism: A perspective article. *Tourism Review*, *75*(1), 267–272. doi:10.1108/TR-06-2019-0258

Buhalis, D., & Amaranggana, A. (2013). Smart Tourism Destinations. In Z. Xiang & I. Tussyadiah (Eds.), *Information and Communication Technologies in Tourism 2014* (pp. 553–564). Springer. doi:10.1007/978-3-319-03973-2_40

Buhalis, D., & Darcy, S. (2010). *Accessible tourism: Concepts and issues. Accessible tourism: Concepts and issues*. Channel View Publications. doi:10.21832/9781845411626

Buhalis, D., & Darcy, S. (2011). *Accessible tourism concepts and issues*. Channel View Publications.

Buhalis, D., & Darcy, S. (2011). *Accessible Tourism*. Channel View Publications.

Buhalis, D., & Darcy, S. (Eds.). (2011). *Accessible tourism: Concepts and issues*. Channel View Publications.

Buhalis, D., & Law, R. (2008). Progress in information technology and tourism management: 20 years on and 10 years after the Internet-The state of eTourism research. *Tourism Management*, *29*(4), 609–623. doi:10.1016/j.tourman.2008.01.005

Buhalis, D., & Michopoulou, E. (2011). Information-enabled tourism destination marketing: Addressing the accessibility market. *Current Issues in Tourism*, *14*(2), 145–168. doi:10.1080/13683501003653361

Buhalis, D., & Sinarta, Y. (2019). Real-time co-creation and nowness service: Lessons from tourism and hospitality. *Journal of Travel & Tourism Marketing*, *36*(5), 563–582. doi:10.1080/10548408.2019.1592059

Bule, J., & Peer, P. (2013). *Interactive augmented reality marketing system in World Usability Day 2013*. Retrieved from http://eprints.fri.uni-lj.si/2505/

Bulearca, M., & Tamarjan, D. (2010). Augmented reality: A sustainable marketing tool. *Global Business and Management Research: An International Journal*, *2*(2), 237–252.

Burnett, J. J., & Baker, H. (2001). Assessing the Travel-related Behaviors of the Mobility-Disabled Customer. *Journal of Travel Research*, *40*(4), 4–11. doi:10.1177/004728750104000102

Burton, H. (2003). Visual cortex activity in early and late blind people. *The Journal of Neuroscience: The Official Journal of the Society for Neuroscience*, *23*(10), 4005–4011. doi:10.1523/JNEUROSCI.23-10-04005.2003 PMID:12764085

ByrneD. M.CorradoC. A. (2017). *ICT services and their prices: What do they tell us about productivity and technology?* (FEDS Working Paper No. 2017-015). https://ssrn.com/abstract=2920451

Cacho, A., Estaregue, D., Figueredo, M., Lucas, J., Aurelio, M., Farias, H., . . . Alves, C. (2015). A smart destination initiative: The case of a 2014 FIFA world cup host city. *2015 IEEE 1st International Smart Cities Conference, ISC2 2015.* 10.1109/ISC2.2015.7366223

Calver, S. J., & Page, S. J. (2013). Enlightened hedonism: Exploring the relationship of service value, visitor knowledge and interest, to visitor enjoyment at heritage attractions. *Tourism Management, 39*, 23–36. doi:10.1016/j.tourman.2013.03.008

Camisão, V. (2006). Turismo e Acessibilidade. In W. M. Guia, A. Pereira, T. Brizolla, M. F. Krahl, M. D. Pinto, & A. C. Santos (Eds.), *Turismo Social: Diálogos do turismo - Uma viagem de inclusão* (pp. 320–356). Ministério do Turismo e Instituto Brasileiro de Administração Municipal.

Camprubí, R., & Coromina, L. (2016). Content analysis in tourism research. *Tourism Management Perspectives, 18*, 134–140. doi:10.1016/j.tmp.2016.03.002

Capriotti, P. (2013). Managing Strategic Communication in Museums. The case of Catalan museums. *Communicatio Socialis, 26*(3), 98–116.

Carmigniani, J., & Furht, B. (2011). Augmented Reality: An Overview. In B. Furht (Ed.), *Handbook of Augmented Reality* (pp. 3–46). Springer New York. doi:10.1007/978-1-4614-0064-6_1

Carmigniani, J., Furht, B., Anisetti, M., Ceravolo, P., Damiani, E., & Ivkovic, M. (2011). Augmented reality technologies, systems and applications. *Multimedia Tools and Applications, 51*(1), 341–377. doi:10.100711042-010-0660-6

Casanova Ferro, G. (2011). *Turismo: El camino de la Discapacidad a la Accesibilidad.* Retrieved from: https://www.academia.edu/594410/Turismo_El_camino_de_la_discapacidad_a_la_accesibilidad

Cassia, F., Castellani, P., Rossato, C., & Baccarani, C. (2020). Finding a way towards high-quality, accessible tourism: The role of digital ecosystems. *The TQM Journal.* Advance online publication. doi:10.1108/TQM-03-2020-0062

Cavinato, J. L., & Cuckovich, M. L. (1992). Transportation and Tourism for the Disabled: An Assessment. *Transportation Journal, 31*(3), 46–53.

Čejka, J., Zsíros, A., & Liarokapis, F. (2020). A hybrid augmented reality guide for underwater cultural heritage sites. *Personal and Ubiquitous Computing*, 1–14.

Çeltek, E. (2016). Augmented Reality Advertisements in Tourism Marketing. In Mobile Computing and Wireless Networks: Concepts, Methodologies, Tools, and Applications (pp. 1082–1102). IGI Global.

Celtek, E. (2015). Augmented Reality Advertisements in Tourism Marketing. In N. O. Taşkıran & R. Yılmaz (Eds.), *Handbook of Research on Effective Advertising Strategies in the Social Media Age* (pp. 125–146). IGI Global. doi:10.4018/978-1-4666-8125-5.ch007

Çeltek, E. (2019). Advantages of Augmented Reality, Virtual Reality, QR Code, Near Field Communication, Geo-Tagging, Geo-Fencing, and Geo-Targeting for Marketing Tourism. In D. Simões (Ed.), *Smart Marketing With the Internet of Things* (pp. 94–113). IGI Global. doi:10.4018/978-1-5225-5763-0.ch005

Cervantes, O., Gutiérrez, E., Gutiérrez, F., & Sánchez, J. A. (2016). SOCIAL METRICS APPLIED to SMART TOURISM. *ISPRS Annals of the Photogrammetry, Remote Sensing and Spatial Information Sciences, 4*(4W1), 117-124. 10.5194/isprs-annals-IV-4-W1-117-2016

Chalkia, E., Bekiaris, E., & Madrid, R. I. (2015). Evaluating all-inclusive ICT with developers, end users and stakeholders. In *International Conference on Human Aspects of IT for the Aged Population* (pp. 157-165). Springer. 10.1007/978-3-319-20892-3_16

Chanana, P., Paul, R., Balakrishnan, M., & Rao, P. V. M. (2017). Assistive technology solutions for aiding travel of pedestrians with visual impairment. *Journal of Rehabilitation and Assistive Technologies Engineering*, *4*, 2055668317725993. doi:10.1177/2055668317725993 PMID:31186934

Chang, F., Wang, Y., Jang, Y., & Wang, C. (2012). Factors Associated With Quality of Life Among People With Spinal Cord Injury: Application of the International Classification of Functioning, Disability and Health Model. *Archives of Physical Medicine and Rehabilitation*, *93*(12), 2264–2271. doi:10.1016/j.apmr.2012.06.008 PMID:22728700

Chang, G., & Caneday, L. (2011). Web-based GIS in tourism information search: Perceptions, tasks, and trip attributes. *Tourism Management*, *32*(6), 1435–1437. doi:10.1016/j.tourman.2011.01.006

Chang, Y., & Chen, C. (2012). Meeting the needs of disabled air passengers: Factors that facilitate help from airline and airports. *Tourism Management*, *33*(3), 529–536. doi:10.1016/j.tourman.2011.06.002

Chanjaraspong, T. (2016). Acceptance factors for the use of video call via smartphones by blind people. *Kasetsart Journal of Social Sciences*, 1–7. doi:10.1016/j.kjss.2016.02.001

Charara, S. (2016). *Hussein Chalayan has turned emotion sensing into a thing of beauty: Biometric wearables at the Design Museum's new Fear and Love exhibition*. Accessed March 7, 2020, from https://www.wareable.com/wearable-tech/hussein-chalayan-emotions-design-museum-3603

Charlier, R., & Chaineux, M. (2009). The healing sea: A sustainable coastal ocean resource: Thalassotherapy. *Journal of Coastal Research*, *25*(4), 838–856. doi:10.2112/08A-0008.1

Chauhan, S.; Agarwal, N.; Kar, A. K. (2016). Addressing big data challenges in smart cities: a systematic literature review. *Info, 18*(4), 73-90.

Chelini, M. (2012). Novas tecnologias para... novas (?) expografias. *Revista Museologia & Interdisciplinaridade*, *1*(2), 59–71. doi:10.26512/museologia.v1i2.12655

Chiou, C. K., Tseng, J. C., Hwang, G. J., & Heller, S. (2010). An adaptive navigation support system for conducting context-aware ubiquitous learning in museums. *Computers & Education*, *55*(2), 834–845. doi:10.1016/j.compedu.2010.03.015

Choi, I. Y., Ryu, Y. U., & Kim, J. K. (2019). A recommender system based on personal constraints for smart tourism city. *Asia Pacific Journal of Tourism Research*, 1–14. Advance online publication. doi:10.1080/10941665.2019.1592765

Chun, R. (2016). *The SFMOMA's New App Will Forever Change How You Enjoy Museums*. Retrieved October 29, 2016, from https://www.wired.com/2016/05/sfmoma-audio-tour-app/

City of Sydney. (2002). *Disability Action Plan 2002-2005*. Retrieved from https://www.cityofsydney.nsw.gov.au/AboutSydney/CBDDisabledAccess/Default.asp

City of Sydney. (2012). *City of Sydney workforce strategy 2012-2015*. Retrieved from: https://www.cityofsydney.nsw.gov.au/surveys-case-studies-reports/community-indicators-report

City of Sydney. (2015). *City of Sydney Wellbeing survey 2015*. Retrieved from: https://www.cityofsydney.nsw.gov.au/surveys-case-studies-reports/community-indicators-report

City of Sydney. (2016). *Digital Strategy* Retrieved from: https://www.cityofsydney.nsw.gov.au/__data/assets/pdf_file/0005/288167/Digital-Strategy.pdf

City of Sydney. (2017a). *A city for all: inclusion (disability) action plan 2017-21*. Retrieved from Sydney:https://www.cityofsydney.nsw.gov.au/__data/assets/pdf_file/0003/287121/Inclusion-Disability-Action-Plan-2017-2021_Accessible.pdf

City of Sydney. (2017b). *Sustainable Sydney 2030: community strategic plan 2017-2021*. Retrieved from Sydney: https://www.cityofsydney.nsw.gov.au/__data/assets/pdf_file/0011/288173/Adopted-Sustainable-Sydney-2030_Accessible-Version.pdf

City of Sydney. (2019a). *Economic development strategy - Sydney's economy: global City, local action*. Author.

City of Sydney. (2019b). *Technology start-ups action plan*. Retrieved from: https://www.cityofsydney.nsw.gov.au/business-economy/tech-startups

City of Sydney. (2019c). *Accessibility Map*. Retrieved from: https://www.cityofsydney.nsw.gov.au/explore/getting-around/accessibility/accessibility-map

City of Sydney. (2020). *Sydney maps*. Retrieved from: https://www.cityofsydney.nsw.gov.au/learn/maps

City of Sydney. (2020a). *Smart City Strategic Framework*. Retrieved from Sydney: https://www.cityofsydney.nsw.gov.au/council/your-say/smart-city-strategic-framework

City of Sydney. (2020b). *Wayfinding signage (for blind and vision impaired)*. Retrieved from https://www.cityofsydney.nsw.gov.au/vision/sustainable-sydney-2030/transport-and-access/liveable-green-network/wayfinding-signage

City of Sydney. (2020c). *Accessible Parking map*. Retrieved from: https://cityofsydney.maps.arcgis.com/apps/Minimalist/index.html?appid=c6d11d61dc994bce8715335e924f5151

Claudio, G., Luca, G., & Luce, L. M. (2017). Interaction design for cultural heritage. A robotic cultural game for visiting the museum's inaccessible areas. *The Design Journal, 20*(sup1), S3925-S3934.

Clift, S., & Page, S. (1996). *Health And The International Tourist* (1st ed.). Routledge.

Cloquet, I., Palomino, M., Shaw, G., Stephen, G., & Taylor, T. (2018). Disability, social inclusion and the marketing of tourist attractions. *Journal of Sustainable Tourism, 26*(2), 221–237. doi:10.1080/09669582.2017.1339710

Cohen, S. A., Prayag, G., & Moital, M. (2014). Consumer behaviour in tourism: Concepts, influences and opportunities. *Current Issues in Tourism, 17*(10), 872–909. doi:10.1080/13683500.2013.850064

Colace, F., de Santo, M., Lombardi, M., & Santaniello, D. (2019). Chars: A cultural heritage adaptive recommender system. *TESCA 2019- 1st ACM International Workshop on Technology Enablers and Innovative Applications for Smart Cities and Communities, Co-Located with the 6th ACM International Conference on Systems for Energy-Efficient Buildings, Cities, and Transportation, ACM BuildSys 2019*, 58-61. 10.1145/3364544.3364830

Colace, F., Santo, M. D., Lemma, S., Lombardi, M., Rossi, A., Santoriello, A., ... Vigorito, M. (2016). How to describe cultural heritage resources in the web 2.0 era? *11th International Conference on Signal-Image Technology and Internet-Based Systems, SITIS 2015*, 809-815. 10.1109/SITIS.2015.50

Colorado Pástor, B. A., Fois Lugo, M. M., Leyva Vázquez, M., & Hechavarría Hernández, J. R. (2020). Proposal of a technological ergonomic model for people with disabilities in the public transport system in guayaquil. In *International Conference on Applied Human Factors and Ergonomics* (pp. 831-843). Springer. 10.1007/978-3-030-19135-1_81

Comisión Europea (2014). *Economic impact and travel patterns of accessible tourism in Europe – Final report*. Service Contract SI2. ACPROCE052481700 – European Commission, DG Enterprise and Industry.

Connolly, D. J., Olsen, M. D., & Moore, R. G. (1998). The internet as a distribution channel. *The Cornell Hotel and Restaurant Administration Quarterly, 39*(4), 42–54. doi:10.1177/001088049803900408

Coromina, L., & Camprubí, R. (2016). Analysis of tourism information sources using a Mokken Scale perspective. *Tourism Management*, *56*, 75–84. doi:10.1016/j.tourman.2016.03.025

Cortés Bello, R., & Vargas Martínez, E. E. (2018). Prospectiva en Agencias de Viajes: Una Revisión de la Literatura (Prospective in Travel Agencies: A Revision of Literature). *Turismo e Sociedade*, *22*, 45–64. doi:10.18601/01207555.n22.03

Coughlan, J., Manduchi, R., & Shen, H. (2006). Computer vision-based terrain sensors for blind wheelchair users. In *International Conference on Computers for Handicapped Persons* (pp. 1294-1297). Springer. 10.1007/11788713_186

Craig, A., Tran, Y., & Middleton, J. (2009). Psychological morbidity and spinal chord injury: A systematic review. *Spinal Cord*, *47*(2), 108–114. doi:10.1038c.2008.115 PMID:18779835

Crawford, D., & Godbey, G. (1987). Reconceptualizing barriers to family leisure. *Leisure Sciences*, *9*(2), 119–127. doi:10.1080/01490408709512151

Crawford, J. L., & Stodolska, M. (2008). Constraints experienced by elite athletes with disabilities in Kenya, with implications for the development of a new hierarchical model of constraints at the societal level. *Journal of Leisure Research*, *40*(1), 128–155. doi:10.1080/00222216.2008.11950136

Crismariu, O. (2017). Accessible Tourism – An Expression of Social Responsibility. *Cactus*, *2*, 33.

Crow, K. L. (2008). Four types of disabilities: Their impact on online learning. *TechTrends*, *52*(1), 51–55. doi:10.100711528-008-0112-6

Cunliffe, D., Kritou, E., & Tudhope, D. (2001). Usability evaluation for museum web sites. *Museum Management and Curatorship*, *19*(3), 229–252. doi:10.1080/09647770100201903

Cupples, M. E., Hart, P. M., Johnston, A., & Jackson, A. J. (2012). Improving healthcare access for people with visual impairment and blindness. *BMJ (Clinical Research Ed.)*, *344*(1), 1–5. doi:10.1136/bmj.e542 PMID:22290101

Curran, K., Walters, N., & Robinson, D. (2007). Investigating the problems faced by older adults and people with disabilities in online environments. *Behaviour & Information Technology*, *26*(6), 447–453. doi:10.1080/01449290600740868

Czikk, J. (2013). *New ScopifyROM App Gives Royal Ontario Museum Visitors an Augmented Experience*. Retrieved March 15, 2020, from https://betakit.com/new-scopifyrom-app-gives-royal-ontario-museum-visitors-an-augmented-experience/

D'Andrea, F. M., & Siu, Y. T. (2015). Students with visual impairments: Considerations and effective practices for technology use. In D. Edyburn (Ed.), Efficacy of Assistive Technology Interventions (Advances in Special Education Technology, Volume 1) (pp. 111-138). Emerald Group Publishing Limited.

D'aniello, G., Gaeta, M., & Reformat, M. Z. (2017). Collective perception in smart tourism destinations with rough sets. *2017 3rd IEEE International Conference on Cybernetics, CYBCONF 2017*. 10.1109/CYBConf.2017.7985765

da Costa Liberato, P. M., Alén-González, E., & de Azevedo Liberato, D. F. V. (2018). Digital technology in a smart tourist destination: The case of Porto. *Journal of Urban Technology*, *25*(1), 75–97. doi:10.1080/10630732.2017.1413228

Dadwal, S. S., & Hassan, A. (2016). The Augmented Reality Marketing. In Mobile Computing and Wireless Networks: Concepts, Methodologies, Tools, and Applications (pp. 63–80). IGI Global. doi:10.4018/978-1-4666-8751-6.ch004

Dalli, A., & Bri, S. (2017). Design of electronic ticket system for smart tourism. *12th International Conference on Signal Image Technology and Internet-Based Systems, SITIS 2016*, 490-492. 10.1109/SITIS.2016.82

Daniels, M. S., Drogen-Rodgers, E. B., & Wiggins, B. P. (2005). 'Travel Tales' an interpretive analysis of constraints and negotiations to pleasure travel as experienced by persons with physical disabilities. *Tourism Management*, *26*(6), 919–930. doi:10.1016/j.tourman.2004.06.010

Darcy, S. (2003). *Riding the Wave of Tourism and Hospitality Research*. Paper presented at the CAUTHE Conference - Disabling Journeys: The tourism patterns of people with impairments in Australia, Lismore: Southern Cross University.

Darcy, S. (2010). Inherent complexity: Disability, accessible tourism and accommodation information preferences. *Tourism Management, 31*(6), 816–826. doi:.tourman.2009.08.010 doi:10.1016/j

Darcy, S. (2014). Sydney for All. In UNWTO (Ed.), Manual on Accessible Tourism for All: Principles, Tools and Good Practices (pp. 50-66). Madrid: UNWTO, the Spanish ONCE Foundation and the European Network on Accessible Tourism (ENAT).

Darcy, S., & Buhalis, D. (2011). Introduction: from disabled tourists to accessible tourism. *Accessible tourism: Concepts and issues*, 1-20.

Darcy, S., & Cameron, B. (2009a). *Uniquest Award - Highly Commended Sydney for All Web portal tourism innovation*. https://www.uniquest.com.au/index.php?sectionID=863

Darcy, S., & Cameron, B. (2009b). *Vision Australia - Making a Difference award for Sydney for All Web portal*. http://www.visionaustralia.org/info.aspx?page=2267#TourismNSW

Darcy, S., & Cameron, B. (2013). Sydney for All - Collaborative Marketing of Accessible Attractions within the Sydney Destination Management Precinct. In I. Ambrose & A. Barragán Iturriaga (Eds.), Exemplars in Global Accessible Tourism Management. Geneva: UN World Tourism Organisation in conjunction with the European Network For Accessible Tourism and the ONCE Foundation for Cooperation and Social Inclusion For PWD.

Darcy, S., Cameron, B., Dwyer, L., & Taylor, T. (2008b). *Research Note: Understanding accessible spaces and places*. www.sydneyforall.com

Darcy, S., Cameron, B., Dwyer, L., Wong, E., & Thomson, A. (2008a). *Visitor Accessibility in Urban Centres (Project # 90064)*. Retrieved from Gold Coast: https://www.researchgate.net/publication/228716596_Visitor_Accessibility_in_Urban_Centres

Darcy, S. (1998). *Anxiety to access: Tourism patterns and experiences of New South Wales people with a physical disability*. Tourism New South Wales.

Darcy, S. (2001). A games for everyone?: Planning for disability and access at the Sydney 2000 Games. *Disability Studies Quarterly, 21*(3), 70–84. doi:10.18061/dsq.v21i4.319

Darcy, S. (2003). The politics of disability and access: The Sydney 2000 Games experience. *Disability & Society, 18*(6), 737–757. doi:10.1080/0968759032000119497

Darcy, S. (2010). Inherent Complexity: Disability, accessible tourism and accommodation information preferences. *Tourism Management, 31*(6), 816–826. doi:10.1016/j.tourman.2009.08.010

Darcy, S. (2011). Developing Sustainable Approaches to Accessible Accommodation Information Provision: A Foundation for Strategic Knowledge Management. *Tourism Recreation Research, 36*(2), 141–157. doi:10.1080/02508281.2011.11081315

Darcy, S. (2012). (Dis)Embodied Air Travel Experiences: Disability, Discrimination and the Affect of a Discontinuous Air Travel Chain. *Journal of Hospitality and Tourism Management, 19*(1), E8. doi:10.1017/jht.2012.9

Darcy, S. (2013). Sydney for All: developing an equality of experience for tourists with disabilities. In I. Cooper, J. Smith, & K. Moss (Eds.), *The World Leisure International Innovation Prize Winning Projects 2006-2012* (pp. 97–103). World Leisure Organisation.

Darcy, S., & Almond, B. (2016). Case study: Sydney's Cultural Ribbon. In F. Aragall, P. Neumann, & S. Sagramola (Eds.), *Design for All in Tourism Destinations - European concept for accessibility* (pp. 76–81). European Concept for Accessibility.

Darcy, S., & Buhalis, D. (2011). Conceptualising disability. In D. Buhalis & S. Darcy (Eds.), *Accessible tourism: Concepts and issues* (pp. 21–45). Channel View Publications.

Darcy, S., & Burke, P. F. (2018). On the road again: The barriers and benefits of automobility for PWD. *Transportation Research Part A, Policy and Practice*, *107*(Supplement C), 229–245. doi:10.1016/j.tra.2017.11.002

Darcy, S., & Cameron, B. (2009/2010). *Competitive Grant Application $30,000 - Urban accessibility and hotel accommodation*. Academic Press.

Darcy, S., Cameron, B., & Pegg, S. (2010). Acessible Tourism and Sustainability: A discussion and case study. *Journal of Sustainable Tourism*, *18*(4), 515–537. doi:10.1080/09669581003690668

Darcy, S., & Dickson, T. J. (2009). A Whole-of-Life approach to tourism: The case for accessible tourism experiences. *Journal of Hospitality and Tourism Management*, *16*(1), 32–44. doi:10.1375/jhtm.16.1.32

Darcy, S., McKercher, B., & Schweinsberg, S. (2020). From tourism and disability to accessible tourism: A perspective article. *Tourism Review*, *75*(1), 140–144. doi:10.1108/TR-07-2019-0323

Darcy, S., & Pegg, S. (2010). Towards Strategic Intent: Perceptions of disability service provision amongst hotel accommodation managers. *International Journal of Hospitality Management*, *30*(2), 468–476. doi:10.1016/j.ijhm.2010.09.009

Darcy, S., Yerbury, H., & Maxwell, H. (2019). Disability citizenship and digital capital: The case of engagement with a social enterprise telco. *Information Communication and Society*, *22*(4), 538–553. doi:10.1080/1369118X.2018.1548632

Daruwalla, P., & Darcy, S. (2005). Personal and Societal Attitudes to Disability. *Annals of Tourism Research*, *32*(3), 549–570. doi:10.1016/j.annals.2004.10.008

Dattolo, A., Luccio, F. L., & Pirone, E. (2016). Web accessibility recommendations for the design of tourism websites for people with autism spectrum disorders. *International Journal on Advances in Life Sciences*, *8*(3–4), 297–308.

Davies, A., & Christie, N. (2017). An exploratory study of the experiences of wheelchair users as aircraft passenger – implications for policy and practice. *IATSS Research*, *4*(2), 89–93. doi:10.1016/j.iatssr.2017.05.003

Davies, D. K., Stock, S. E., Holloway, S., & Wehmeyer, M. L. (2010). Evaluating a GPS-based transportation device to support independent bus travel by people with intellectual disability. *Intellectual and Developmental Disabilities*, *48*(6), 454–463. doi:10.1352/1934-9556-48.6.454 PMID:21166550

De Lemos, F. F., Vieira, D., & Pinho, T. (2020). Digital Marketing Strategies for Accessible Tourism. In Digital Marketing Strategies for Tourism, Hospitality, and Airline Industries (pp. 71-92). IGI Global. doi:10.4018/978-1-5225-9783-4.ch004

De Oliveira, L. C., Andrade, A. O., De Oliveira, E. C., Soares, A., Cardoso, A., & Lamounier, E. (2017). Indoor navigation with mobile augmented reality and beacon technology for wheelchair users. *2017 IEEE EMBS International Conference on Biomedical & Health Informatics (BHI)*, 37–40. 10.1109/BHI.2017.7897199

Decrop, A. (2006). *Vacation Decision Making*. Cabi Publishing. doi:10.1079/9781845930400.0000

Dekelver, J., Daems, J., Solberg, S., Bosch, N., Van de Perre, L., & De Vliegher, A. (2015). Viamigo: A digital travel assistant for people with intellectual disabilities: Modeling and design using contemporary intelligent technologies as a support for independent traveling of people with intellectual disabilities. In *2015 6th International Conference on Information, Intelligence, Systems and Applications (IISA)* (pp. 1-6). IEEE.

Del Chiappa, G., & Baggio, R. (2015). Knowledge transfer in smart tourism destinations: Analyzing the effects of a network structure. *Journal of Destination Marketing & Management*, *4*(3), 145–150. doi:10.1016/j.jdmm.2015.02.001

Deloitte and Touche. (2003). *Access to Assistive Technology in the European Union*. European Commission.

DEO. (2005). *Disability definitions and related links*. https://www.doe.mass.edu/sped/definitions.html

Department for Work and Pensions. (2018). *Spending power of disabled people and their families estimated at £249 billion*. https://www.gov.uk/government/news/purple-tuesday-is-the-uks-first-accessible-shopping-day

Deque Systems. (2020a). *1.3.6 Identify Purpose (AAA)*. Retrieved from https://dequeuniversity.com/resources/wcag2.1/1.3.6-identify-purpose

Deque Systems. (2020b). *1.4.10 Reflow (AA)*. Retrieved from https://dequeuniversity.com/resources/wcag2.1/1.4.10-reflow

Design for all foundation. (2019). *Design for all*. Retrieved from http://www.designforall.org/

Devigne, L., Babel, M., Nouviale, F., Narayanan, V. K., Pasteau, F., & Gallien, P. (2017). Design of an immersive simulator for assisted power wheelchair driving. *2017 International Conference on Rehabilitation Robotics (ICORR)*, 995–1000. 10.1109/ICORR.2017.8009379

Deville, E., & Kastenholz, E. (2018). Accessible tourism experiences: The voice of people with visual disabilities. *Journal of Policy Research in Tourism, Leisure & Events*, *10*(3), 265–285. doi:10.1080/19407963.2018.1470183

Dhalwal, S. K., & Juyal, S. L. (2017). Assistive technology as a tool of making difference in the life of persons with visual impairment. *Indian Journal of Health and Wellbeing*, *8*(4), 319–321.

Didehbani, N., Allen, T., Kandalaft, M., Krawczyk, D., & Chapman, S. (2016). Virtual Reality Social Cognition Training for children with high functioning autism. *Computers in Human Behavior*, *62*, 703–711. doi:10.1016/j.chb.2016.04.033

Dillenburg, E. (2011). What, if anything, is a museum. *The Exhibitionist*, *30*(1), 8–13.

Dinis, M. G., Eusébio, C., & Breda, Z. (2020). Assessing social media accessibility: The case of the Rock in Rio Lisboa music festival. *International Journal of Event and Festival Management*, *11*(1), 26–46. doi:10.1108/IJEFM-02-2019-0012

Disability Accessibility Centre. (2020). *Audit and Accreditation*. http://digitalaccessibilitycentre.org/index.php/audit-and-accreditation

Disability Rights Commission. (2004). *The Web: Access and Inclusion for Disabled People*. London: TSO. Available at: http://www.drc-gb.org/publicationsandreports/report.asp 11.

Disabled People's Association (DPA). (2015). *Glossary of disability terminology*. https://www.dpa.org.sg/wp-content/uploads/2015/10/DPA-Disability-Glossary-FINAL.pdf

Disabled World. (2019). *Disabilities: Definition, Types and Models of Disability*. Accessed on April 18, 2020, from https://www.disabled-world.com/disability/types/

Domingos, T., & Braga, E. (2014). Meaning of aromatherapy massage in mental health. *Acta Paulista de Enfermagem*, *27*(6), 579–584. doi:10.1590/1982-0194201400094

Domínguez Vila, T., Alén González, E., & Darcy, S. (2017). Website accessibility in the tourism industry: An analysis of official national tourism organization websites around the world. *Disability and Rehabilitation*, *40*(24), 2895–2906. doi:10.1080/09638288.2017.1362709 PMID:28793789

Domínguez Vila, T., Alén González, E., & Darcy, S. (2018). Accessible tourism online resources: A Northern European perspective. *Scandinavian Journal of Hospitality and Tourism*, *19*(2), 140–156. doi:10.1080/15022250.2018.1478325

Domínguez Vila, T., Alén González, E., & Darcy, S. (2020). Accessibility of tourism websites: The level of countries' commitment. *Universal Access in the Information Society*, *19*(2), 331–346. doi:10.100710209-019-00643-4

Domínguez Vila, T., & Brea, J. A. F. (2009). A new challenge: The content and the accessibility on the Spanish tourist web [Un nuevo desafío: El contenido y la accesibilidad al contenido de las web turísticas españolas]. *Revista Galega de Economia*, *18*(1), 1–23.

Domínguez Vila, T., Darcy, S., & Alén González, E. (2015). Competing for the disability tourism market – A comparative exploration of the factors of accessible tourism competitiveness in Spain and Australia. *Tourism Management*, *47*, 261–272. doi:10.1016/j.tourman.2014.10.008

Domínguez, T., Fraiz, J. A., & Alén, E. (2013). Economic profitability of accessible tourism for the tourism sector in Spain. *Tourism Economics*, *19*(6), 1385–1399. doi:10.5367/te.2013.0246

Doolin, B., Burgess, L., & Cooper, J. (2002). Evaluating the use of the Web for tourism marketing: A case study from New Zealand. *Tourism Management*, *23*(5), 557–561. doi:10.1016/S0261-5177(02)00014-6

Drews, W., & Schemer, C. (2010). eTourism for All? Online Travel Planning of Disabled People. *Information and Communication Technologies in Tourism*, *2010*, 507–518. doi:10.1007/978-3-211-99407-8_42

Duckett, P. S., & Pratt, R. (2001). The Researched Opinions on Research: Visually impaired people and visual impairment research. *Disability & Society*, *16*(6), 815–835. doi:10.1080/09687590120083976

Dwyer, L., & Darcy, S. (2008). Economic contribution of disability to tourism in Australia. In S. Darcy, B. Cameron, L., Dwyer, T., Taylor, E., Wong, & A. Thomson (Eds.), Technical report 90040: Visitor accessibility in urban centres (pp. 15–21). Gold Coast, Australia: Sustainable Tourism Cooperative Research Centre.

Dyson, M. C., & Moran, K. (2000). Informing the design of web interfaces to museum collections. *Museum Management and Curatorship*, *18*(4), 391–406. doi:10.1080/09647770000501804

ECLAC. (2020). *Recovery measures for the tourism sector in Latin America and the Caribbean present an opportunity to promote sustainability and resilience*. Retrieved from: https://repositorio.cepal.org/bitstream/handle/11362/45767/4/S2000440_en.pdf

Economic Commission for Latin America and the Caribbean (ECLAC). (2013). *Social Panorama of Latin America 2012*. Retrieved from: https://www.cepal.org/en/publications/1248-social-panorama-latin-america-2012

Eichhorn, V., & Buhalis, D. (2011). Accessibility: A key objective for the tourism industry. Accessible tourism: Concepts and issues. In Accessible tourism: Concepts and issues. Bristol: Channel View Publications.

Eichhorn, V., & Buhalis, D. (2011). Accessibility: A Key Objective for the Tourism Industry. In D. Buhalis & S. Darcy (Eds.), *Accessible tourism: Concepts and issues* (pp. 46–62). Channel View Publications.

Eichhorn, V., Miller, G., Michopoulou, E., & Buhalis, D. (2008). Enabling access to tourism through information schemes? *Annals of Tourism Research*, *35*(1), 189–210. doi:10.1016/j.annals.2007.07.005

Eichhorn, V., Miller, G., & Tribe, J. (2013). Tourism: A site of resistance strategies of individuals with a disability. *Annals of Tourism Research*, *43*, 578–600. doi:10.1016/j.annals.2013.03.006

Emrouzeh, M. P., Dewar, K., Fleet, G., & Bourgeois, Y. (2017). Implementing ICT for tourists with disabilities. In *Proceedings of the 2017 International Conference on E-Education, E-Business and E-Technology* (vol. 6, pp. 50–53). 10.1145/3141151.3141152

ENAT - European Network for Accessible Tourism. (2018). *What is "Accessible Tourism"?* Retrieved from https://www.accessibletourism.org/?i=enat.en.faq.744

Encalada, L., Boavida-Portugal, I., Ferreira, C. C., & Rocha, J. (2017). Identifying tourist places of interest based on digital imprints: Towards a sustainable smart city. *Sustainability (Switzerland), 9*(12). doi:.. doi:10.3390u9122317

Ente de Turismo de la Ciudad de Buenos Aires. (2018). *Noticia: Buenos Aires presentó sus avances como destino inteligente*. Retrieved from: https://turismo.buenosaires.gob.ar/es/article/noticia-buenos-aires-present%C3%B3-sus-avances-como-destino-inteligente

Equality Act 2010–UK Equality legislation. https://www.legislation.gov.uk/ukpga/2010/15/contents

ESI. (2018). *Smarter Cities 2025*. Retrieved from https://econsultsolutions.com/esi-thoughtlab/smarter-cities-2025/

Espadinha, C., Pereira, L. M., Da Silva, F. M., & Lopes, J. B. (2011). Accessibility of Portuguese Public Universities' sites. *Disability and Rehabilitation, 33*(6), 475–485. doi:10.3109/09638288.2010.498554 PMID:20594034

Esper, F. (2020). Enfoques, modelos y buenas prácticas internacionales en DTI: los casos de Buenos Aires y la Red Argentina. In *Sistemas de inteligencia y modelos de trabajo en red para una gestión turística inteligente en Argentina*. INVAT-TUR, Webinar.

EUR-Lex. (2018). *Accessibility of public sector websites and mobile apps*. Retrieved from https://eur-lex.europa.eu/eli/dir/2016/2102/oj

European Commision (EU). (2015). *Mapping and Performance Check of the Supply of Accessible Tourism Services*. Retrieved from: https://op.europa.eu/en/publication-detail/-/publication/f0948779-1303-481a-a2d8-5aa043035b81

European Commission. (2014). *Economic Impact and Travel Patterns of Accessible Tourism in Europe – Final Report*. Retrieved from: http://ec.europa.eu/DocsRoom/documents/7221/attachments/1/translations/en/renditions/native

European Commission. (2015). *Economic impact and travel patterns of accessible tourism in Europe Final report*. Retrieved From: https://op.europa.eu/en/publication-detail/-/publication/d04fbb3d-8d15-4fe3-b554-844bde75c171/language-en/format-PDF/sourcesearch#

European Commission. (2018). *Access city award*. Retrieved from https://ec.europa.eu/social/main.jsp?catId=1141

European Network for Accessible Tourism. (2008). Rights of Tourists with Disabilities in the European Union Framework. *European Network for Accessible Tourism.*

European Network for Accessible Tourism. (2020). *eAccess+. A Thematic Network for Improving Implementation of eAccessibility throughout Europe*. https://www.accessibletourism.org/?i=enat.en.enat_projects_and_good_practices.1038

European Telecommunications Standards Institute. (2018). *EN 301 549 V2.1.2 (2018-08) - Accessibility requirements for ICT products and services*. Author.

European Union. (2020). Guide for applicants. *Competition For The European Capital Of Smart Tourism 2020.*

Eurostat. (2018). *1 in 4 people in the EU have a long-term disability*. Retrieved from https://ec.europa.eu/eurostat/web/products-eurostat-news/-/EDN-20181203-1?inheritRedirect=true

Falk, J. H., & Dierking, L. D. (2016). *The museum experience revisited*. Routledge.

Fernandes, H., Costa, P., Filipe, V., Paredes, H., & Barroso, J. (2019). A review of assistive spatial orientation and navigation technologies for the visually impaired. *Universal Access in the Information Society, 18*(1), 155–168. doi:10.100710209-017-0570-8

Figueiredo, E., Eusébio, C., & Kastenholz, E. (2012). How diverse are tourists with disabilities? A pilot study on accessible leisure tourism experiences in Portugal. *International Journal of Tourism Research*, *14*(6), 531–550. doi:10.1002/jtr.1913

Figueredo, M., Cacho, N., Thome, A., Cacho, A., Lopes, F., & Araujo, M. (2017). Using social media photos to identify tourism preferences in smart tourism destination. *2017 IEEE International Conference on Big Data, Big Data 2017*, 4068-4073. 10.1109/BigData.2017.8258423

Finkelstein, V. (2004). Representing Disability. In J. Swain, S. French, C. Barnes, & C. Thomas (Eds.), *Disabling Barriers – Enabling Environments* (2nd ed., pp. 13–20). Sage.

Fleischer, A., & Pizam, A. (2002). Tourism constraints among Israeli seniors. *Annals of Tourism Research*, *29*(1), 106–123. doi:10.1016/S0160-7383(01)00026-3

Florim, de L.M., Viera, D., & Pinho, T. (2019). Digital Marketing Strategies for Accessible Tourism. In J.D. Santos & Ó. Lima Silva (Eds.), *Digital Marketing Strategies for Tourism, Hospitality and Airline Industries*. IGI Global.

Fogli, D., Sansoni, D., Trivella, E., Arenghi, A., & Passamani, I. (2018). Advanced Interaction Technologies for Accessible and Engaging Cultural Heritage. In *Lecture Notes of the Institute for Computer Sciences, Social-Informatics and Telecommunications Engineering* (pp. 364–373). LNICST. doi:10.1007/978-3-319-76111-4_36

Fontes, F., Martins, B. S., & Hespanha, P. (2014). The emancipation of disability studies in Portugal. *Disability & Society*, *29*(6), 849–862.

Françoso, M. T., Costa, D. C., Valin, M. M., & Amarante, R. R. (2013). Free software for development of Web GIS in tourism accessibility. *Applied Mechanics and Materials*, *256*, 2953–2956.

Freeman, A., Becker, S. A., Cummins, M., McKelroy, E., Giesinger, C., & Yuhnke, B. (2016). *Nmc horizon report: 2016 museum edition*. The New Media Consortium.

Freund, P. (2001). Bodies, Disability and Spaces: The social model and disabling spatial organisations. *Disability & Society*, *16*(5), 689–706. doi:10.1080/09687590120070079

Fundación, O. N. C. E. (2017). *Observatorio de Accesibilidad Universal del turismo en España*. Retrieved from: http://www.drtturismo.com/sites/default/files/observatorio_de_turismo_110917_4.pdf

G3ict & World Enabled. (2017). *Smart Cities for All: Results of the 2016 Global Expert Survey*. Retrieved from: https://smartcities4all.org/wp-content/uploads/2017/06/Smart-Cities-for-All-A-Vision-for-an-Inclusive-Accessible-Urban-Futur...-min.pdf

Gamble, M. J., Dowler, D. L., & Orslene, L. E. (2006). Assistive technology: Choosing the right tool for the right job. *Journal of Vocational Rehabilitation*, *24*(2), 73–80.

Garau, C. (2014). Smart paths for advanced management of cultural heritage. *Regional Studies, Regional Science*, *1*(1), 286–293. doi:10.1080/21681376.2014.973439

García-Crespo, Á., González-Carrasco, I., López-Cuadrado, J. L., Villanueva, D., & González, Á. (2016). CESARSC: Framework for creating cultural entertainment systems with augmented reality in smart cities. *Computer Science and Information Systems*, *13*(2), 395–425. doi:10.2298/CSIS150620006G

Gass, E., & Gass, G. (2001). Thermoregulatory responses to repeated warm water immersion in subjects who are paraplegic. *Spinal Cord*, *39*(3), 149–155. doi:10.1038j.sc.3101117 PMID:11326325

Gautam, B. P., Asami, H., Batajoo, A., & Fujisaki, T. (2016). Regional revival through IoT enabled smart tourism process framework (STPF): A proposal. *2016 Joint 8th International Conference on Soft Computing and Intelligent Systems and 2016 17th International Symposium on Advanced Intelligent Systems, SCIS-ISIS 2016,* 743-748. 10.1109/SCIS-ISIS.2016.0162

Geehan, T., & Suen, L. (1993). User acceptance of advanced traveler information systems for elderly and disabled travellers in Canada. In *Proceedings of VNIS'93-Vehicle Navigation and Information Systems Conference* (pp. 389-392). IEEE. 10.1109/VNIS.1993.585655

Gell, N. M., Rosenberg, D. E., Carlson, J., Kerr, J., & Belza, B. (2015). Built environment attributes related to GPS measured active trips in mid-life and older adults with mobility disabilities. *Disability and Health Journal, 8*(2), 290–295. doi:10.1016/j.dhjo.2014.12.002 PMID:25637503

Gerino, A., Picinali, L., Bernareggi, C., Alabastro, N., & Mascetti, M. (2015). Towards large scale evaluation of novel sonification techniques for non-visual shape exploration. In *Proceedings of the 17th International ACM SIGACCESS Conference on Computers & Accessibility*. ACM. 10.1145/2700648.2809848

GfK. (2015). Economic Impact on Travel Patterns of Accessible Tourism in Europe – Final Report. GfK, University of Surrey, The Neumann Consulting and PRO Solutions, Surrey.

Ghapanchi, A. H., & Aurum, A. (2011). Antecedents to IT Personnel's Intentions to Leave: A Systematic Literature Review. *Journal of Systems and Software, 84*(2), 238–249. doi:10.1016/j.jss.2010.09.022

Giannoulis, G., Coliou, M., Kamarinos, G. S., Roussou, M., Trahanias, P., Argyros, A., . . . Haehnel, D. (2001). Enhancing museum visitor access through robotic avatars connected to the web. Proc of museums and the web.

Gibson, A., & O'Rawe, M. (2018). Virtual Reality as a Travel Promotional Tool: Insights from a Consumer Travel Fair. In Augmented Reality and Virtual Reality (pp. 93–107). Springer.

Giffinger, R., & Gudrun, H. (2010). Smart cities ranking: an effective instrument for the positioning of the cities? *ACE: Architecture, City and Environment, 4*(12), 7-26.

Giffinger, R., Fertner, C., Kramar, H., Meijers, E., & Pichler-Milanovic, N. (2007). *Smart cities: Ranking of European medium-sized cities*. Centre of Regional Science. Retrieved from: http://www.smart-cities.eu/download/smart_cities_final_report.pdf

Gilbert, C. (1999). Les musées ouvrent leurs portes à tous les publics. *Espaces (Paris), 161*, 26–29.

Giovannella, C., Iosue, A., Moggio, F., Rinaldi, E., & Schiattarella, M. (2013). User experience of kinect based applications for smart city scenarios integrating tourism and learning. *2013 IEEE 13th International Conference on Advanced Learning Technologies, ICALT 2013,* 459-460. 10.1109/ICALT.2013.142

Gladwell, N., & Bedini, L. (2004). In search of lost leisure: The impact of caregiving on leisure travel. *Tourism Management, 25*(6), 685–693. doi:10.1016/j.tourman.2003.09.003

Global Wellness Institute. (2018a). *Global Wellness Economy Monitor – October 2018.* Available at: https://globalwellnessinstitute.org/wp-content/uploads/2019/04/GWIWellnessEconomyMonitor2018_042019.pdf

Global Wellness Institute. (2018b). *Global Wellness Tourism Economy – November 2018.* Available at: https://globalwellnessinstitute.org/industry-research/global-wellness-tourism-economy/

Global Wellness Summit. (2019). *2019 Wellness Trends.* Available at: https://www.globalwellnesssummit.com/2019-global-wellness-trends/download-report/

Goggin, G., & Newell, C. (2005). Disability. In *Australia: Exposing a Social Apartheid*. University of New South Wales Press.

Gomez, C., Carretero, M., Maraver, F., Cantista, P., Armijo, F., Legido, J., ... Delgado, R. (2013). Peloids and Pelotherapy: Historical Evolution, classification and glossary. *Applied Clay Science*, *75-76*, 28–38. doi:10.1016/j.clay.2013.02.008

Gonçalves, M. J. A., Camarinha, A. P., Abreu, A. J., Teixeira, S., & Ferreira, A. (2019). Web Accessibility in the Tourism Sector: an analsis of the most used Websites in Portugal. *Advances in Tourism, Technology and Smart Systems, Proceedings of ICOTTS 2019*.

Gonçalves, M. J. A., Camarinha, A. P., Abreu, A. J., Teixeira, S., & Da Silva, A. F. (2020). Web accessibility in the tourism sector: An analysis of the most used websites in Portugal. In Á. Rocha, A. Abreu, J. de Carvalho, D. Liberato, E. Gonzâlez, & P. Liberato (Eds.), *Advances in Tourism, Technology and Smart Systems. Smart Innovation, Systems and Technologies* (Vol. 171, pp. 141–150). Springer. doi:10.1007/978-981-15-2024-2_13

Gondos, B., & Nárai, M. (2019). The opportunities in accessible tourism. *The USV Annals of Economics and Public Administration*, *19*(1), 48–57.

Goodall, B. (2006). Disabled acess and heritage attractions. *Tourism Review International*, *7*, 57–78.

Goodall, B., Pottinger, G., Dixon, T., & Russell, H. (2004). Heritage property, tourism and the UK Disability Discrimination Act. *Property Management*, *22*(5), 345–357. doi:10.1108/02637470410570734

Graham, M., Zook, M., & Boulton, A. (2013). Augmented reality in urban places: Contested content and the duplicity of code. *Transactions of the Institute of British Geographers*, *38*(3), 464–479. doi:10.1111/j.1475-5661.2012.00539.x

Gray, D., Gould, M., & Bickenbach, J. (2003). Environmental barriers and disability. *Journal of Architectural and Planning Research*, *20*(1), 29–37.

Green, J., Pridmore, T., & Benford, S. (2014). Exploring attractions and exhibits with interactive flashlights. *Personal and Ubiquitous Computing*, *18*(1), 239–251. doi:10.100700779-013-0661-3

Gretzel, U. (2011). Intelligent Systems in Tourism. *Annals of Tourism Research*, *38*(3), 757–779. doi:10.1016/j.annals.2011.04.014

Gretzel, U., Koo, C., Sigala, M., & Xiang, Z. (2015). Special issue on smart tourism: Convergence of information technologies, experiences and theories. *Electronic Markets*, *25*(3), 175–177. doi:10.100712525-015-0194-x

Gretzel, U., Sigala, M., Xiang, Z., & Koo, C. (2015). Smart tourism: Foundations and developments. *Electronic Markets*, *25*(3), 179–188. doi:10.100712525-015-0196-8

Gretzel, U., & Yoo, K. H. (2008). Use and Impact of Online Travel Reviews. In P. O'Connor, W. Höpken, & U. Gretzel (Eds.), *Information and Communication Technologies in Tourism*. Springer. (Original work published 2008)

Griffin, T., Hwan, S., Lee, M., & Guttentag, D. (2017, June). *Virtual Reality and Implications for Destination Marketing* (Paper presentation). Travel and Tourism Research Association: Advancing Tourism Research Globally, Quebec, Canada.

Gronflaten, O. (2009). Predicting Travelers' Choice of Information Sources and Information Channels. *Journal of Travel Research*, *48*(2), 230–244. doi:10.1177/0047287509332333

Guerra, J., Pinto, M., & Beato, C. (2015). Virtual Reality-Shows A new vision for tourism and heritage. *European Scientific Journal*, *11*(9), 49–54.

Guo, R., Samaraweera, G., & Quarles, J. (2014). A unique way to increase presence of mobility impaired users—Increasing confidence in balance. *2014 IEEE Virtual Reality (VR)*, 77–78.

Gutenbrunner, C., Bender, T., Cantista, P., & Karagulle, Z. (2010). A proposal for a worldwide definition on health resort medicine, balneotherapy, medical hydrology and climatology. *International Journal of Biometeorology*, *54*(5), 495–507. doi:10.100700484-010-0321-5 PMID:20532921

Gutierrez, C. F., Loucopoulos, C., & Reinsch, R. W. (2005). Disability-accessibility of airlines' websites for US reservations online. *Journal of Air Transport Management*, *11*(4), 239–247. doi:10.1016/j.jairtraman.2005.01.006

Guttentag, D. A. (2010). Virtual reality: Applications and implications for tourism. *Tourism Management*, *31*(5), 637–651. doi:10.1016/j.tourman.2009.07.003

Hackett, S., Parmanto, B., & Zeng, X. (2005). A retrospective look at website accessibility over time. *Behaviour & Information Technology*, *24*(6), 407–417. doi:10.1080/01449290500066661

Haegele, J., & Hodge, S. (2016). Disability Discourse: Over view and Critiques of the Medical and Social Models. *Quest*, *68*(2), 193-206.

Hale, K. S., & Stanney, K. M. (2019). *Virtual Reality*. Cambridge University Press., http://vr.cs.uiuc.edu/vrbookbig.pdf

Hamann, L. M. A., Airaldi, L. L., Molinas, M. E. B., Rujana, M., Torre, J., & Gramajo, S. (2015). Smart doorbell: An ICT solution to enhance inclusion of disabled people. In 2015 ITU Kaleidoscope: Trust in the Information Society (K-2015) (pp. 1-7). IEEE.

Hammerle, J. (2005). The rise of stadium seating in movie theatres and the disabled's fight for a comparable seat. *De Paul Law Review*, *2*, 589.

Han, D.-I., Tom Dieck, M. C., & Jung, T. (2018). User experience model for augmented reality applications in urban heritage tourism. *Journal of Heritage Tourism*, *13*(1), 46–61. doi:10.1080/1743873X.2016.1251931

Handa, K., Dairoku, H., & Toriyama, Y. (2010). Investigation of priority needs in terms of museum service accessibility for visually impaired visitors. *British Journal of Visual Impairment*, *28*(3), 221–234. doi:10.1177/0264619610374680

Hano, K. (2011). Accessible Tourism: Concepts and Issues. *Tourism Planning & Development*, *8*(4), 467–468. doi:10.1080/21568316.2011.603886

Harper, K. A., & DeWaters, J. (2008). A quest for website accessibility in higher education institutions. *The Internet and Higher Education*, *11*(3-4), 160–164. doi:10.1016/j.iheduc.2008.06.007

Harpur, P., Connolly, U., & Blanck, P. (2017). Socially Constructed Hierarchies of Impairments: The Case of Australian and Irish Workers' Access to Compensation for Injuries. *Journal of Occupational Research*, *27*(4), 507–519. doi:10.100710926-017-9745-7 PMID:29181808

Harris, F., Yang, H., & Sanford, J. (2015). Physical Environmental Barriers to Community Mobility in Older and Younger Wheelchair Users. *Topics in Geriatric Rehabilitation*, *31*(1), 42–51. doi:10.1097/TGR.0000000000000043

Harrison, E. (2018, Aug 21). Whiteboards used as Gatwick flight information screens fail. *University Wire*. Retrieved from http://ezproxy.cardiffmet.ac.uk/login?url=https://search-proquest-com.ezproxy.cardiffmet.ac.uk/docview/2090168288?accountid=15588

Hattan, A. (2004). Stadium-Style Seating Movie Theatres: Does the Cornerstone of the Theatre Industry's Recent Transformation Violate the Americans with Disabilities Act? *Seton Hall Journal of Sports & Entertainment Law*, *14*(1), 267–303.

Hawkey, R. (2004). *Learning with digital technologies in museums, science centres and galleries*. Institute of Museum and Library Services 2006. Status of Technology and Digitization in the Nation's Museums and Libraries. Accessed 29 April 2020 at https://telearn.archives-ouvertes.fr/hal-00190496/

Haworth, A., & Williams, P. (2012). Using QR codes to aid accessibility in a museum. *Journal of Assistive Technologies*, *6*(4), 285–291. doi:10.1108/17549451211285771

Henriques, M. H., Canales, M. L., García-Frank, A., & Gomez-Heras, M. (2019). Accessible Geoparks in Iberia: A challenge to promote geotourism and education for sustainable development. *Geoheritage*, *11*(2), 471–484. doi:10.100712371-018-0300-5

Hersh, M. A., & Johnson, M. A. (2008). On modelling assistive technology systems–Part I: Modelling framework. *Technology and Disability*, *20*(3), 193–215. doi:10.3233/TAD-2008-20303

Hersh, M., Johnson, M., & Keating, D. (2008). *Assistive technology for visually impaired and blind people*. Springer. doi:10.1007/978-1-84628-867-8

Herzog, D., Sikander, S., & Worndl, W. (2019). Integrating route attractiveness attributes into tourist trip recommendations. *The Web Conference 2019 - Companion of the World Wide Web Conference, WWW2019*, 96-101. 10.1145/3308560.3317052

He, Z., Wu, L., & Li, X. (2018). When art meets tech: The role of augmented reality in enhancing museum experiences and purchase intentions. *Tourism Management*, *68*, 127–139. doi:10.1016/j.tourman.2018.03.003

Hicks, S. L., Wilson, I., Muhammed, L., Worsfold, J., Downes, S. M., & Kennard, C. (2013). A depth-based head-mounted visual display to aid navigation in partially sighted individuals. *PLoS One*, *8*(7), 667–695. doi:10.1371/journal.pone.0067695 PMID:23844067

Hinch, T. D., & Jackson, E. L. (2000). Leisure Constraints Research: Its Value as a Framework for Understanding Tourism Seasonability. *Current Issues in Tourism*, *3*(2), 87–106. doi:10.1080/13683500008667868

Hjalager, A. M. (2002). Repairing innovation defectiveness in tourism. *Tourism Management*, *23*(5), 465–474. doi:10.1016/S0261-5177(02)00013-4

Ho, C.-I., Lin, M.-H., & Chen, H.-M. (2012). Web users' behavioural patterns of tourism information search: From online to offline. *Tourism Management*, *33*(6), 1468–1482. doi:10.1016/j.tourman.2012.01.016

Holsti, O. R. (1968). Content analysis. The handbook of social psychology, 2, 596-692.

Holsti, O. R. (1969). *Content Analysis for the Social Sciences and Humanities. Reading. MA: Addison-Wesley (content analysis)*. Publishing Company.

Hu, P. Y., & Tsai, P. F. (2017). Mobile outdoor augmented reality project for historic sites in Tainan. *2016 International Conference on Advanced Materials for Science and Engineering (ICAMSE)*, 509-511. 10.1109/ICAMSE.2016.7840184

Huang, H. (2018). Blind users' expectations of touch interfaces: Factors affecting interface accessibility of touchscreen-based smartphones for people with moderate visual impairment. *Universal Access in the Information Society, 17*(2), 291-304. doi:http://dx.doi.org.ezproxy.cardiffmet.ac.uk/10.1007/s10209-017-0550-z

Huang, L., Yung, C. Y., & Yang, E. (2011). How do travel agencies obtain a competitive advantage?: Through a travel blog marketing channel. *Journal of Vacation Marketing*, *17*(2), 139–149. doi:10.1177/1356766710392737

Huber, D., Milne, S., & Hyde, K. F. (2018). Constraints and facilitators for senior tourism. *Tourism Management Perspectives*, *27*, 55–67. doi:10.1016/j.tmp.2018.04.003

Hudson, S., Thal, K., Cardeas, D., & Meng, F. (2017). Wellness tourism: Stress alleviation or indulging healthful habits? *International Journal of Culture, Tourism and Hospitality Research*, *11*(1), 35–52. doi:10.1108/IJCTHR-09-2015-0111

Hung, K., & Lu, J. (2016). Active living in later life: An overview of aging studies in hospitality and tourism journals. *International Journal of Hospitality Management*, *53*, 133–144. doi:10.1016/j.ijhm.2015.10.008

Hung, K., & Petrick, J. (2009). How Do We Get Baby Boomers And Future Seniors On Motorcoaches? *Tourism Analysis*, *14*(5), 665–675. doi:10.3727/108354209X12597959359338

Hunter-Jones, P. (2004). Young People, holiday-taking and cancer – an exploratory analysis. *Tourism Management*, *25*(2), 249–258. doi:10.1016/S0261-5177(03)00094-3

ICOM. (2016). *Museum Definition.* International Council of Museum http://archives.icom.museum/definition.html

ICOM. (2018). https://icom.museum/wp-content/uploads/2018/07/ICOM-code-En-web.pdf

ICT. (2019). In *Cambridge dictionary*. https://dictionary.cambridge.org/pt/dicionario/ingles/ict

IESE. (2019). *Índice IESE. Cities in Motion.* Universidad de Navarra. Retrieved from: https://media.iese.edu/research/pdfs/ST-0509.pdf

IMD. (2019). *Smart City Index 2019.* Retrieved 04/05/2019 from: www.imd.org/research-knowledge/reports/imd-smart-city-index-2019

INDEC. (2019). *Informes técnicos Turismo Internacional.* Retrieved from: https://www.indec.gob.ar/indec/web/Nivel4-Tema-3-13-55

INDEC. (2020). *Proyecciones y estimaciones poblacionales.* Retrieved from: https://www.indec.gob.ar/indec/web/Nivel3-Tema-2-24

Instituto Ciudades del Futuro. (2020). *Red Argentina de Destinos turísticos inteligentes.* Retrieved from: https://www.google.com/search?q=red+argentina+dti&oq=red+argentina+dti&aqs=chrome.0.69i59j0l4j69i60l3.1410j0j7&sourceid=chrome&ie=UTF-8

Instituto Nacional De Estadísticas y Censos (INDEC). (2018). *Estudio Nacional sobre el Perfil de las Personas con Discapacidad.* Retrieved from: https://www.indec.gob.ar/ftp/cuadros/poblacion/estudio_discapacidad_12_18.pdf

International Federation of Robitics (IFR). (2020). *Service Robots.* https://ifr.org/service-robots

International Organization for Standardization. (2012). ISO/IEC 40500:2012 Information technology - W3C Web Content Accessibility Guidelines (WCAG) 2.0. International Organization for Standarization, 2012.

International Standard Organization. (2016). *ISO 9999:2016 Assistive products for persons with disability — Classification and terminology.* https://www.iso.org/standard/60547.html

Internet World Stats. (2020). http://www.Internetworldstats.com/stats.htm

Ioannidisa, C., & Vozikisb, K. T. (2007). *Application of a GIS for the accessibility of archaeological sites by visitors with disability and visitors with reduced mobility.* In *XXI International CIPA Symposium*, Athens, Greece.

Ip, C., Lee, H., & Law, R. (2012). Profiling the Users of Travel Websites for Planning and Online Experience Sharing. *Journal of Hospitality & Tourism Research (Washington, D.C.)*, *36*(3), 418–426. doi:10.1177/1096348010388663

Irshad, M. (2016). A systematic review of information security frameworks in the internet of things (iot). In *IEEE 18th International Conference on High Performance Computing and Communications; IEEE 14th International Conference on Smart City; IEEE 2nd International Conference on Data Science and Systems (HPCC/SmartCity/DSS)* (pp. 1270-1275). IEEE.

Ismail, A., & Kuppusamy, K. S. (2016). Accessibility of Indian universities' homepages: An exploratory study. *Journal of King Saud University - Computer and Information Sciences*. doi:10.1016/j.jksuci.2016.06.006

ISO 26800:2011. (2011). *ISO 26800:2011 - Ergonomics - General approach, principles and concepts.*

ISO 9241-112:2017. (2017). *ISO 9241-112:2017 Ergonomics of human-system interaction — Part 112: Principles for the presentation of information.*

ISO 9241-171. (2008). *ISO 9241-171:2008 Ergonomics of human-system interaction - Part 171 : Guidance on software accessibility. International Organization.* International Organization for Standardization, 2008.

ISO/IEC 30071-1:2019. (2019). *ISO/IEC 30071-1:2019 Information technology — Development of user interface accessibility — Part 1: Code of practice for creating accessible ICT products and services.*

ISO/TR 22411:2008. (2008). *ISO/TR 22411:2008 Ergonomics data and guidelines for the application of ISO/IEC Guide 71 to products and services to address the needs of older persons and persons with disabilities.*

Israeli, A. (2002). A Preliminary investigation of the importance of site accessibility factors for disabled tourists. *Journal of Travel Research*, *41*(1), 101–104. doi:10.1177/004728750204100114

Istanbullu Dincer, F., Ozcit, H., Cifci, I., Sezer, B., Kahraman, O. C., & Sahinoglu, S. (2019). Accessible Museums for Visually Impaired: A Case Study from Istanbul. *Journal of Tourismology*, *5*(2), 113–126. doi:10.26650/jot.2019.5.2.0032

Jackson, E. L. (2000). Will Research on Leisure Constraints Still Be Relevant in the Twenty-first Century? *Journal of Leisure Research*, *32*(1), 62–68. doi:10.1080/00222216.2000.11949887

Jackson, E., & Searle, M. (1985). Recreation Non-Participation And Barriers To Participation: Concepts, And Models. *Loisir et Societe. Society and Leisure*, *8*(2), 693–707. doi:10.1080/07053436.1985.10715236

Jackson, M. (2018). Models of Disability and Human Rights: Informing the Improvement of Built Environment Accessibility for People with Disability at Neighborhood Scale? *Laws*, *7*(1), 10. doi:10.3390/laws7010010

Jang, S. S., Bai, B., Hu, C., & Wu, C. (2009). Affect, travel motivation, and travel intention: A senior market. *Journal of Hospitality & Tourism Research (Washington, D.C.)*, *33*(1), 51–73. doi:10.1177/1096348008329666

Jennings, G. R. (2005). Interviewing: a Focus on Qualitative Techniques. In B. W. Ritchie, P. Burns, & C. Palmer (Eds.), *Tourism Research Methods - Integrating Theory with Practice* (pp. 99–117). Cabi Publishing. doi:10.1079/9780851999968.0099

Jensen, J. M., & Hjalager, A.-M. (2013). The role of demographics and travel motivation in travellers' use of the internet before, during, and after a trip. *International Journal of Tourism Policy*, *5*(1-2), 34–58. doi:10.1504/IJTP.2013.054051

Jhajharia, S., Pal, S. K., & Verma, S. (2014). Wearable computing and its application. *International Journal of Computer Science and Information Technologies*, *5*(4), 5700–5704.

Johnson, L., Adams, S., & Witchey, H. (2011). *The NMC Horizon Report: 2011 Museum Edition.* http://www.nmc.org/pdf/2011-horizon-report-museum.pdf

Johnson, J. (2011). *Deep Tissue Massage.* Human Kinetics.

Johnson, L., Becker, S., Estrada, V., & Freeman, A. (2015). *NMC Horizon Report: 2015 Higher Education Edition.* The New Me.

Johnson-Migalski, L., & Drout, M. O. (2018). Using the International Classification of Functioning, Disability, and Health (ICF) in Adlerian approaches. *Journal of Individual Psychology*, *74*(1), 38–54. doi:10.1353/jip.2018.0003

Jung, T., & Han, D. (2014). Augmented Reality (AR) in Urban Heritage Tourism. *E-Review of Tourism Research*, *5*.

Jung, T., Chung, N., & Leue, M. C. (2015). The determinants of recommendations to use augmented reality technologies: The case of a Korean theme park. *Tourism Management*, *49*, 75–86. doi:10.1016/j.tourman.2015.02.013

Jung, T., tom Dieck, M. C., Lee, H., & Chung, N. (2016). Effects of virtual reality and augmented reality on visitor experiences in museum. In *Information and communication technologies in tourism 2016* (pp. 621–635). Springer. doi:10.1007/978-3-319-28231-2_45

Kaganek, K., Ambrozy, T., Mucha, D., Jurczak, A., Bornikowska, A., Ostrowski, A., Janiszewska, R., & Mucha, T. (2017). Barriers To Participation In Tourism In The Disabled. *Sport Tourism, 24*(2), 121–129. doi:10.1515/pjst-2017-0013

Karagulle, M., & Karagulle, M. (2015). Effectiveness of balneotherapy and spa therapy for the treatment of chronic low back pain: A review on latest evidence. *Clinical Rheumatology, 34*(2), 207–214. doi:10.100710067-014-2845-2 PMID:25535198

Kasnauskiene, G., & Michnevic, K. (2017). Contribution of increased life expectancy to economic growth: Evidence from CEE countries. *International Journal of Economic Sciences, 2*, 82.

Kastenholz, E., Eusébio, C., Figueiredo, E., & Lima, J. (2012). Accessibility as Competitive Advantage of a Tourism Destination: The Case of Lousã. In K. F. Hyde, C. Ryan, & A. Woodside (Eds.), Field Guide for Case Study Research in Tourism, Hospitality, and Leisure (Advances in Culture, Tourism and Hospitality Research, Volume 6) (pp. 369-385). Bingley: Emerald Publishers.

Kastenholz, E., Eusébio, C., & Figueiredo, E. (2015). Contributions of tourism to social inclusion of persons with disability. *Society and Disability, 30*(8), 1259–1281. doi:10.1080/09687599.2015.1075868

Kastenholz, E., Eusébio, M. C., Moura, A., & Figueiredo, E. (2010). Animação turística acessível: Do sonho à realidade. In N. Abranja, A. Marques, F. Coelhoso, & I. Carneiro (Eds.), *Turismo acessível: Estudos e experiências* (pp. 171–192). Edições Pedago.

Kaur, M. J., & Maheshwari, P. (2016). Smart tourist for Dubai city. *2016 2nd International Conference on Next Generation Computing Technologies, NGCT 2016*, 30-34. 10.1109/NGCT.2016.7877385

Keeratitanont, K., Jensen, M., Chatchawan, U., & Auvichayapat, P. (2015). The efficacy of traditional Thai massage for the treatment of chronic pain: A systematic review. *Complementary Therapies in Clinical Practice, 22*(1), 26–32. doi:10.1016/j.ctcp.2015.01.006 PMID:25682523

Kempiak, J., Hollywood, L., Bolan, P., & McMahon-Beattie, U. (2017). The heritage tourist: An understanding of the visitor experience at heritage attractions. *International Journal of Heritage Studies, 23*(4), 375–392. doi:10.1080/13527258.2016.1277776

Kennard C (2013) A depth-based head-mounted visual display to aid navigation in partially sighted individuals. *PLoS One, 8*(7).

Khalil, T. (2000). *Management of technology -The key to competitiveness and wealth creation.* Prentice Hall.

Kim, E., Chiange, L., & Tang, L. (2016). Investigating wellness tourists' motivation, engagement and loyalty: In search of the missing link. *Journal of Travel & Tourism Marketing, 34*(7), 867–879. doi:10.1080/10548408.2016.1261756

Kim, H. K., Han, S. H., Park, J., & Park, J. (2016). The interaction experiences of visually impaired people with assistive technology: A case study of smartphones. *International Journal of Industrial Ergonomics, 55*, 22–33. doi:10.1016/j.ergon.2016.07.002

Kim, H., & Fesenmaier, D. R. (2008). Persuasive design of destination web sites: An analysis of first impression. *Journal of Travel Research, 47*(1), 3–13. doi:10.1177/0047287507312405

Kim, H., Xiang, Z., & Fesenmaier, D. R. (2015). Use of the internet for trip planning: A generational analysis. *Journal of Travel & Tourism Marketing, 32*(3), 276–289. doi:10.1080/10548408.2014.896765

Kim, S. E., & Lehto, X. Y. (2012). The voice of tourists with mobility disabilities: Insights from online customer complaint websites. *International Journal of Contemporary Hospitality Management*, *24*(3), 451–476. doi:10.1108/09596111211217905

Kiss, G., Jansen, H., Castaldo, V. L., & Orsi, L. (2015). The 2050 City. *Procedia Engineering*, *118*, 326–355. doi:10.1016/j.proeng.2015.08.434

Kitchin, R. (2000). The researched opinions on research: Disabled people and disability research. *Society and Disability*, *15*(1), 25–47. doi:10.1080/09687590025757

Kluitenberg, E. (2008). *Delusive Spaces: Essays on Culture, Media and Technology. NAi Publishers*. Institute of Network Cultures.

Knochel, A. D., Hsiao, W. H., & Pittenger, A. (2018). Touching to see: Tactile learning, assistive technologies, and 3-D printing. *Art Education*, *71*(3), 7–13. doi:10.1080/00043125.2018.1436320

Kolbe, R. H., & Burnett, M. S. (1991). Content-analysis research: An examination of applications with directives for improving research reliability and objectivity. *The Journal of Consumer Research*, *18*(2), 243–250. doi:10.1086/209256

Kołodziejczak, A. (2019). Information as a factor of the development of Accessible Tourism for People with Disabilities. *Quaestiones Geographicae*, *38*(2), 67–73. doi:10.2478/quageo-2019-0014

Koo, C., Gretzel, U., Hunter, W. C., & Chung, N. (2015). The Role of IT in Tourism. *Asia Pacific Journal of Information System*, *25*(1), 99–104. doi:10.14329/apjis.2015.25.1.099

Koog, Y., Jin, S., Yoon, K., & Min, B. (2010). Interventions for hemiplegic shoulder pain: Systematic review of randomised controlled trials. *Disability and Rehabilitation*, *32*(4), 282–291. doi:10.3109/09638280903127685 PMID:20055567

Kotler, P., Bowen, J. T., Makens, J. C., & Baloglu, S. (2006). Bradford & Bingley audits age diversity in the workplace. *Human Resource Management International Digest*, *14*(6), 27–28. doi:10.1108/09670730610690358

Kotler, P., Kartajaya, H., & Hooi, D. H. (2019). Marketing 4.0: Moving From Traditional to Digital. In P. Kotler, H. Kartajaya, & H. D. Huan (Eds.), *Asian Competitors:Marketing for Competitiveness in the Age of Digital Consumers* (pp. 99–123). World Scientific. doi:10.1142/9789813275478_0004

Kounavis, C. D., Kasimati, A. E., & Zamani, E. D. (2012). Enhancing the Tourism Experience through Mobile Augmented Reality: Challenges and Prospects. *International Journal of Engineering Business Management*, *4*, 10. doi:10.5772/51644

Kous, K., Kuhar, S., Pušnik, M., & Šumak, B. (2019). Comparative analysis of faculties ' websites accessibility based on an automatic evaluation. In *42nd International Convention on Information and Communication Technology, Electronics and Microelectronics (MIPRO)* (pp. 1498–1502). Croatian Society MIPRO. 10.23919/MIPRO.2019.8757202

Kous, K., Kuhar, S., Rajšp, A., & Šumak, B. (2020). Investigation of the accessibility of non-text content published on websites. In *MIPRO 2020 : 43nd International Convention* (pp. 1953–1958). Retrieved from http://docs.mipro-proceedings.com/proceedings/mipro_2020_proceedings.pdf

Kous, K., Kuhar, S., Pavlinek, M., Heričko, M., & Pušnik, M. (2020). Web accessibility investigation of Slovenian municipalities' websites before and after the adoption of European Standard EN 301 549. *Universal Access in the Information Society*, (0123456789). Advance online publication. doi:10.100710209-020-00732-9

Kous, K., & Pavlinek, M. (2017). Longitudinalna raziskava dostopnosti spletnih mest slovenskih občin. *Uporabna Informatika*, *25*(3), 135–152.

Kous, K., & Polančič, G. (2019). Empirical insights of individual website adjustments for people with dyslexia. *Sensors (Basel)*, *19*(10), 1–25. doi:10.339019102235 PMID:31091827

Ktena, S. I., Abbott, W., & Faisal, A. A. (2015). A virtual reality platform for safe evaluation and training of natural gaze-based wheelchair driving. *2015 7th International IEEE/EMBS Conference on Neural Engineering (NER)*, 236–239.

Kusayama, K. (2005). Access to museums for visually challenged people in Japan. *International Congress Series, 1282*, 877–880. doi:10.1016/j.ics.2005.05.068

Kuzma, J., Yen, D., & Oestreicher, K. (2009). Global e-government Web Accessibility : An Empirical Examination of EU, Asian and African Sites. In *International Conference on Information and Communication Technologies and Accessibility*, (pp. 1–6). Hammamet, Tunisia: Academic Press.

Kwok, A. O. J., & Koh, S. G. M. (2020). COVID-19 and Extended Reality (XR). *Current Issues in Tourism*, 1–6.

La Rosa, G. (2020). Enfoques, modelos y buenas prácticas internacionales en DTI: los casos de Buenos Aires y la Red Argentina. In Sistemas de inteligencia y modelos de trabajo en red para una gestión turística inteligente en Argentina. INVAT-TUR, Webinar.

Lacalle, L. (2013). Agencias de viajes en España. Una industria convulsa. *Papers de turismo*, (54), 122–138.

Lai, K. W. (2014). The role of service quality, perceived value, and relationship quality in enhancing customer loyalty in the travel agency sector. *Journal of Travel & Tourism Marketing*, *31*(3), 417–442. doi:10.1080/10548408.2014.883346

Lam, K. L., Chan, C.-S., & Peters, M. (2020). Understanding technological contributions to accessible tourism from the perspective of destination design for visually impaired visitors in Hong Kong. *Journal of Destination Marketing & Management*, *17*, 1–9. doi:10.1016/j.jdmm.2020.100434

Lamont, D., Kenyon, S., & Lyons, G. (2013). Dyslexia and mobility-related social exclusion: The role of travel information provision. *Journal of Transport Geography*, *26*, 147–157. doi:10.1016/j.jtrangeo.2012.08.013

Lamsfus, C., Martin, D., Alzua-Sorzabal, A., & Torres-Manzanera, E. (2015). Smart Tourism Destinations: An Extended Conception of Smart Cities Focusing on Human Mobility. In L. Tussyadiah & A. Inversini (Eds.), *Information and Communication Technologies in Tourism 2015* (pp. 363–375). Springer. doi:10.1007/978-3-319-14343-9_27

Lancioni, G. E., O'Reilly, M. F., Oliva, D., & Bracalent, S. (1998). Guiding a person with blindness and intellectual dis- ability in indoor travel with fewer auditory cues. *Journal of Visual Impairment & Blindness*, *92*(9), 609–614. doi:10.1177/0145482X9809200904

Lancioni, G. E., Singh, N. N., O'Reilly, M. F., Alberti, G., Chiariello, V., Campanella, C., Grillo, G., & Tagliente, V. (2019). A Program Based on Common Technology to Support Communication Exchanges and Leisure in People With Intellectual and Other Disabilities. *Behavior Modification*, *43*(6), 879–897. doi:10.1177/0145445519850747 PMID:31104482

Langa, L., Herman, A., Hillsgrove, D., Jones, B., Li, C. M., Matthai, Z., Schlaffer, J., Slack, J., Sizemore, B., Weir, R. Jr, Fagan, J., Lazar, J., Henderson, H., Knauff, S., Adesina, T., & Baumgart, D. M. (2012). Museum web accessibility: A Pilot assessment of Mid-Atlantic museums. *International Journal of the Inclusive Museum*, *4*(1), 15–28. doi:10.18848/1835-2014/CGP/v04i01/44360

Laurìa, A. (2016). Accessibility to cultural heritage as a design challenge. In Accessibility as a key enabling knowledge for enhancement of cultural heritage. Franco Angeli Edizioni.

Lauría, A. (2016). "The Florence Experience": A multimedia and multisensory guidebook for cultural towns inspired by Universal Design approach. *Work (Reading, Mass.)*, *53*(4), 709–727. doi:10.3233/WOR-162256 PMID:26890596

Lavalle, S. M. (2019). Virtual reality. In *Virtual Reality*. Cambridge University Press. http://vr.cs.uiuc.edu/vrbookbig.pdf

Law N° 25.643 (2002). Accessible Tourism Law.

Law, R., Qi, S., & Buhalis, D. (2010). Progress in tourism management: A review of website evaluation in tourism research. *Tourism Management, 31*(3), 297–313. doi:10.1016/j.tourman.2009.11.007

Lazar, J., Jaeger, P. T., Olalere, A., Algarne, M., Augustine, Z., Brown, D'Erasmo, F., Dotson, B., Endiape, G., Govender, L., Hagos, N., Hunt, G., Lawrence, W., Plummer, R., Richardson, A., Saval, J., Schickling, C., Schley, C., Szoka, D., Torrie, R., & Walton, M. (2012). Still up in the air: government regulation of airline websites and continuing price inequality for persons with disabilities online. In *Proceedings of the 13th Annual International Conference on Digital Government Research* (pp. 240-245). ACM. 10.1145/2307729.2307768

Lazaroiu, G. C., & Roscia, M. (2012). Definition methodology for the smart cities model. *Energy, 47*(1), 326–332. doi:10.1016/j.energy.2012.09.028

Lee, H., Chung, N., & Nam, Y. (2019). Do online information sources really make tourists visit more diverse places? Based on the social networking analysis. *Information Processing & Management, 56*(4), 1376–1390. doi:10.1016/j.ipm.2018.01.005

Leong, I., & Higgins, S. (2010). Public Library Services For Wheelchair-Bound Young People in Singapore. *Public Library Quarterly, 29*(3), 210–229. doi:10.1080/01616846.2010.502033

Leporini, B., & Norscia, I. (2008). "Fine-Tuning" image accessibility for museum websites. *Journal of Universal Computer Science, 14*(19), 3250–3264.

Lestari, N. S., Wiastuti, R. D., & Triana, I. (2019). An overview of accessible Tourism information at Taman Mini Indonesia Indah Jakarta. *African Journal of Hospitality, Tourism and Leisure, 8*(4), 1–11.

Lijing, Z., Yanrong, P., & Jianhua, H. (2015). The development strategy for the tourism in Hebei under the background of smart city based on data mining. *7th International Conference on Intelligent Computation Technology and Automation, ICICTA 2014*, 991-994. 10.1109/ICICTA.2014.238

Lima Devile, E., Kastenholz, E., & Santiago, R., (2012). Inibidores, facilitadores e estratégias de negociação associadas às experiências turísticas das pessoas com incapacidade. *RDT/Journal of Tourism and Development, 17/18*, 1417-1430.

Lim, Y., Park, Y., Heo, J., Yang, J., Kang, M., & Byun, Y.-C. (2011). A Smart Phone Application Based on AR for Jeju Tourism. *2011 First ACIS/JNU International Conference on Computers, Networks, Systems and Industrial Engineering*, 271–272. 10.1109/CNSI.2011.101

Lisney, E., Bowen, J. P., Hearn, K., & Zedda, M. (2013). Museums and Technology: Being Inclusive Helps Accessibility for All. *Curator (New York, N.Y.), 56*(3), 353–361. doi:10.1111/cura.12034

Littrell, M. A., Paige, R. C., & Song, K. (2004). Senior travellers: Tourism activities and shopping behaviours. *Journal of Vacation Marketing, 10*(4), 348–362. doi:10.1177/135676670401000406

Liu, Z., Sokhn, M., Le Calvé, A., & Schegg, R. (2016). City Eye: Accessibility for All. In *ICEGOV '15-16 Proceedings of the 9th International Conference on Theory and Practice of Electronic Governance* (pp. 350-351). ACM.

Liu, J. N. K., & Zhang, E. Y. (2014). An investigation of factors affecting customer selection of online hotel booking channels. *International Journal of Hospitality Management, 39*, 71–83. doi:10.1016/j.ijhm.2014.01.011

Liu, P., Zhang, H., Zhang, J., Sun, Y., & Qiu, M. (2019). Spatial-temporal response patterns of tourist flow under impulse pre-trip information search: From online to arrival. *Tourism Management, 73*, 105–114. doi:10.1016/j.tourman.2019.01.021

Liu, Z., Sokhn, M., Le Calvé, A., & Schegg, R. (2016). City Eye: Accessibility for all. In *Proceedings of the 9th International Conference on Theory and Practice of Electronic Governance* (pp. 350-351). 10.1145/2910019.2910020

Livingston, M. A., Rosenblum, L. J., Brown, D. G., Schmidt, G. S., Julier, S. J., Baillot, Y., Swan, J. E., Ai, Z., & Maassel, P. (2011). Military Applications of Augmented Reality. In *Handbook of Augmented Reality* (pp. 671–706). Springer New York. doi:10.1007/978-1-4614-0064-6_31

Li, Y., Hu, C., Huang, C., & Duan, L. (2017). The concept of smart tourism in the context of tourism information services. *Tourism Management, 58*, 293–300. doi:10.1016/j.tourman.2016.03.014

Llodrà-Riera, I., Martínez-Ruiz, M. P., Jiménez-Zarco, A. I., & Izquierdo-Yusta, A. (2015). A multidimensional analysis of the information sources construct and its relevance for destination image formation. *Tourism Management, 48*, 319–328. doi:10.1016/j.tourman.2014.11.012

Loi, K. L., & Kong, W. H. (2017). Tourism for All: Challenges and Issues Faced by People with Vision Impairment. *Tourism Planning & Development, 14*(2), 181–197. doi:10.1080/21568316.2016.1204357

López Sintas, J., García Álvarez, E., & Pérez Rubiales, E. (2014). Art museum visitors: Interaction strategies for sharing experiences. *Museum Management and Curatorship, 29*(3), 241–259. doi:10.1080/09647775.2014.919175

Lorenz, M., Knopp, S., & Klimant, P. (2018, October). Industrial Augmented Reality: Requirements for an Augmented Reality Maintenance Worker Support System (Paper presentation). *2018 IEEE International Symposium on Mixed and Augmented Reality Adjunct (ISMAR-Adjunct)*, 151-153. 10.1109/ISMAR-Adjunct.2018.00055

Lorenzo, G., Lledó, A., Pomares, J., & Roig, R. (2016). Design and application of an immersive virtual reality system to enhance emotional skills for children with autism spectrum disorders. *Computers & Education, 98*, 192–205. doi:10.1016/j.compedu.2016.03.018

Loureiro, S. M. C., Guerreiro, J., & Ali, F. (2020). 20 years of research on virtual reality and augmented reality in tourism context: A text-mining approach. *Tourism Management, 77*, 104028. doi:10.1016/j.tourman.2019.104028

Louvre. (2020). https://www.louvre.fr/en/touch-gallerya-new-tour-sculpting-body

Louw, C. (2017). Good, better, best practise - a comparative analysis of official destination websites. *African Journal of Hospitality, Tourism and Leisure, 6*(3).

Lozynska, O., Savchuk, V., & Pasichnyk, V. (2019). Individual sign translator component of tourist information system. In *International Conference on Computer Science and Information Technology* (pp. 593-601). Springer.

Luberg, A., Tammet, T., & Järv, P. (2011). Smart City: A Rule-based Tourist Recommendation System. In R. Law, M. Fuchs, & F. Ricci (Eds.), *Information and Communication Technologies in Tourism 2011*. Springer. doi:10.1007/978-3-7091-0503-0_5

Luccio, F. L., & Beltrame, L. (2018). Accessible tourism for users with hearing loss. In *Proceedings of the 4th EAI International Conference on Smart Objects and Technologies for Social Good* (pp. 243-248). 10.1145/3284869.3284909

Luo, M., Feng, R., & Cai, L. A. (2004). Information search behavior and tourist characteristics. *Journal of Travel & Tourism Marketing, 17*(2), 269–193.

Lyu, S. (2017). Which accessible travel products are people with disabilities willing to pay more? A choice experiment. *Tourism Management, 59*, 404–412. doi:10.1016/j.tourman.2016.09.002

Mandal. (2019). https://www.news-medical.net/health/Types-of-visual-impairment.aspx

Mangani, A., & Bassi, L. (2019). Web information, accessibility and museum ownership. *International Journal of Tourism Policy, 9*(4), 265–281. doi:10.1504/IJTP.2019.105486

Mankoff, J., Fait, H., & Tran, T. (2005). Is your web page accessible?: A comparative study of methods for assessing web page accessibility for the blind. In *Proc. of the SIGCHI Conf. on Human factors in Computing Systems*, (pp. 41–50). 10.1145/1054972.1054979

Marasco, A., & Balbi, B. (2019). Designing Accessible Experiences for Heritage Visitors Through Virtual Reality. *Ereview of Tourism Research, 17*(3), 426–443.

Marjury, D. H., Karen, B. C., Diana, M., & Gabriel, L. F. (2017). Offline mobile application for places identification with augmented reality. *2017 4th International Conference on eDemocracy and eGovernment, ICEDEG 2017,* 261-264. . doi:10.1109/ICEDEG.2017.7962546

Marshall, C., & Rossman, G. (2011). *Designing Qualitative Research* (5th ed.). SAGE Publications Inc.

Martin, J., Bohuslava, J., & Igor, H. (2018). Augmented Reality in Education 4.0. *2018 IEEE 13th International Scientific and Technical Conference on Computer Sciences and Information Technologies (CSIT)*, 231–236.

Marty, P. F. (2007). Museum Websites and Museum Visitors: Before and After the Museum Visit. *Museum Management and Curatorship, 22*(4), 337–360. doi:10.1080/09647770701757708

Marty, P. F., & Twidale, M. B. (2004). Lost in gallery space: A conceptual framework ofr analysing the usability flaws of museum Web sites. *First Monday, 9*(9), 1–13. doi:10.5210/fm.v9i9.1171

Masala, C., & Petretto, D. (2008). From disablement to enablement: Conceptual models of disability in the 20th Century. *Disability and Rehabilitation, 30*(17), 1233–1244. doi:10.1080/09638280701602418 PMID:18821191

Mascetti, S., Gerino, A., Bernareggi, C., & Picinali, L. (2017). On the Evaluation of Novel Sonification Techniques for Non-Visual Shape Exploration. *ACM Transactions on Accessible Computing, 9*(4), 13. doi:10.1145/3046789

Mascho, E., & Singh, N. (2014). Virtual tourism: Use of "second life" for destination marketing. *Anatolia, 25*(1), 140–143. doi:10.1080/13032917.2013.815644

Maswera, T., Dawson, R., & Edwards, J. (2005). Analysis of Usability and Accessibility Errors of E-Commerce Websites of Tourist Organisations in Four African Countries Tonderai Maswera. *World Wide Web Internet And Web Information Systems*. Retrieved from http://www.ifitt.org/admin/public/uploads/Maswera.pdf

Mayordomo-Martínez, D., Sánchez-Aarnoutse, J. C., Carrillo-de-Gea, J. M., García-Berná, J. A., Fernández-Alemán, J. L., & García-Mateos, G. (2019). Design and Development of a Mobile App for Accessible Beach Tourism Information for People with Disabilities. *International Journal of Environmental Research and Public Health, 16*(12), 2131. doi:10.3390/ijerph16122131 PMID:31208146

McKercher, B., & Darcy, S. (2018). Re-conceptualizing barriers to travel by people with disabilities. *Tourism Management Perspectives, 26*, 59–66. doi:10.1016/j.tmp.2018.01.003

McKercher, B., Packer, T., Yau, M., & Lam, P. (2003). Travel agents as facilitators or constraints of travel: Perceptions of people with disabilities. *Tourism Management, 24*(4), 465–474. doi:10.1016/S0261-5177(02)00107-3

McMillen, R., & Alter, F. (2017). Social media, social inclusion, and museum disability access. *Museums & Social Issues, 12*(2), 115–125. doi:10.1080/15596893.2017.1361689

McMulla. (2015). *Shake up the gallery: how iPads are changing the way we visit museums*. Accessed on March 27, 2020, from https://www.alphr.com/apple/1001182/shake-up-the-gallery-how-ipads-are-changing-the-way-we-visit-museums

Meliones, A., & Sampson, D. (2018). Blind MuseumTourer: A system for self-guided tours in museums and blind indoor navigation. *Technologies, 6*(1), 1–31. doi:10.3390/technologies6010004

Mesároš, P., Mandicák, T., Mesárošová, A., Hernandez, M. F., Kršák, B., Sidor, C., Štrba, L., Molokác, M., Hvizdák, L., Blišt'an, P., & Delina, R. (2016). Use of Augmented Reality and Gamification techniques in tourism. *Ereview of Tourism Research, 13*(1/2), 366–381.

Mesquita, S., & Carneiro, M. J. (2016). Accessibility of European museums to visitors with visual impairments. *Disability & Society, 31*(3), 373–388. doi:10.1080/09687599.2016.1167671

Metatla, O., Serrano, M., Jouffrais, C., Thieme, A., Kane, S., Branham, S., . . . Bennett, C. L. (2018). Inclusive education technologies: Emerging opportunities for people with visual impairments. *Extended Abstracts of the 2018 CHI Conference on Human Factors in Computing Systems*, 1-8.

Michopoulou, E., & Buhalis, D. (2011). Technology platforms and challenges. Accessible Tourism Concepts and Issues, 287-299.

Michopoulou, E. (2017). Marketing for the Spa Industry. In S. Rawlinson & T. Heap (Eds.), *International Spa Management*. Goodfellow Publishers.

Michopoulou, E., & Buhalis, D. (2013). Information provision for challenging markets: The case of the accessibility requiring market in the context of tourism. *Information & Management, 50*(5), 229–239. doi:10.1016/j.im.2013.04.001

Michopoulou, E., Darcy, S., Ambrose, I., & Buhalis, D. (2015). Accessible tourism futures: The world we dream to live in and the opportunities we hope to have. *Journal of Tourism Futures, 1*(3), 179–188. doi:10.1108/JTF-08-2015-0043

Miesenberger, K., Klaus, J., Karshmer, A., & Zagler, W. (Eds.). (2006). *Computers Helping People with Special Needs: Proceedings of the 10th International Conference* (Vol. 4061). Springer Science & Business Media. 10.1007/11788713

Mighali, V., Del Fiore, G., Patrono, L., Mainetti, L., Alletto, S., Serra, G., & Cucchiara, R. (2015). Innovative IoT-aware services for a smart museum. *WWW 2015 Companion 24th International Conference on World Wide Web*, 547-550. 10.1145/2740908.2744711

Millan, A., & Esteban, A. (2004). Development of a multiple-item scale for measuring customer satisfaction in travel agencies services. *Tourism Management, 25*(5), 533–546. doi:10.1016/j.tourman.2003.07.002

Miller, G., & Kirk, E. (2002). The disability discrimination act: Time for a stick? *Journal of Sustainable Tourism, 10*(1), 82–88. doi:10.1080/09669580208667154

Mills, J. E., Han, J.-H., & Clay, M. J. (2008). Accessibility of hospitality and tourism websites: A challenge for visually impaired persons. *Cornell Hospitality Quarterly, 49*(1), 28–41. doi:10.1177/1938965507311499

Mintel. (2015). *Complementary and Alternative Medicine – UK – December 2015*. Available at: https://academic.mintel.com/display/716196/

Moher, D., Liberati, A., Tetzlaff, J., & Altman, D. G. (2009). Preferred reporting items for systematic reviews and meta-analyses: The PRISMA statement. *Annals of Internal Medicine, 151*(4), 264–269. doi:10.7326/0003-4819-151-4-200908180-00135 PMID:19622511

Moreno Gil, S., & Ritchie, J. R. B. (2009). Understanding the museum image formation process: A comparison of residents and tourists. *Journal of Travel Research, 47*(4), 480–493. doi:10.1177/0047287508326510

Morris, J., Mueller, J., & Jones, M. (2010). Tomorrows Elders with Disabilities: What the Wireless Industry Needs to Know. *Journal of Engineering Design, 21*(2/3), 131–146. doi:10.1080/09544820903303431

Morrison, A. M., Taylor, S., Morrison, A. J., & Morrison, A. D. (1999). Marketing small hotels on the World Wide Web. *Information Technology & Tourism, 2*(2), 97–113.

Moura, A. (2018). *Benefícios do Turismo para Lidar com o Stress da Incapacidade: Turismo Acessível, Stress-Coping e Incapacidade Motora e Sensorial*. Editorial Académica Española - OmniScriptum Publishing Group.

Moura, A. F. A., Kastenholz, E., & Pereira, A. M. S. (2018). Accessible tourism and its benefits for coping with stress. *Journal of Policy Research in Tourism, Leisure & Events, 10*(3), 241–264. doi:10.1080/19407963.2017.1409750

Moura, A., Mónico, L., & Mira, M. R. (2019). Measuring quality regarding destination marketing: Perceptions from local public stakeholders in Portugal. *Tourism & Management Studies, 15*(1), 44–53. doi:10.18089/tms.2019.150104

Muller, L., & Laburda, A. (2012). *ENAT Reaching all customers: How do European NTOs compare on online Accessibility*. ENTER 12 IFITT Conference, Helsingborg, Sweden. https://www.accessibletourism.org/resources/enter2012-helsingborg_enat_final_ia--2.pdf

Müller, K. (2002). Museums and Virtuality. *Curator (New York, N.Y.), 45*(1), 21–33. doi:10.1111/j.2151-6952.2002.tb00047.x

Murray, M., & Sproats, J. (1990). The disabled traveller: Tourism and disability in Australia. *Journal of Tourism Studies, 1*(1), 9–14.

Nayyar, A., Mahapatra, B., Nhuong Le, D., & Suseendran, G. (2018). Virtual Reality (VR) & Augmented Reality (AR) technologies for tourism and hospitality industry. *International Journal of Engineering & Technology, 7*(2.21), 156.

Neumann, P., & Reuber, P. (2004). Economic Impulses of Accessible Tourism for All. Berlin: Federal Ministry of Economics and Technology.

Neumann, P., & Reuber, P. (2004). Economic Impulses of Accessible Tourism for All. Study commissioned by the Federal Ministry of Economics and Technology and Federal Ministry of Economics and Labour, Berlin.

Neumann, P., & Reuber, P. (2004). *Estímulos económicos do turismo acessível para todos*. Munique e Berlim: Ministério Federal da Economia e do Trabalho Alemão (BMWA) – versão portuguesa, não publicada, disponibilizada pelo Secretariado Nacional de Reabilitação e Integração da Pessoa com Deficiência.

Neven, A., Vanrompay, Y., Declercq, K., Janssens, D., Wets, G., Dekelver, J., Daems, J., & Bellemans, T. (2017). Viamigo: Monitoring tool to support independent travel by persons with intellectual disabilities. *Transportation Research Record: Journal of the Transportation Research Board, 2650*(1), 25–32. doi:10.3141/2650-04

Next Tourism Generation. (2020). *About the Next Tourism Generation Alliance*. https://nexttourismgeneration.eu/about/

Nichol, P. B. (2017). Healthcare VR innovations are healing patients. *CIO (Framingham, Mass.)*.

Niininen, O., Buhalis, D., & March, R. (2007). Customer Empowerment in Tourism through Consumer Centric Marketing (CCM). *Qualitative Market Research, 10*(3), 265–281. doi:10.1108/13522750710754308

Nikitina, O., & Vorontsova, G. (2015). Aging population and tourism: Socially determined model of consumer behavior in the "senior tourism" segment. *Procedia: Social and Behavioral Sciences, 214*, 845–851. doi:10.1016/j.sbspro.2015.11.736

Nitti, M., Giusto, D., Zanda, S., Di Francesco, M., Casari, C., Clemente, M. L., Lai, C., Milesi, C., & Popescu, V. Using IoT for Accessible Tourism. In A. R. Garcia Ramirez & M. Gitirana Gomes Ferreira (Eds.), *Smart Cities, Assistive Technologies in Smart Cities* (pp. 31–50). IntechOpen.

Nowak Da Costa, J., & Bielski, C. (2018). Towards ↜tourism for all↝ - improving maps for persons with reduced mobility. *International Archives of the Photogrammetry, Remote Sensing and Spatial Information Sciences - ISPRS Archives, 42*(4) 547-550. 10.5194/isprs-archives-XLII-4-475-2018

NSW. (2019). *Sydney for All*. Retrieved from: https://sydneyforall.com/

Nunez, A., Moquillaza, A., & Paz, F. (2019). Web Accessibility Evaluation Methods: A Systematic Review. In A. Marcus & W. Wang (Eds.), *Design, User Experience, and Usability, Practice and Case Studies, 21st HCI International Conference, HCII 2019*. Orlando, FL: Academic Press.

Nurhasan, U., Pradibta, H., Suryadi, S. B., & Alfinda, A. (2019). Smart tourist guide application for the introduction of Malang city tourism potential using hybrid technology. *Journal of Physics: Conference Series*, *1402*(6) 10.1088/1742-6596/1402/6/066051

Nyanjom, J., Boxall, K., & Slaven, J. (2018). Towards inclusive tourism? Stakeholder collaboration in the development of accessible tourism. *Tourism Geographies*, *20*(4), 675–697. doi:10.1080/14616688.2018.1477828

Nyaupane, G. P., & Andereck, K. L. (2008). Understanding travel constraints: Application and extension of a leisure constraints model. *Journal of Travel Research*, *46*(4), 433–439. doi:10.1177/0047287507308325

Nyman, E., Westin, K., & Carson, D. (2018). Tourism destination choice sets for families with wheelchair-bound children. *Tourism Recreation Research*, *43*(1), 26–38. doi:10.1080/02508281.2017.1362172

O'Neill, M., & Ali Knight, J. (2000). Disability tourism dollars in Western Australia hotels. *FIU Hospitality Review*, *18*(2), 72–88.

Ochoggia, R. (2003). Persons with Disabilities Bill 2002: Implications concerning visual disabilities for academic library and Information Services in Kenya. Kenya. *New Library World*, *104*(7/8), 307–312. doi:10.1108/03074800310488086

OECD. (2018). *OECD Tourism Trends and Policies 2018*. OECD Publishing. https://www.oecd.org/cfe/tourism/2018-Tourism-Trends-Policies-Highlights-ENG.pdf

Oertel, B., Haße, C., Scheermesser, M., Thio, S. L., & Feil, T. (2004). Accessibility of Tourism Web sites within the European Union. In *Information and communication technologies in tourism 2004* (pp. 358–368). Springer-Verlag New York Inc. doi:10.1007/978-3-7091-0594-8_34

Olsen, J. (2018). Socially disabled: The fight disabled people face against loneliness and stress. *Disability & Society*, *33*(7), 1160–1164. doi:10.1080/09687599.2018.1459228

ONCE. (2020). http://museo.once.es/home.cfm?id=5&nivel=1

Öörni, A. (2004). Consumer objectives and the amount of search in electronic travel and tourism markets. *Journal of Travel & Tourism Marketing*, *17*(2/3), 3–14. doi:10.1300/J073v17n02_01

Open Doors Organization. (2015). *Marketing Outlook Forum 2015, New ODO Market Study Finds Positive Trends in Accessible Travel*. Retrieved From: https://opendoorsnfp.org/wpcontent/uploads/2016/05/ODOMarketingForecast2015.pdf

Ortega, E., Galvez, I., Hinchado, M., Guerro, J., Martin-Cordero, L., & Torres-Piles, S. (2017). Anti-inflammatory effect as a mechanism of effectiveness underlying the clinical benefits of Pelotherapy in osteoarthritis patients: Regulation of the altered inflammatory and stress feedback response. *International Journal of Biometeorology*, *61*(10), 1777–1785. doi:10.100700484-017-1361-x PMID:28456879

Ospina, J. P. (2018). *Ciudades inteligentes en América Latina*. Banco Interamericano de Desarrollo (BID). Retrieved from: https://conexionintal.iadb.org/2018/11/27/267_e_ideas6/

Ozkuk, K., Uysal, B., Ates, Z., Okmen, B., Refia, S., & Gamze, D. (2018). The effects of inpatient versus outpatient spa therapy on pain, anxiety, and quality of life in elderly patients with generalized osteoarthritis: A pilot study. *International Journal of Biometeorology*, *62*(10), 1823–1832. doi:10.100700484-018-1584-5 PMID:30022244

Özogul, G., & Baran, G. G. (2016). Accessible tourism: The golden key in the future for the specialized travel agencies. *Journal of Tourism Futures*, *2*(1), 79–87. doi:10.1108/JTF-03-2015-0005

Packer, T. L., McKercher, B., & Yau, M. (2007). Understanding the complex interplay between tourism, disability and environmental contexts. *Disability and Rehabilitation*, *29*(4), 281–292. doi:10.1080/09638280600756331 PMID:17364778

Pagán, R. (2015). The contribution of holiday trips to life satisfaction: The case of people with disabilities. *Current Issues in Tourism*, *18*(6), 524–538. doi:10.1080/13683500.2013.860086

Pallud, J., & Straub, D. W. (2014). Effective website design for experience-influenced environments: The case of high culture museums. *Information & Management*, *51*(3), 359–373. doi:10.1016/j.im.2014.02.010

Panou, M., Bekiaris, E., & Robledo, M. G. (2007). Towards an accessible Europe. In *International Conference on Universal Access in Human-Computer Interaction* (pp. 527-535). Springer.

Park, Y. A., Gretzel, U., & Sirakaya-Turk, E. (2007). Measuring web site quality for online travel agencies. *Journal of Travel & Tourism Marketing*, *23*(1), 15–30. doi:10.1300/J073v23n01_02

Parsons, S., & Cobb, S. (2011). State-of-the-art of virtual reality technologies for children on the autism spectrum. *European Journal of Special Needs Education*, *26*(3), 355–366. doi:10.1080/08856257.2011.593831

Paterno, F., & Schiavone, A. G. (2015). The role of tool support in public policies and accessibility. *Interactions (New York, N.Y.)*, *22*(3), 60–63. doi:10.1145/2745395

Patterson, I., & Balderas, A. (2020). Continuing and Emerging Trends of Senior Tourism: A Review of the Literature. *Population Ageing*, *13*(3), 385–399. doi:10.100712062-018-9228-4

Paulo Lima, J., Roberto, R., Simões, F., Almeida, M., Figueiredo, L., Marcelo Teixeira, J., & Teichrieb, V. (2017). Markerless tracking system for augmented reality in the automotive industry. *Expert Systems with Applications*, *82*, 100–114. doi:10.1016/j.eswa.2017.03.060

Peamruetai, B., Khunnara, N., Angkana, A., Sukree, K., Suton, J., Premsuda, J., ... Pravit, A. (2016). The Effect of the Posture of the "Hermit Doing Body Contortion" on Relief of Shoulder and Scapular Pain Caused by Chronic Myofascial Pain Syndrome: A Randomised, Parallel Group, Controlled Trial. *Siriraj Medical Journal*, *68*(6), 350–357.

Pearn, M. (2011). Heritage Sites: Attitudinal and Experimental Differences of Disabled and Able-Bodied Visitors. In S. Darcy & D. Buhalis (Eds.), *Acessible Tourism: Concepts and Issues* (pp. 201–213). Channel View Publications.

Peng, L., & Lai, L. (2014). A service innovation evaluation framework for tourism ecommerce in China based on BP neural network. *Electronic Markets*, *24*(1), 37–46. doi:10.100712525-013-0148-0

Penton, V. M. (2015). Assistive Technology Provision for People with Disabilities in Newfoundland and Labrador, Canada. In D. Edyburn (Ed.), Efficacy of Assistive Technology Interventions (Advances in Special Education Technology, Vol. 1), (pp. 139-162). Emerald Group Publishing Limited.

Pereira, L. (2018). Seaweeds as Source of Bioactive Substances and Skin Care Therapy – Cosmeceuticals, Algotherapy and Thalassotherapy. *Cosmetics*, *5*(4), 68. doi:10.3390/cosmetics5040068

Pereira, R. L., Sousa, P. C., Barata, R., Oliveira, A., & Monsieur, G. (2015). CitySDK tourism API - building value around open data. *Journal of Internet Services and Applications*, *6*(1), 1–13. doi:10.118613174-015-0039-z

Perez-Jimenez, R., Rabadan, J., Rufo, J., Solana, E., & Luna-Rivera, J. M. (2015). Visible light communications technologies for smart tourism destinations. *2015 IEEE 1st International Smart Cities Conference, ISC2 2015*. 10.1109/ISC2.2015.7366193

Perles Ribes, J. F., Ramón Rodriguez, A. B., Vera Rebollo, J. F., & Ivars Baidal, J. (2017). The End of Growth in Residential Tourism Destinations: Steady State or Sustainable Development? The Case of Calpe. *Current Issues in Tourism.* Retrieved from: https://www.tandfonline.com/loi/rcit20

Perna, G., Varriale, L., & Ferrara, M. (2020). Internet for Supporting and Promoting Accessible Tourism: Evidence from Italy. In Y. Baghdadi & A. Harfouche (Eds.), *ICT for an Inclusive World* (Vol. 35, pp. 149–160). Lecture Notes in Information Systems and Organisation. doi:10.1007/978-3-030-34269-2_12

Petrie, H., King, N., & Weisen, M. (2005). The accessibility of museum web sites: Results from an English investigation and international comparisons. In J. Trant & D. Bearman (Eds.), *Museums and the Web 2005: Proceedings. Archives & Museum Informatics, 2005.* Available https://www.archimuse.com/mw2005/papers/petrie/petrie.html

Petticrew, M., & Roberts, H. (2008). *Systematic reviews in the social sciences: A practical guide.* John Wiley & Sons.

Pierdicca, R., Frontoni, E., Zingaretti, P., Malinverni, E. S., Galli, A., Marcheggiani, E., & Costa, C. S. (2016). *Cyberarchaeology: Improved way findings for archaeological parks through mobile augmented reality.* . doi:10.1007/978-3-319-40651-0_14

Pimentel, K., & Teixeira, K. (1994). Virtual Reality: Through the New Looking Glass. *Bulletin of Science, Technology & Society, 14*(4), 224–224.

Pine, B. J., & Gilmore, J. H. (1998). Welcome to the experience economy. *Harvard Business Review, 76*(4), 97–105. PMID:10181589

Pine, B. J., & Gilmore, J. H. (2013). *The experience economy: Past, present and future. Handbook on the experience economy.* Edward Elgar Publishing., doi:10.4337/9781781004227.00007

Plimmer, F., Pottinger, G., & Goodall, B. (2006, October). *Accessibility Issues for Heritage Properties: a Frame of Mind?* Shaping the Change. *XXIII FIG Congress*, Munich, Germany.

PN-Paraplegia News. (2005). Pool Party: Why Swelter in Summers Heat? Whether for therapy or just cooling off, residential swimming pools and spas can be accessed by just about everyone. *PN- Paraplegia News, 5*, 55.

Poon, A. (1988). Tourism and information technologies. *Annals of Tourism Research, 15*(4), 531–549. doi:10.1016/0160-7383(88)90048-5

Pope, C., & Mays, N. (2000). Qualitative research in Health Care. In *Analysing Qualitative Data.* https://www.bmj.com/content

Poria, Y., Reichel, A., & Brandt, Y. (2009). People with disabilities visit art museums: An exploratory study of obstacles and difficulties. *Journal of Heritage Tourism, 4*(2), 117–129. doi:10.1080/17438730802366508

Poria, Y., Reichel, A., & Brandt, Y. (2010). The flight experience of people with disabilities: An exploratory study. *Journal of Travel Research, 49*(2), 216–227. doi:10.1177/0047287509336477

Poria, Y., Reichel, A., & Brandt, Y. (2011). Blind People's Tourism Experiences: An exploratory study. In D. Buhalis & S. Darcy (Eds.), *Accessible Tourism: Concepts and Issues.* Channel View Publications.

Poria, Y., Reichel, A., & Brandt, Y. (2011). Dimensions of hotel experience of people with disabilities: An exploratory study. *International Journal of Contemporary Hospitality Management, 23*(5), 571–591. doi:10.1108/09596111111143340

Porto, N., Rucci, A. C., Darcy, S., Garbero, N., & Almond, B. (2019). Critical elements in accessible tourism: Principal component analysis from Oceania and South America. *Tourism Management, 75*, 169–185. doi:10.1016/j.tourman.2019.04.012

Posada, R. H., Cano, T. C., & Rodríguez, J. A. (2014). Accessibility in eGovernment. In *Proceedings of the XV International Conference on Human Computer Interaction* (pp. 1-2). ACM.

Presley, I., & D'Andrea, F. M. (2009). *Assistive technology for students who are blind or visually impaired: A guide to assessment*. American Foundation for the Blind.

Pritchard, A., Morgan, N., & Ateljevic, I. (2011). Hopeful Tourism: A New Transformative Perspective. *Annals of Tourism Research, 38*(3), 941–963. doi:10.1016/j.annals.2011.01.004

Puhretmair, F., & Nussbaum, G. (2011). *Web design, assistive technologies and accessible tourism. Accessible Tourism Concepts and Issues*. Channel View Publications.

Puhretmair, F., & Nussbaum, G. (2011). Web Design, Assistive Technologies and Accessible Tourism. In D. Buhalis & S. Darcy (Eds.), *Accessible Tourism: Concepts and Issues* (pp. 274–285). Channel View Publications.

Pühretmair, F., & Nussbaum, G. (2011). Web design, assistive technologies and accessible tourism. In D. Buhalis & S. Darcy (Eds.), *Accessible tourism: Concepts and issues* (pp. 274–285). Channel View Publications.

Pühretmair, F., & Wöß, W. (2008). A flexible concept to establish accessibility information in tourism web-pages. In *International Conference on Computers for Handicapped Persons* (pp. 981-988). Springer. 10.1007/978-3-540-70540-6_147

Pulina, F., Broccia, G., Paternò, F., Manca, M., & Schiavone, A. G. (2020). *Multiguideline Accessibility and Usability Validation Environment.* Retrieved from https://mauve.isti.cnr.it/index.jsp

Pulugurtha, S. S., Hirshorn, B. A., Krishnakumar, V. K., & Stewart, J. E. (2006). Transportation network analysis to access community resources by older populations. In Applications of Advanced Technology in Transportation (pp. 677-682). doi:10.1061/40799(213)108

Puma, P. (2019). *The Cortona heritage project: Digital applications in education to architectural and archaeological heritage* . doi:10.1007/978-3-030-12240-9_71

Purnomo, F., & Prabowo, H. (2016). Smart city indicators: A systematic literature review. *Journal of Telecommunication. Electronic and Computer Engineering, 8*(3), 161–164.

Puyuelo, M., Higón, J. L., Merino, L., & Contero, M. (2013). Experiencing Augmented Reality as an Accessibility Resource in the UNESCO Heritage Site Called "La Lonja", Valencia. *Procedia Computer Science, 25*, 171–178. doi:10.1016/j.procs.2013.11.021

Quarati, A., Clematis, A., Roverelli, L., Zereik, G., D'Agostino, D., Mosca, G., & Masnata, M. (2017). Integrating heterogeneous weather-sensors data into a smart-city app. *2017 International Conference on High Performance Computing and Simulation, HPCS 2017*, 152-159. 10.1109/HPCS.2017.33

Quivy, R., & Campenhoudt, L. (1992). *Manual de Investigação em Ciências Sociais*. Gradiva.

Rachid-Kandvani, F., Nicolau, B., & Bedos, C. (2015). Access to Dental Services for People Using a Wheelchair. *American Journal of Public Health, 105*(11), 2312–2317. doi:10.2105/AJPH.2015.302686 PMID:26378836

Racz, A., & Zilizi, G. (2019). Virtual reality aided tourism. *2019 Smart Cities Symposium Prague, SCSP 2019*. 10.1109/SCSP.2019.8805727

Radosav, D., Karuovic, D., Markoski, B., & Ivankovic, Z. (2011). Guidelines on accessible web portal design. In *12th International Symposium on Computational Intelligence and Informatics (CINTI)* (pp. 297-302). IEEE.

Rae, J. (2020). As technology giveth, technology taketh away. In The Routledge Handbook of Disability Activism 2020. Taylor and Francis.

Raigada, J. L. P. (2002). Epistemología, metodología y técnicas del análisis de contenido. *Sociolinguistic Studies*, *3*(1), 1–42.

Rajšp, A., Kous, K., Kuhar, S., Šumak, B., & Šorgo, A. (2019). Preliminary review of jobs, skills and competencies for implementation of Digital Accessibility. In *Proceedings of the Central European Conference on Information and Intelligent Systems* (pp. 93–99). Retrieved from http://archive.ceciis.foi.hr/app/public/conferences/2019/Proceedings/ETICT/ETICT3.pdf

Ramón-Rodríguez, A. B., & Such-Devesa, M. J. (2018). Is hotel efficiency necessary for tourism destination competitiveness? An integrated approach. *Tourism Economics*, *24*(1), 3–26. doi:10.5367/te.2016.0555

Rancati, E., Scuotto, V., & Codignola, F. (2016). *Augmented Reality for Enhancing the Customer Experience in Museums : An Exploratory Analysis. Toulon-Verona Conference. Excellence in Services.*

Randle, M., & Dolnicar, S. (2019). Enabling people with impairments to use Airbnb. *Annals of Tourism Research*, *76*, 278–289. doi:10.1016/j.annals.2019.04.015

Rawlinson, S., & Heap, T. (2017). *International Spa Management: Principles and practice*. Goodfellow Publishers Ltd.

Ray, N. M., & Ryder, M. E. (2003). Ebilities'' tourism: An exploratory discussion of the travel needs and motivations of the mobility-disabled. *Tourism Management*, *24*(1), 57–72. doi:10.1016/S0261-5177(02)00037-7

Reich, C., Lindgren-Streicher, A., Beyer, M., Levent, N., Pursley, J., & Mesiti, L. A. (2011). *Speaking out on art and museums: A study on the needs and preferences of adults who are blind or have low vision*. Boston: Museum of Science, Boston and Art Beyond Sight.

Ribeiro, F., Silva, A., Barbosa, F., Silva, A. P., & Metrôlho, J. C. (2018b). Mobile applications for accessible tourism: Overview, challenges and a proposed platform. *Information Technology & Tourism*, *19*(1–4), 29–59. doi:10.100740558-018-0110-2

Ribeiro, F., Silva, A., Metrôlho, J. C., Silva, A. P., & Barbosa, F. S. (2018a). A new framework for accessible tourism mobile application development. *International Journal of Mobile Computing and Multimedia Communications*, *9*(2), 31–46. doi:10.4018/IJMCMC.2018040103

Richards, V. (2013). *(Re) Envisioning the Tourism Experiences of vision impaired people* (PhD Thesis). Cardiff Metropolitan University.

Richards, V., Pritchard, A., & Morgan, N. (2010). (Re)Envisioning tourism and vision impairment. *Annals of Tourism Research*, *37*(4), 1097–1116. doi:10.1016/j.annals.2010.04.011

Richards, V., Pritchard, A., Morgan, N., & Sedgley, D. (2010). Tourism and Visual Impairment. In S. Cole & N. Morgan (Eds.), *Tourism and Inequality: Problems and Prospects*. CABI. doi:10.1079/9781845936624.0021

Rocha, N. P., Dias, A., Santinha, G., Rodrigues, M., Queirós, A., & Rodrigues, C. (2019). Smart Cities and Public Health: A Systematic Review. *Procedia Computer Science*, *164*, 516–523. doi:10.1016/j.procs.2019.12.214

Rocha, N., Dias, A., Santinha, G., Rodrigues, M., Queirós, A., & Rodrigues, A. (2020). A Systematic Literature Review of Smart Cities' Information Services to Support the Mobility of Impaired People. *International Conference on Health and Social Care Information Systems and Technologies.*

Rodrigues, J. M. F., Sousa, N., Martins, M., & Rosa, M. (2018). IoE accessible bus stop: An initial concept. *ACM International Conference Proceeding Series*, 137-143. 10.1145/3218585.3218659

Rosa, M. P., & Lopes, J. D. C. (2019). Senior tourists' perceptions of bus stop environments used in collaborative design. *WIT Transactions on the Built Environment*, 121-132. 10.2495/UT190111

Rouse, M. (2014). *ICT (information and communications technology, or technologies)*. https://searchcio.techtarget.com/definition/ICT-information-and-communications-technology-or-technologies

Royal National Institute for Blind People. (2015). *My Voice Research: The views and experiences of blind and partially sighted people in the UK*. https://www.rnib.org.uk/sites/default/files/My%20Voice%202015%20-%20Full%20report%20-%20Accessible%20PDF_0.pdf

Royal National Institute for Blind People. (2017). *State of the Nation Eye Health Annual Review*. https://www.rnib.org.uk/state-nation-2017

Royal National Institute for Blind People. (2020). *Key Information and Statistics*. https://www.rnib.org.uk/aboutus/Reserach/statistics/Pages/statistics.aspx

Royal National Institute for Blind People. (2020a). *Calling out inaccessible coronavirus information*. https://www.rnib.org.uk/news/campaigning/inaccessible-coronavirus-information

Royal National Institute for Blind People. (2020b). *Quick tips for digital accessibility*. https://www.rnib.org.uk/campaigning/priority-campaigns/accessible-health-information/coronavirus-and-accessible-online-information

Royal National Institute for Blind People. (2020c). *RNIB asks Number 10 for leadership on accessible coronavirus information*. https://www.rnib.org.uk/news/campaigning/letter-to-prime-minister

Rucci, A. C. (2018). *Discapacidad en el MERCOSUR: accesibilidad turística como una propuesta de desarrollo regional* (Master thesis). Universidad Nacional de La Plata, Argentina.

Rucci, A. C., & Ravlic, I. (2020 in press). La accesibilidad en turismo desde 2005: una mirada desde la LNT. In E. Amadasi & J. L. López Ibáñez (Eds.), El turismo en la Argentina desde 2005: una mirada desde la Ley Nacional de Turismo. Academic Press.

Rumetshofer, H., & Wöß, W. (2004). Tourism information systems promoting barrier-free tourism for people with disabilities. In *International Conference on Computers for Handicapped Persons* (pp. 280–286). 10.1007/978-3-540-27817-7_42

Rumetshofer, H., & Woss, W. (2005). Semantic maps and meta-data enhancing e-accessibility in tourism information systems. In *16th International Workshop on Database and Expert Systems Applications (DEXA'05)* (pp. 881-885). IEEE. 10.1109/DEXA.2005.176

Sahib, N., Tombros, A., & Stockman, T. (2012). A Comparative Analysis of the Information-Seeking Behavior of Visually Impaired and Sighted Searchers. *Journal of the American Society for Information Science and Technology*, *63*(2), 377–391. doi:10.1002/asi.21696

Sahib, N., Tombros, A., & Stockman, T. (2015). Evaluating a Search Interface for Visually Impaired Searchers. *Journal of the Association for Information Science and Technology*, *63*(11), 2235–2248. doi:10.1002/asi.23325

Sánchez-Corcuera, R., Nuñez-Marcos, A., Sesma-Solance, J., Bilbao-Jayo, A., Mulero, R., Zulaika, U., & Almeida, A. (2019). Smart cities survey: Technologies, application domains and challenges for the cities of the future. *International Journal of Distributed Sensor Networks*, *15*(6), 1–36. doi:10.1177/1550147719853984

Santinha, G., Dias, A., Rodrigues, M., Queirós, A., Rodrigues, C., & Rocha, N. P. (2019). How Do Smart Cities Impact on Sustainable Urban Growth and on Opportunities for Entrepreneurship? Evidence from Portugal: The Case of Águeda. In *New Paths of Entrepreneurship Development* (pp. 31–53). Springer. doi:10.1007/978-3-319-96032-6_3

Sapey, B. (2001). From stigma to the social exclusion of disabled people. In T. Mason, C. Carlisle, C. Watkins, & E. Whitehead (Eds.), *Stigma and social exclusion in healthcare*. Routledge.

Schiavone, A. G., & Paterno', F. (2015). An extensible environment for guideline-based accessibility evaluation of dynamic Web applications. *Universal Access in the Information Society*, *14*(1), 111–132. doi:10.100710209-014-0399-3

Schmalstieg, D., & Hollerer, T. (2016). *Augmented reality: principles and practice*. Addison-Wesley Professional.

Scholl-Grissemann, U., & Schnurr, B. (2016). Room with a view: How hedonic and utilitarian choice options of online travel agencies affect consumers' booking intentions. *International Journal of Culture, Tourism and Hospitality Research*, *10*(4), 361–376. doi:10.1108/IJCTHR-06-2016-0062

Schwab, K. (2017). *The Fourth Industrial Revolution*. Crown Business.

Seale, J., Garcia Carrizosa, H., Rix, J., Sheehy, K., & Hayhoe, S. (2020). A participatory approach to the evaluation of participatory museum research projects. *International Journal of Research & Method in Education*, 1–21. Advance online publication. doi:10.1080/1743727X.2019.1706468

Sedlar, U., Kos, A., Pustisek, M., Bester, J., Pogacnik, M., Mali, L., & Stojmenova Duh, E. (2018). *Tackling the challenges of ICT innovation and talents for Industry 4.0*. https://ltfe.org/reference/tackling-the-challenges-of-ict-innovation-and-talents-for-industry-4-0/

Serra, R., Metrôlho, J., & Ribeiro, F. (2019). App for Inclusive Tourism: A case study. In *2019 14th Iberian Conference on Information Systems and Technologies (CISTI)* (pp. 1-6). IEEE. 10.23919/CISTI.2019.8760623

Sert, A. N. (2019). Senior Tourism in the aging world. In V. Krystev, R. Efe & E. Atasoy (Eds.), Theory and Practice in Social Sciences (pp. 488-498). Sofía, Bulgaria: St. Kliment Ohridski University Press.

SFMOMA. (2016). *The New SFMOMA Announces Transformed Digital Strategy: Supported by Bloom- berg Philanthropies, SFMOMA Puts Mobile at the Forefront*. San Francisco Museum of Modern Art. Retrieved from https://s3-us-west-2.amazonaws.com/sfmomamedia/media/pressreleases/downloads/SFMOMA_Digital_Engagement_Press_Release_5_2_2016.pdf

Shabani, N., Munir, A., & Hassan, A. (2019). E-Marketing via Augmented Reality: A Case Study in the Tourism and Hospitality Industry. *IEEE Potentials*, *38*(1), 43–47. doi:10.1109/MPOT.2018.2850598

Shakespeare, T. (2018). *Disability: the basics*. Routledge.

Shakespeare, T., Lezzoni, L. I., & Groce, N. E. (2009). Disability and the training of health professionals. *Lancet*, *374*(9704), 1815–1816.

Shanker, D. (2008). ICT and Tourism: Challenges and Opportunities. In *International Conference "Tourism in India-Challenges Ahead"*. Indian Institute of Management Kozhikode.

Sharpley, R. (2018). *Tourism, Tourists and Society*. Routledge.

Shaw, G., & Coles, T. (2004). Disability, holiday making and the tourism industry in the UK: a preliminary survey. *Tourism Management, 25*(3), 397-403. doi:(03)00139-0 doi:10.1016/S0261-5177

Shaw, G., & Coles, T. (2004). Disability, holiday making and the tourism industry in the UK: A preliminary survey. *Tourism Management*, *25*(3), 397–403. doi:10.1016/S0261-5177(03)00139-0

Shawn, L. H., & McGe, L. (2020). *Accessibility, World Wide Web Consortium*. Available: https://www.w3.org/standards/webdesign/accessibility

Sheehy, K., Garcia Carrizosa, H., Rix, J., Seale, J., & Hayhoe, S. (2019). Inclusive Museums and Augmented Reality: Affordances, Participation, Ethics, and Fun. *The International Journal of the Inclusive Museum*, *12*(4), 67–85. doi:10.18848/1835-2014/CGP/v12i04/67-85

Sheldon, A. (2004). Women and Disability. In J. Swain, S. French, C. Barnes, & C. Thomas (Eds.), *Disabling Barriers – Enabling Environments* (2nd ed.). Sage Publications.

Sheng, J., & Lo, A. (2010). Evaluating the tourism potential of public museums in Hangzhou: A supply-side perspective. *Journal of Travel & Tourism Marketing*, *27*(3), 287–305. doi:10.1080/10548401003744719

Shih, N.-J., Diao, P.-H., & Chen, Y. (2019). ARTS, an AR Tourism System, for the Integration of 3D Scanning and Smartphone AR in Cultural Heritage Tourism and Pedagogy. Sensors, 19(17), 3725.

Shi, Y. (2006). The accessibility of Queensland visitor information centres' websites. *Tourism Management*, *27*(5), 829–841. doi:10.1016/j.tourman.2005.05.012

Shoemaker, S. (2000). Segmenting the mature market: 10 years later. *Journal of Travel Research*, *39*(1), 11–26. doi:10.1177/004728750003900103

Shwab, K. (2016). *The Fourth Industrial Revolution*. World Economic Forum.

Siddaway, A. (2014). *What is a systematic literature review and how do I do one?* University of Stirling.

Silva, F. M., & Borges, I. (2019). Digital Accessibility on Institutional Websites of Portuguese Tourism. In V. Ratten (Ed.), *Technological Progress, Inequality and Entrepreneurship* (pp. 67–85). Springer.

Silveiro, A., Eusébio, C., & Teixeira, L. (2019). Heterogeneity in accessibility of travel agency websites: A study in the central Portugal region. *RISTI - Revista Iberica de Sistemas e Tecnologias de Informação, 35*, 18–34.

Silveiro, A., Eusébio, C., & Teixeira, L. (2019). Heterogeneity in accessibility of travel agency websites: A study in the central Portugal region. *RISTI - Revista Ibérica de Sistemas e Tecnologias de Informação, 35*, 18-34.

Silveiro, A., Eusébio, C., & Teixeira, L. (2019). Heterogeneity in accessibility of travel agency websites: A study in the central Portugal region. *Revista Ibérica de Sistemas y Tecnologías de la Información*, *35*, 18–34. doi:10.17013/risti.35.18-34

Silverman, D. (2000). *Doing Qualitative Data: A Practical Handbook*. Sage Publications.

Small, J. (2014). Interconnecting mobilities on tour: Tourists with vision impairment partnered with sighted tourists. *Tourism Geographies, An International Journal of Tourism Space. Place and Environment*, *17*(1), 76–90.

Small, J., & Darcy, S. (2010). Tourism, Disability and Mobility in Tourism and Inequality. In S. Cole & N. M. Morgan (Eds.), *Tourism and Inequality: Problems and Prospects*. CABI.

Small, J., Darcy, S., & Packer, T. (2011). The embodied tourist experiences of people with vision impairment: Management implications beyond the visual gaze. *Tourism Management*. Advance online publication. doi:10.1016/j.tourman.2011.09.015

Smith, E., Sakakibara, B., & Miller, W. (2016). A review of factors influencing participation in social and community activities for wheelchair users. *Disability and Rehabilitation. Assistive Technology*, *11*(5), 361–374. doi:10.3109/17483107.2014.989420 PMID:25472004

Smith-Jackson, T., Nussbaum, M., & Mooney, A. (2003). Accessible cell phone design: Development and application of a needs analysis framework. *Disability and Rehabilitation*, *25*(10), 549–560.

Smith, M., & Puczko, L. (2017). *The Routledge Handbook of Health Tourism*. Routledge.

Smith, R. W. (1987). Leisure of disabled tourists: Barriers to participation. *Annals of Tourism Research*, *14*(3), 376–389.

Smith, R. W. (1987). Leisure of disabled tourists: Barriers to travel. *Annals of Tourism Research*, *14*(3), 376–389. doi:10.1016/0160-7383(87)90109-5

Smithsonian American Art Museum. (2020). https://americanart.si.edu/visit/accessibility

Sniadek, J. (2006). Age of seniors–a challenge for tourism and leisure industry. *Studies in the Physical Culture and Tourism, 13*, 103–105.

Sobek, A. D., & Miller, H. J. (2006). U-Access: A web-based system for routing pedestrians of differing abilities. *Journal of Geographical Systems, 8*(3), 269–287. doi:10.100710109-006-0021-1

Sociedad Estatal para la Gestión de la Innovación y las Tecnologías Turísticas (SEGITTUR). (2015). *Informe destinos turísticos: construyendo el futuro.* Madrid: Ministerio de Industria, Energía y Turismo.

Sorrell, M., Norton, D., McAdams, J., Winterling, R., & Dipple, K. (2017). Creating an online scientific art exhibit formatted for people with a visual impairment. *Journal of Web Librarianship, 11*(2), 105–123. doi:10.1080/19322909.2017.1300788

Spiel, K. K., Werner, K., Hödl, O., Ehrenstrasser, L., & Fitzpatrick, G. (2017). Creating community fountains by (re-) designing the digital layer of way-finding pillars. *19th International Conference on Human-Computer Interaction with Mobile Devices and Services, MobileHCI 2017*, 10.1145/3098279.3122135

Sritoomma, N., Moyle, W., Cooke, M., & O'Dwyer, S. (2012). The effectiveness of Swedish massage and traditional Thai massage in treating chronic low back pain: A review of the literature. *Complementary Therapies in Clinical Practice, 18*(4), 227–234. doi:10.1016/j.ctcp.2012.07.001 PMID:23059437

Standen, P. J., & Brown, D. J. (2005). Virtual reality in the rehabilitation of people with intellectual disabilities [review]. *Cyberpsychology & Behavior, 8*(3), 272–282. doi:10.1089/cpb.2005.8.272 PMID:15971976

Stangl, B., & Weismayer, C. (2008). Websites and virtual realities: a useful marketing tool combination? An exploratory investigation. In Information and communication technologies in tourism 2008 (pp. 141–151). Springer. doi:10.1007/978-3-211-77280-5_13

Stillman, M., Williams, S., Bertocci, G., Smalley, C., & Frost, K. (2017). Healthcare utilization and associated barriers experienced by wheelchair users: A pilot study. *Disability and Health Journal, 10*(4), 502–508. doi:10.1016/j.dhjo.2017.02.003 PMID:28245968

Stone, E., & Priestley, M. (1996). Parasites, pawns and partners: Disability research and the role of non-disabled researchers. *The British Journal of Sociology, 47*, 699–716.

Stumbo, N. J., & Pegg, S. (2005). Travelers and tourists with disabilities: A matter of priorities and loyalties. *Tourism Review International, 8*(3), 195–209. doi:10.3727/154427205774791537

Stylianou-Lambert, T., & Stylianou, E. (2010). A third space: reconsidering issues of neutrality and accessibility in the virtual art museum. *EVA London 2010: Electronic visualisation & the arts. Proceedings of a conference held in London*, 66-71. 10.14236/ewic/EVA2010.12

Suarez, M., Gonzalez, P., Dominguez, R., Bravo, A., Melian, C., Perez, M., & Fagundo, J. (2011). Identification of organic compounds in San Diego de los Banos Peliod (Pinar del Rio, Cuba). *Journal of Alternative and Complementary Medicine (New York, N.Y.), 17*(2), 155–165. doi:10.1089/acm.2009.0587 PMID:21309710

Sun, Y., Song, H., Jara, A. J., & Bie, R. (2016). Internet of Things and Big Data Analytics for Smart and Connected Communities. *IEEE Access: Practical Innovations, Open Solutions, 4*, 766–773. doi:10.1109/ACCESS.2016.2529723

Swain, J., & French, S. (2000). Towards an Affirmation Model of Disability. *Disability & Society, 15*(4), 569–582.

Sylaiou, S., & Papaioannou, G. (2019). ICT in the Promotion of Arts and Cultural Heritage Education in Museums. In *Strategic Innovative Marketing and Tourism* (pp. 363–370). Springer. doi:10.1007/978-3-030-12453-3_41

Szcygiel, A., & Piotrowicz, M. (1995). Leisure activities of people with special needs, their selfconcept and social interactions. In A. Slezynski & M. Petrrynski (Eds.), *Progress in Tourism of People with Special Needs* (pp. 280–289). AGAT.

Tadić, V., Hundt, G. L., Keeley, S., & Rahi, J. S. (2015). Seeing it my way: Living with childhood onset visual disability. *Child: Care, Health and Development, 41*(2), 239–248. doi:10.1111/cch.12158 PMID:24913734

Taheri, B. (2011). *Unpacking visitor engagement: Examining drivers of engagement in museums* (Doctoral dissertation). University of Strathclyde.

Tao, B. C., Goh, E., Huang, S., & Moyle, B. (2019). Travel constraint perceptions of people with mobility disability: A study of Sichuan earthquake survivors. *Tourism Recreation Research, 44*(2), 203–216. doi:10.1080/02508281.2019.1589085

Tarasoff, L. (2017). Sexual and Reproductive Health of Women with Disability: "We don't know. We've never had anybody like you before. Barriers to perinatal care for women with physical disabilities. *Disability and Health Journal, 10*(3), 426–433. doi:10.1016/j.dhjo.2017.03.017 PMID:28404229

Tecau, A. S., Bratucu, G., Tescaşiu, B., Chiţu, I. B., Constantin, C. P., & Foris, D. (2019). Responsible tourism-integrating families with disabled children in tourist destinations. *Sustainability (Switzerland), 11*(16), 4420. doi:10.3390u11164420

Teixeira, L., Eusébio, C., & Silveiro, A. (2019). Website Accessibility of Portuguese Travel Agents. 14th Conferência Ibérica de Sistemas y Tecnologías de Información, 1–6.

Teixeira, L., Eusébio, C., & Silveiro, A. (2019). Website Accessibility of Portuguese Travel Agents. In *14th Iberian Conference on Information Systems and Technologies (CISTI)* (pp. 1–6). 10.23919/CISTI.2019.8760949

Teixeira, L., Eusebio, C., & Silveiro, A. (2019). Website accessibility of portuguese travel agents. In *14th Iberian Conference on Information Systems and Technologies (CISTI)* (pp. 1–6). IEEE.

Thangaraj, M., & Gomathi, K. S. (2019). Design and development of MDOE for virtual tourism management for elderly people. *International Journal of Recent Technology and Engineering, 8*(2), 3558–3562. doi:10.35940/ijrte.B3067.078219

The Centre for Excellence in Universal Design. (2019). *What is Universal Design?* Retrieved from: http://universaldesign.ie/What-is-Universal-Design/

The World Bank. (2016). *Disability-overview.* Retrieved from The World Bank: http:// www.worldbank.org/en/topic/disability/overview#1

Thomas, W., & Carey, S. (2005). Actual/Virtual Visits: What Are The Links? In J. Trant & D. Bearman (Eds.), *Museums and the Web 2005: Proceedings*. Archives & Museum Informatics.

Thompson, N. (2016). *Anti-discriminatory Practice, Equality, Diversity and Social Justice* (6th ed.). Palgrave.

Thomson, D. J., Gylseth, M., McGarry, R., & Garcia, C. V. (2007). The VVIP system: encouraging the use of public transport in Edinburgh. In CHI'07 Extended Abstracts on Human Factors in Computing Systems (pp. 2091-2096). doi:10.1145/1240866.1240955

Tom Dieck, M. C., & Jung, T. (2018). A theoretical model of mobile augmented reality acceptance in urban heritage tourism. *Current Issues in Tourism, 21*(2), 154–174. doi:10.1080/13683500.2015.1070801

Tom Dieck, M. C., Jung, T., & Michopoulou, E. (2019). Experiencing Virtual Reality in Heritage Attractions: Perceptions of Elderly Users. In M. C. Tom Dieck & T. Jun (Eds.), *Augmented Reality and Virtual Reality The Power of AR and VR for Business* (pp. 89–98). Springer. doi:10.1007/978-3-030-06246-0_7

TRA - Tourism Research Australia. (2019). *Latest International Visitor Survey - Results Year ending December 2019*. Retrieved from: https://www.tra.gov.au/International/international-tourism-results

Tregaskis, C. (2003). Towards inclusive practice: An insider perspective on leisure provision for disabled people. *Managing Leisure*, *8*(1), 28–40.

Tsang, K., Lai, M. T., & Law, R. (2010). Measuring e-service quality for online travel agencies. *Journal of Travel & Tourism Marketing*, *27*(3), 306–323. doi:10.1080/10548401003744743

Tsiotsou, R., & Hudson, S. (2010). Wooing zoomers: Marketing to the mature traveller. *Marketing Intelligence & Planning*, *28*(4), 444–461. doi:10.1108/02634501011053568

Tung, V., & Law, R. (2017). The potential for tourism and hospitality experience research in human-robot interactions. *International Journal of Contemporary Hospitality Management*, *29*(10), 2498–2513. doi:10.1108/IJCHM-09-2016-0520

Tussyadiah, I. P., Jung, T. H., & tom Dieck, M. C. (2018). Embodiment of Wearable Augmented Reality Technology in Tourism Experiences. *Journal of Travel Research*, *57*(5), 597–611. doi:10.1177/0047287517709090

Tussyadiah, I. P., Wang, D., & Jia, C. (2017). Virtual Reality and Attitudes Toward Tourism Destinations. In R. Schegg & B. Stangl (Eds.), *Information and Communication Technologies in Tourism 2017* (pp. 229–239). Springer International Publishing. doi:10.1007/978-3-319-51168-9_17

Tussyadiah, I. P., Wang, D., Jung, T. H., & tom Dieck, M. C. (2018). Virtual reality, presence, and attitude change: Empirical evidence from tourism. *Tourism Management*, *66*, 140–154. doi:10.1016/j.tourman.2017.12.003

Ucok, K., Mollaoglu, H., Demirel, R., Ucok, K., Mollaoglu, H., Kavuncu, V., ... Evcik, D. (2008). Effects of balneotherapy with exercise in patients with low back pain. *Journal of Back and Musculoskeletal Rehabilitation*, *21*(4), 263–272. doi:10.3233/BMR-2008-21407

UN. (2019). *Perspectivas de la población mundial 2019*. Retrieved from: https://population.un.org/wpp/

UNE 178501 and UNE 178502 (2016). *Destinos Turísticos Inteligentes*. AENOR, Spain.

UNESCO. (2019). *ICTs for Persons living with Disabilities*. https://en.unesco.org/partnerships/partnering/icts-persons-living-disabilities

United Nations News. (2016). Accessible tourism will benefit everyone, say senior UN officials on World Day. *UN News*. Accessed May 2020 https://news.un.org/en/story/2016/09/541232-accessible-tourism-will-benefit-everyone-say-senior-un-officials-world-day#.V-vdOPkrKUl

United Nations World Tourism Organisation. (2020). *Why Tourism? Tourism – an economic and social phenomenon*. https://www.unwto.org/why-tourism

United Nations, Department of Economic and Social Affairs, Population Division. (2015). *World Population Prospects: The 2015 Revision, Key Findings and Advance Tables*. Working Paper No. ESA/P/WP.241.

United Nations, Department of Economic and Social Affairs, Population Division. (2019). *World Population Prospects 2019: Highlights*. Working Paper No. ST/ESA/SER.A/423.

United Nations. (2006). *Convención sobre los Derechos de las Personas con Discapacidad y Protocolo Facultativo*. Nueva York: UN.

United Nations. (2006a). *Convention on the Rights of Persons with Disabilities – Articles*. Convention on the Rights of Persons with Disabilities (CRDP). Available at: https://www.un.org/development/desa/disabilities/convention-on-the-rights-of-persons-with-disabilities/convention-on-the-rights-of-persons-with-disabilities-2.html

United Nations. (2006b). *Article 19 – Living independently and being included in the community.* Convention on the Rights of Persons with Disabilities (CRDP). Available at: https://www.un.org/development/desa/disabilities/convention-on-the-rights-of-persons-with-disabilities/article-19-living-independently-and-being-included-in-the-community.html

United Nations. (2006c). Article 30 – Participation in cultural life, recreation, leisure and sport. *Convention on the Rights of Persons with Disabilities (CRDP).* Available at: https://www.un.org/development/desa/disabilities/convention-on-the-rights-of-persons-with-disabilities/article-30-participation-in-cultural-life-recreation-leisure-and-sport.html

United Nations. (2015). *World population ageing 2015.* Retrieved from: https://www.un.org/en/development/desa/population/publications/pdf/ageing/WPA2015_Report.pdf

United States Environmental Protection Agency. (2019). *What is Section 508?* Retrieved from https://www.epa.gov/accessibility/what-section-508

UNWTO. (2014). *Manual sobre Turismo Accesible para todos. Alianzas público-privadas y buenas prácticas sobre Turismo Accesible.* UNWTO.

UNWTO. (2016). *Manual on accessible tourism for all: Principles, tools and best practices.* Retrieved from: https://www.e-unwto.org/doi/pdf/10.18111/9789284418077

UNWTO. (2016). *Manual on Accessible Tourism for All: Principles, Tools and Best Practices.* UNWTO Publications.

UNWTO. (2018). *Official Celebration of World Tourism Day 2018 - Tourism and digital transformation.* Retrieved from: http://media.unwto.org/es/event/celebracion-oficial-dia-mundial-del-turismo-2018-turismo-y-la-transformacion-digital

UNWTO. (2019). *UNWTO Tourism Definitions.* UNWTO Publications.

Uslan, M., Peck, A., & Waddell, W. (1988). Audible Traffic Signals: How Useful Are They. *ITE Journal*, 37–43.

Vacas, T. (2005). Los museos del "Paseo del Arte: Espacio turístico cultural de ámbito internacional. *Estudios Turísticos, 165*, 97–129.

Vanolo, A. (2014). Smartmentality: The smart city as disciplinary strategy. *Urban Studies (Edinburgh, Scotland), 51*(5), 883–898. doi:10.1177/0042098013494427

Varga, J., Gaal, J., Varga, J., Szekanecz, Z., Kurko, J., Ficzere, A., ... Bender, T. (2008). Balneotherapy in elderly patients: Effect on pain from degenerative knee and skin conditions and on quality of life. *The Israel Medical Association Journal, 10*(5), 365–369. PMID:18605361

Var, T., Yesiltas, M., Yayli, A., & Öztürk, Y. (2011). A study on the travel patterns of physically disabled people. *Asia Pacific Journal of Tourism Research, 16*(6), 599–618. doi:10.1080/10941665.2011.610143

Vaz, R. (2020). Co-Creating an Integrative Framework to Enhance the Museum Experience of Blind and Visually Impaired Visitors. In J. M. F. Rodrigues, C. M. Q. Ramos, P. J. S. Cardoso, & C. Henriques (Eds.), *Handbook of Research on Social Media Applications for the Tourism and Hospitality Sector* (pp. 164–191). IGI Global. doi:10.4018/978-1-7998-1947-9.ch011

Vaz, R., Fernandes, P., & Veiga, A. (2018). Interactive technologies in museums: How digital installations and media are enhancing the visitors' experience. In *Handbook of Research on Technological Developments for Cultural Heritage and eTourism Applications* (pp. 30–53). IGI Global. doi:10.4018/978-1-5225-2927-9.ch002

Velasco, M. (2009). Gestión Turística del Patrimonio Cultural: Enfoques para un desarrollo sostenible del Turismo cultural. *Cuadernos de Turismo, 23*, 237–253.

Verma, S., Venugopala, N., Rao, P., & Balasubramanian, K. (2019). *Snapping, Virtual Inking, and Accesibility in Augmented Reality*. https://patentscope.wipo.int/search/zh/detail.jsf;jsessionid=111B01279B2DD3C7C81BC846BEA6C84A.wapp1nA?docId=US298181772&_cid=P10-KCM60G-37278-23

Victoria and Albert Museum. (2020). https://www.vam.ac.uk/info/disability-access

Visit England. (2015). *The Purple Pound. Volume and Value of Accessible Tourism in England.* Available at: https://www.visitbritain.org/sites/default/files/vb-corporate/Documents-Library/ve_purplepound2015.pdf

VisitEngland. (2018). *Accessibility – non participation*. VisitEngland. https://www.visitbritain.org/sites/default/files/vb-corporate/Documents-Library/documents/England-documents/april_2018_accessibility_non_participation_reportv3.pdf

Vitaliev, V. (2016). VR around the... world. *Engineering & Technology*, *11*(3), 33. doi:10.1049/et.2016.0312

Voigt, C., Brown, G., & Howat, G. (2011). Wellness tourists: In search of transformation. *Tourism Review*, *66*(1), 16–30. doi:10.1108/16605371111127206

Voigt, C., & Pforr, C. (2014). *Wellness Tourism – A destination perspective*. Routledge.

W3C WAI. (2018a). *Introduction to Web Accessibility*. Retrieved from https://www.w3.org/WAI/fundamentals/accessibility-intro/

W3C WAI. (2018b). *Techniques for WCAG 2.1*. Retrieved from https://www.w3.org/WAI/WCAG21/Techniques/#techniques

W3C WAI. (2018c). *Understanding WCAG 2.1*. Retrieved from https://www.w3.org/WAI/WCAG21/Understanding/

W3C WAI. (2018d). *Web Accessibility Laws & Policies*. Retrieved from https://www.w3.org/WAI/policies/

W3C WAI. (2018e). *Web Content Accessibility Guidelines (WCAG) 2.1*. Retrieved from https://www.w3.org/TR/WCAG21/

W3C WAI. (2019a). *Web Accessibility Evaluation Tools List*. Retrieved from https://www.w3.org/WAI/ER/tools/?q=wcag-21-w3c-web-content-accessibility-guidelines-21&q=groups-of-web-pages-or-web-sites

W3C WAI. (2019b). *What's New in WCAG 2.1*. Retrieved from https://www.w3.org/WAI/standards-guidelines/wcag/new-in-21/

W3C. (2016). *How to Meet WCAG 2 (Quick Reference)*. Available: https://www.w3.org/WAI/WCAG21/quickref

W3C. (2016). *Understanding WCAG 2.0*. https://www.w3.org/TR/UNDERSTANDING-WCAG20/

W3C. (2018). *Web Content Accessibility Guidelines (WCAG) 2.1*. Available: https://www.w3.org/TR/2018/REC-WCAG21-20180605

W3C. (2018). *Web Content Accessibility Guidelines (WCAG) 2.1*. Retrieved from: https://www.w3.org/TR/2018/REC-WCAG21-20180605

W3C. (2019). *Web Accessibility Evaluation Tools List*. Available: https://www.w3.org/WAI/ER/tools

W3C. (2020). *Web Content Accessibility Guidelines (WCAG) 2.2 in Development*. Available: https://www.w3.org/blog/2020/02/wcag-22-in-development/

Wales Council for the Blind. (2008). Statistics in Wales. In *Re) Envisioning the Tourism Experiences of vision impaired people*. Cardiff Metropolitan University.

Waligo, V. M., Clarke, J., & Hawkins, R. (2013). Implementing sustainable tourism: A multistake- holders involvement management framework. *Tourism Management*, *36*, 342–353. doi:10.1016/j.tourman.2012.10.008

Walsh, D., Hall, M. M., Clough, P., & Foster, J. (2020). Characterising online museum users: A study of the National Museums Liverpool museum website. *International Journal on Digital Libraries*, *21*(1), 75–87. doi:10.100700799-018-0248-8

Wang, D., & Fesenmaier, D. R. (2013). Transforming the Travel Experience: The Use of Smartphones for Travel. In Information and Communication Technologies in Tourism 2013 (pp. 58–69). Springer Berlin Heidelberg.

Wang, D., Li, R., & Li, Y. (2013). China's "smart tourism destination initiative: A state of the service-dominant logic. *Journal of Destination Marketing & Management*, *2*(2), 59–61. doi:10.1016/j.jdmm.2013.05.004

Wan, Y. (2013). Barriers for people with disabilities in visiting casinos. *International Journal of Contemporary Hospitality Management*, *25*(5), 660–682. doi:10.1108/IJCHM-Jul-2012-0112

Waruwu, A. F., Agung Bayupati, I. P., & Gede Darma Putra, I. K. (2015). Augmented Reality Mobile Application of Balinese Hindu Temples: DewataAR. *International Journal of Computer Network and Information Security*, *7*(2), 59–66. doi:10.5815/ijcnis.2015.02.07

Waschke, S. (2004). *Labeling im barrierefreien tourismus in deutschland–Vergleichende analyse auf basis europäischer beispiele* [Labelling in barrier-free tourism in germany comparative analysis based on european examples]. University of Lüneburg.

Web Content Accessibility Guidelines. (2018). http://www.w3.org/WAI/standards-guidelines/wcag

WebAIM. (2018). *Section 508 Checklist*. Retrieved from https://webaim.org/standards/508/checklist

Weiermair, K., & Mathies, C. (2004). *The Tourism and Leisure Industry: Shaping the Future*. The Haworth Press, Inc.

Weisen, M., Petrie, H., King, N., & Hamilton, F. (2005). Web Accessibility Revealed: The Museums, Libraries and Archives Council Audit. *Ariadne*, (44), 1–10.

Wei, W. (2019). Research progress on virtual reality (VR) and augmented reality (AR) in tourism and hospitality. *Journal of Hospitality and Tourism Technology*, *10*(4), 539–570. doi:10.1108/JHTT-04-2018-0030

Werthner, H., & Klein, S. (1999). *Information technology and tourism: A challenging relationship. Viena*. Springer-Verlag. doi:10.1007/978-3-7091-6363-4

Werthner, H., & Ricci, F. (2004). E-commerce and tourism. *Communications of the ACM*, *47*(12), 101–105. doi:10.1145/1035134.1035141

WHO - World Health Organization. (2020). *10 facts on disability*. Retrieved from: https://www.who.int/features/factfiles/disability/en/

WHO. (2015). *World report on ageing and health 2015*. Retrieved from: https://www.who.int/ageing/events/world-report-2015-launch/en/

Wiener, W., Welsh, R., & Blasch, B. (2010). Foundations of orientation and mobility (3rd ed. Vol. 1). New York: AFB (American Foundation for the Blind) Press.

Williams, O. J. (2019). *Low Vision in the 21st century – keeping up with the technology: Wales Optometry Postgraduate Education Centre*. Wales Council of the Blind.

Williams, R., & Brownlow, S. (2019). *The Click-Away Pound Report*. Freeney Williams Ltd.

Williams, R., & Rattray, R. (2005). UK hotel web page accessibility for disabled and challenged users. *Tourism and Hospitality Research*, *5*(3), 255–268. doi:10.1057/palgrave.thr.6040024

Williams, R., Rattray, R., Grimes, A., Darcy, S., Mills, J. E., Han, J. H., & Clay, M. J. (2006). Accessibility of hospitality and tourism websites: A challenge for visually impaired persons. *International Journal of Tourism Research*, *5*(1), 28–41.

Williams, V. (2017). The 'Purple Pound' and Breaking down Barriers in the Workplace in the UK. *International Journal of Therapy and Rehabilitation*, *24*(3), 97. doi:10.12968/ijtr.2017.24.3.97

Wilson, R. J. (2011). Behind the scenes of the museum website. *Museum Management and Curatorship*, *26*(4), 373–389. doi:10.1080/09647775.2011.603934

Winkler, M., & Wöß, W. (2006). Accessibility add-on box enabling barrier-free tourism information systems (TIS). In *International Conference on Computers for Handicapped Persons* (pp. 298-305). Springer. 10.1007/11788713_45

Winkowska, J., Szpilko, D., & Pejić, S. (2019). Smart city concept in the light of the literature review. *Engineering Management in Production and Services*, *11*(2), 70–86. doi:10.2478/emj-2019-0012

Wobbrock, J. O., Kane, S. K., & Ladner, R. (2011). Usable Gestures for Blind People: Understanding Preference and Performance. CHI 2011, Vancouver, Canada

World Bank (2018). *Disability Inclusion and Accountability Framework*. Retrieved from: file:///D:/Docs%20Andreia%202019_2020/2.%20ARTIGOS/1.%20ARTIGOS%202019_2020/ICOTTS_2020/126977-WP-PUBLIC-DisabilityInclusionAccountabilitydigital.pdf

World Bank. (2020). *Argentina Data*. Retrieved from: https://data.worldbank.org/country/argentina

World Blind Union. (n.d.). *White Cane Safety Day*. http://www.worldblindunion.org

World Economic Forum. (2017). *Digital Transformation Initiative: Aviation, Travel and Tourism Industry White Paper*. http://reports.weforum.org/digital-transformation

World Health Organisation. (2017). *10 facts on disability*. Available at: https://www.who.int/features/factfiles/disability/en/

World Health Organisation. (2020). *Blindness and Vision Impairment*. https://www.who.int/news-room/fact-sheets/detail/blindness-and-visual-impairment

World Health Organization (WHO). (2011). *Informe Mundial sobre la Discapacidad*. Ginebra: WHO.

World Health Organization. (2010). *New estimates of visual impairment and blindness*. Retrieved from: http://www.who.int/blindness/publications/globaldata/ en/

World Health Organization. (2011). *Relatório Mundial sobre a Deficiência*. Retrieved from: https://apps.who.int/iris/bitstream/handle/10665/70670/WHO_NMH_VIP_11.01_por.pdf;jsessionid=5BBD98488DC0CF3AD13E7C780D460A2F?sequence=9

World Health Organization. (2016). *Disability and health*. Accessed on December 5, 2016, from https://www.who.int/mediacentre/factsheets/fs352/en/

World Health Organization. (2016). *World Report on Disability*. Retrieved from https://apps.who.int/iris/bitstream/handle/10665/70670/WHO_NMH_VIP_11.01_eng.pdf?sequence=1WHO

World Health Organization. (2017). *Vision impairment and blindness*. Accessed on March 24, 2020, from http://who.int/mediacentre/factsheets/fs282/en/

World Health Organization. (2018). *Disability and health*. Retrieved from https://www.who.int/news-room/fact-sheets/detail/disability-and-health

World Health Organization. (2018). *International classification of diseases: the global standard for diagnostic health information.* Accessed on March 24, 2020 from https://icd.who.int/en/

World Health Organization. (2019). *Blindness and vision impairment.* Accessed on March 24, 2020. from https://www.who.int/news-room/fact-sheets/detail/blindness-and-visual-impairment

World Tourism Organization Business Council. (1999). *Marketing Tourism Destinations Online: Strategies for the Information Age.* Madrid: World Tourism Organization. https://www.e-unwto.org/doi/book/10.18111/9789284403288

World Tourism Organization. (2013). *Recommendations on Accessible Tourism.* Madrid: World Tourism Organization. Retrieved from: http://business.turismodeportugal.pt/SiteCollectionDocuments/all-for-all/recommendations-accesible-tourism.pdf

World Wide Consortium. (2020). *Web Content Accessibility Guidelines 1.0.* Accessed on April 1, 2020 from https://www.w3.org/TR/WCAG21/#orientation

Wu, Y. C. J., Chang, C. L., & Hsieh, Y. J. (2014). Enhancing learning experience of the disabled: An accessible tourism platform. *Journal of Universal Computer Science, 20*(15), 2080–2095.

Xiang, Z., Gretzel, U., & Fesenmaier, D. (2009). Semantic Representation of Tourism on the Internet. *Journal of Travel Research, 47*(4), 440–453.

Xiang, Z., Magnini, V. P., & Fesenmaier, D. R. (2015). Information technology and consumer behavior in travel and tourism: Insights from travel planning using the internet. *Journal of Retailing and Consumer Services, 22*, 244–249. doi:10.1016/j.jretconser.2014.08.005

Yamamoto, K., & Ikeda, T. (2016). Social recommendation GIS for urban tourist spots. *18th IEEE International Conference on High Performance Computing and Communications, 14th IEEE International Conference on Smart City and 2nd IEEE International Conference on Data Science and Systems, HPCC/SmartCity/DSS 2016*, 50-57. 10.1109/HPCC-SmartCity-DSS.2016.0018

Yau, M. K., McKercher, B., & Packer, T. L. (2004). Traveling with a Disability More than an Access Issue. *Annals of Tourism Research, 31*(4), 946–960.

Yau, M. K., McKercher, B., & Packer, T. L. (2004). Traveling with a Disability: More than an Access Issue. *Annals of Tourism Research, 31*(4), 946–960. doi:10.1016/j.annals.2004.03.007

Yerbury, H., Darcy, S., & Burridg, N. (2019). Accessing learning resources: experiences of students with disability. *Proceedings of the Tenth International Conference on Conceptions of Library and Information Science, Ljubljana, Slovenia, June 16-19, 2019. Information Research, 24*(4), paper colis 1929. Retrieved from http://InformationR.net/ir/24-4/colis/colis1929.html

Yung, R., & Khoo-Lattimore, C. (2019). New realities: A systematic literature review on virtual reality and augmented reality in tourism research. *Current Issues in Tourism, 22*(17), 2056–2081. doi:10.1080/13683500.2017.1417359

Zajadacz, A. (2014). Sources of tourist information used by deaf people. Case study: The Polish Deaf community. *Current Issues in Tourism, 17*(5), 434–454. doi:10.1080/13683500.2012.725713

Zajadacz, A., & Szmal, P. (2017). Accessible tourism for deaf people in Poland: The SITur and SITex programs as proposals for accessible urban information. In *International Conference on Universal Access in Human-Computer Interaction* (pp. 348-359). Springer. 10.1007/978-3-319-58700-4_29

Zakaria, N. N. (2020). Barriers to social inclusion with the Egyptian museums; New approach towards disability. *Curator, 63*(1), 115–130. doi:10.1111/cura.12353

Zanella, A., Bui, N., Castellani, A., Vangelista, L., & Zorzi, M. (2014). Internet of things for smart cities. *IEEE Internet of Things Journal*, *1*(1), 22–32. doi:10.1109/JIOT.2014.2306328

Zhang, Y. J., Driscoll, D. A., Lazarony, P. J., Lin, B., & Zhang, R. (2006). What affects travellers' preference of e-travel or traditional travel agencies: An empirical study. *International Journal of Services and Standards*, *2*(4), 385–399. doi:10.1504/IJSS.2006.010471

Zhang, Y., & Cole, S. T. (2016). Dimensions of lodging guest satisfaction among guests with mobility challenges: A mixed-method analysis of web-based texts. *Tourism Management*, *53*, 13–27. doi:10.1016/j.tourman.2015.09.001

Zhang, Y., Roxburgh, R., Huang, L., Parsons, J., & Davies, T. (2014). The effect of hydrotherapy treatment on gait characteristics of hereditary spastic paraperesis patients. *Gait & Posture*, *39*(4), 1074–1079. doi:10.1016/j.gaitpost.2014.01.010 PMID:24556467

Zolotas, M., Elsdon, J., & Demiris, Y. (2018). Head-Mounted Augmented Reality for Explainable Robotic Wheelchair Assistance. *2018 IEEE/RSJ International Conference on Intelligent Robots and Systems (IROS)*, 1823–1829. 10.1109/IROS.2018.8594002

About the Contributors

Celeste Eusébio is an Associate Professor of Tourism and a researcher at the GOVCOPP Research Unit at the University of Aveiro (Portugal). She holds a Degree in Tourism Management and Planning from the University of Aveiro (1995), a Master in Economics from the University of Coimbra (1998) and a Doctoral Degree in Tourism, also from the University of Aveiro (2006). Her research interests include tourism economics, tourism impacts, tourism forecasts and consumer behavior in tourism.

Leonor Teixeira graduated in Industrial Engineering and Management, received a M.Sc. degree in Information Management, and a PhD in Industrial Management (Information Systems area), in 2008, from the University of Aveiro, Portugal. She is currently an Associate Professor of the Department of Economics, Management, Industrial Engineering and Tourism at the University of Aveiro and teaching subjects related to Information Systems, Information Technologies to undergraduate and post-graduate studies. She is also a researcher at the Institute of Electronics and Telematics Engineering (IEETA) of University of Aveiro. Her current research interests include Information Systems applied to Industry, focusing on Requirement Engineering, Software Engineering and Usability Engineering. She has several publications in peer-reviewed journals, book chapters and proceedings, and oral communications at international scientific conferences. She serves as a member of Program Board and Organizing Committees for several Scientific Committees of International Conferences. She is associated with IIIS, IEEE and APSI/PTAIS.

Maria João Carneiro is Assistant Professor of Tourism at the University of Aveiro (UA) (in Portugal) and researcher at the Governance, Competitiveness, and Public Policy (GOVCOPP) research unit at this University. She holds a Degree in Tourism Management and Planning from UA, an MBA from New University of Lisbon and a PhD in Tourism from UA. She is the Vice-Director of the degree program (licenciatura) in Tourism of the UA. She has published papers in several journals and conference proceedings, and has also presented several works in conferences and seminars. Her research interests are competitiveness of tourism destinations, consumer behaviour in tourism and destination marketing.

* * *

Meltem Altinay Özdemir is a senior lecturer of the Department of Travel and Tourism at the University of Mugla Sitki Kocman in Turkey. She holds a Degree in Travel Management and Tour Guiding from the University of Gazi (2013), a Master in Recreation Management at the same university (2016), and a PhD in Tourism Management from the University of Istanbul (2020). She teaches several courses in tourism marketing, tourist behavior, tour guiding, and recreation management. She is also an official tour guide. After she worked in different travel agencies and hospitality businesses between 2008 and 2013, she worked as a lecturer at the University of Istanbul Arel and an official tour guide between 2015 and 2018. She has published papers in several journals and conference proceedings. Meltem has published over 30 academic journal articles, book chapters, and conference papers. Her research interests are tourism marketing, destination management, consumer behavior in tourism, and leisure studies.

Leandro F. Becka is a researcher at the Tourism Research Institute of the Faculty of Economics, Universidad Nacional de La Plata (UNLP), Argentina. Bachelor in Tourism in the same university, his research interests include event tourism, accessible tourism, and impacts of tourism in world heritage cases. Currently, he is in the process of becoming a Master in Business Management in the National University of La Plata.

Raquel Camprubí was lecturer at University of Girona from 2004 to 2010 and has been Associate Professor at University of Girona, since then. She completed her PhD in 2009, focusing on tourism image formation and relational networks, and she was awarded with the PhD Extraordinary Award on Management and Business Administration 2011. She graduated in Tourism at University of Girona and Maîtrise en Ingenerie et Commercialisation des Produits Touristiques from University of Toulouse - Le Mirail. She teaches several courses in research methods, communication and strategic management. She is coordinator of the Master in Tourism Management and Planning at the University of Girona. Her research interests cover destination management, tourism image and branding, and consumer behavior. She is member of the research group Organizational Networks and Innovation in Tourism (ONIT), and she has published in several academic journals, such as Annals of Tourism Research and Tourism Management.

Simon Darcy is a Professor at the UTS Business School, University of Technology Sydney. He specialises in developing inclusive organisational approaches for diversity groups and understanding the social impact of organisations and individuals. Simon's research has spanned sport, tourism, events, volunteers, transport, the built environment and disability services. His research and industry collaboration on accessible tourism has been recognised for its outstanding contribution to the field through the World Leisure Organisation's Innovation Prize, the United Nations World Tourism Organization's publication on best practice and the Asociación Española de Expertos Científicos en Turismo best research paper published in 2015. In 2017 Simon presented the Richard Jones Oration for the Queensland Anti-Discrimination Commission examining issues of transport, travel, sport and tourism as it relates to the UN Convention on the Rights of Persons with Disabilities. He is co-author of Benchmark Games: Sydney 2000 Paralympic Games; Accessible Tourism: Concepts and Issues (2011); Best Practice in Accessible Tourism: Inclusion, Disability, Ageing Population and Tourism (2012); Australian Leisure (2013); Research Methods for Sport Studies and Sport Management (2014); and Managing the Paralympiics (2017).

Hugo García-Andreu (Alicante, 1976) is an Associate Professor in sociology and tourism at University of Alicante, being part of the University Institute for Tourism Research and the Interuniversity Doctorate Program in Tourism. He has participated in numerous R&D projects, some of them international, in the field of tourism aimed at analyzing social impacts and designing reorientation proposals. In the impact assessment, he has applied and developed innovative methodologies that combine causal maps and social network analysis. He has investigated the impact of forms of social organization on the success of community-based tourism projects (Chile and Brazil). He has experience in technical advice on tourism, highlighting the collaboration with the regional government to improve the management of visits to Natural Parks. He has completed training stays in the United States of America and the United Kingdom. He has taught courses, seminars and conferences in Brazil, Chile and Cuba. He is currently investigating the drivers and brakes on accessible tourism on the Costa Blanca (Spain).

Ariadna Gassiot Melian is a part-time lecturer at University of Girona and EU Mediterrani, Barcelona. She has a PhD in Tourism (University of Girona, 2016), Master's Degree in Tourism Management and Planning (University of Girona, 2012), Bachelor's Degree in Translation and Interpreting and Bachelor's Degree in Linguistics (University Pompeu Fabra, 2011). Her teaching activities are related to strategic management, statistics in marketing, among other subjects. Her research interests are linked to tourists' behavior, especially focusing on accessible tourism. She has published in academic journals, such as the International Journal of Management Cases, Documents d'Anàlisi Geogràfica, Tourism Review and Journal of Revenue and Pricing Management and has participated in both international conferences and projects related to her research.

Sarah Hilton is a researcher and holds a BSc in International Spa Management and an MSc in Tourism Management. With almost twenty years experience in the spa industry, she has a personal interest in the subject of accessibility and a desire to make a meaningful contribution towards making the spa experience accessible for all.

Katja Kous is a teaching assistant at the Institute of Informatics, Faculty of Electrical Engineering and Computer Science, at the University of Maribor. She received her Ph.D. in Computer Science in 2016 from the University of Maribor, Slovenia. Her research interests include the investigations of human-computer interaction and user experiences with a focus on web accessibility and usability evaluation to improve the integration of people with disabilities into an information society. She has appeared as an author and co-author in several peer-reviewed journals and has presented her work at several international conferences.

Susana Mesquita holds a Degree in Tourism Management and Planning from ISAG -Instituto Superior de Administração in Porto e Gestão, a Master in Tourism Management and Development and a Doctoral Degree in Tourism, also from the University of Aveiro. She is currently working on her PhD in Tourism in the University of Aveiro. Presently she teaches several courses on Tourism and on the Tourism and Hospitality Management degrees at ISAG – Instituto Superior de Administração e Gestão and in other private and public entities. Her research interests include accessible tourism, people with visual impairments, consumer behavior in museums, co creation of experiences and experience economy. She is a researcher at the Competitiveness, Governance and Public Policy (GOVCOPP) research unit at the University of Aveiro, Portugal.

Eleni (Elina) Michopoulou is an Associate Professor in Business Management at University of Derby, UK. Her research interests include information systems in tourism, online consumer behaviour and technology acceptance. She is particularly interested in the field of accessible and wellness tourism, which she has actively been researching for over fifteen years. Eleni has published over 60 academic journal articles, book chapters and conference papers. She is the co-founder of THEINC International

Conference, sits on the editorial board of over 10 high impact academic journals and is the European Editor for the International Journal of Spa and Wellness.

Andreia Moura has a PhD in Tourism from the University of Aveiro, Portugal, specifically versing about Accessible Tourism/ Tourism for All. Until November 2019, she was the Coordinator for the Polytechnic Institute of Coimbra (co-promotor) of the project HCTourism - Profile and Trends in Human Capital in the Tourism Sector, co-financed by the European Regional Development Fund (ERDF) through the Portugal 2020 Operational Program for Competitiveness and Internationalization (COMPETE 2020). Nowadays, she is Researcher of the project ACTION - Accessible tourism: Co-creation of tourism experiences through intelligent web-based systems, financed by the FCT (Portuguese public agency that supports science, technology and innovation). She is Adjunct Professor at the Polytechnic Institute of Coimbra, Portugal; Integrated researcher at CiTUR-Centre for Research, Development &Innovation in Tourism, Portugal; Collaborating researcher at GOVCOPP-Research Unit on Governance, Competitiveness & Public Policies, University of Aveiro, Portugal; and Researcher of the Alliance on Training and Research in Social and Fair Tourism, at ISTO-International Social Tourism Organization.

Gregor Polančič is an associate professor in Informatics at the University of Maribor. He received his Ph.D. in 2008 in the fields of software engineering and information systems. His main research interests include empirical investigations of Business Process Modeling and Management, IT management, IT acceptance, e-communication, and e-collaboration. He has appeared as an author or co-author in more than 30 peer-reviewed scientific journals. In all, his bibliography contains more than 300 records.

Natalia Porto is a Ph.D. in Economics from Universidad Nacional de La Plata (UNLP), Argentina, and she is the director of the Institute of Economics Investigations (UNLP). She is Titular Professor of International Economics and Economic Growth, Tourism and the Environment at the same university. She is a full-time researcher, has published various articles in national and international journals and has worked on many research project related to tourism.

Nelson Pacheco Rocha is Full Professor of the University of Aveiro. He received his BSc degree in Electronics and Telecommunications Engineering in 1983 and his PhD in Electronics Engineering in 1992, from the University of Aveiro. He was the Head of Health Sciences School (2001-2011), the Head of the Health Sciences Department of the University of Aveiro (2001-2014) and Pro-Rector of the University of Aveiro (2005-2010). His current research interests include the application of information and communications technologies to healthcare and social services, the secondary use of electronic health records, and the interconnection of human functionality and ambient assisted living services. He has been involved in various European and national funded research projects, has supervised several PhD and MSc students and has a patent and more than one hundred research publications distributed by books, book chapters, international journals and proceedings of international conferences.

Lucía Rubio-Escuderos (Alicante, 1992) is a Sociologist with a Master's degree in Tourism Planning and Management. She is currently a PhD candidate in Tourism at the University of Alicante, having done a research stay at the University of Derby. She follows a line of research in accessible tourism, having analysed implications of tourists with disabilities' experiences, market provision for tourists with disabilities and stakeholder relationships for the development of accessibility in tourist destinations.

Ana Clara Rucci is a researcher at the Tourism Research Institute of the Faculty of Economics, Universidad Nacional de La Plata (UNLP), Argentina. She has a Master Degree focused in accessible tourism in the Latin American countries from the UNLP. She is a PhD candidate at the Tourism Program in the University of Alicante, Spain, and her research interests include accessible tourism, inclusion, destination competitiveness and smart destinations. She has published articles in national and international journals related to tourism and accessibility.

Gonçalo Santinha is an Assistant Professor, Department of Social, Political and Territorial Sciences, University of Aveiro.

António Teixeira has a MsC in Electronics and Telecomm. and a Ph.D. in Electrical Engineering from the University of Aveiro, Portugal. Research interests include Speech Processing, Human Language Technologies and Multimodal Human-Machine Interaction. Involved in several projects (e.g. IRIS, AL4ALL, AAL Paelife) and in the organization of several events (iberSpeech 2016, 2014, 2012; TPC Chair PROPOR 2012; Co-Chair FALA 2010, Special Sessions Chair InterSpeech 2005, Chair PROPOR 2008).

Pedro Teixeira has a master's degree in Industrial Engineering in Industrial Engineering (2019). Currently is a PhD student in Industrial Engineering, at the University of Aveiro. The main research preferences are in the area of Information Systems, Industry 4.0, Accessible Tourism and investigation related to technological solutions to real-world problems.

Index

E

I

L

M

N

O

P

R

S

T

U

V

W

www.ingramcontent.com/pod-product-compliance
Lightning Source LLC
LaVergne TN
LVHW060412300926
849162LV00006B/1403